The Method Framework
for Engineering
System Architectures

The Method Framework for Engineering System Architectures

Donald G. Firesmith
with
Peter Capell
Dietrich Falkenthal
Charles B. Hammons
DeWitt Latimer
Tom Merendino

CRC Press
Taylor & Francis Group
Boca Raton   London   New York

CRC Press is an imprint of the
Taylor & Francis Group, an **informa** business

AN AUERBACH BOOK

Auerbach Publications
Taylor & Francis Group
6000 Broken Sound Parkway NW, Suite 300
Boca Raton, FL 33487-2742

© 2009 by Taylor & Francis Group, LLC
Auerbach is an imprint of Taylor & Francis Group, an Informa business

International Standard Book Number-13: 978-1-4200-8575-4 (Hardcover)

Library of Congress Cataloging-in-Publication Data

The method framework for engineering system architectures / Donald G. Firesmith ... [et al.].
    p. cm.
  Includes bibliographical references and index.
  ISBN 978-1-4200-8575-4 (alk. paper)
  1. Computer architecture. 2. System design. I. Firesmith, Donald G., 1952-

QA76.9.A73M46 2008
004.2'2--dc22
                                   2008043271

Visit the Taylor & Francis Web site at
http://www.taylorandfrancis.com

and the Auerbach Web site at
http://www.auerbach-publications.com

Concise Table of Contents

Preface xxvii

1   Introduction 1

2   System Architecture Engineering Challenges 13

3   System Architecture Engineering Principles 39

4   MFESA: An Overview 49

5   MFESA: The Ontology of Concepts and Terminology 81

6   Task 1: Plan and Resource the Architecture Engineering Effort 137

7   Task 2: Identify the Architectural Drivers 153

8   Task 3: Create the First Versions of the Most Important Architectural Models 171

9   Task 4: Identify Opportunities for the Reuse of Architectural Elements 191

10 Task 5: Create the Candidate Architectural Visions 205

11 Task 6: Analyze Reusable Components and Their Sources 219

12 Task 7: Select or Create the Most Suitable Architectural Vision 233

13 Task 8: Complete the Architecture and Its Representations 245

14 Task 9: Evaluate and Accept the Architecture 257

15 Task 10: Maintain the Architecture and Its Representations 279

16 MFESA Method Components: Architectural Workers 293

17 MFESA: The Metamethod for Creating Endeavor-Specific Methods 339

18 Architecture and Quality 355

19 Conclusions 397

Appendix A: Acronyms and Glossary..415

Appendix B: MFESA Method Components 431

Appendix C: List of Guidelines and Pitfalls...441

Appendix D: Decision-Making Techniques 449

Annotated References/Bibliography 459

Index 467

Contents

List of Figures xvii

List of Tables xxi

Foreword xxiii

Preface...xxvii

1 Introduction 1
   1.1   To Begin 1
   1.2   Why This Book Is Needed...2
   1.3   Why System Architecture Is Critical to Success..............................4
   1.4   Why System Architecture Engineering Is Critical to Architecture..........................9
   1.5   A Common System Architecture Engineering Method Is Insufficient 11

2 System Architecture Engineering Challenges 13
   2.1   Introduction 13
   2.2   General Systems Architecture Engineering Challenges....................13
   2.3   Challenges Observed in System Architecture Engineering Practice 23
        2.3.1   Industry Has a Poor Architecture Engineering Track Record 24
        2.3.2   Many Architecture Defects Are Found during Integration and Testing........................ 24
        2.3.3   Processes Are Inconsistent in Practice...25
        2.3.4   Architectural Representations Are Often Missing, Incomplete, Incorrect, or Out of Date.................................25
        2.3.5   Architectural Models Are Treated as the Sole Architectural Representations.......................... 26
        2.3.6   Architectural Models Are Often *Not* Understandable.............................. 26
        2.3.7   Architecture Engineering Underemphasizes Specialty Engineering Focus Areas 27
        2.3.8   How Good Is "Good Enough"?... 27
        2.3.9   Because We Lack Sufficient Adequately Trained and Experienced Architects, They Must Sometimes Perform Tasks for Which They Are Unqualified.............................. 28

      2.3.10 Architects Use Multiple Inconsistent Architecture Engineering
             Methods 29

      2.3.11 Architecture Engineering Methods Are Incompletely Documented 29

      2.3.12 Architects Rely Too Much on Architectural Engineering Tools................ 29

      2.3.13 The Connection between the Architecture and the Design It Drives
             Is Weak.. 30

  2.4   Challenges Observed in Systems Architecture Engineering Methods 30

      2.4.1   Current System Architecture Engineering Methods Are Incomplete 31

      2.4.2   Current Methods Do Not Scale Up..31

      2.4.3   Current Methods Assume "Greenfield" Development 31

      2.4.4   Current Methods Overemphasize Architecture Development over
            Other Tasks..31

      2.4.5   Current Methods Overemphasize Functional Decomposition for
            Logical Structures... 32

      2.4.6   Current Methods Overemphasize Physical Decomposition for
            Physical Structures 32

      2.4.7   Current Methods Are Weak on Structure, View, and Focus Area
            Consistency 32

      2.4.8   Current Methods Codify Old Processes 33

      2.4.9   Current Methods Emphasize the Waterfall Development Cycle 33

      2.4.10 Current Methods Confuse Requirements Engineering with
             Architecture Engineering...33

      2.4.11 Current Methods Underemphasize Support for the Quality
             Characteristics 34

      2.4.12 Current Methods Assume That "One Size Fits All" 35

      2.4.13 Current Methods Produce Only a Single Architectural Vision 35

      2.4.14 Current Methods Overly Rely on Local Interface Specifications 36

      2.4.15 Current Methods Lack an Underlying Ontology.................................... 36

      2.4.16 Current Methods Confuse Architecture and Architecture
             Representations.. 36

      2.4.17 Current Methods Excessively Emphasize Architectural Models over
             Other Architectural Representations 36

      2.4.18 Current Methods Overemphasize the Points of View of Different
             Types of Experts 37

  2.5   Reasons for Improved Systems Architecture Engineering Methods 37

  2.6   Summary of the Challenges.. 38

3   System Architecture Engineering Principles 39

  3.1   The Importance of Principles.. 39

  3.2   The Individual Principles.. 40

  3.3   Summary of the Principles...47

4   MFESA: An Overview 49

  4.1   The Need for MFESA.. 49

  4.2   MFESA Goal and Objectives 51

  4.3   What Is MFESA? 51

      4.3.1   Ontology 52

       4.3.2   Metamodel 52

       4.3.3   Repository 56

             4.3.3.1   Method Components: Tasks.............................57

             4.3.3.2   Method Components: Architectural Workers.......................... 60

       4.3.4   Metamethod 61

4.4   Inputs 61

4.5   Outputs 66

4.6   Assumptions 66

       4.6.1   The Number and Timing of System Architecture Engineering

             Processes 67

4.7   Relationships with Other Disciplines 67

       4.7.1   Requirements Engineering.. 68

       4.7.2   Design 71

       4.7.3   Implementation 71

       4.7.4   Integration 71

       4.7.5   Testing.. 71

       4.7.6   Quality Engineering.. 72

       4.7.7   Process Engineering.. 72

       4.7.8   Training 72

       4.7.9   Project/Program Management....................................... 73

       4.7.10 Configuration Management 73

       4.7.11 Risk Management 74

       4.7.12 Measurements and Metrics..74

       4.7.13 Specialty Engineering Disciplines 75

4.8   Guidelines 76

4.9   Summary.. 78

       4.9.1   MFESA Components 79

       4.9.2   Goal and Objectives 79

       4.9.3   Inputs 79

       4.9.4   Tasks 79

       4.9.5   Outputs 80

       4.9.6   Assumptions 80

       4.9.7   Other Disciplines... 80

       4.9.8   Guidelines 80

5   MFESA: The Ontology of Concepts and Terminology........................81

5.1   The Need for Mastering Concepts and Their Ramifications 81

5.2   Systems...81

5.3   System Architecture... 92

5.4   Architectural Structures 95

5.5   Architectural Styles, Patterns, and Mechanisms 100

5.6   Architectural Drivers and Concerns 102

5.7   Architectural Representations...106

5.8   Architectural Models, Views, and Focus Areas........................109

5.9   Architecture Work Products 122

5.10 Architectural Visions and Vision Components 125

5.11 Guidelines 126

5.12	Pitfalls	131
5.13	Summary	135

6 Task 1: Plan and Resource the Architecture Engineering Effort............137
6.1	Introduction	137
6.2	Goal and Objectives	137
6.3	Preconditions	138
6.4	Inputs	138
6.5	Steps	140
6.6	Postconditions	141
6.7	Work Products	142
6.8	Guidelines	144
6.9	Pitfalls	146
6.10	Summary	151
	6.10.1 Steps	151
	6.10.2 Work Products	151
	6.10.3 Guidelines	152
	6.10.4 Pitfalls	152

7 Task 2: Identify the Architectural Drivers 153
7.1	Introduction	153
7.2	Goal and Objectives	153
7.3	Preconditions	154
7.4	Inputs	155
7.5	Steps	156
7.6	Postconditions	159
7.7	Work Products	159
7.8	Guidelines	160
7.9	Pitfalls	162
7.10	Summary	168
	7.10.1 Steps	168
	7.10.2 Work Products	168
	7.10.3 Guidelines	169
	7.10.4 Pitfalls	169

8 Task 3: Create the First Versions of the Most Important Architectural Models.......171
8.1	Introduction	171
8.2	Goal and Objectives	173
8.3	Preconditions	174
8.4	Inputs	174
8.5	Steps	175
8.6	Postconditions	176
8.7	Work Products	177
8.8	Guidelines	177
8.9	Pitfalls	183
8.10	Summary	187
	8.10.1 Steps	187
	8.10.2 Work Products	188

8.10.3   Guidelines 188
8.10.4   Pitfalls 188

9   Task 4: Identify Opportunities for the Reuse of Architectural Elements.................191
9.1   Introduction 191
9.2   Goal and Objectives 192
9.3   Preconditions..192
9.4   Inputs 193
9.5   Steps...193
9.6   Postconditions 195
9.7   Work Products...197
9.8   Guidelines 197
9.9   Pitfalls 198
9.10   Summary..202
9.10.1   Steps..203
9.10.2   Work Products...203
9.10.3   Guidelines 203
9.10.4   Pitfalls 203

10   Task 5: Create the Candidate Architectural Visions 205
10.1   Introduction 205
10.2   Goal and Objectives 206
10.3   Preconditions.. 206
10.4   Inputs 206
10.5   Steps... 207
10.6   Postconditions 208
10.7   Work Products... 208
10.8   Guidelines 209
10.9   Pitfalls 211
10.10   Summary..215
10.10.1   Steps..216
10.10.2   Work Products...216
10.10.3   Guidelines 216
10.10.4   Pitfalls 216

11   Task 6: Analyze Reusable Components and Their Sources 219
11.1   Introduction 219
11.2   Goal and Objectives 220
11.3   Preconditions.. 220
11.4   Inputs 221
11.5   Steps...221
11.6   Postconditions 222
11.7   Work Products... 223
11.8   Guidelines 223
11.9   Pitfalls 224
11.10   Summary.. 230
11.10.1   Steps..231
11.10.2   Work Products...231

      11.10.3 Guidelines 231
      11.10.4 Pitfalls 231

12 Task 7: Select or Create the Most Suitable Architectural Vision 233
   12.1 Introduction 233
   12.2 Goal and Objectives 234
   12.3 Preconditions...234
   12.4 Inputs 234
   12.5 Steps 235
   12.6 Postconditions 237
   12.7 Work Products 237
   12.8 Guidelines 238
   12.9 Pitfalls 240
   12.10 Summary 243
      12.10.1 Steps 243
      12.10.2 Work Products 243
      12.10.3 Guidelines 244
      12.10.4 Pitfalls 244

13 Task 8: Complete the Architecture and Its Representations 245
   13.1 Introduction 245
   13.2 Goals and Objectives 246
   13.3 Preconditions...246
   13.4 Inputs 247
   13.5 Steps 247
   13.6 Postconditions 250
   13.7 Work Products 250
   13.8 Guidelines 251
   13.9 Pitfalls 252
   13.10 Summary 254
      13.10.1 Steps 255
      13.10.2 Work Products 255
      13.10.3 Guidelines 255
      13.10.4 Pitfalls 255

14 Task 9: Evaluate and Accept the Architecture 257
   14.1 Introduction 257
   14.2 Goals and Objectives 257
   14.3 Preconditions...259
   14.4 Inputs 259
   14.5 Steps 259
   14.6 Postconditions 262
   14.7 Work Products 263
   14.8 Guidelines 263
   14.9 Pitfalls 267
   14.10 Summary 275
      14.10.1 Steps 275
      14.10.2 Work Products 276

14.10.3 Guidelines 276
14.10.4 Pitfalls 277

15 Task 10: Maintain the Architecture and Its Representations 279
15.1 Introduction 279
15.2 Goals and Objectives 280
15.3 Preconditions.. 280
15.4 Inputs 281
15.5 Steps 282
15.6 Invariants 283
15.7 Work Products.. 284
15.8 Guidelines 286
15.9 Pitfalls 288
15.10 Summary...291
     15.10.1 Steps 291
     15.10.2 Work Products..291
     15.10.3 Guidelines 292
     15.10.4 Pitfalls 292

16 MFESA Method Components: Architectural Workers.................................293
16.1 Introduction 293
16.2 System Architects 295
     16.2.1 Definitions 296
     16.2.2 Types of System Architect................................ 296
     16.2.3 Responsibilities 297
     16.2.4 Authority.. 300
     16.2.5 Tasks 300
     16.2.6 Profile...301
          16.2.6.1 Personal Characteristics.....................301
          16.2.6.2 Expertise.. 302
          16.2.6.3 Training 303
          16.2.6.4 Experience 303
          16.2.6.5 Interfaces 303
     16.2.7 Guidelines 305
     16.2.8 Pitfalls 305
16.3 System Architecture Teams....................................... 307
     16.3.1 Types of Architecture Teams 307
     16.3.2 Responsibilities.. 309
     16.3.3 Membership..310
     16.3.4 Collaborations 311
     16.3.5 Guidelines 313
     16.3.6 Pitfalls 315
16.4 Architectural Tools...317
     16.4.1 Example Tools 317
     16.4.2 Types of Architecture Tools 318
     16.4.3 Relationships 328
     16.4.4 Guidelines 328

16.4.5 Pitfalls 331
16.5 Architecture Worker Summary..335
    16.5.1 System Architects 335
    16.5.2 System Architecture Teams......................................336
    16.5.3 Architecture Tools 336

17 MFESA: The Metamethod for Creating Endeavor-Specific Methods......................339
17.1 Introduction 339
17.2 Metamethod Overview..340
17.3 Method Needs Assessment 341
17.4 Number of Methods Determination....................................346
17.5 Method Reuse Type Determination 346
17.6 Method Reuse 346
17.7 Method Construction..346
17.8 Method Documentation..347
17.9 Method Verification 348
17.10 Method Publication..348
17.11 Guidelines 348
17.12 Pitfalls 350
17.13 Summary..352

18 Architecture and Quality 355
18.1 Introduction 355
18.2 Quality Model Components and Their Relationships............356
18.3 Internal Quality Characteristics 360
18.4 External Quality Characteristics..363
18.5 Quality Requirements 373
    18.5.1 Example Quality Requirements 374
18.6 Architectural Quality Cases 375
    18.6.1 Quality Case Components 376
    18.6.2 Architectural Quality Case Components..................376
    18.6.3 Example Architectural Quality Case 378
18.7 Architectural Quality Case Evaluation Using QUASAR 380
    18.7.1 Work Products 386
18.8 Guidelines 388
18.9 Pitfalls 389
18.10 Summary..394

19 Conclusions 397
19.1 Introduction 397
19.2 Summary of MFESA..397
    19.2.1 MFESA Components 397
    19.2.2 Overview of the MFESA Tasks 398
19.3 Key Points to Remember 400
    19.3.1 System Architecture and System Architecture Engineering Are
           Critical 400
    19.3.2 MFESA Is Not a System Architecture Engineering Method................... 400
    19.3.3 Quality Is Key 401

19.3.4 Architectural Quality Cases Are Important............................... 402
19.3.5 Capture the Rationales 403
19.3.6 Stay at the Right Level.. 403
19.3.7 Reuse Significantly Affects Architecture Engineering............. 403
19.3.8 Architecture Is Never Finished 404
19.3.9 Beware of Ultra-Large Systems of Systems.......................... 404
19.4 Future Directions 405
19.4.1 The Future Directions of System Architecture Engineering.......... 405
19.4.1.1 Trends in Systems and System Engineering 405
19.4.1.2 Trends in System Architecture Engineering, Architects, and Tools... 407
19.4.2 The Future Directions of MFESA....................................... 410
19.4.2.1 MFESA Organization.. 410
19.4.2.2 Informational Web Site 410
19.4.2.3 Method Engineering Tool Support......................... 411
19.5 Final Thoughts 412

Appendix A: Acronyms and Glossary..**415**

Appendix B: MFESA Method Components **431**

Appendix C: List of Guidelines and Pitfalls..**441**

Appendix D: Decision-Making Techniques **449**

Annotated References/Bibliography **459**

Index **467**

List of Figures

Figure 1.1   Architecture capabilities versus project performance. 10

Figure 2.1   Challenge of system size and complexity. 15

Figure 2.2   Software size in high-end television sets. 18

Figure 2.3   Software size increase in military aircraft. 19

Figure 2.4   Air Force and NASA software size increase from 1960 to 1995. 20

Figure 2.5   Increasing functionality implemented by software. 20

Figure 4.1   The four components of the MFESA method engineering framework. 52

Figure 4.2   Methods and processes. 54

Figure 4.3   System architecture engineering methods and processes. 56

Figure 4.4   The primary contents of the MFESA repository. 57

Figure 4.5   The logical ordering of MFESA tasks. 58

Figure 4.6   MFESA tasks by life-cycle phase. 59

Figure 4.7   Plan, prepare, check, and act cycle for a single architectural element. 60

Figure 4.8   Interactions between concurrent Tasks 3, 4, and 5. 61

Figure 4.9   How architectural visions are created, selected, and iterated. 62

Figure 4.10   Primary MFESA inputs and outputs. 63

Figure 4.11   The MFESA metamethod tasks. 64

Figure 4.12   A generic system aggregation structure. 68

Figure 4.13   Interleaving of requirements engineering and architecture engineering............. 69

Figure 4.14   Incremental requirements and architecture engineering over multiple
releases. 70

Figure 4.15   Architecture engineering effort as a function of phase. 70

Figure 5.1   Example aircraft system of systems. 85

Figure 5.2   System architecture.. 93

Figure 5.3   Architectural structures. 96

Figure 5.4   Architectural styles, patterns, and mechanisms. 100

Figure 5.5   Architectural concerns and drivers. 103

Figure 5.6   Architectural representations. 107

Figure 5.7   Example block diagram. 115

Figure 5.8   Example configuration diagram. 117

Figure 5.9   Views versus models versus structures versus focus areas............................119

Figure 5.10   Some example quality characteristics. 121

Figure 5.11   The natural flow from architectural concerns to architecture tools.................. 122

Figure 5.12   Multiple views of multiple structures of a single multifaceted architecture. 123

Figure 5.13   Structure of architecture quality cases... 124

Figure 5.14   Architecture visions composed of architectural vision components. 126

Figure 5.15   Complete ontology of architectural work product concepts and terminology....135

Figure 6.1   Summary of Task 1 inputs, steps, and outputs...138

Figure 6.2   The optimum amount of architecture engineering. 145

Figure 7.1   Summary of Task 2 inputs, steps, and outputs. 154

Figure 8.1   Summary of Task 3 inputs, steps, and outputs...170

Figure 8.2   General and example model creation from concerns..171

Figure 9.1   Summary of Task 4 inputs, steps, and outputs...188

Figure 9.2   Potential sources of architectural reuse. 192

Figure 10.1   Summary of Task 5 inputs, steps, and outputs... 202

Figure 10.2   Architecting OTS subsystems.. 208

Figure 11.1   Summary of Task 6 inputs, steps, and outputs..216

Figure 12.1   Summary of Task 7 inputs, steps, and outputs.. 230

Figure 13.1   Summary of Task 8 inputs, steps, and outputs... 242

Figure 14.1   Summary of Task 9 inputs, steps, and outputs...256

Figure 14.2   Three example evaluation scopes. 265

Figure 15.1   Summary of Task 10 inputs, steps, and outputs...276

Figure 16.1   The three types of MFESA method components. 288

Figure 16.2   Three types of system architecture workers... 288

Figure 16.3   Types of architects.. 289

Figure 16.4   Types and memberships of architecture teams. 302

Figure 16.5   Architecture repositories. 322

Figure 17.1   The four primary MFESA components. 332

Figure 17.2   The MFESA metamodel of reusable abstract method component types. 333

Figure 17.3   MFESA metamethod tasks. 334

Figure 18.1   The components of a quality model. 348

Figure 18.2   Performance as an example quality characteristic with associated attributes. 349

Figure 18.3   Safety and security as example quality characteristics with associated attributes. 350

Figure 18.4   An example partial hierarchy of important internal quality characteristics. 352

Figure 18.5   An example partial hierarchy of important external quality characteristics. 356

Figure 18.6   Quality requirements are based on a quality model. 365

Figure 18.7   The three components of a general quality case. 367

Figure 18.8   The three components of architectural quality cases. 368

Figure 18.9   Architectural quality case diagram notation. 369

Figure 18.10   Example architectural quality case diagram. 371

Figure 18.11   The three phases of the QUASAR method. 372

Figure 18.12   QUASAR tasks. 373

Figure 18.13   QUASAR team responsibilities. 374

Figure 19.1   The four primary components of MFESA. 388

Figure 19.2   MFESA tasks. 389

Figure 19.3   Future integrated MFESA toolset. 398

Figure B.1   Reusable method components in the MFESA repository. 418

Figure D.1   A generic decision-making method. 436

List of Tables

Table 5.1 Differences between Architecture and Design 94

Table 10.1 Architectural Vision Component versus Vision Matrix 205

Table 10.2 Example Partial Architectural Concern versus Architectural Component Matrix 206

Table 12.1 Example Architectural Concern versus Candidate Architectural Vision Matrix 233

Table 18.1 QUASAR Assessment Results Matrix 375

Foreword

One of the biggest sources of pain in system development is "system integration and test." This is frequently where projects sailing along with all-green progress reports and Earned Value Management System status summaries start to see these indicators increasingly turn to yellow and then to red. Projects that were thought to be 80 percent complete may be found to still have another 120 percent to go, increasing the relative costs of integration and test from 20 percent of the total to 120/200 = 60 percent of the total.

Managers often look at this 60 percent figure and say, "We need to find a way to speed up integration and test," and invest in test tools to make testing go faster. But this is not the root cause of the cost escalation. That happened a lot earlier in the definition and validation (or more often the lack of these) of the system's architecture. Components that were supposed to fit together did not. Unsuspected features in commercial off-the-shelf (COTS) products were found to be incompatible, with no way to fix them and little vendor interest in doing anything about the problems. Nominal-case tests worked beautifully but the more frequent off-nominal cases led to system failures. Readiness tests for safety and security certification were unacceptable. Defect fixes caused regression tests to fail due to unanticipated side effects. Required response times were impossible to meet. And award fees for on-time delivery and expected career promotions faded away.

Suppose, however, that you could do most of this integration before you bought or developed the components. An increasing number of projects have been able to do this. Some good examples are the Boeing 777 aircraft, which developed and validated a digital version of the aircraft before committing to its production, and the TRW CCPDS-R command and control system, well documented in Walker Royce's book, *Software Project Management*. These and other successful projects concurrently engineered their system's architecture along with its concept of operations, requirements, life-cycle plans, and prototypes or early working versions of its high-risk elements. And they also concurrently prepared for and developed the evidence that if the system were developed to the given architecture, it would support the operational concept, satisfy the requirements, and be buildable within the budgets and schedules in the plans. Further, they checked the consistency of the interfaces of the elements so that if the developers complied with the interface specifications, the developed elements would plug-and-play together (well, almost).

Thus, the managers proceeding into development had much more than a set of blueprints and software architecture diagrams upon which to base their decision to proceed. They had strong technical evidence of the feasibility of the specifications and plans, and often a business case showing that the costs to be invested in the system would provide a positive return on investment (ROI). The costs of doing all this up-front work are higher, but as we show for software-intensive systems in an upcoming paper in the INCOSE journal *Systems Engineering* (B. Boehm, R. Valerdi, and E. Honour, "The ROI of Systems Engineering: Some Quantitative Results for Software-Intensive Systems," 2008), the ROI is generally quite positive and becomes increasingly large as the systems become increasingly large. For example, consider a software-intensive system with one million equivalent source lines of code, on which the time spent in systems engineering for the project before proceeding into development increases from 5 to 25 percent of the nominal project duration. Based on the Constructive Cost Model (COCOMO II) calibration to 161 project data points, an additional 13.5 percent of the nominal project budget will be invested in the project in doing so, but 41.4 percent of the budget will be saved by avoiding rework due to weak architecting and risk resolution, for a return on investment of over 2:1.

This book, *The Method Framework for Engineering System Architectures (MFESA)*, is the first book of a new generation of books that will provide guidelines for how to develop systems in this way. The book strongly emphasizes, as have others, that there is no one-size-fits-all set of architecting guidelines and representations. But this book breaks new ground (while being practical and useful) by providing an architectural framework that can be tailored to a project's particular situation. It provides a ten-task process (in which steps can be performed concurrently) that enables one to evaluate a project's architectural options with respect to its situation; to synthesize a solution; to verify and validate its feasibility; and to elaborate it into a solid build-to (or acquire-to) set of architectural representations. The ten tasks are formulated as reusable and tailorable method components, and are described in Chapters 6 through 15 in the book:

Task 1:   Plan and resource the architecture engineering effort.
Task 2:   Identify the architectural drivers.
Task 3:   Create the first versions of the most important architectural models.
Task 4:   Identify opportunities for the reuse of architectural elements.
Task 5:   Create the candidate architectural visions.
Task 6:   Analyze reusable components and their sources.
Task 7:   Select or create the most suitable architectural vision.
Task 8:   Complete the architecture and its representations.
Task 9:   Evaluate and accept the architecture.
Task 10: Maintain the architecture and its representations.

Each chapter describing a task is organized in the same way, presenting the task's goal and objectives, preconditions, inputs, steps, postconditions, work products, guidelines, pitfalls, and summary. These provide a uniformity of coverage and a readily understandable organization of the content.

If there is one thing that I wish the book had done more of, it would be to address the interplay between architecture tasks and other interdependent project tasks such as operational concept formulation, requirements determination, and project planning, budgeting, and scheduling. The book is extremely thorough about how architects go about their function of architecting. But an integrated product team involving users, acquirers, requirements engineers, and planners can get into a great deal of trouble without the services of a good architect to collaborate with and identify as early as possible which of the users' wishes, acquirers' constraints, requirements engineers' assertions, and planners' increments are architecturally insupportable and need to be reworked early and cheaply rather than late and expensively.

But other books can come along and do this, and the later chapters in this book address some of the key aspects of this interaction. Chapter 16 on architectural workers emphasizes that architects should be stakeholder advocates; should know requirements engineering; should interface with management, systems engineering, and integration and test; and should employ tools including requirements and business process engineering tools. Chapter 18 on architecture and quality emphasizes the need for architectural validation of quality requirements, and often the need for iteration of quality requirements if no architecture can support the desired quality levels. Chapter 18 is particularly good at addressing the critical role that quality requirements levels play in determining architectural solutions, and in presenting the QUASAR quality-case approach for assessing the architecture's support for the system's quality requirements. Finally, Chapter 19 summarizes the book's content, addresses future trends such as integrated architecting tool support, and provides a set of points-to-remember that is valuable for everyone involved in systems engineering and development:

- System architecture and system architecture engineering are critical to success.
- MFESA is not a system architecture engineering method, but rather a framework for constructing appropriate, project-specific system architecture engineering methods.
- Architectural quality cases make the architects' case that their architecture sufficiently supports the architecturally significant requirements.
- It is critical to capture the rationale for architectural decisions, inventions, and trade-offs.
- Architects should keep their work at the right level of abstraction.
- Reuse has a major impact on system architecture engineering.
- Architecture engineering is never finished.

As a bottom line, I would say that anyone wishing to keep pace with the job of architecting the systems of the future should consider buying, understanding, and applying the framework and insights in this book. If you do, you will reap a very strong return on your investment, and help produce the stronger and more flexible architectures that our world is going to need.

Barry Boehm
TRW Professor of Software Engineering
Director of the Center for Systems and Software Engineering (CSSE)
University of Southern California

Preface

Goals and Objectives

The goals of this reference book are to:

- Document the Method Framework for Engineering System Architectures (MFESA*) repository of reusable architecture engineering method components† for creating methods for effectively and efficiently engineering high-quality architectures for software-intensive systems and their subsystems.‡
- Provide a more complete look at system architecture engineering than that which is commonly seen in industry and academia.
- Thereby open readers' eyes to the very large scale of architecture engineering, including the numerous potential architectural:
  - Workers (e.g., architects, architecture teams, and architecture tools)
  - Work units they perform
  - Work products they produce

The subordinate objectives of this reference book are to document:

- The *major challenges* facing the architects of today's large and complex systems
- A *consistent set of principles* that should underlie system architecture engineering

---

* MFESA is pronounced as em-fay-suh.

† Method components are also known as *method fragments* and *method chunks* in the situational method engineering community. The term *component* was selected to emphasize the close relationship between method engineering and component-based engineering, which is well known within the system architecture engineering community. The term *chunk* was rejected as being too informal, and the term *fragment* was rejected because it implied destructive decomposition rather than constructive composition.

‡ Although MFESA was primarily developed to produce methods for engineering the *system* architectures of software-intensive systems, it can also be used to engineer the *software* architectures of systems, their subsystems, and their software architectural components.

- The *components* of the MFESA:
  - A *cohesive and consistent ontology* defining the fundamental concepts and terminology underlying system architecture engineering
  - The *metamodel* defining the types of reusable method components
  - The *repository* of reusable method components, including:
    - A *cohesive and effective set of tasks* and component steps for producing associated architectural work products
    - The *architectural workers* who perform architectural work units to produce architectures and their representations
- A *recommended set of industry best practices* and guidelines for engineering system architectures
- The *common architecture engineering pitfalls* and the means to avoid or mitigate them
  - A *metamethod* for creating project-specific system architecture engineering methods
- The close *relationship between quality and architecture* in terms of a quality model, quality requirements, and architectural quality cases

Scope

The scope of MFESA and this reference book is the engineering of system architectures. This includes systems ultimately consisting of one or more of the following architectural types of components: data, equipment, facilities, firmware, hardware, human roles, manual procedures, materials, and software. This also includes the engineering of the architecture of new systems as well as the maintenance of the architectures of existing systems, as well as the architecture of individual systems, their subsystems, and systems of systems.

Note that this book is about *system* architectures, not enterprise architectures. It is also about software architectures to the extent that they are part of and significantly affect system architectures.

The following three terms and their definitions will help the reader understand the scope of this book:

1. *System architecture:* the set of all of the *most important, pervasive, higher-level, strategic decisions, inventions, engineering trade-offs, assumptions,* and their associated *rationales* concerning *how* the system meets its allocated and derived product and process requirements.

    Note that the preceding definition includes more than just the system structure (i.e., the major components of the system, their relationships, and how they collaborate to meet the requirements).

2. *System architecture engineering (SAE):* the subdiscipline of systems engineering consisting of all architectural work units performed by architecture workers to develop and maintain architectural work products (including system or subsystem architectures and their representations).

    Note that system architecture engineering is part of system engineering and not a totally independent discipline.

3. *System architect:* the highly specialized role played by a system engineer when performing architecture engineering work units to produce system architectural work products.

Thus, you are a system architect if you are a system engineer who performs system architecture engineering to create a system architecture and its representations.

MFESA Applicability

The MFESA reusable architecture engineering method components have been designed for wide applicability:

- Because a system's architecture grows and evolves from the system's earliest concept until the system is retired, MFESA has been designed to apply during the entire system life cycle from conception through development, initial small-scale production, full-scale production, utilization, and sustainment to retirement.
- MFESA can be applied to acquired systems as well as systems developed in-house.
- MFESA has been designed for both new "greenfield" development as well as the evolving of legacy systems.
- MFESA has been designed for the development of a system's new built-from-scratch components as well as development involving the heavy reuse (e.g., commercial off-the-shelf [COTS], government off-the-shelf [GOTS], military off-the-shelf [MOTS], open source, and freeware) of existing components.
- In addition to individual systems, MFESA can also be applied to the architecting of systems of systems (SOS), including product lines, families of systems, and networks of systems.*
- However, MFESA is neither designed nor intended for the development of enterprise architectures.

Intended Audiences

Although primarily intended for system architects, MFESA and this reference book are also intended for all other system architecture engineering stakeholders. This includes stakeholders in system architectures and their representations, as well as stakeholders in how system architecture engineering is performed. This also includes all stakeholders who may be sources of architecturally significant requirements. Specifically, the intended audience includes:

- *System, subsystem, software, and hardware architects*, who will perform architectural tasks, use architectural techniques, and develop the associated system, subsystem, software, and hardware architectures, their representations, and other architectural work products
- *Process engineers*, who will collaborate with the architects to determine and define how system architecture engineering will be performed and therefore develop:
  - Appropriate, project-specific, MFESA-compliant architecture engineering methods
  - Engineering conventions (e.g., standards, procedures, guidelines, templates, and tool manuals) affecting how to perform architecture engineering
- *Customers* and *owners*, who will acquire or own the system or its components and may thus need to perform oversight of or visibility into the work performed by the architects
- *Marketers and sellers*, who must market and sell the system or its components
- *Policy makers*, who will develop policies affecting the architecture or architecture engineering

---

* Unfortunately, *no* architecture engineering methods or method frameworks have yet been shown to be effective and efficient for the architecting of ultra-large systems of systems. While we feel that the best practices incorporated within MFESA will help with such unprecedented systems of systems, no one knows for sure how to best architect such systems and this is an area of active research.

- *Requirements engineers*, who will engineer the architecturally significant requirements that drive the development of the architecture and its representations
- *Technical* and *administrative managers*, who will manage, staff, and resource the architecture teams that perform the architectural work units and develop the associated architectural work products
- *Designers* and *implementers*, who will design and implement the system's components
- *Integrators*, who will either integrate the system's components or integrate the system into a larger system of systems
- *Evaluators*, who evaluate (e.g., review, inspect, audit, or independently assess) the correctness, quality, and maturity of the architecture and its representations
- *Testers*, who will perform integration testing of the architecture's components
- *Certifiers* and *accreditors*, who will certify that the system or its components have the necessary behavior and properties and are therefore ready and authorized to be placed into operation
- *Operators* and *administrators*, who will operate and administer the system and its components
- *Maintainers*, who will maintain and service the system and its components
- *Trainers* and *educators*, who will train architects and other stakeholders in (1) how to perform architecture engineering, (2) how to create and use architectural work products, and (3) how to use, operate, and maintain the system and its components
- *Academic researchers*, who research system architecture engineering
- *Tool developers*, who will develop, market, and maintain architecture engineering tools
- All other stakeholders who are interested in how a system's architecture is engineered

MFESA Flexibility

MFESA is intended to be very flexible so that it can be used to construct appropriate architecture engineering methods that meet the specific needs of different projects — whether it be constructing a new system or doing a major upgrade of an existing system. Therefore, MFESA does not mandate (and can be used with) any specific development and life cycle, although MFESA does assume a modern concurrent, iterative, and recursively incremental development and life cycle as default. Similarly, MFESA does not mandate any specific set of architectural work products, and does not mandate specific names, formats, explicit content, or recording media for architecture representations. Finally, MFESA does not require compliance with any specific architecture process or product standards. If specific standards have been contractually imposed or if specific standards have been selected and mandated by the development organization, then the project architects and process engineers may choose to tailor MFESA to accommodate the contents of the required standard.

Organization of This Reference Book

As illustrated in the following figure, this reference book is organized into the following chapters and appendices:

- Chapter 1 provides a brief introduction to this reference book.
- Chapter 2 documents the common challenges that must be addressed by system architecture engineering methods.
- Chapter 3 documents the major principles that should underlie system architecture engineering, that address the preceding challenges, and that guided the development of the MFESA method framework.
- Chapter 4 provides an introduction to the MFESA method framework, including its primary goals, inputs, tasks, outputs, and assumptions.
- Chapter 5 documents an ontology of the fundamental concepts and terminology on which systems architecture engineering is founded.
- Chapters 6 through 15 document the individual tasks comprising the MFESA method, including their goals, preconditions, steps, postconditions, work products including examples, and associated guidelines and pitfalls.
- Chapter 16 documents the architectural workers, including architects, their teams, and their tools.
- Chapter 17 describes the MFESA metamethod for constructing system architecture engineering methods from the repository of reusable method components.
- Chapter 18 documents the relationship between quality and architecture, including the quality model underlying MFESA, quality requirements, and architectural quality cases.
- Chapter 19 is the conclusion, providing a summary of the MFESA method framework, a list of points to remember, and future directions planned for MFESA.
- Appendix A is a glossary of the technical acronyms and terms used in this book.
- Appendix B is a brief list of all of the MFESA reusable method components.
- Appendix C is a comprehensive listing of MFESA guidelines to follow and pitfalls to avoid.

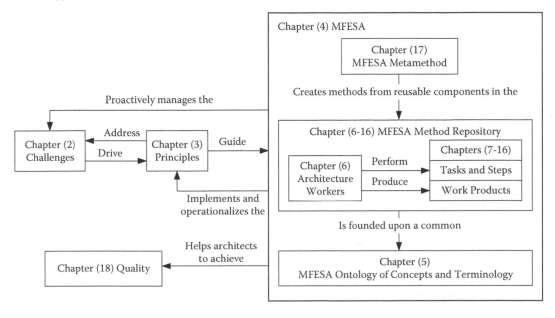

- Appendix D provides an overview of decision-making techniques that can be used during the selection of both reusable components and architectural visions.
- References and bibliography.

Industry Best Practices

This reference book documents industry best practices based on such sources as industry and international standards, industry society handbooks, documented architecture engineering methods, and other similar bodies of knowledge. It is also based on the extensive experience of its contributors acquired while both engineering and evaluating the architectures of real-world systems. However, it is not intended to be the final word on how to engineer system architectures. Instead, we intend this reference book to be a living document and thus present a single snapshot in time of a large body of work in progress. In making it available to the system and software engineering communities, we are looking to practicing architects, educators, and researchers for their support, input, and guidance. We actively solicit your comments and recommendations to help advance the contents of this book, and based on your inputs and future usage, we intend to publish updated versions of this book as it evolves.

How to Read and Use This Reference Book

This is the primary reference book documenting the Method Framework for Engineering System Architectures (MFESA). This book has been intentionally organized and formatted to make it quick and easy for its readers to find the relevant information they need to create and use appropriate, effective, and efficient project-specific system architecture engineering methods. To do this, a significant part of the contents of this book is organized in the form of lists, the entries of which are identified and summarized using short descriptive titles in boldface, followed by more detailed paragraphs. As such, it will seem familiar to anyone who has used informational Web sites and, in fact, it is the authors' intent that the contents of this book will eventually be republished as a Web site with significant hyperlinks among related concepts.

Different readers may choose to read different parts of this book for different purposes:

- Those wishing to obtain a quick overview of MFESA can jump straight to Chapter 4 ("MFESA: An Overview").
- Everyone should read Chapter 2 ("System Architecture Engineering Challenges"), Chapter 4 ("MFESA: An Overview"), Chapter 5 ("MFESA: The Ontology of Concepts and Terminology"), and Chapter 18 ("Architecture and Quality").
- New architects and students should read the entire book to learn about the challenges they face, the principles they should follow, system architecture engineering methods including the work products to perform to engineer system architectures and their representations, and the important relationships between architecture and quality.
- Experienced architects who are responsible for performing a specific architectural task or create one or more related specific architectural work products may wish to turn directly to the specific chapter describing the relevant MFESA task and associated work products.
- Architects, methodologists, and process engineers who are responsible for developing and documenting project-specific system architecture engineering methods may wish to read

Chapter 17 ("MFESA: The Metamethod for Creating Endeavor-Specific Methods") before reading Chapters 7 through 10 on the individual MFESA tasks and work products.

■ Managers should pay special attention to Chapter 2 ("System Architecture Engineering Challenges"), Chapter 4 ("MFESA: An Overview"), Chapter 6 ("Task 1: Plan and Resource the Architecture Engineering Effort"), and Chapter 16 ("MFESA Method Components: Architectural Workers").

Acknowledgments

We cannot begin to properly thank our many technical reviewers, whose insightful observations and recommendations greatly improved the final manuscript of this book. We are very grateful to Richard Barbour (The Software Engineering Institute [SEI]), Stephen Blanchette (SEI), Jørgen Bøegh (Terma A/S), Mike Bossert (U.S. Navy — Civilian), Dan Cocks (Lockheed Martin), Adriano Comai (Private Consultant — Italy), Joe Elm (SEI), Summer Fowler (SEI), Jon Hall (The Open University), Brian Henderson-Sellers (University of Technology Sydney), Harry Levinson (SEI), Grace Lewis (SEI), Azad Madni (ISTI), Abe Meilich (Lockheed Martin), Ira Monarch (SEI), Peter G. Neumann (SRI International), Ken Nidiffer (SEI), Binh Ninh (NAVAIR Systems Engineering), Mary Popeck (SEI), Samuel Redwine (James Mason University), Linda Roush (SEI), Lui Sha (University of Illinois at Urbana-Champaign [UIUC]), Greg Spaulding (The MITRE Corp.), Andras Szakal (IBM), and Carol Woody (SEI).

We would also like to thank Gerald Miller (SEI) and Mary Jamieson, who copyedited this book. Their work has made it considerably more readable.

Finally, we would like to thank John Wyzalek, Senior Acquisitions Editor at Auerbach Publications and Andrea Demby, Project Editor, for their constant support.

Donald G. Firesmith (SEI)
Peter Capell (SEI)
Dietrich Falkenthal (The MITRE Corp.)
Charles B. Hammons (SEI)
DeWitt T. Latimer IV (U.S. Air Force)
Tom Merendino (SEI)

Chapter 1

Introduction

1.1 To Begin ...

This is the reference book for the Method Framework for Engineering System Architectures (MFESA), and as its name implies, MFESA helps *system architects* use *system architecture engineering* to create *system architectures*. However, there are no universally accepted definitions for these three terms, and the lack of standard definitions has been the cause of considerable confusion and disagreement. Therefore, to begin at the beginning, we start by defining these three foundational concepts.

Most common definitions equate *system architecture* with the overall top-level structure of the system in terms of its major components, the relationships between them, and the principles guiding how they collaborate to meet the system's requirements. This definition has the advantages of being intuitive given the original meaning of the word "architecture," and it captures the most obvious aspect of the word's meaning. However, this definition is too restrictive. An architect makes many architectural decisions, inventions, trade-offs, and assumptions when architecting a system, and they are not all restricted to the system's overall structure. In fact, a system does not have a single overall structure, but rather a great number of architecturally important, inter-related, and interwoven structures that are logical or physical as well as static or dynamic. The system architecture must also capture nonstructural strategic decisions and inventions such as the system-wide use of design paradigms, technologies, and programming languages. Architects also have important rationales for these decisions, inventions, trade-offs, and assumptions. Therefore, MFESA uses the following more general definition:

> **System architecture:** the set of all of the *most important, pervasive, higher-level, strategic decisions, inventions, engineering trade-offs, assumptions,* and their associated *rationales* concerning *how* the system meets its allocated and derived product and process requirements

The scope of MFESA is the engineering of system (and by implication subsystem) architectures. There are many books about enterprise architectures but this is not one of them, even

1

though businesses, government, and military enterprises can be construed as extremely large systems. Similarly, there are many fine books on software architectures; but although software is a critical component of most systems, this is a software architecture book only to the extent that system architecture must include and properly address software architecture.*

Like requirements engineering, implementation and fabrication, integration, testing, and manufacturing, the development of system architectures is an *engineering* subdiscipline of system engineering and should not be thought of as merely a subjective art or craft. When engineering architectural work products such as the architecture and its representations, architectural workers (such as architects and architecture teams) perform multiple architectural work units (including architectural tasks and techniques). Thus, MFESA uses the following definition:

> **System architecture engineering:** the subdiscipline of systems engineering consisting of all architectural work units performed by architecture workers to develop and maintain architectural work products.

In MFESA, the term *system architect* is a role that is played rather than a job title. A person who performs system architecture engineering tasks may have the job title of system architect, but he or she may also have the job title of system engineer, chief engineer, lead engineer, software architect, or hardware architect. Similarly, and although these are two different disciplines with different training and expertise, a person playing the role of system architect may also perform requirements engineering, technical leadership, or other tasks. Finally, to be successful as a system architect, a person must also be a system engineer just as a doctor who is a specialist in a subdiscipline of medicine (e.g., cardiologist or surgeon) must first be a doctor to be successful.

> **System architect:** the highly specialized role played by a system engineer when performing architecture engineering work units to produce system architectural work products.

1.2 Why This Book Is Needed

One of the objectives of this book is to open readers' eyes to the true scale of system architecture engineering. Architecture engineering is a much larger and more difficult activity and discipline than many development and acquisition organizations currently realize. When reading this book, ask yourself the following questions:

- How many of the architectural work products (such as architectural models, views, and other architectural representations) and work units (such as architectural tasks) described in this book was I previously aware of?
- How many of these work products do I currently produce, and how many of these work units do I currently perform?
- What is the value and cost-effectiveness of those work products that I do not yet produce?

---

* Systems (and their subsystems) can be very heterogeneous, having many different types of components including data, equipment, facilities, hardware, manual procedures, personnel, and software. Thus, the scope of MFESA is much greater than pure software.

- Which of this book's many guidelines am I currently following, and into which of its pitfalls am I currently stumbling?
- How do I measure the performance of my system architecture engineering effort, and how successful are my current system architecture engineering efforts?
- Which of these work products and work units should I incorporate into my next project?

Although some national and international standards exist for performing systems engineering, these standards typically include only a very small number of pages describing system *architecture* engineering. Systems engineering handbooks (such as the International Council on Systems Engineering's *System Engineering Handbook*) also tend to provide only a relatively brief, high-level overview of system architecture engineering [INCOSE, 2006]. In practice, the architecture engineering sections of system engineering management plans are also frequently very brief and shallow, often consisting of little more than a short description of architecture models to be produced and the tools to be used to generate them. While the system architecture plans developed for large complex programs are often considerably more complete, they still seldom adequately describe how the system architecture is to be engineered in terms of an architecture engineering method's tasks, steps, techniques, and intermediate work products.

Whereas significantly more has been written concerning software architecture and software engineering being increasingly critical to systems engineering, software architecture engineering *by itself* does not adequately cover system architecture engineering. In addition to including purely system-level and hardware issues, system architecture engineering must also address the effect of relevant specialty engineering areas such as interoperability, performance, reliability, reuse, safety, and security on system architecture engineering.

Unfortunately, there currently seems to be no relatively comprehensive, integrated description of system architecture engineering, because what one finds in practice is that the different sources of this information provide only partial views of system architecture engineering, covering only those aspects considered important by their authors. In essence, we have descriptions of the elephant from the viewpoints of several blind men.* A major goal of this reference book is to show the whole elephant by providing a much more complete look at system architecture engineering.

It is becoming widely understood that the architecture of a system has a fundamental impact on the quality of the system. The decisions, inventions, and engineering trade-offs made by the system's architects have a critical effect on the achievement of the system's performance and quality as well as the cost and schedule of the system's development and maintenance. It is also becoming widely understood that the documented method used to engineer a system's architecture has a significant impact on the effectiveness and efficiency of the corresponding process performed to engineer the system's architecture, the resulting quality of the system architecture and its representations, and the corresponding quality of the resulting system.

A good system architecture is critical to project success, and system architecture engineering is critical to engineering a good system architecture. MFESA will help system architects use effective, efficient, and appropriate methods for engineering their system's architectures.

---

* A famous parable describes how six blind men react upon encountering an elephant for the first time. The first man touches the trunk and feels a snake, while the second man touches the tail and feels a rope. A third man touches a leg and feels a tree trunk, while the fourth man touches an ear and feels a fan. The fifth man touches a tusk and feels a spear, while the sixth man touches the elephant's side and feels a wall.

1.3 Why System Architecture Is Critical to Success

Given the above informal and incomplete definition of system architecture, the next question is, *"Why is it important?"* Why should you (or any system stakeholder) care whether your system has an adequate, well-documented architecture? Why should you invest in the production, usage, and maintenance of a system architecture as part of the system development process? Why not divert your project's limited budget away from architecture modeling and documentation to more "important" project tasks, especially when a project begins to experience budget and schedule pressures? In fact, why should you buy and read this book?

Ultimately, the answers to these kinds of questions emerge from examining historical trends for large system development projects over the past 25 years. Such trends indicate that the system architecture is becoming critical to the success of the system and the project that produces it. A good architecture* and associated architectural representations improve the probability of success as measured in the following terms:

- **Cost.** To be successful, a system must be affordable in terms of its development, maintenance, and operational costs. A good architecture can decrease a system's cost — not only its development and operational costs, but also, and especially, its maintenance cost. Good architectural decisions can promote a system's maintainability.
- **Schedule.** To be successful, a system must be producible or updatable within given schedule constraints. A good architecture supported by proper architectural representations provides a solid foundation on which to produce or refine the overall development or maintenance schedule as well as the more limited development schedules of the individual architectural components (e.g., subsystems). A good architecture can therefore also increase confidence in meeting a system's planned development and maintenance schedule.

  Significant architectural defects typically have major negative effects on the project schedule, especially when they are discovered late in the development process after significant design, implementation, and testing have been based on the defective architecture. Thus, a good architecture will significantly decrease the overall project schedule by limiting the amount of rework needed to identify and fix architectural defects, as well as to regression-test the system once these defects have been removed.
- **Functionality.** To be successful, a system must perform its required functions. A good architecture promotes a system's functionality, enabling it to perform its necessary functions. In fact, a good system architecture can also promote a system's extensibility, making it easier to modify the system to perform new functions in the future.
- **Quality.** Finally and perhaps most importantly, a system must have the necessary quality to be successful.

  But what is quality? It is far more than merely meeting requirements — and thus not having any defects. After all, requirements are far too often ambiguous, incomplete, and incorrect. As documented in Chapter 18, a quality model defines a system's quality in terms

---

* By "good" architecture, we mean an architecture that sufficiently supports the meeting of its associated architecturally significant requirements, especially the quality requirements. By quality factors, we mean a nonfunctional requirement that specifies that the system exhibit at least a specific threshold level of quality in terms of the quality characteristics (or "ilities") as defined by the project quality model.

of the degree to which it exhibits required levels of relevant quality characteristics* and their associated quality attributes. For example, availability, capacity, efficiency, extensibility, interoperability, maintainability, performance, portability, reliability, safety, security, stability, and usability are some of the quality characteristics, whereas event schedulability, jitter, latency, response time, and throughput are the quality attributes of performance. Using a quality model makes it clear that we should speak in terms of a system's *qualities* instead of its quality.†

Requirements engineers use quality requirements to specify the system's required levels of these quality characteristics or quality attributes. A good architecture promotes a system's quality by enabling the system to achieve its quality requirements. And how well the system's quality requirements are met is a major factor in determining a system's ultimate success.

In summary, the project's quality model defines the meaning of quality in terms of quality characteristics and attributes, and the system's quality requirements mandate its required levels of the relevant quality characteristics and attributes. A system's architecture largely determines its qualities and therefore largely determines its success.

It is important to remember that architecture is not the only major determinant of system quality and system success. Another of the most important root causes of system and associated development project failure is poor requirements. Requirements engineering,‡ after all, is the earliest technical discipline that can be poorly performed, resulting in an immense negative influence on all downstream technical disciplines including architecture engineering. Poorly identified, analyzed, specified, and managed requirements have a direct negative impact on the quality of the system architecture because the architecturally significant quality requirements are often the least-well engineered requirements. As previously noted, a good architecture is critical to achieving a quality system. If the architects are not given good quality requirements that specify the required levels of the different qualities that their architecture must support, it should not come as a surprise to anyone when the system architectures does not support adequate levels of the associated quality characteristics.

■ **Communication.** The representations of the system architecture are the primary means used to communicate the overall vision of the system among its stakeholders. They are used to ensure mutual understanding of (and form a consensus as to what) the system is and how the system will perform. Although the size of the architectural representations for a major system (such as an aircraft, chemical plant, or nuclear reactor) can be huge and extremely daunting, the architectural representations are nevertheless at a higher level and are much more concise than the design or implementation of the system. Specifically, the architectural models concisely capture how the system is organized and how its components collaborate to meet their most important system requirements. The architectural representations also

---

* Quality characteristics are often also called quality attributes, quality factors, and "ilities" (because many of them end with the letters *ility*). MFESA follows the ISO quality model by using the terms *quality characteristic* and *quality attribute.*.

† Trying to roll all of these different system qualities into the single term *quality* is roughly the same kind of error as trying to roll all of the different types of intelligence (e.g., academic, athletic, musical, and social) into a single intelligence quotient (IQ).

‡ Requirements engineering and system architecture engineering are two different but related disciplines. Although poor requirements are a major cause of poor architectures, system architecture engineering should *not* be improperly expanded to include requirements engineering. The requirements engineers are responsible for the requirements and the architects are responsible for the architecture. Expertise in one discipline does not imply expertise in the other, and each should only be performed by those who are qualified to do so.

communicate the architects' other architectural decisions, inventions, trade-offs, assumptions, and rationales.

■ **Driver of downstream disciplines.** The system's architectural representations are developed early in the system development cycle and capture the strategic decisions, inventions, and engineering trade-offs concerning how the system will meet its requirements. Because of this, the architecture and its representations will drive all the downstream effort, including design, implementation, integration, testing, and manufacturing of the system. The system's static physical aggregation structure can also be used to largely structure the development organization using Conway's law.*

■ **Reuse.** Architecture engineering supports reuse in four ways: First, a good system architecture supports the identification, reuse, and integration of off-the-shelf (OTS) architectural components as part of the system. Second, a good system architecture often produces architectural components that can be reused as part of other systems. Third, the representations of the system architecture form a highly valuable set of work products that can be reused on similar systems. Not only can individual system architectures be reused on similar individual systems, but reference architectures can be reused within families of related systems (such as product lines of similar systems). Finally, the personal experience engineering of a specific architect may be applicable to the engineering of future architectures.

■ **Organizational balance.** Architecture engineering captures, integrates, and balances the contributions of system engineers and other stakeholders having a diverse set of skills and experience in such domains as command and control logic, communications, human factors, performance modeling, reliability, sensor and actuator technologies, safety, and security. Architects must balance and contrast competing interests, and optimize the architecture across multiple contradictory stakeholder needs and requirements.

There are typically a very great number of ways that a project can fail, whereas there are many fewer ways that a project can succeed. While having a good† system architecture will not guarantee project success, having a bad system architecture will greatly increase the probability of system failure. The following are a few examples of systems that either have failed (or are subject to failure), largely due to having poor architectures:

■ **Failure of Ariane 5 flight 501.** On June 4th, 1996, the inaugural flight 501 of the European Space Agency's (ESA) Ariane 5 launch vehicle performed nominally until 37 seconds after launch, at which time the active Inertial Reference System failed [Lions, 1996]. This was immediately followed by the failure of the backup Inertial Reference System. These failures caused the nozzles of the solid booster rockets to steer to an extreme position, thereby causing the rocket to suddenly veer off of its planned flight path. The solid booster rockets broke off the core stage, thereby triggering the self-destruction and explosion of the launch vehicle.

Each inertial reference system was of standard design, using an internal computer, laser gyroscopes, and accelerometers to calculate angles and velocities. To increase redundancy

---

* Conway's law states that a system's aggregation structure tends to mirror the aggregation structure of the organization that produces it.

† By "good" architecture, we mean an architecture that sufficiently supports the meeting of its associated architecturally significant requirements, especially the quality requirements. By quality requirement, we mean a nonfunctional requirement that specifies that the system exhibit at least a specific threshold level of quality in terms of the quality characteristics (or "ilities") as defined by the project quality model.

at the equipment level, the launch vehicle included two identical inertial reference systems operating in parallel, whereby one was active and one was in hot standby. The software in these systems was reused essentially without modification and requirements verification from the previous Ariane 4 Inertial Reference System.

The Inertial Reference System software contained a horizontal alignment function that computed meaningful results only prior to liftoff, but that nevertheless continued to operate approximately 40 seconds into the flight. The function returned an unexpectedly high value because the trajectory of the Ariane 5 differs from that of the Ariane 4 in having considerably higher horizontal values. The software did not properly handle the exceptional values calculated and instead raised a generic operand error causing the active Inertial Reference System to fail. Because the hot standby Inertial Reference System used the same software, it also failed. The Inertial Reference Systems then passed a diagnostic bit value to the on-board computer, which then misinterpreted the data and ordered the engine nozzles into extreme position.

The software caused system failure in several ways: (1) the alignment function continued to operate after liftoff when it did not provide meaningful data; (2) data conversion functions in the Ada software did not properly handle the error condition, although other modules were properly protected from out of range values; and (3) the on-board computer did not check the input data (diagnostic bit value) and therefore misinterpreted it. Nevertheless, it is important to note that the Ariane 5 failure was *not* merely due to software coding defects. The primary causes of the failure were in the engineering of the architecture: (1) the redundant Ariane 4 Inertial Reference Systems were reused in Ariane 5 without proper verification that their behavior met requirements, (2) interfaces between the Inertial Reference System and nozzle control software running on the on-board computer were not verified and may not have been properly specified, and (3) use of identical Inertial Reference Systems with identical software (one active and one on hot standby) was not an adequate architectural pattern to use to obtain system reliability.

Given the magnitude of the changes to the Ariane 5, both physically and in software, the decision to reuse software wholesale without detailed analysis led to a $1.2 billion USD loss for the ESA in that year, of which $370 million USD was the direct cost of the payload. This failure is a good example of the following pitfalls:

- Architects should not uncritically believe that reuse automatically saves money and improves schedule.
- Architects should not uncritically believe that when a component works in one system, it will also work in another similar system. This is a fundamental unsolved problem because there is no automated mechanism for determining if an existing software component will continue to function correctly when ported to a new system, possibly in a new environment. Current development efforts depend on architecture representations, design documentation, and human evaluations to ensure that the software component is truly reusable in the new system and its environment.
- Architects should not uncritically believe that identical redundant software provides the same improvement in reliability as identical redundant hardware.

■ **Cascading network system failures.** Many business- or mission-critical systems (of systems) can be modeled as heterogeneous networks in which information, power, or substances must continuously flow along the connections between the nodes. An example of each type is a telecommunications system, an electrical power supply system, and a petrochemical pipeline system. To maintain the flow and to ensure that the capacity of the nodes and connections are not exceeded, these systems must be dynamically load balanced so that the flow of information, power, or substances remains continuous.

When developing such systems, architects must make appropriate engineering trade-offs to ensure that the system architecture adequately supports the following quality characteristics:

- **Availability.** Typically, such systems must be available 24/7. Any discontinuity in service can be catastrophic in terms of cost and may even lead to large losses of property and life.
- **Capacity.** Typically, such systems must maintain a very large capacity. Although any significant loss of capacity may have the most serious of consequences, it is nevertheless important for the system to degrade gracefully. When the system fails, it should maintain as much of its capacity as practical.
- **Cost.** Typically, such systems are very large, complex, and expensive. Such systems typically have large capital costs limiting the number of nodes and connections that can practically be built, and thereby limiting the amount of redundancy and excess capacity that it is practical for the architects to include in the system architecture.
- **Reliability.** Typically, such systems must be highly reliable. The mean time between failures must be very high, and the mean time to fix failures must be very low.
- **Robustness.** Typically, such systems must be highly robust. When parts of such systems fail, they should nevertheless continue to provide their most important services as long as is practical. They must typically exhibit a high degree of environmental, error, fault, and failure tolerance.
- **Safety.** Typically, such systems are safety critical. Such systems may be subject to the existence of hazards that can lead to accidents that may cause *unintentional* unauthorized harm to valuable assets. This can include damage to or loss of property, injury, illness, and death, as well as environmental degradation.
- **Security.** Typically, such systems are security critical. Such systems may be subject to the existence of threats that can lead to attacks (e.g., denial-of-service attacks) that may cause *intentional* unauthorized harm to valuable assets. This can include loss of integrity and confidentiality, loss of service, and successful repudiation of transactions that have actually occurred.

In most such networks, the loads carried by each node or connection within the network are dynamically balanced. If a node or connection is lost, then the load that was carried by that node or connection is rapidly redistributed to other nodes and connections within the network. This enables the networked nodes and connections to continue to provide the necessary flow of information, power, or substances despite the failure.

This process of maintaining service by redistributing the load has its limitations. When the load is redistributed from the failed high-load node or connection to lower-load nodes and connections, it is possible that the increased flow exceeds their capacity. To protect these newly overloaded nodes and connections from damage, many networked systems automatically force the overloaded nodes and connections to either shut down or slow down, causing the failures or increased congestion to "cascade" to other nodes and connections. This cascade can eventually lead to a total system failure unless the failing sections of the network are not properly isolated from the rest of the network. Such dynamic systems tend to be vulnerable to cascading network failures when high load nodes or connections fail due to either accident or attack.

To avoid single point catastrophic failures, such systems need to be architecturally engineered with sufficient hardware and software redundancy for nodes and connections so that the loss of any single node or connection will not by itself cause a cascading series of failures, even when the network is under its maximum expected loads. Such systems should also be sufficiently self-monitoring to identify the occurrence of cascading failures and have the ability to disconnect the failing part of the network in such a manner that the loads on the remaining parts of the network do not cause failure (i.e., the surviving part of the network must be able to degrade gracefully). Unfortunately, due to cost, poor architecting, lack of functionality, and poor human interfaces, several such systems have failed in the past.

Power grids are a good example of such systems. An interesting fact about power grids is that they cannot store any electrical power. At any one point in time, there are millions of customers consuming megawatts of power that is being generated by dozens of power plants producing the necessary power. Too much power can damage components of the grid, whereas too little electricity can cause blackouts and brownouts because it takes significant time for additional amounts of power to be generated. The following are examples of power grid failures:

- **13 March 1989 Quebec blackout.** On March 13, 1989, massive induced currents in power lines caused by a major space magnetic storm caused a transformer failure on one of the main power transmission lines in the HydroQuebec system. The failure of the transformer caused a series of events that resulted in the catastrophic collapse of the entire power grid in only 90 seconds. This in turn caused 6 million people to lose electrical power for 9 hours or more.
- **14 August 2003 blackout.** The blackout on August 14, 2003, began at 1:58 p.m. when the power generating plant in Eastlake, Ohio, shut down. A little over an hour later at 3:06 p.m., a 345-kV transmission line failed south of Cleveland, Ohio. At 3:17 p.m., the voltage on the Ohio portion of the grid dipped temporarily but controllers took no action. Because of the first failure, electrical power was shifted to another power line, at which point the resulting heat caused that line to sag into a tree and go offline. Local controllers attempted to understand the failures but did not inform the system controllers in nearby states. At 3:41 and 3:46 p.m., breakers tripped that connected the affected grid to a neighboring grid. At 4:09 p.m., voltage dropped sharply as Ohio drew 2 gigawatts of power from Michigan. At 4:10 p.m., many transmission lines tripped out, starting in Michigan and later in Ohio, thereby blocking the flow of electricity eastward. This caused generators to go down, thereby creating a huge deficit of electrical power. Within seconds, power surges that could damage East Coast generators caused them to shut down, resulting in the biggest blackout in U.S. history.

1.4 Why System Architecture Engineering Is Critical to Architecture

Why is it important to engineer system architectures? Is it not sufficient to architect systems as we traditionally have, using an informal mixture of personal experience and art? Why is the architectural equivalent of flying-by-the-seat-of-the-pants or hacking not sufficient?

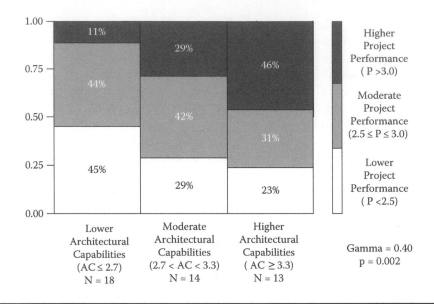

Figure 1.1 Architecture capabilities versus project performance. (*Source:* From Elm et al., 2007.)

During 2006 and 2007, the Systems Engineering Effectiveness Committee (SEEC) of the Systems Engineering Division (SED) of the National Defense Industrial Association (NDIA) performed a survey of defense industry contractors to identify the best system engineering practices used on defense projects [Elm et al., 2007]. In the survey, a project's system architecture engineering capability was operationalized in terms of the respondent's replies as to the degree to which the following six survey statements applied to their projects:

1. This project maintains accurate and up-to-date descriptions (e.g., interface control documents, models, etc.) defining interfaces in detail.
2. Interface definition descriptions are maintained in a designated location, under configuration management, and accessible to all who need them.
3. For this project, the product high-level structure is documented, kept up-to-date, and managed under configuration control.
4. For this project, the product high-level structure is documented using multiple views (e.g., functional views, module views, etc.).
5. For this project, the product high-level structure is accessible to all relevant project personnel.
6. This project has defined and documented guidelines for choosing COTS product components.

As illustrated in Figure 1.1, those projects that exhibited higher levels of system architecture engineering capabilities (as measured in terms of positive answers to the six preceding architecture-related survey questions) had better project performance in terms of schedule, cost, and functionality delivered. The figure was developed using information obtained from 45 projects, divided roughly into three groups based on their relative level (i.e., lower, moderate, and high) of architectural capability. It is interesting to note that of all the systems engineering areas surveyed and analyzed, the largest positive relationship between best practices and project performance was found in the area of architecture.* Despite the relatively small sample size, the probability that one would observe such a positive relationship by chance alone (p = 0.002) was also the smallest found for any of the areas of system engineering surveyed and analyzed.

1.5 A Common System Architecture Engineering Method Is Insufficient

Systems vary greatly in terms of size, complexity, criticality, incorporation of reusable components, and application domain. A few systems are new "greenfield" developments, many systems are new versions of existing legacy systems, and some systems are members of product lines of similar systems. Similarly, system development projects vary widely in terms of size, complexity, organizational structure and culture, and staffing expertise and geographical distribution. Different individuals, teams, and organizations may have different goals and may thus need different architecture engineering methods, even within the same endeavor. Therefore, no single system architecture engineering method, no matter how tailorable, is sufficiently flexible to support all of architecture engineering. Although flexibility is critical, extremely important and obvious benefits can also be obtained by providing the appropriate amount of standardization with regard to methods for engineering system architectures.

This is the reason why this reference book does not document and recommend any individual method for engineering system architectures. Instead of being restricted by the limitations of a single generic system architecture engineering method, architects need a *framework* to provide the advantages of both flexibility and standardization. System architects and process engineers need a framework to enable them to construct system architecture engineering methods that are appropriate for the situation at hand. The Method Framework for Engineering System Architectures (MFESA) was developed specifically to meet this need.

Situational method engineering is the engineering of methods that are appropriate for the situation at hand, such as the system being developed and the organization doing the developing [Henderson-Sellers, 2003]. Situational method engineering involves the selection of appropriate method components from a repository of reusable method components, tailoring these method components to meet the specific situation, and integrating these components to produce an appropriate method. MFESA is based on the application of situational method engineering to engineering system architectures. MFESA provides a repository of free, open source architecture engineering method components as documented in this reference book. When using MFESA, the architects:

- Select the appropriate architecture engineering method components (i.e., work products, work units, and performers of work units)
- Tailor these selected method components to meet the needs of the project
- Integrate the selected and tailored method components to form an appropriate cohesive and consistent method for engineering system architectures

---

* Note that Gamma can vary from +1 to −1, whereby positive values represent positive correlations, 0 represents no correlation, and negative values represent negative (or inverse) correlations. In this case, Gamma was equal to 0.40.

Chapter 2

System Architecture Engineering Challenges

2.1 Introduction

This chapter discusses the major challenges of system architecture engineering, including:

- General challenges associated with system architecture engineering
- Challenges commonly observed in practice with the performance of system architecture engineering
- Challenges observed with major current system architecture engineering methods
- Reasons for improved system architecture engineering methods

Both system architecture engineering (as a subdiscipline of systems engineering) and system architecture engineering methods should help system architects rise to meet these challenges. The Method Framework for Engineering System Architectures (MFESA) was developed to produce such system architecture engineering methods.

2.2 General Systems Architecture Engineering Challenges

When developing the architecture of large, complex, software-intensive systems, the system architects currently face the following very difficult challenges:

- Systems are becoming increasingly critical to their users.
- The maximum size and complexity of systems and their architectures are exponentially increasing.
- Systems have complex development cycles.
- Software has an increasingly critical and pervasive influence on system architectures.

■ Systems are hierarchically decomposed.
■ Systems reuse architectural components.
■ System architectures are never finished.
■ Different stakeholders need different architectural representations.
■ There are many different system structures that need to be architected.
■ It is difficult to maintain architectural integrity.
■ Many different architecture teams must collaborate.
■ Architectural decisions are sometimes made prematurely.

The following provides a more detailed description of these challenges.

■ **Systems are becoming increasingly critical to their stakeholders.** As software-intensive systems become ever more pervasive and as humanity grows increasingly dependent on them, such systems (and their quality) are becoming increasingly critical to individual people, their businesses and other organizations, and society as a whole. One reason why system architecture engineering is becoming more and more critical is because it greatly improves the quality of the resulting system architectures, and these architectures largely determine whether the system achieves its necessary levels of system quality.

  Thus, system architecture engineering helps the architect produce more dependable systems that are critical to users, organizations, governments, and society.

■ **Maximum size and complexity of systems and their architectures are exponentially increasing.** One may not need to architect a dog house as a simple sketch may do. However, a huge skyscraper cannot be built without a complete and high-quality set of architectural representations, and the same applies to any significant system. A massive and complex system has a massive and complex architecture that unfortunately needs a massive and complex set of architectural representations. For example, a modern commercial or military aircraft may have such a large associated set of architectural models and related architectural documentation that they will not easily fit on a standard compact disk.* The complexity of the architecture of modern software-intensive systems is due to more than just the size of such systems. Complexity is also greatly increased by the inherent complexity both within and between architectural components. Complexity is increased by the many interactions between subsystems that are often developed by geographically distributed organizations and teams within organizations.

  The maximum size and complexity of systems is steadily increasing at an exponential rate, whereby the size and complexity can be measured in terms of the:

  – **Maximum system size and complexity.** The maximum size and complexity of systems are increasing in terms of:
    • Numbers and types of requirements including functional, quality, data, and interface requirements as well as architecture, design, and implementation constraints
    • Interrelationships between requirements
    • Number, types, size, and complexity of system components

---

* At least it is difficult to store such a set of models and documentation on an average CD as of 2007. Although the size of the architectures of large systems is steadily increasing, the amount of storage available is also increasing rapidly, and it is difficult to know which rate of increase is higher.

- **Maximum project size and complexity.** The maximum size and complexity of projects are increasing in terms of:
  - Project development and sustainment costs, both actual and needed
  - Project development schedules (duration and intermediate milestones), both actual and needed
  - Number (and type) of contracts involved
  - Number and types of system stakeholders, who may have inconsistent needs and desires
  - Number of organizations involved (e.g., acquisition, prime contractor/integrator, subcontractors, suppliers, and vendors)
  - Number of architects and architecture teams
  - Number of interfaces between individuals, teams, and organizations

The corresponding maximum size and complexity of system architectures is also steadily increasing at an exponential rate, whereby the size and complexity can be measured in terms of the number of architectural components, the number of architectural views, and the number of different types of architectural models to be engineered. The resulting complexity is overwhelming and has outgrown the ability of any single system architect to master in their entirety. Such size and complexity demand proper architecture engineering if a complete and correct architecture and associated architectural representations are to be created and maintained so that project risk remains acceptable.

As illustrated in Figure 2.1, the maximum size and complexity of systems has increased exponentially over the years. Initially, systems and their architectures were sufficiently small and simple that existing individual generic architecture engineering standards and methods were reasonably effective and efficient. We will label these existing standards and methods as "first generation." In the past few years as the size and complexity curve has turned more rapidly upward, the combination of best industry practices has enabled the engineering of much larger and more complex systems and their architectures, but current approaches have

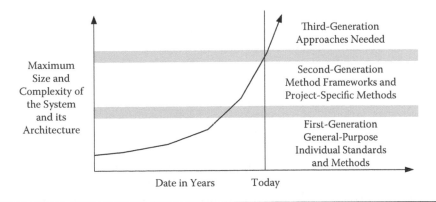

Figure 2.1 Challenge of system size and complexity. Note that the lines separating the three categories of systems in terms of their size and complexity are wide and gray, symbolizing that the demarcations are imprecise and the categories shade into one another. Although in that sense the three categories are not well defined, they are nevertheless useful for communicating the points made in the text. The diagram is also neither drawn to scale nor based on hard data using clear definitions of size, complexity, and success. Instead, the figure is based on many years of experience. Regardless of the exact shape of the curve, the basic arguments hold.

been neither effective nor efficient. It is common for system development costs to be overrun, schedule milestones to slip or be exceeded, system functionality to be postponed to future releases, quality to decrease, and risks to increase. Thus, the size and complexity of many systems and their architectures now exceed the current ability of our architecture engineering methods and tools to effectively and efficiently handle. That is, today's methods and standards are woefully inadequate for coping with the size and complexity of modern systems.

Only by moving from existing generic first-generation standards and architecture engineering methods to method frameworks incorporating industry best practices will there be a reasonable probability of effectively and efficiently engineering the architectures of such larger and more complex systems. Such method frameworks and the appropriate project-specific architecture engineering methods derived from them will be labeled "second generation."

Although method frameworks such as MFESA will be very useful for engineering the architecture of today's large and complex systems of systems, architects now face the challenge of engineering ultra-large-scale systems of systems [Northrop et al., 2006]. Industry has very little experience in architecting such systems, and the track record is very poor. No one yet knows what third-generation architecture engineering approaches for such ultra-large and complex systems will look like, and therefore those unknown approaches are unfortunately beyond the scope of this book. But as a profession, architecture engineering cannot wait until such time when third-generation approaches are developed and proven effective and efficient. Such systems are beginning to be developed now, and today's architects must do their best to engineer the associated architectures.

Therefore, the aim of this book is not to present a silver bullet panacea for the engineering of all systems, no matter how large and complex. Its aims are more modest and achievable:

- To help architects and their organizations move forward from using less effective and efficient *first-generation* general-purpose standards and methods to using a much more powerful *second-generation* method framework to create more appropriate project-specific architecture engineering methods
- To provide architects, methodologists, process engineers, and academic researchers a strong foundation on which to eventually build *third-generation* methods and standards for the architecting of ultra-large and complex systems

■ **Systems have complex development cycles.** The system being architected is sufficiently large and complex to require a recursively incremental, iterative, parallel, and time-boxed development cycle:

- The system architecture will largely be developed incrementally in a top-down, hierarchical manner in terms of subsystems at multiple lower tiers in the hierarchy.
- The "final" system and/or subsystem architecture will not be obvious so that multiple candidate architectural visions should be created and/or initial potential architectures will need to be iterated to produce the as-build architecture.
- Schedule constraints will not allow the architects to wait until all architecturally significant requirements have been engineered and baselined before beginning to architect. Similarly, lower-level architectures may need to be started before the higher-level architectures are completed and baselined.

■ **Software has an increasingly critical and persuasive influence on system architectures.** Software is having an ever-increasing influence on system architectures in terms of the system architects' decisions, inventions, trade-offs, assumptions, and rationales. Software greatly influences the architectural structures, views, and models that must be engineered, as well as their contents. Software significantly influences system architectures in at least the following ways:

- **Functionality.** Most systems are very software intensive, and software implements an ever increasing amount of the functionality of software-intensive systems. See Figure 2.5. It is no longer uncommon for software to implement significantly more than half of a system's functionality.

- **Data and data flow.** Software is almost universally used to create, update, and delete the data that systems input, store, and output. Software is thus responsible for the existence and flow of data within and between systems and subsystems.

- **Behavior and control flow.** Software is almost universally used to control the behavior of systems as well as implement control flows within and between systems and subsystems. Analog and mechanical control has largely been replaced by digital control as software has become responsible for ever larger amounts of system functionality.

- **Fault tolerance and exception flow.** In modern systems, software is used for the implementation of the vast majority of fault and failure tolerance. It is also used for implementing the flow of exceptions, a special type of control flow, including the identification, raising or throwing, catching, and handling of exceptions.

- **Primary integration glue for subsystem intra-operability.** Decades ago, subsystems were primarily connected by physical connections (e.g., bolts and welds), by energy transfer (e.g., electricity and mechanical energy), as well as by mechanical, hydraulic, or pneumatic linkages. But from a behavioral standpoint, software is now the primary integration mechanism used to glue interacting and collaborating subsystems together to support subsystem intra-operability. Examples of software-mitigated subsystem intra-operability via data, control, and exception flow include communication of sensor information, communication of actuator control (e.g., fly-by-wire), and the intra-subsystem communication of large amounts of application data and messages.

- **System interoperability.** Systems are rarely isolated. Most systems must exist in a network or system of systems. Thus, it is critical for systems to be interoperable with other systems. The primary ways that systems interoperate is via data and control flows, and these are implemented primarily via software. Architectural structures must often contain logical and physical architectural elements that exist to support software-based interoperability.

- **System quality.** As described in Chapter 18, the quality of a system is defined in terms of the system's quality model, which consists of the important quality characteristics (a.k.a. quality attributes, quality factors, or "ilities"), quality attributes (parts of quality characteristics), and associated quality measures. The system (and especially the software) architecture are largely determined by their need to address the system's quality concerns.

- **Increasingly responsible for safety and security.** Despite the continued widespread use of hardware interlocks, more and more of the safety and security controls (safeguards and countermeasures) are being implemented in the form of software components or software-intensive components. Although often appropriate and sometimes necessary to provide needed functionality, these uses of software have intrinsic risks and have resulted in several high-profile accidents [Perrow, 1999].

- **Cross-cutting architectural structures.** According to some documents, there are only two system architecture structures, which are often mislabeled as system architectures: (1) a static logical functional decomposition structure and (2) the static physical system into subsystem hierarchical decomposition structure. Although many system components may perform a single major system function, many will not [INCOSE, 2006]. However, these are not the only two system structures. As will be noted in Chapter 5, there are many logical and physical, static and dynamic architectural structures that system architects need to consider and model. These other structures typically cut across numerous architectural components, which is one reason why multiple subsystems must typically collaborate to implement many system requirements. It is difficult for system architects to determine which of the potential architectural structures is important to model, and it is very difficult to keep all of these cross-cutting structures and their models consistent with one another.

– **Disruptive technology.** Because of the previously listed impacts of software on system architectures, the increasing use of software within systems has become a disruptive technology with regard to system architecture engineering methods. For example, the traditional, relatively straightforward mapping from system functions to functionally decomposed architectural components is neither always possible nor appropriate. Thus, the system community is beginning to learn the same lesson learned by the software community back in the 1980s: that functional decomposition often increases data coupling, the number of interfaces, and the amount of message traffic between architectural components. System architecture engineering methods must therefore adequately address software and support the engineering of the architectures of software-intensive systems.

We live in a time of rapid change, and the technologies incorporated into systems are rapidly evolving. One of the most influential changes is the rapid increase in the amount of software in today's systems. As illustrated in Figures 2.2 through 2.4, systems are rapidly incorporating increasing amounts of software, and this software incorporates increasingly sophisticated software technologies. As illustrated in Figure 2.5, software is also being used to provide or control ever-increasing percentages of system functionality. One need only consider the safety, control, and communications subsystems that have added to aircraft and automobiles to see that software is enabling systems to do many things that were impossible only a very few years ago. Software is now the most critical

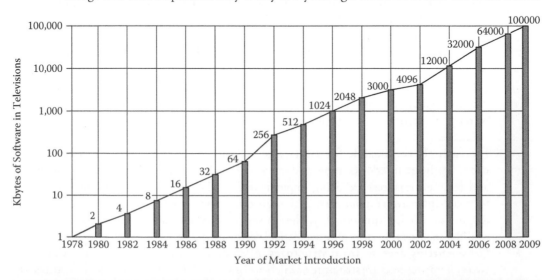

Figure 2.2   Software size in high-end television sets. (*Source:* From van Ommering, 2004.)

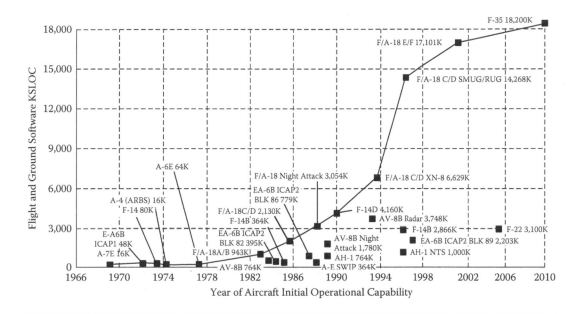

Figure 2.3 Software size increase in military aircraft. Note that the means used to count lines of code have varied from project to project and over the years. The scope of the software counted (e.g., avionics, all flight software, or both flight and ground software) may also have varied from project to project. Nevertheless, the trend of ever-increasing amounts of software remains clear as software has been used to provide ever-increasing amounts of functionality. (*Source:* From Saint-Amand and Hodgins, 2007, corrected and updated with F-22 and F-35 data from their respective programs.)

glue used to integrate subsystems together, and software is also the means by which systems are made interoperable with other systems. In fact, software currently implements a majority of the architecturally significant requirements (such as quality requirements specifying minimum levels of quality characteristics and quality attributes) as well as the functional requirements. For these reasons, system architecture engineering must incorporate and be consistent with software architecture engineering.

Software is not the only "disruptive" technology having a fundamental effect on system architectures. Other such technologies include nanotechnology and new strong and lightweight materials such as carbon-fiber composites.

- **Systems are hierarchically decomposed.** Although software-intensive, the system being architected will consist of a hierarchy of architectural elements (e.g., subsystems) that may contain significant amounts of data, hardware, software, procedural, and potentially other types of components. Although these architectural components may technically be subsystems, assemblies, subassemblies, and the like, we will use the generic term *subsystem* in the remainder of this book to mean any of these different types of architectural components. As will be seen in Figure 4.12, systems can be quite large, consisting of many subsystems in many tiers of decomposition.

- **Systems reuse architectural components.** Systems (and many of their subsystems) are rarely developed from scratch without any type of reuse. Thus, subsystems may contain architectural components that are developed specifically for the system, reused with modification (e.g., tailoring or configuration), and reused without modification. Reuse without modification of off-

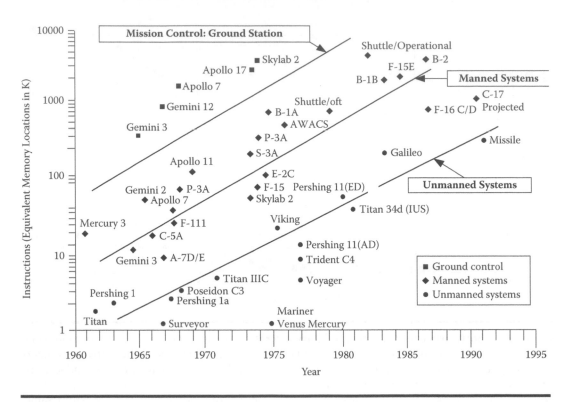

Figure 2.4 Air Force and NASA software size increase from 1960 to 1995. (*Source:* From AFSTCS, 1996).

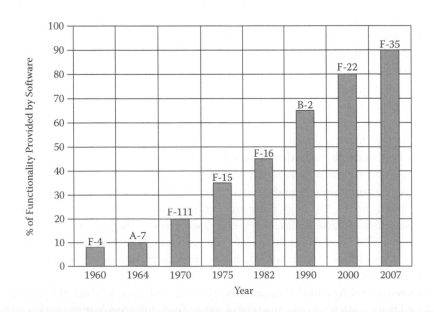

Figure 2.5 Increasing functionality implemented by software. (*Source:* From Etter, 2000, updated with F-35 data from the F-35 Program Office.)

the-shelf (OTS) architectural components may include the reuse of commercial off-the-shelf (COTS), government off-the-shelf (GOTS), open source, freeware, etc., components. These reused architectural components may also be systems, subsystems, hardware components, software components, data components, procedural components, facilities, etc. Logical components may be services as in a service-oriented architecture (SOA). There are also many programmatic constraints associated with the source of the reusable component (such as vendor, subcontractor, government, organizational reuse repository, and others).

■ **System architectures are never finished.** A system's architecture begins with its initial concept consisting of one or more candidate competing architectural visions. The architecture then grows through the selection of the winning vision and completion of the architecture of the initial version of the system. During the life cycle of the system, architecturally significant system requirements are added, modified, or removed and the system architecture evolves to meet these changes. For a system's architecture to retain its value, architectural integrity must be ensured and architectural representations must be maintained during the life of the system. All or part of the system architecture may be incorporated into reusable reference architectures and live on, even after the original system is retired. Thus, although architects strive to produce a stable architecture that provides a solid foundation supporting extensibility and modifiability, the architecture is never completely finished and the architects' work is never done. Neither is the work of the system architects, even if the individual architects making up the system architecture team also change over time.

The system architecture will evolve over time, both within a single block of development and across blocks of development.

  − **New system architectures.** During pure "greenfield" development of a new system, an initial, vague, highly incomplete vision of the system architecture typically will be developed during a concept evaluation phase. Then, the acquisition organization may, by itself, develop a more complete architecture or else the acquisition organization may contract out the development of a more complete system architecture for the request for proposals (RFP). Then a set of competing prime contractors may develop competing proposed architectures. Once the contract is let, then the winning prime contractor will incrementally and iteratively develop the system architecture during the initial releases and their phases of the development of the initial production version of the system. As part of this, subcontractors will probably develop architectures of the subsystems they will deliver.

  − **Reused architectures.** Systems are most often developed as variations of existing legacy systems. Systems are possibly developed as new additions to existing families of systems or product lines of systems. In such cases, the system architecture is typically a new version or variant of existing, reusable system architectures. There are often system- and subsystem-level reference architectures that provide a foundation for (and therefore greatly constrain) the architectures of the system under development. Similarly, the architecture of the system under development may be highly constrained by existing enterprise architectures that limit architectural choices, especially with regard to achieving interoperability.

    There is sometimes a strong tendency to place too much reliance on the architecture of the existing legacy systems, thereby risking the loss of opportunity to make a great leap forward. Sometimes what is needed is a revolutionary rather than evolutionary change in the architecture. A similar potential problem is that of accidentally inheriting architectural risks when new requirements imply the need for significant changes to the existing architecture.

- **Incremental development.** As previously mentioned, the size of most systems forces architecture engineering to be recursively performed in an incremental manner. The system architecture is often developed in a top-down fashion, initially at the overall system or system-of-systems level, then for each architectural element (subsystem, assembly, subassembly, etc.) at the next tier down in the system architectural hierarchy, and so on downward until only system design and implementation remains because of the small size and simplicity of the leaf node architectural elements.

- **Iterative development.** Because architecture engineering is a human activity and we all make mistakes and learn over the course of engineering an architecture, the initial architecture proposed is probably not the optimal one that should be incorporated into the system and its subsystems. As will be illustrated in Figure 4.9, an architecture team will typically develop a small set of candidate competing architectural visions, of which an optimal (although not optimum) one is selected. Even then, the selected architectural vision will evolve through several versions as the architectural vision is completed and implemented.

■ **Different stakeholders need different architectural representations.** As indicated in the Preface (*Intended Audiences*), there are many different kinds of stakeholders of the system architecture and its architectural representations. Because these different stakeholders have different interests and needs and use the architecture for different purposes, they need different architectural representations. The architect must meet their needs. That is, they need different architectural documentation consisting of different text, diagrams, and models showing different architectural views and architectural focus areas. To support their different needs, these different architectural representations should be at different levels of abstraction and detail. Also, different stakeholders use different terminology and the architect must communicate with them in terms they understand; not everyone in the world will learn UML.

To properly document a large and complex system architecture requires the engineering of a significant number of large and complex architectural representations. These representations and their relationships must be kept consistent. These representations will consist largely of multiple textual documents and static/dynamic and logical/physical models. The models will require both model-specific views as well as focus areas for different quality characteristics, specialty engineering groups, and disciplines. It is very difficult and expensive to decide which representations should be engineered, to create them with the correct level of abstraction and detail, to ensure their quality and consistency, and to maintain them and their integrity as the system evolves over time.

■ **There are many different system structures to be architected.** System architects need to engineer a system's numerous architectural structures. These logical and physical as well as static and dynamic structures are often large and complex. As will be illustrated in Figure 5.12, these structures, their relationships, and their associated views and models must be kept consistent.

■ **It is difficult to maintain architectural integrity.** A major task of system architecture engineering is to maintain the integrity of the architecture as it evolves. It is very difficult to keep the architectural decisions, inventions, trade-offs, and assumptions consistent during both the development and the maintenance of the architecture. The many different architectural representations including both models and documents must remain consistent with one another. The system architecture may also need to be consistent with organizational enterprise architectures as well as product line reference architectures.

- **Many different system architecture teams must collaborate.** Systems are developed by a large number of teams of various types. These teams have cross-functional memberships, relationships between teams, and multiple stakeholders. The organizations will also tend to be divided into acquisition organizations and development organizations consisting of both prime contractor (a.k.a. system integrator) and multiple subcontractors. It is difficult to properly staff these teams and to get them to properly collaborate.

  The size and complexity of many software-intensive systems and their architectures have outgrown the ability of any single architect to completely master. Typically, one or more architecture teams are required to provide adequate expertise and personnel to produce the architectural work products. When developing very large systems with a large number of architectural components, it is not at all unusual to set up multiple architecture teams to develop the architecture for either individual architectural components or sets of related architectural components.

  Proper architecture engineering is needed to provide the benefits of standardization and improve the communication and collaboration between multiple architects and architecture teams, to create and maintain a consistent architecture and associated architectural representations, and to keep project architectural risk acceptable.

  System architecture engineering helps the architects effectively and efficiently collaborate in large architecture teams and organizations.

- **Architectural decisions are sometimes made prematurely.** There are times when non-architects prematurely make major architectural decisions. For example, multiple contracts or subcontracts may be let based on organizational expertise, traditional decomposition structures, project schedule, and other reasons before the requirements and resulting overall system architecture have been adequately engineered. This can result in a suboptimal top-level system decomposition that is almost impossible to change.

As previously illustrated in Figure 2.1, systems engineering has passed a critical flash-over point to where existing architecture engineering methods have failed to scale up because of the extent of these challenges. Additional complications have arisen due to increasing volatility and incremental understanding of requirements that accompany such large and complex systems. The situation is further exacerbated by difficult-to-predict emergent properties that accompany the creation of systems of systems (SOS).

A major advance in addressing these challenges, MFESA captures industry best practices in a standardized way while retaining the flexibility needed to fulfill project-specific needs. Although no *method* should be viewed as a panacea or silver bullet, the MFESA *method framework* is a major step forward in the continuing evolution of system architecture engineering, and we believe that it is a proper stepping stone on the way to our ability to architect ultra-large-scale systems of systems.

2.3 Challenges Observed in System Architecture Engineering Practice

The previous general system architectural engineering challenges are not the only ones faced by system architects. While working on numerous systems development projects, the authors have also observed and experienced the following significant architectural engineering challenges reoccurring repeatedly in practice despite near-heroic efforts made by many dedicated and professional architects:

- Industry has a poor architecture engineering track record.
- Many architecture defects are found during integration and testing.
- Architecture engineering processes are used inconsistently in practice.
- Architecture representations are often confused with the underlying architecture.
- Architectural models are sometimes treated as the sole architectural representations.
- Architectural models are often incomprehensible to their stakeholders.
- Architecture engineering typically underemphasizes specialty engineering focus areas such as reliability, safety, security, and usability.
- It is difficult for architects to know how good is *good enough*.
- Because we lack sufficient adequately trained and experienced architects, they must sometimes perform tasks for which they are unqualified.
- Large projects often use multiple inconsistent architecture engineering methods.
- Architecture engineering methods are typically incompletely documented.
- Architects sometimes rely too much on architectural engineering tools.
- The connection between the architecture and the design it drives is weak.

The following provides a more detailed description of each of the above challenges observed in system architecture engineering practice.

2.3.1 Industry Has a Poor Architecture Engineering Track Record

Despite near-heroic efforts by experienced system architects, industry historically has a poor track record for developing large, complex, software-intensive systems on time, within budget, with the necessary functionality, and with adequate quality (i.e., with adequate levels of the relevant quality characteristics such as availability, interoperability, maintainability, portability, reliability, safety, security, stability, and usability). Studies and experience clearly demonstrate that inadequate requirements and inadequate architectures are two of the most important reasons for these failures [Elbert, 2006; Elm et al., 2007; Jones, 2008; Johnson, 2003; NASA, 2004; and Standish, 2008]. Inadequate architectural decisions, inventions, trade-offs, and assumptions are being made. They are not being adequately documented, and many of the resulting architectural defects and weaknesses are not being recognized until system integration and test time, when they are very difficult and very expensive to fix. In fact, in practice, many architectural defects are often not even recognized as such, but rather ascribed to general design deficiencies or developer errors.

2.3.2 Many Architecture Defects Are Found during Integration and Testing

Currently, a common problem with system architectures is that many of the defects in these architectures and their representations are often not found until system integration and integration testing. By having requirements engineers and architects closely collaborate, the critical need to properly engineer the architecturally significant requirements can receive proper emphasis. For example, when quality requirements that should be driving the system architecture are inadequately specified, it is often not until late in the project that the stakeholders recognize that the achieved levels of quality are inadequate. By then, architectural trade-offs have already been made, the architecture has been essentially completed, and the system design and implementation have been based on the inadequate architecture. At this point, making architectural changes to increase support for one quality characteristic is likely to decrease support for competing quality characteristics. For example, increasing reliability may decrease performance, and increasing security may decrease usability. Thus, architectural defects and inadequacies found during system integration can be very difficult and expensive to fix. Evaluation of architectural work products is a critical part of any complete architectural engineering method, and it is a responsibility of the architects to ensure that their architecture and associated architectural representations are adequately evaluated and any defects are fixed in a timely manner. An actual example of late discovery of architectural defects occurred as a result of Boeing's selection of Pratt and Whitney's PW2037 engine for the Boeing 757 [Womack and Jones, 1996].

System architecture engineering helps eliminate architectural defects and helps ensure that those defects that occur will be found earlier during the development and life cycle.

2.3.3 Processes Are Inconsistent in Practice

The as-performed architecture engineering processes actually used in practice tend to be inconsistent with the as-planned method. They vary within and between projects, between primes and subcontractors, and between different organizations. These often unnecessary differences make communication and reuse of training and experience more difficult than necessary. These discrepancies also cause defects and failures during integration with other systems.

2.3.4 Architectural Representations Are Often Missing, Incomplete, Incorrect, or Out of Date

The architecture of a system *is* the collection of the most important, pervasive, top-level, strategic decisions, inventions, engineering trade-offs, assumptions, and their associated rationales concerning *how* the system and its components collaborate to meet their derived and allocated requirements. The system's architectural representations describe the architecture, and they are frequently the only representations of the system. To be useful to their many stakeholders, elements of an architecture must be captured in the form of architectural representations such as models, text, diagrams, graphs, tables, drawings, and photographs.

Note that the architecture resides in the minds and memories of the architects. Although the architecture may be implicit in the system, the architectural representations exist independently of the architects and are explicit in the system and its documentation, models, and prototypes. When stakeholders (including architects) confuse the architecture with its architectural representations, they overlook that the two can be inconsistent, which necessitates updating the representations.

Architects are fond of saying that the architecture always exists, whether or not it is documented in the architectural representations. While this is true in one sense (because the architecture is implicit in the system), it is false in another sense. When the architects move on to new projects and forget the architectural decisions, inventions, trade-offs, assumptions, and rationales they made, then in an important sense, the architecture ceases to exist, which is an important reason why architectural representations are so critical. These representations become the architects' external durable memory of the architecture, enabling the architects and others to remember the architecture so that it can be maintained, its integrity can be maintained, and it can be reused by other architects on future systems.

These are reasons why a considerable amount of a project's limited time, effort, and funding is invested in producing the architectural representations. These are also reasons for verifying the completeness, correctness, and timeliness of the architectural representations prior to their acceptance by management and acquisition representatives. Finally, these are reasons for properly maintaining the architectural representations once they have been verified and accepted so that they do not become shelfware.

2.3.5 Architectural Models Are Treated as the Sole Architectural Representations

There is currently a strong trend to replace architecture documents containing free-form text and cartoon diagrams with more formally defined architecture models. For example, these models are often developed using version 2 of the Unified Modeling Language (UML), an updated version of UML reinterpreted for systems, the System Modeling Language (SysML) [OMG, 2007]; the Architecture Analysis and Design Language (AADL) [Feiler et al., 2006]; and the Department of Defense Architecture Framework (DODAF) [DODAF, 2007a–c]. While these models provide useful formality and have associated tool support for model production and consistency checking, there are many architectural decisions, inventions, engineering trade-offs, assumptions, and rationales that are not easily captured using the graphical models. Also, a huge model is typically not the best way to communicate architectural elements with stakeholders, who often lose sight of the architectural forest because of all the model element trees.

Unfortunately, when using frameworks such as the DODAF or modeling languages such as UML, there is a temptation to assume that "if there is not a view or notation defined for it, then it must not be needed." Similarly, on projects where a specific framework or modeling language is mandated by contract, there is a tendency to think that "if it was important, the customer would have required it and (explicitly) paid for it."

2.3.6 Architectural Models Are Often Not Understandable

It is important to remember that the primary purpose of architectural representations is to communicate with the architecture's many human stakeholders. Although architectural models are good for many reasons (such as organizing information, analysis and simulation, communicating structures and behaviors to engineers, and identifying redundancies), they are not appropriate for communicating with nontechnical stakeholders. One should also remember that different representations have different uses, whereby some are more oriented toward human readers, whereas others are more oriented toward computer-based simulation, automated checking, and automated generation of software.

While providing the ability to analyze and unambiguously communicate the architecture, the trend away from cartoon diagrams to semiformal graphical and formal textual models has also made it much more difficult to communicate the architecture to all of its stakeholders, most of whom are not trained and experienced in the chosen modeling language. The complete set of architectural models for a nontrivial system can also be huge, especially if one includes most of the important architectural structures (i.e., not just a functional decomposition logical structure and a physical decomposition structure of the system into a hierarchy of subsystems). When presented with multiple CDs containing a massive set of UML models in the form of an extensive HTML Web site, many stakeholders are beginning to rebel and demand traditional textual architecture documents, thereby making it even more difficult to ensure the consistency of the architectural representations.

The modeling notations themselves present barriers to understanding large models. Not only are some popular notations not understood by many stakeholders, they are not sufficiently well formed even to be fully understood by the modeling tools that use them. The DODAF [DODAF, 2007a–c], SysML [OMG, 2007], and UML [ISO/IEC, 2005] models at their core are supported by informally specified semantics, so that machine processing of large models for completeness and/or consistency is not entirely achievable. AADL [Feiler et al., 2006] has a somewhat more carefully specified underpinning, but there remains some question regarding its ability to scale up to capture and communicate very large models.

The point is to use the appropriate models appropriately, but not to rely on them for everything.

2.3.7 Architecture Engineering Underemphasizes Specialty Engineering Focus Areas

Different architectural models provide different views of the system architecture. These architecture models are often in the form of UML allocation diagrams, class diagrams, collaboration diagrams, sequence diagrams, and statecharts. While these diagrams provide an overall view of the associated model elements, they are often quite inappropriate for documenting the localized parts of the architecture that address specialty engineering concerns. For example, it is difficult to see how a wall-sized class diagram and network diagram clearly document how an architecture's decisions, inventions, and trade-offs support the achievement of minimum acceptable levels of such quality characteristics such as availability, configurability (e.g., internationalization and personalization), interoperability, maintainability, performance, portability, reliability, safety, security, and usability. The architectural elements resulting from some commonly existing analysis techniques (such as hazard analysis fault trees or security attack trees) are hidden in the architectural modeling representations noted above, and thus escape the notice of the architects despite their significance. Thus, existing architectural modeling languages and documentation as used in practice do not provide good ways of focusing on the relevant bits of the architecture. In turn, this makes it difficult to verify and validate that the architecture adequately supports any specialty engineering area or associated quality requirements.

This is a major weakness of current architecture engineering tools and most architecture engineering methods. It can be overcome by the consistent use of architecture quality cases [Firesmith et al., 2006], as documented in Chapter 18, "Architecture and Quality."

2.3.8 How Good Is "Good Enough"?

Because many, if not most, architecturally significant product requirements (especially quality requirements) and programmatic process requirements are very poorly engineered and specified, it is very difficult for the architect to know when the architecture is complete and when the architectural decisions, inventions, and trade-offs are good enough. Because different qualities are often inconsistent (i.e., architectural decisions that increase one quality characteristic tend to decrease other quality characteristics), it is also very difficult for architects to objectively know how to perform architectural trade-offs between the different quality characteristics.

Because in practice there is currently a strong emphasis on functionality over quality, a functional decomposition architectural structure is often the only logical view modeled, and performance is often the only quality characteristic given adequate weight during architecture engineering. However, many systems have been built that provide the required functionality but were not acceptable to their stakeholders precisely because of their lack of quality. Just because quality may not be emphasized by the stakeholders and adequately captured during requirements engineering does not necessarily mean that quality is not critically important. One of the major responsibilities of architects is to review the requirements for their adequacy with regard to architecturally significant requirements, especially the quality requirements.

Architectural quality cases are a way for the architects to make their case (e.g., to assessors and architecture stakeholders) that their architecture adequately supports the meeting of quality-related requirements. Quality cases consist of the following three components: (1) claims (that the architecture is good enough — sufficiently supports a specific type of quality requirement); (2) clear and compelling arguments (e.g., architectural decisions, inventions, trade-offs, assumptions, and rationales); and (3) sufficient supporting evidence (e.g., architectural representations such as architecture documents and architectural models). Inadequate requirements and an overemphasis on functional and physical decomposition make it very difficult for architects to produce proper architectural quality cases as a natural part of the architecture engineering process, which in turn makes it difficult and expensive to assess the quality of the architecture and to achieve safety and security certification and accreditation.

2.3.9 Because We Lack Sufficient Adequately Trained and Experienced Architects, They Must Sometimes Perform Tasks for Which They Are Unqualified

There is a significant lack of highly experienced system architects, especially those with experience architecting very large and complex software-intensive systems. Because of the multiyear duration of such development projects, there is a natural limitation to the number of such systems that can be architected by any single architect. It is not unusual for such projects to have a very small number of experienced architects and to give "field promotions" to a large number of general systems engineers and designers who must then perform the role of architect even if they lack the necessary training, experience, aptitude, and attitude to effectively and efficiently perform their new duties. Whereas highly experienced architects may have internalized reasonably complete and effective system architecture engineering methods, the beginning architects need sufficiently more in the way of methodological guidance. This guidance is incorporated into the architecture engineering methods created using the MFESA architecture engineering framework.

Not every designer has the necessary mindset, attitude, aptitude, training, and experience to make a good architect. University curricula do not provide an adequate breadth of material to adequately teach a nascent architect, nor is the faculty ordinarily steeped in sufficient practical experience to recognize and mentor people with the intellectual equipment to grow into the role. Additionally, architects need practical experience that can only be obtained by working on the architectures of real systems. Unfortunately, the industries that need such talent do not in general have apprenticeship programs to cultivate promising individuals by placing them in close contact with experienced architects.

As a profession and industry, we need to ensure that more systems engineering students learn architecture engineering in their universities; more systems engineers learn architecture engineering on the job, through both classroom training and apprenticeship programs; and that development organizations do a better job of hiring, rewarding, and retaining system architects.

There is a chronic shortage of qualified architects needed to support large projects. This results in the Peter Principle "field promotion" of many designers to architects [Peter and Hull, 2001].

Because the architecturally significant requirements are so often poorly and incompletely engineered, architects often find themselves informally performing requirements engineering tasks for which they are untrained and unqualified and for which they have inadequate input (e.g., access to stakeholders). Architects are also often called on to make architectural decisions to support specialty engineering disciplines (e.g., safety engineering and security engineering) for which they are unqualified. There is a general lack of adequate collaboration between architects and other teams. Even when the architects are performing architectural tasks, if they are inadequately trained and experienced, they will naturally make a relatively large number of architectural mistakes.

System architecture engineering helps inexperienced architects to more effectively and efficiently perform architectural work units and produce better quality architectural work products.

2.3.10 Architects Use Multiple Inconsistent Architecture Engineering Methods

Different architecture teams tend to use different and inconsistent versions of architecture methods, even when working on the same system. When lower-level subsystem architecture teams cannot wait for their top-level system architecture team to specify an endeavor-wide architecture engineering method, they develop variants of their own. Individual development programs often use different architecture engineering methods than the official common organizational architecture engineering method, which is sometimes used to support proposal efforts rather than to actually architect systems.

System architects must also do a better job during planning and proposal preparation to ensure the consistent usage of standardized, yet project specific, architecture-engineering methods. Method standardization can be achieved via reuse of standardized method components, which is a major benefit of MFESA and other method frameworks based on method engineering.

2.3.11 Architecture Engineering Methods Are Incompletely Documented

Architecture engineering methods are often very briefly documented in short sections of system architecture plans or in short procedure documents. Often, they are documented in one or two brief Web pages as part of an overall system engineering method Web site. Generally, the method documentation briefly states what architectural work products to produce but provides very little detail about how to produce those work products in terms of the architectural tasks, techniques, guidelines, patterns, principles, and heuristics to perform and use. When there is more detail, it usually concerns architectural models, the associated architectural modeling languages (e.g., UML, SysML, and AADL), and how to use any associated architectural engineering tools.

2.3.12 Architects Rely Too Much on Architectural Engineering Tools

Most architecture engineering tools only support architectural modeling, leaving many architectural engineering decisions, inventions, trade-offs, rationales, and assumptions undocumented. The tools may enable architects to develop huge individual models but they often make it almost impossible to restrict viewing to only the relevant parts of individual relevant models. They also make it very difficult, if not impossible, to see relevant bits of multiple models. And although architectural tools often support many fields for capturing information about model elements, the tools do not typically make it easy to see which fields have not yet been completed. Yet in practice, these fields are far too often left empty so that the boxes on diagrams are inadequately documented and the diagrams become little more than elaborately drawn cartoons.

2.3.13 The Connection between the Architecture and the Design It Drives Is Weak

Sometimes an architecture is explicitly created only because everyone knows that is how development is started. Because the architectural development is a *pro forma* activity, the resulting architectural representations lack sufficient detail to clearly capture all the architecture. As a result, the design of architectural components proceeds within the loose confines described by the architectural representations but is not driven by a need to satisfy the constraints of the architecture. Thus, the resulting components conform to the loosely defined architecture (e.g., the system is segmented as defined by the architectural representations). However, none of the advantages of a clearly documented architecture are realized. Attention to quality characteristics is omitted, and inter-component interfaces are incompatible or are not harmonized through negotiation among architectural component development teams, etc.

2.4 Challenges Observed in Systems Architecture Engineering Methods

The previously listed challenges with architecture engineering we have observed in practice suggest that either existing system architecture engineering methods need improvement, the methods need to be better applied, or both [ANSI/EIA, 2003; Albert et al., 2002; Chrissis et al., 2007; Clements et al., 2003; Hatley et al., 2000; IEEE, 2000; INCOSE, 2006; ISO/IEC, 2002; Maier and Rechtin, 2002; NAVY, 2004; Rechtin, 1991; and Wojcik et al., 2006]. However, when reviewing existing systems architecture engineering standards, methods, and guidebooks, we often discovered the following significant challenges with architectural engineering methods. Specifically, in most cases the following apply:

- Current system architecture engineering methods are incomplete.
- Current methods do not scale up.
- Current methods typically assume "greenfield" development.
- Current methods overemphasize architecture development over other tasks.
- Current methods overemphasize functional decomposition for logical structures.
- Current methods overemphasize physical decomposition for physical structures.
- Current methods are weak on structure, view, and focus area consistency.
- Current methods commonly codify old processes.
- Current methods emphasize the waterfall development cycle.
- Current methods confuse requirements engineering with architecture engineering.
- Current methods underemphasize support for the quality characteristics.
- Current methods assume that "one size fits all."
- Current methods produce only a single architectural vision.
- Current methods overly rely on local interface specifications.
- Current methods lack an underlying ontology.
- Current methods confuse architecture with architectural representations.
- Current methods excessively emphasize architectural models over other architectural representations.
- Current methods overemphasize the points of view of different types of experts.

The following provides a more detailed description of each of these challenges observed in system architecture engineering methods:

2.4.1 Current System Architecture Engineering Methods Are Incomplete

Many system architecture engineering methods are inadequately documented in practice. For example, the ISO/IEC 15288 (Life-Cycle Management — System Life Cycle Processes) devotes less than two pages to "architectural design," while the *INCOSE System Engineering Handbook* devotes 31 pages to a summary of "functional analysis/allocation" and "system architecture synthesis." The mere size of this book should make it clear that there is far more to system architecture engineering than that.

Many of the system architecture engineering methods in practice are very incomplete. They tend to concentrate on architectural work products, although they seldom address all architectural work products. Many methods do not provide much detail about the work units to be performed, such as tasks and techniques, leaving it up to the individual architects to decide how they are going to develop the work products. Often, the methods address the architecture tools but are relatively weak with regard to describing the different types of architects and architecture teams in terms of responsibilities, authorities, and required expertise and training.

The exact amount of information needed to adequately document a system architecture engineering method will vary greatly in practice because the size and complexity of the method are a complex function of the system (e.g., size, complexity, and criticality); the organizations involved in its development (e.g., experience, culture, and geographical distribution); and the project (e.g., contractual requirements). Therefore, we will not recommend a page count or even page count range for the documentation. However, by the time you have finished this book, we hope that you will recognize the inadequacy of much of the method documentation you have read in the past.

2.4.2 Current Methods Do Not Scale Up

Current methods work best on small and mid-sized systems. Unfortunately, experience clearly shows that they have not been effective and efficient when applied to large and complex systems, systems of systems, or systems composed of many OTS components, which often make incompatible assumptions or use incompatible underlying technologies. When it comes to the architecture engineering of ultra-large-scale systems, current methods are both untried and unproven [Northrop et al., 2006].

2.4.3 Current Methods Assume "Greenfield" Development

Although most system engineering efforts involve the iterative and incremental improvement of existing systems, many of the methods for engineering system architectures assume "greenfield" development of entirely new systems. Yet even new systems developed from scratch have many constraints and typically exist within the context of other systems, both internal and external to the user, operator, and maintainer organizations.

2.4.4 Current Methods Overemphasize Architecture Development over Other Tasks

Most system architecture engineering methods tend to concentrate only on the tasks needed to develop the architecture. They often lack any discussion and inclusion of steps needed for planning the architectural tasks, assessing the quality of the architecture, maintaining the architecture, and ensuring the continued integrity of the architecture over time.

One reason for this is that different organizations are often used to develop and maintain systems. In government contracting, for example, development organizations (and their architects) are typically requested to write proposals for conducting development through initial system delivery. Any sustainment (i.e., maintenance, support, and operations) is typically managed by a different contracting office and must be bid separately. Because extra work is not free and proposals are competitive, this separation of development and sustainment contracts has strongly discouraged development organizations from addressing the post-delivery system architecture engineering tasks.

2.4.5 Current Methods Overemphasize Functional Decomposition for Logical Structures

Many system architecture engineering methods equate logical architecture with a functional decomposition of the system into logical functions. There are actually many logical structures in addition to the functional decomposition one. For example, architects should be interested in class structures (domain modeling), control and data flows, data and information structures, logical concurrency, as well as modes and states.

2.4.6 Current Methods Overemphasize Physical Decomposition for Physical Structures

Many system architecture engineering methods equate physical architecture with a decomposition of the system into subsystems, sub-subsystems, and so on down to individual configuration items (such as data, hardware, software, manual procedures, and human roles). There are actually many physical structures in which the architect should be interested — for example, data, hardware, networks, and technology allocation.

2.4.7 Current Methods Are Weak on Structure, View, and Focus Area Consistency

A view of an architectural structure consists of one or more architectural models that tend to show the entire system or subsystem being architected. Yet only a few parts of any one model are relevant to show how the system addresses any given architectural concern (such as interoperability, performance, reliability, and security), and a focus area documenting architectural support for such a concern typically involves a relatively small number of relevant bits from parts of multiple models and views (as well as other architectural representations). Thus, a comprehensive set of architectural representations include views, models, and focus areas.

Perhaps because they are overly simplistic as far as the number and types of architectural structures and views, system architecture methods do not tend to address how to keep these multiple views of multiple architectural structures consistent with one another. Because most architecture methods do not explicitly address architectural focus areas, they also do not address how to keep the focus areas consistent with the models and views. Because this architectural consistency challenge will only become more difficult as architects recognize the value of additional architectural views and focus areas, architectural integrity needs to be an explicit part of architecture engineering methods.

Many current system architecture engineering methods rely heavily on a hierarchical functional decomposition of systems into subsystems. These methods fail to capture focus areas as well as the cross-cutting interactions that drive the complexity of interfaces between subsystems and other architectural components. Only a small portion of the cross-cutting interactions become sufficiently visible that associated program risks are identified. Unfortunately, these unacknowledged interactions drive many of the surprises that most projects encounter during system integration and testing.

2.4.8 Current Methods Codify Old Processes

Many current system architecture methods and associated standards codify processes that are some 10 to 15 years old, or older. This is exacerbated by the architects' natural tendency to use the same method with which they have previously had some success. However, traditional methods do not have excellent track records and are even less likely to be successful when applied to programs of today's size and complexity. While enabling architects to incorporate best practices of the past when they are still appropriate, architecture engineering methods may need to enable the architects to apply newer approaches when previous approaches are inappropriate.

For example, allocating functions performed by software to their own individual computational hardware boxes may no longer work if the number of functions increases and there are power, weight, air conditioning, and space constraints that prohibit incorporating additional computing hardware. This often requires individual computer platforms to perform multiple functions, and multiple platforms to collaborate to perform individual functions. Thus, the allocation of software to computing hardware is becoming more complex and architecture engineering methods must not assume a simplistic one-function-one-box allocation.

2.4.9 Current Methods Emphasize the Waterfall Development Cycle

Many current methods still place too much emphasis on the use of an impractical waterfall approach, thereby forcing the as-performed process to deviate from the as-documented method. The waterfall model assumes an unrealistic degree of determinism and comprehensive understanding of requirements, constraints, and the consequences of engineering decisions that is not realized in practice, particularly for complex systems that require the collaboration of multiple disciplines. The architecturally significant product and process requirements drive the architecture, but the architecture also drives the requirements, both in terms of the appropriate level of requirements when reusing OTS components, as well as the derivation and allocation of requirements down through the static physical decomposition structure of the architecture. To be successful, architecture engineering is typically performed as part of a concurrent, iterative, recursively incremental, and time-boxed development cycle.

To prevent inconsistencies between the planned method and the performed process, the method should explicitly cover modern development cycles. The architecture engineering method should clearly address how the architecting tasks will be performed in such a development cycle because platitudes and reverse arrows on development cycle diagrams are insufficient.

2.4.10 Current Methods Confuse Requirements Engineering with Architecture Engineering

The engineering of a system's architecture is quite distinct from the engineering of the system's requirements. Proper crafting of requirements entails specifying requirements that are:

- Unambiguous
- Verifiable
- Ideally neutral with respect to application of particular technology, materials, components
- Focused on items of interest to end users and related stakeholders for the system being specified

By contrast, architectural engineering produces engineering models that take requirements considered architecturally significant, life-cycle considerations, technological constraints, and future adaptation that may be expected of conforming systems to produce a set of architectural artifacts that are well-formed, sufficiently comprehensive, and sufficiently complete to enable the design and implementation of one or more conforming systems.

Systems architecture engineering often includes steps that are properly a part of requirements engineering and should therefore be performed by requirements engineers. Although the allocation of requirements to architectural elements is an architecture task, deriving new requirements for the architectural elements is a requirements engineering task. Although requirements engineering and architecture engineering are interleaved as part of a concurrent, iterative, and recursively incremental development cycle, this is no excuse to incorporate requirements engineering as a part of architecture engineering. Although their methods must be consistent and closely coordinated, requirements and architecture engineering are distinct disciplines with distinct tasks, techniques, and tools requiring different expertise and skill sets of their practitioners.

Requirements engineering and architecture engineering are two distinct but interwoven disciplines and activities. They should collaborate closely but not be allowed to merge together. In theory, the same person can play the roles of requirements engineer and architect on the same project and be a member of both the requirements and architecture teams. In practice, however, few people have the qualifying training, expertise, experience, and mindset to play both roles effectively and efficiently. Thus, even if cross-functional integrated product teams (IPTs) are used to develop the system, they should be subdivided (at least logically) into requirements and architecture teams. However, because of the critical differences between requirements and architecture, we recommend that separate — but closely collaborating — requirements and architecture teams be staffed, in many ways like other specialty engineering teams such as safety and security. To ensure adequate collaboration, an effective compromise is to have an architect be a member of the requirements team so that the architect can:

- Ensure that the requirements are appropriate, feasible, consistent with the architectural work being performed concurrently with the requirements work, and not improperly constraining (i.e., unnecessary architectural constraints)
- Act as a liaison between the requirements team and the architecture team

2.4.11 Current Methods Underemphasize Support for the Quality Characteristics

There is often an insufficient emphasis on ensuring that the system architecture adequately supports the achievement of the quality requirements that specify minimum amounts of quality characteristics* such as availability, interoperability, performance, reliability, safety, security, and

---

* Also known as quality attributes or "ilities."

usability. These are system-level properties, not just software properties, and some of them (e.g., manufacturability) only make sense at the system level.

2.4.12 Current Methods Assume That "One Size Fits All"

Many system architecture engineering methods exhibit a propensity to emphasize a "one size fits most or all" approach. They have either too much emphasis (or too little emphasis) on:

- A small and simplistic set of views and models*
- *Greenfield development,* although many (most) systems:
  - Are enhanced versions of existing systems
  - Must fit into (and interoperate with the components of) an existing system or network of systems
  - Must plug into an existing infrastructure
- *One of a kind systems,* although there are benefits to viewing many systems as members of a product line and conforming to a product line reference architecture
- *Stand-alone systems,* although the system may need to fit within a system of systems and be consistent with an enterprise architecture
- *Reuse* (e.g., COTS), whereby some methods ignore reuse and others are built around COTS, reuse, and product lines
- *Software or hardware,* when the physical decomposition architectural structures include components consisting of one or more of the following: data, equipment, facilities, firmware, hardware, materials, procedures, and software
- *One life-cycle phase,* with too much emphasis on the initial development of the architecture and insufficient emphasis on its evaluation and maintenance

2.4.13 Current Methods Produce Only a Single Architectural Vision

Although the first architectural vision is unlikely to be optimal, many system architecture engineering methods do not include the development of multiple competing architectural visions (i.e., incomplete draft architectures capturing the most important architectural decisions, inventions, and trade-offs). Because the methods do not explicitly include the development of multiple architectural visions followed by the selection of the most suitable vision, planning contains neither sufficient budget, schedule, and availability of architects to develop the multiple visions nor the resources to fix the negative consequences for having to go with the first suboptimal vision developed.

The architecture teams should produce multiple visions that provide a reasonable sampling of the architecture space so that the most appropriate can be selected or so that a better architectural vision can be created by combining the best architectural decisions from the competing visions. MFESA makes this creation of multiple candidate architectural visions and selection of the most suitable of these competing architectural visions explicit system architecture engineering tasks.

---

* If a person only knows how to use a hammer, everything looks like a nail. Thus, although traditional views are often the most commonly useful ones, there are many times when a specialty view is critical when dealing with an architectural concern (i.e., a cohesive set of architecturally significant requirements) such as those related to a specialty engineering area like reliability, safety, and security.

2.4.14 Current Methods Overly Rely on Local Interface Specifications

A major part of architecture engineering is the decomposition of a system into its component subsystems. However, this static physical decomposition structure is only one of many. While it is important to properly capture the local subsystem-to-subsystem interfaces in terms of their syntax and semantics, such interface specifications do not do justice to the many other static and dynamic, logical and physical structures that overlay and crosscut this primary static structure.

Architecture engineering methods and the architecture representations they produce should properly address architectural support for quality requirements and emergent, holistic system properties and behaviors that cut across multiple architectural structures and components. Therefore, they should address global as well as local views of the interfaces. These views should capture dependencies induced by the interfaces, including technical dependencies and those created among the teams that will develop and maintain the interacting subsystems. For example, interactions across multiple architectural components, including both requests for services and the raising of exceptions, may well need to include information about the overall mode of the system and the relevant states of the intermediary architectural components.

2.4.15 Current Methods Lack an Underlying Ontology

Most current architecture engineering methods lack a well-defined, consistent, complete, ontology defining the foundational concepts underlying architectural engineering. Some terms are not defined, whereas other terms are not clearly defined and differentiated. When terms are defined, they differ from method to method. While an industry-wide standard ontology may well be optimal, a minimum requirement should be that all of the architecture engineering methods used during the development and maintenance of a single system or system of systems should be based on a common understanding of architectural terms and concepts.

2.4.16 Current Methods Confuse Architecture and Architecture Representations

Many current system architecture engineering methods tend to confuse a system's architecture with representations of the system's architecture. Because they tend to consider architectural structures to be the only parts of the architecture, methods therefore tend to overemphasize views of these structures and the models comprising these views. Less and less emphasis is being given to other types of architectural representations, including trade-offs, assumptions, rationales, and more traditional documentation of the architecture.

2.4.17 Current Methods Excessively Emphasize Architectural Models over Other Architectural Representations

Most current architecture engineering methods tend to emphasize the development of architectural models capturing architectural structures to the exclusion of all else. However, the architecture includes numerous architectural decisions, trade-offs, assumptions, and rationales that are not easily or best captured and communicated in the form of architectural models.

2.4.18 Current Methods Overemphasize the Points of View of Different Types of Experts

Based on their past experience and areas of expertise, different groups of experts develop different architecture engineering methods (or systems engineering methods including architecture engineering) that emphasize their interests while remaining silent on other important aspects of architecture engineering. Thus, there are architecture engineering methods inspired by traditional hardware engineering, methods based on software engineering, methods emphasizing COTS-based development, and methods centered on product line development. Some standards only address architectural documentation, whereas others address architectural engineering tasks, techniques, tools, and teams. The situation is analogous to the old parable of the six blind men who happen upon an elephant for the first time and describe it quite differently, depending on the part of the elephant they happen to touch. No current system architecture engineering method adequately addresses the interdisciplinary nature of the discipline, and this can lead, for example, to separate architecture engineering efforts neither being adequately coordinated nor coordinated sufficiently early. Some of these methods delay software architecting tasks until well after system architecting tasks are done, even when applied to the development of software-intensive systems. Such methods defeat any iterative architectural refinements that may be needed due to suboptimal interdependencies of decisions made within the two separated sets of architectural tasks.

2.5 Reasons for Improved Systems Architecture Engineering Methods

The quality of system architectures depends on many factors in addition to the system architecture engineering method used, including project resources, management support, and the availability of experienced architects. Nevertheless, the preceding lists of challenges make it clear that there is significant room for improvement. A major goal of MFESA is to support process improvement by significantly improving current system architecture engineering methods.

The challenges due to the shortcomings of current methods must be addressed if the systems and software engineering communities are to successfully architect systems that satisfy current and envisioned operational needs. As capabilities of individual components extend our potential reach, our ability to efficiently incorporate those capabilities into ever larger, ever more complex structures is a necessary prerequisite to extending our grasp and ultimate mastery of future operational needs. An improved method for engineering systems architectures provides a concrete means to augment our ability to meet current and future expectations.

Improved methods will leverage the best of the existing methods, guidebooks, and standards, while filling the "holes" that have been observed in practice.

To deal with the increasing scale and complexity of software-intensive acquisitions and development programs, improved systems architecture engineering methods should enable the architects to focus their attention on those critical system quality characteristics that are identified as truly driving the architecture. Improved methods need to place a much greater emphasis on identifying, analyzing, and properly addressing these system qualities, by giving them at least as much emphasis as the existing methods have given to functional decomposition.

Requirements that are not identified as being architecturally significant need to be "abstracted out" to avoid becoming unnecessary baggage for the architects to carry around during architecture engineering. The development of these nonarchitecturally interesting requirements needs to continue but must be off the critical path of the architects' incremental iteration loops with requirements development activities for the architecturally significant requirements.

In addition to the above, any improved method should focus on permitting:

- Quicker and more numerous iteration cycles between requirements and systems architecture engineering.
- Quicker and more numerous engineering iteration cycles between systems architecture numerous engineering and software architecture engineering (particularly for software-intensive systems).
- Quicker and more iteration cycles between higher-level systems architecture decisions and subsystems-level architecture decisions.
- An overall net effect of reducing the "time constant" of iterations between requirements engineering and system/software architecture engineering.
- Greater emphasis on identifying and producing the system architecture views and decisions well beyond "traditional" functional decomposition and static hierarchical aggregation views. These views will illustrate the broad system qualities that were deemed important to the architecture and that cross-cut the system functions that are typically shown in the functional hierarchy view.
- Clear identification of the primary audience for, and intended use of, the various architectural work products produced by the methods.
- Greater emphasis on explaining and recording the "allocation" of the important system qualities to architectural decisions, inventions, and structures. This will likely involve close collaboration among the system architects and the system requirements engineers.
- Greater emphasis on capturing the architectural rationales that promote those system qualities that have been identified as important to or driving the architecture.
- Greater emphasis on keeping the relevant architectural views and models consistent.
- Greater emphasis on providing consistent architectural engineering methods across different architectural teams.
- Ease of explaining and understanding how the system architectural artifacts are used to constrain downstream design, implementation, integration, testing, and deployment efforts.
- Ease of understanding how the architectural work products are used and updated over the entire system acquisition life cycle.

2.6 Summary of the Challenges

This chapter listed and summarized the major challenges facing system architecture engineering. The first list contained *general challenges* associated with systems architecture engineering itself. The second list contained challenges commonly *observed in practice* during the performance of architecture engineering. The third list contained challenges often observed with major current *systems architecture engineering methods*. Finally, the fourth list summarized important reasons for *improved system architecture engineering methods*. These challenges and reasons have led us to develop the MFESA for generating effective and efficient system architecture engineering methods. They are also reasons why you should seriously consider using MFESA in your organization.

Chapter 3

System Architecture Engineering Principles

3.1 The Importance of Principles

System architecture engineering (SAE) should be founded on a consistent set of underlying principles that drive how it should be performed and the work products that it should produce. These principles should be reasonably small in number and at a high level of abstraction. These principles should also be clearly identifiable, well defined, and have rationales based on experiences and practical lessons learned during the engineering of real system architectures. Because we have found the following SAE principles to be fundamental and a foundation on which SAE should be built, we have therefore based MFESA (Method Framework for Engineering System Architectures) upon them:

- Support for situational method engineering
- Scalability
- Architecture evolution
- Enabling system quality
- All types of architecture components
- Component-based development
- Well-defined concepts and terminology
- Architectures and their representations are valuable resources
- Both architectural views and focus areas
- Sufficient architectural representations
- Completeness of methods produced
- Teamwork
- Independence of architecture assessments
- Analysis impact before architectural changes

3.2 The Individual Principles

The following is a description of the preceding principles together with an associated rationale:

■ **Support for situational method engineering.** SAE should properly support situational method engineering to produce *appropriate* architecture engineering methods, whether these methods will be specific to the organization, endeavor, system, or team.

Rationale: Individual systems, endeavors such as individual projects and programs of projects for developing product lines of systems, and both acquisition and development organizations can vary widely in terms of size, complexity, and criticality. Methods that work well in one context are not guaranteed to be effective and efficient when used in different contexts. Although these architecture engineering methods may be produced by tailoring existing methods, often that does not provide adequate flexibility. Producing new situation-specific methods often is better accomplished by selecting and integrating reusable method components from a repository of such components. In general, it tends to be easier to tailor out (or not select) tasks, steps, and work products that are not necessary or not cost-effective than it is to add or invent missing method components either before or during the performance of system architecture engineering tasks. To support the creation of appropriate situation-specific architecture engineering methods, it is better that generic architecture engineering methods and repositories of reusable architecture engineering method components are relatively complete.

■ **Scalability.** SAE must be able to properly address both the size and associated unavoidable complexity of the systems being architected.

Rationale: Systems come in all sizes and associated inherent (as opposed to accidental) complexity from the nearly trivial to the intractable. Remember that system architectures consist of *all* strategic, top-level decisions, inventions, trade-offs, assumptions, and rationales related to how the system meets its requirements. Typically, system architectures primarily consist of a large number of logical and physical static and dynamic structures that should be represented using appropriate models.* As the number of system components increase, the number of relationships between the system components also increases in an even more rapid manner.

■ **Architecture evolution.** SAE must recognize that architectures are never done and architecture engineering never stops. System architectures can and will evolve over time. This evolution occurs both during the initial development of the initial system and its architecture, as well as during the following development of future versions of the system as the system's context and requirements change. Thus, architecture engineering must support the successive refinement of system architectures during the system development and life cycle.

Rationale: The final architecture is highly unlikely to be the first architecture that an architecture team envisions. Architecture engineering must successively refine the architecture to meet its evolving architectural drivers.

Although SAE must properly address systems of all sizes and complexities, the trend is clearly toward ever larger and more complex systems. Architectures must typically be iterated because they are typically too large and complex to be correctly developed without

---

* Remember that there are far more important architectural structures than just the single static logical functional decomposition structure and the single static physical hierarchical decomposition of the system into a tree of subsystems. As the number of these structures increases, the number of views and models naturally increases.

defects and opportunities for improvement. Architectures must be recursively and incrementally engineered because they are typically too large (many architectural elements, many subsystems, many structures, and many decisions, inventions, trade-offs, and assumptions) to be engineered all at once. Architectures must evolve with the systems and their associated systems because the systems themselves tend to evolve over time as stakeholders become more proficient with them and recognize opportunities for new and enhanced features. Architectures also evolve from one version of the system to the next. Architectures are developed in parallel with the requirements, design, and implementation, which themselves evolve. Therefore, architects should develop multiple candidate competing architectural visions, select or create the most suitable architectural vision, and use the resulting most suitable vision as a foundation on which to complete and then maintain the architecture. From this, it is clear that architecture engineering should not assume sequential waterfall development cycles and milestones.

■ **Enabling system quality.** The system architects should concentrate on ensuring that the system architecture adequately supports the achievement of the system's required levels of quality.

Rationale: Not all requirements are architecturally significant. Major functional requirements (especially when grouped into major feature sets) do influence the system architecture. Similarly, development and sustainment process requirements (such as mandated maximum development schedule, maximum development and life-cycle costs, and mandated usage of available development staffing and manufacturing facilities) also influence the system architecture. However, the primary architecturally significant requirements are the quality requirements that mandate minimum acceptable levels of individual quality characteristics (the "ilities") or their component quality attributes. The quality requirements are thus the primary architectural drivers and concerns. Because the quality requirements drive the development of the structures of the system architecture, the quality requirements are also used as a sieve for selecting among competing candidate architectural visions. The system architecture is also verified against these requirements.

Given adequate quality requirements, the architects' primary responsibility is to engineer an architecture that adequately supports the achievement of the system's required levels of quality. Although a good architecture *by itself* cannot guarantee that the system will meet its quality goals and requirements, a poor architecture can make achieving them essentially impossible. Thus, architecture is necessary — but not sufficient by itself — to achieve success.

■ **All types of architectural components.** Systems can be composed of many different types of architectural components. System architecture engineering should adequately address all relevant types of system architectural components from which systems may be composed. This includes components composed of data, equipment, facilities, firmware, hardware, manual procedures, roles played by people and organizations, and software. System architecture engineering should neither be excessively driven by hardware engineering nor software engineering, but address all components in a manner commensurate with their importance to the system.*

---

* Note that the need to properly represent all the different types of architectural components does not imply the need to use all the different types of architectural views. A single model may well be able to represent many (if not all) of the different types of components.

That being said, there can be no doubt that many if not the vast majority of modern systems and subsystems are software-intensive and have a large amount of their requirements, behavior, characteristics, and internal and external interfaces implemented by software. Software has become arguably the biggest disruptive technology in systems engineering during the past quarter century. Thus, system architecture engineering should properly address and include software architecture engineering. A major consequence of this is that architects should *not* wait to address software's critical influence on the system architecture until after the system has been decomposed through multiple levels of subsystems until pure software architectural components and configuration items are identified.

Rationale: Most modern systems are quite complex, physically consisting of subsystems composed of combinations of data, equipment, facilities, firmware, hardware, manual procedures, roles played by people and organizations, and software. These different components and the emergent properties and behaviors that result from their combination must properly collaborate to implement the system's requirements and fulfill the system's mission. Yet a system architecture engineering method that addresses (or primarily addresses) only a single type of component will fail to properly address the other types of components and the emergent properties and behaviors that result from their combination. Unfortunately, many systems architecture engineering methods exhibit one of the following weaknesses:

- Systems engineers who were originally hardware (or electrical) engineers tend to overemphasize the architecting of hardware components and thus tend to develop and use system architecture engineering methods that have been derived from hardware development methods.
- Systems engineers who were originally software engineers tend to underemphasize the architecting of non-software components and thus tend to develop and use system architecture engineering methods that have been derived from software architecture engineering methods.

On the other hand, software is a disruptive technology that is having an increasingly critical and persuasive influence on system architectures. Therefore, as identified in the associate software architecture engineering challenges in section 2.2, software often does require special emphasis in the architecture engineering methods because software is the primary:

- Means used to provide needed functionality and implement the majority of functional requirements
- Means used to achieve the necessary level of quality characteristics and attributes
- Implementer of data and controller of data flow
- Implementer of behavior and control flow
- Provider of fault tolerance and exception flow
- Provider of system interoperability
- Glue to integrate subsystems together and provide intra-operability
- Implementer of cross-cutting architectural structures
- Component responsible for safety and security

A software architectural decision, invention, or trade-off made in one architectural component often has a major influence on the software architecture of other components. Interface, infrastructure, and modeling incompatibilities between software in different architectural components is a major source of problems during integration and system testing.

- Systems engineers who have neither extensive hardware nor software backgrounds tend to develop and use system architecture engineering methods that are very high level and generic such that they do not adequately address the hardware and software components.

– Because of the typical emphasis on either software or hardware components, most systems architecture engineering methods do not adequately address architectural components such as data, equipment, facilities, and manual procedures.

■ **Component-based development.** System architecture engineering must include the task steps, techniques, and work products needed to support the identification, evaluation, and selection of reused architectural components. Architecture engineering must include determining the often significant impact of their reuse on development and life-cycle costs, schedule, and risks. As part of the evaluation of reusable architectural components (and their sources), architecture engineering should also include the testing of these components within the context of the proposed architectural vision(s) and actual system architecture.

Rationale: Most systems include at least one reused architectural component, and many systems result from component-based development (CBD) involving the incorporation of multiple reused components to provide a significant portion of the system's functionality and quality characteristics. Such system architectures tend to reuse a significant number of both hardware and software components.

Significant reuse influences many disciplines, including requirements engineering, integration, testing, and management, as well as architecture engineering. For example, requirements engineering must be sufficiently flexible to include negotiation, license issues and support the trade-offs needed to enable the reuse of significant COTS (commercial off-the-shelf) components. Note that a challenge associated with the pervasive effect of the reuse of architectural elements is that it sometimes becomes difficult to determine whether a task, technique, or work product supporting CBD belongs to architecture engineering or some other discipline (such as requirements engineering or management).

■ **Well-defined concepts and terminology.** System architecture engineering methods should be based on an ontology of related engineering concepts and terminology. The elements of the ontology should be well defined and mutually consistent, capturing not only the definitions of the terms but also the important relationships between the underlying concepts. This standard glossary should be applied consistently across the project, organization, and architecture teams.

Rationale: Such an ontology of concepts and terminology promotes communication and understanding among architects and other stakeholders of architecture engineering. The substance of any engineering discipline resides in the strength of its commonly applied terms and the degree to which these concepts and terminology are applied globally and consistently by practitioners and stakeholders. Conversely, an engineering discipline may be viewed as weak, or not in keeping with good engineering practices if these definitions are missing or have unnecessary variability from one stakeholder group to another.

■ **Architectures and their representations as valuable resources.** System architecture engineering methods must recognize the critical importance of the system architecture as the bridge from the most critical system requirements to the resulting design and implementation that it drives. Similarly, SAE methods must recognize that although the architecture may exist within the system, it is not easily visible until it is communicated among stakeholders and this communication requires the proper production of representations of the architecture. Therefore, SAE methods must provide sufficient tasks, steps, and techniques to properly produce the system architecture and provide sufficient architectural representations. The SAE methods must also ensure that these architectural representations are active, living documents, maintained and kept current and consistent as the requirements and design evolve.

Rationale: The architecture as documented in the architectural representations drives the downstream design, implementation, integration, testing, operation, maintenance, and retirement of the system. The architecture implements the architecturally significant requirements and thus addresses the architectural concerns. The architectural representations also support the traceability of requirements to the design, implementation, and tests.

Unless the architecture is properly captured in the architectural representations, it cannot be properly analyzed, tested, and communicated with its stakeholders. Unless architectural representations are kept current, they cannot be used to evaluate the architectural impact of requirements additions, modifications, and deletions.

A great deal of resources (schedule, budget, and staff effort among others) are, or at least should be, invested in the production and maintenance of the system architecture and its architectural representations. Unless the SAE methods recognize this, it is highly likely that the integrity and value of the architecture and its representations will *not* be maintained.

■ **Both architectural views and focus areas.** System architecture engineering must develop multiple architecture views, whereby each such view is an abstraction of a single architectural structure of a system consisting of one or more related models of that structure. System architecture engineering must also develop appropriate architectural focus areas, whereby each focus area is the cohesive set of all architectural decisions, inventions, engineering trade-offs, and assumptions related to a specific architectural concern (such as interoperability, reliability, safety, and security), regardless of the architectural view, model, or structure where they are documented or found.

Rationale: Because a system architecture includes multiple logical and physical, static and dynamic structures, it will require multiple architecture views to capture the models of these structures. Because a system architecture must also adequately address multiple architecture concerns, and because these concerns are implemented by only a few of the model elements of multiple models comprising multiple views, individual views are *not* sufficient by themselves.

Focus areas are also needed. Thus, for example, if one needs to see how and to what degree the architecture supports its architecturally significant availability, interoperability, safety, and security requirements (i.e., supports these four architectural concerns), then one needs the four focus areas that focus on the relevant bits of the relevant models of the relevant views (and other architectural representations such as relevant architectural white papers). Note that focus areas should be thought of as an integral part of the architectural representations and not merely a way to use existing architectural representations to analyze the architecture. Like the models and views, architectural focus areas should be naturally developed by the architects on an ongoing basis to help them ensure that the architecture adequately supports meeting the architecturally significant requirements and to support later architectural evaluations.

Although architectural views and focus areas are not the only architectural representations, they are nevertheless excellent and powerful ways to document and communicate important architectural decisions and inventions, as well as illustrate important engineering trade-offs and assumptions.

■ **Sufficient architectural representations.** *Sufficient* is the key word in this principle. SAE must produce sufficient architectural representations to adequately document and communicate all important aspects of the system architecture, including architectural decisions, inventions, assumptions, and rationales. SAE must include sufficient architectural views and focus areas to properly document the important architectural structures. SAE should produce sufficient architectural representations to support the generation of proper architectural quality cases that document the architects' clear and compelling arguments that their architectures sufficiently address the architectural concerns and provide sufficient evidence supporting these arguments.

Rationale: Every system has an architecture[*] in the sense that its architects consciously and unconsciously made a large number of architectural decisions, inventions, trade-offs, and assumptions. These architects also had explicit or implicit rationales for their architecture. Thus, every system has an architecture, regardless of whether that architecture is described in the form of one or more architectural representations. Thus, an architecture is not the same as an architectural representation. Architectural representations are needed to communicate the architecture to its stakeholders. Formal and semiformal architectural representations are also needed to model the architecture and to enable simulation and testing of the architecture.

Too often, architectural representations have only a superficial relationship to driving quality characteristics or system functionality that is designed to support the quality characteristics. Because the architecture is a *window* into the system for those having both technical and managerial responsibilities for ultimate system production, representations must have sufficient fidelity *and no more*.

In practice, most sets of architectural representations are currently restricted to some "standard" of models such as UML or DODAF. Because of this, they may not model many architectural structures that are relevant to specific stakeholders in specific areas of the system. This problem is not just due to incomplete methods or architects who may not be aware of some types of models. For example, it sometimes occurs that a system acquisition statement of work (SOW) specifies and funds the development of a specific set of architectural work products. In this case, the complete set of appropriate architectural models may not be funded, and the acquisition organization may be suspicious of the development organization's suggestion for the development of additional views as unnecessary "scope creep." This selection of a specific set of models, especially by the acquisition rather than the development organization, prior to the detailed analysis of the system requirements is an inappropriate constraint on the system architecture engineering method and can result in project cost and schedule overruns if important architectural structures are overlooked and mistakes are therefore made.

These representations also do not make it easy to determine the degree to which the architecture supports the achievement of the architecturally significant requirements, especially the quality requirements.

■ **Completeness of methods produced.** System architecture engineering should enable the production of architecture engineering methods that include:
- All appropriate *architecture workers* (roles, teams, and tools)

---

[*] Note that this commonly stated assertion can be very misleading. Just because a system theoretically has an architecture does not mean that the architecture is good or even adequate. Perhaps the system had no one explicitly playing the architect role or perhaps those who did had insufficient architectural training, experience, and expertise. Similarly, it is possible that many, most, or even all of the architectural decisions, inventions, trade-offs, assumptions, and rationales could have been made implicitly and unconsciously. In such cases, the statement that "the system has an architecture" has very little value or meaning. This commonly stated assertion can also be misused by the development organization to justify passing an acquirer requirement mandating "The system shall have an architecture." In any case, there should instead be requirements mandating that sufficient architectural representations exist to properly document the architecture, and that both the architecture and its representations have sufficient completeness and quality to meet the needs of their stakeholders.

- Performing all appropriate *architectural work units* (tasks and techniques)
- Producing and maintaining all appropriate *architecture work products,* including all appropriate models and documents

Although all projects will not need all these architecture engineering method components and although many of those components may well benefit from being tailored down, projects developing large and complex systems will still need large and complex architecture engineering methods. Because different projects may well need different architecture engineering method components, the entire set of potentially reusable architecture engineering method components can be quite large.

The method components used to generate architecture engineering methods should have a standardized form where possible in support of clarity and ease of use. Because the underlying structure of an architecture engineering method must be clearly understandable and repeatable, their constituent method components should be based on the strong foundation of an underlying metamodel and ontology.

Rationale: The form of systems architecture engineering should promote the ability to "consciously leave out" rather than the propensity to "accidentally overlook." This means that the tools for SAE are intended to be tailored under the presumption that "one size *does not* fit all."

Notes: Different sources include different sets of method components, as simple as possible but no simpler. Do not think just hardware, or just software — think system and subsystem. Address all useful parts of major standards.

■ **Teamwork.** System architecture engineering should emphasize the need for system architectures and architectural representations to be developed and maintained by architecture teams. The membership of these teams should include multiple people playing multiple roles, and the members should have sufficient expertise, experience, and training to be able to effectively and efficiently perform the many architecture engineering tasks, use the architecture engineering tools, and develop and maintain the architecture work products.

As documented more completely in Chapter 16, "MFESA Method Components—Architectural Workers" on architects and teams, the architecture teams should be cross-functional, integrated product teams (IPTs), including not just architects, but also other important stakeholders.

Rationale: Although a simple mention of the importance of stakeholder involvement should be sufficient in defining this principle, it may also be added that every major scholarly and technical source defining best practices in organizational excellence emphasizes the importance of team involvement and understanding of team dynamics as essential to successful project endeavors. Architecture engineering is no exception to this rule.

This principle must be made explicit due to a very human tendency to think "if you want it done right, then do it yourself." The difference is that for systems development, the team must possess this perspective as one. The sum of concerns across any system beyond the trivial is too many for any individual to handle successfully. The management of multiple competencies allocated in appropriate measure according to the identified and explicit requirements of the system is essential.

■ **Independence of architecture assessments.** System architecture engineering should support the use of independent assessments to evaluate the correctness and quality of the system and architectural component architectures and associated architectural representations.

Rationale: Independent assessments identify architectural defects and risks that are not as easily identified using only peer-level evaluations. The critical importance and value of the architecture and its representations justify the added expense of independent assessments.

■ **Impact analysis before architectural changes.** System architecture assessment should (as part of configuration management change control) support the impact analysis of proposed changes to system requirements, architecture, and design for their ramifications on the architecture and its architectural representations.

Rationale: The risk of architectural degradation and loss of integrity significantly increases without proper impact analysis.

Notes: It can be very difficult to explicitly identify architectural baselines because of their size and complexity, the iterative and incremental nature of architecture engineering, and that both models and documents often contain requirements and design, as well as architecture.

3.3 Summary of the Principles

System architectural engineering should be founded on a set of underlying principles that, if followed, enable the effective and efficient construction of system architectures and their representations. Some of these principles should be considered *mandatory* in that they should always apply. If followed, these mandatory principles will almost always be cost-effective and improve the architects' effectiveness and efficiency. On the other hand, certain principles should be considered *recommended* because, if followed, they are typically — but not always — cost-effective and improve the architects' effectiveness and efficiency.

We believe the following principles should be considered mandatory, and system architecture engineering (SAE) should be based on them. This is why MFESA was designed to incorporate them.

■ **Support for situational method engineering.** SAE must support situational method engineering because the architecture engineering methods used need to be appropriate for the system being architected and the organization doing the architecture engineering.

■ **Scalability.** SAE must be scalable because it must be able to properly handle the ever larger and more complex systems being developed and updated.

■ **Architecture evolution.** SAE must support the evolution of architectures and their architectural representations because the architecture must be able to evolve during its development as well as once the system is placed into use and its requirements change.

■ **Enable system quality.** SAE must concentrate on ensuring system quality because it is primarily architectural decisions, inventions and trade-offs that have the greatest impact on quality, and because most architectural concerns are related to achieving necessary levels of important quality characteristics.

■ **All types of architectural components.** SAE must address all types of architectural components because systems consist of many types of components. SAE must adequately address software because software is used to implement the majority of the functional and quality requirements as well as many of the system's most important internal and external interfaces.

■ **Component-based development.** SAE must support component-based development because most systems now contain many off-the-shelf components.

■ **Well-defined concepts and terminology.** SAE must be based on well-defined concepts and terminology because clear communication among stakeholders is critical to engineering a successful architecture.

- **Architectures and their representations as valuable resources.** SAE must treat both architectures and architectural representations as valuable resources because of their great usefulness during the entire system life cycle and the great investment that is made in properly producing them.
- **Both architectural views and focus areas.** SAE must develop both views and focus areas because multiple views are necessary to understand the architecture and focus areas are necessary to show the relevant parts of multiple views when addressing architectural support for specific architectural concerns.
- **Sufficient architectural representations.** SAE must produce a sufficient set of architectural representations because individual documents and views are insufficient to document the entire system architecture and communicate it to all of its many stakeholders.
- **Completeness of methods produced.** SAE must enable the production of architecture engineering methods that address all the important architectural workers, work units, and work products because important work will not be done and important work products will not be produced if the methods are incomplete.
- **Teamwork.** SAE should be a team activity because no single architect is typically qualified or physically able to efficiently and effectively develop the large and complex architectures of today's systems.
- **Independence of architectural assessments.** SAE should support independent architectural assessments because such assessments are likely to identify architectural defects and risks that are not as easily identified using only peer-level evaluations.
- **Impact analysis before architectural changes.** Configuration management change control should apply to baselined architectural representations because the risk of architectural degradation and loss of integrity significantly increases without proper impact analysis.

Chapter 4

MFESA: An Overview

4.1 The Need for MFESA

Historically, most system-development projects have used previously existing, general-purpose methods to specify the architecture engineering process their architects are to perform. These reusable methods are often (1) organizational standard methods, (2) methods largely based on an international or military standard, (3) industry domain specific methods, or (4) methods drawn from a book, article, or conference paper or tutorial on architecture engineering. Often, these methods were chosen because they were methods with which the chief architects were already familiar.

However, there can be significant risks associated with using a general-purpose method for engineering the architecture of a specific system on a specific endeavor by specific organizations for specific stakeholders:

- *Systems* can vary greatly in terms of:
  - Size (from small embedded applications to ultra-large systems of systems)
  - Complexity (from simple to incredibly complex)
  - Criticality (such as business criticality, safety criticality, and security criticality)
  - Cardinality of purpose (from single-purpose systems to multipurpose systems)
  - Application domain (such as communications, information processing, transportation, and weapon systems)
  - Operational dependence on other systems (from stand-alone to part of a highly interoperable system of systems)
  - Originality (new "greenfield" systems to updates of existing legacy systems)
  - Uniqueness (from unique systems to systems within product lines)
  - Amount of reuse (such as COTS, GOTS, MOTS, and open source) mandated or incorporated
  - Technologies incorporated (including diversity, volatility, and maturity)
  - Required functionality, required quality characteristics and attributes, and requirements volatility

- Expendability (from single use to reusable systems)
- Automation (from manned to unmanned and from totally autonomous to human operated)
■ *System development endeavors* can vary greatly in terms of:
  - Type (from individual projects to programs of related projects)
  - Size (from small to vast endeavors)
  - Duration (from a few months to many years)
  - Schedule and funding (from more than adequate to very insufficient)
  - Contractual relationships (ranging from informal in-house development to formal contracts between acquisition organizations, prime contractors, subcontractors, and vendors)
  - Development and life-cycle scope (from initial development through major modification of legacy systems)
■ *Development organizations* can vary greatly in terms of:
  - Engineering culture (from early to late adopters of technology) [Rogers, 1962]
  - Management culture (including degree of control and risk taking versus risk avoidance)
  - Staff training, experience, and expertise
  - Staff localization (from co-located to geographically distributed)
■ *System stakeholders* can vary in terms of:
  - Type (such as acquirer, user, operator, trainer, maintainer, certifier, and member of the public)
  - Authority to set system goals and requirements
  - Numbers (from a single individual to literally tens or hundreds of thousands)
  - Accessibility to the architecture teams
  - Willingness to take risks
  - Preferences in terms of architectures and technologies
  - Past experiences with similar systems

Because of these variables, no *single* system architecture engineering method is likely to be sufficiently general to meet the needs of all endeavors. Thus, many architecture teams have had to tailor the previously existing method to meet the needs of their new endeavors. Unfortunately, the variability of architecture engineering efforts has been so great that it has often been very difficult to choose the appropriate reusable architecture engineering method and properly tailor it to meet the specific needs of the endeavor. This difficulty is one of many reasons why the intended architecture engineering method and the actual architecture engineering process performed in practice have often differed greatly. It is also a reason why both methods and associated processes have had limited success in effectively and efficiently developing and maintaining optimal system architectures and their associated representations.

MFESA takes a fundamentally different approach. Because of the vast potential variability of system engineering endeavors, MFESA assumes that no previously existing architectural engineering method will optimally meet the specific needs of the endeavor. Because existing generic methods often require so much tailoring, MFESA assumes that this is not the best way to achieve the necessary flexibility while retaining the benefits of standardization. MFESA therefore embraces the idea of situational method engineering, which is the generation of an appropriate project-specific method by allowing the architects to:

■ Select suitable method components from a repository containing a consistent class library of preexisting reusable method components

- Tailor these selected method components (if necessary)
- Integrate the selected tailored method components to create the actual method to be used
- Verify the consistency and appropriateness of the resulting method

On the other hand, tailoring an existing architecture engineering method is sometimes preferable to creating a new method. For example, tailoring existing methods is often the best approach when creating a product line of highly similar systems. However, method engineering is valuable even under these circumstances. Instead of trying to reuse an existing general-purpose method that requires a large amount of tailoring, organizations can use method engineering to produce an appropriate domain- or organization-specific method that can then be reused with considerably less tailoring. This is why an MFESA repository can store reusable methods as well as reusable method components. The key differences between this approach and the traditional approach are that the reusable methods have been specifically created for the situation under which they will be reused and they have been constructed out of a standard set of reusable components.

These are the reasons why this book neither documents nor recommends any one general-purpose method for engineering system architectures. Instead, it documents a powerful method framework for effectively and efficiently constructing appropriate situation-specific system architecture engineering methods.

4.2 MFESA Goal and Objectives

The primary goal of the MFESA method framework is to help architects effectively and efficiently engineer a consistent high-quality architecture for a system of systems, a single system, and/or a system's subsystems, whereby the architecture meets the architecturally significant acquirer and derived product and process requirements that are allocated to the system and its subsystems.

The objectives of the MFESA method framework are to:

- Provide an organized introduction to system architecture engineering in terms of its work products, work units, and workers
- Clarify the large scope (size and complexity) of system architecture engineering
- Apply situational method engineering to system architecture engineering in terms of method component reuse via selection, tailoring, and integration
- Provide the benefits of *flexibility* via:
  - Method component selection
  - Method component tailoring
- Provide the benefits of *standardization* via:
  - Use of an underlying ontology of concepts and terminology
  - Use of a common method metamodel
  - Use of a repository of reusable method components
  - Restricting extension to subclassing the existing method components

4.3 What Is MFESA?

As illustrated in Figure 4.1, the Method Framework for Engineering System Architectures (MFESA) consists of the following four primary parts:

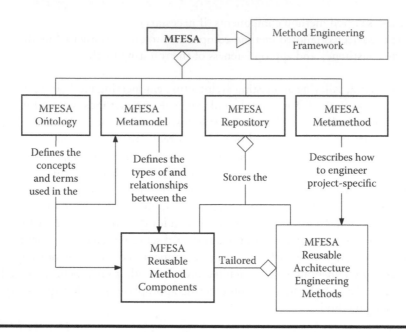

Figure 4.1 The four components of the MFESA method engineering framework.

- **MFESA ontology of concepts and terminology:** an information model defining a consistent set of interrelated concepts and terms underlying system architecture engineering
- **MFESA metamodel:** a model of the types of method components (Figure 4.1) in the repository of reusable method components (i.e., architectural work products, architectural work units, and architectural workers)
- **MFESA repository:** a repository containing (1) a consistent class library of reusable method components for creating situation-specific methods for engineering system architectures, and (2) a set of reusable methods constructed out of the reusable method components
- **MFESA metamethod:** a method for creating effective and efficient project-specific methods for engineering system architectures (i.e., a method for creating methods)

4.3.1 Ontology

MFESA is founded on a common system architecture engineering ontology defining the concepts and terminology used in the MFESA reusable method components and their supertypes specified in the MFESA metamodel. This ontology is more than merely a glossary of system architecture engineering terms. It is an information model that also captures the relationships between these concepts and terms. The MFESA ontology is documented in Chapter 5, "MFESA: The Ontology of Concepts and Terminology."

4.3.2 Metamodel

The Method Framework for Engineering System Architectures (MFESA) is a framework that enables system architects and process engineers to effectively and efficiently create methods for engineering system architectures. MFESA can also be used to create methods for engineering the associated software architectures of the system, its major subsystems, and its software architectural components.

To understand the term *method framework*, it helps to understand the difference between engineering processes and methods [Firesmith and Henderson-Sellers, 2002].* In fact, it is useful to understand the basic differences between processes, process models, and process metamodels, which form the theoretical basis of process engineering. Although these terms will be more formally defined later, it may be useful to provide a discussion of them and their relationships to one another. As illustrated in Figure 1.6, processes, process models, and process metamodels can be viewed as forming three layers. At the lowest level are processes, which are what actually exist in the real world on real projects. Processes can be limited to individual disciplines (e.g., architecture engineering as performed on a project) or include cohesive sets of disciplines (e.g., system engineering as performed on a project). A process (e.g., as performed on a specific project) can be thought of as an integrated set of process components, whereby a process component might be an actual specific architect (e.g., John Smith), the actual performance or execution of an architecture engineering task (e.g., the creation of candidate architectural visions by a specific team on a specific project), or an actual architecture document (e.g., version 1.0 of the architecture document for a specific system on a specific project).

The middle layer consists of methods, which are models of processes. A method documents how a process is intended to be performed, not how the method is actually performed. Like processes, methods can be limited to individual disciplines (e.g., how to perform architecture engineering) or include cohesive sets of disciplines (e.g., how to perform system engineering). Most methods are discipline specific. Methods are typically documented in the form of conventions such as procedures, guidelines, standards, and plans. Two example generic methods from the software community are the Rational Unified Process (RUP) and eXtreme Programming (XP). Similar to a process, a method can be thought of as consisting of method components that specify how the corresponding process components should be performed. Thus, a method (a.k.a. process model) is a model of the processes that are performed in accordance with the method, and each process component can be thought of as an instance of its corresponding method component. The MFESA repository contains reusable method components, not process components that come into existence when the method component is instantiated (used) on a real project. For example, the reusable method component *architecture document* gets instantiated as a series of versions of actual electronic or paper architecture documents (the process components), which describe the actual architecture of an actual system (or subsystem) on an actual project.

The highest level consists of the process metamodel. While architects will be interested in architecture processes and methods, architecture metamodels tend to only be of interest to methodologists, process engineers, and the developers of method supporting tools. Others may feel free to skip over the rest of this paragraph. A process metamodel is a model that provides a higher-level abstraction of a set of related methods. The three MFESA metamodel components are architecture worker, architecture work unit, and architecture work product. In this case, the metamethod components can be viewed as a set of base abstract classes of process components from which the

---

* Note that there is no universal consensus on the meaning of the terms *method* and *process*. Some authors consider these terms to be synonyms. Others consider processes to be large-scale activities and methods to be small-scale ways of implementing the processes. However, from a method engineering standpoint, it is very important to clearly differentiate the documented *as-intended* approach (the *method* documented in plans, procedures, and standards) from the *as-performed* approach (the *process* actually used by people on the endeavor).

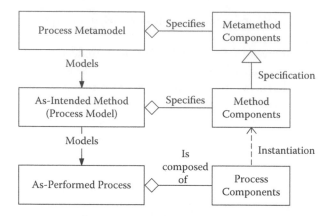

Figure 4.2 Methods and processes. Most of the figures in this book are drawn using the Unified Modeling Language (UML). Rectangles represent either classes of concepts or instances of such classes. Solid lines with small black arrowheads represent associations (general relationships) directed in the direction of dependency. An association will typically be labeled so that it can be read in the direction of the arrow as an English sentence, with the subject of the sentence being the label of the starting rectangle, the verb phrase being the label of the association, and the direct object being the label of the ending rectangle. Lines with a white diamond represent an aggregation (whole-part) relationship between the two rectangles; note that although it is counterintuitive from an object-oriented standpoint, the diamond is at the whole end of the relationship and that there is no arrowhead on the part end of the relationship. The solid line with the large white arrowhead represents a generalization/specialization (inheritance or *is-a-kind-of*) relationship with the line pointing from the subtype (subclass) to the supertype (superclass). Finally, dashed lines with arrowheads represent instance-of (is-a) relationships pointing from the instance to its type or class.

method components (i.e., process component classes) are instantiated.* As with abstract classes in object-oriented programming languages, the primary value of a process metamodel is to ensure the standardization and consistency of the method components (i.e., the concrete subclasses of the abstract superclasses).

The following definitions and examples are provided to clarify Figure 4.2 and discussion:

- **Engineering process:** the actual way that one or more engineering disciplines are performed in practice on a real endeavor. The term "engineering process" is referred to as the "as-performed process" in Figure 4.2 and will be shortened to "process" for the remainder of this book.
- **Engineering process component:** a component part of an engineering process and an instance of a method component. The term "engineering process component" is referred to as "process component" in Figure 4.2 and the remainder of the book.

---

* Actually, the preceding discussion is an oversimplification. Proper modeling of these three layers should actually incorporate powertypes (that is, *clabjects*), which are simultaneously both *cla*sses and *obj*ects. Such an approach is embodied in the recent ISO standard *Software Engineering Metamodel for Development Methodologies* (ISO/IEC, 2007). However, the distinction is relatively technical and primarily of interest only to a subset of methodologists and modeling tool vendors. For the purposes of MFESA, we will stick to the preceding simplification.

■ **Engineering method**: a process model that systematically documents the way that one or more engineering disciplines are intended to be performed. The term "engineering method" is referred to as "as-intended method" and "process model" in Figure 4.2. It will be shortend to "method" in the remainder of the book.

■ **Engineering method component:** a component part of an engineering method. The term "engineering method" is referred to as "method component" in Figure 4.2 and in the remainder of the book.

■ **Engineering process metamodel:** a model of an engineering process model that defines the abstract types of method components and their relationships. The term "engineering process metamodel" is referred to as "process metamodel" in Figure 4.2 and in the remainder of the book.

■ **Metamethod component:** a type of method component that is part of an engineering process metamodel.

Note that processes are the way that things are actually done (as-performed), whereas methods capture the intended way to do something (as-planned). The intent is that the process is performed in accordance with the associated method, and when discrepancies are identified, either the process or the method should be modified accordingly. A major part of process improvement is improving the methods to be followed.

To properly document an engineering method, one must decompose the method into its component parts. A method documents the work products to be produced, the work units to be performed, and the roles to be played by the people, teams, organizations, and tools that perform the work units to produce the work products. The MFESA metamethod consists of the following three metamethod components:

1. **Architectural work product:** a method component modeling a work product (e.g., document, model, system, or system component) that is produced during the performance of an architecture engineering method.

    MFESA models architectural work product as an abstract class (in the object-oriented sense) that must be subclassed prior to instantiation on a project. That is, instances of architectural workers perform instances of specific subtypes of architectural work units to produce instances of architectural work products; they do not instantiate a generic abstract architectural work product.

2. **Architectural work unit:** a method component modeling a unit of work (e.g., activity, task, or technique) that is performed during architecture engineering to create, access, modify, or remove an architectural work product.

    MFESA models architectural work product as an abstract class (in the object-oriented sense) that must be subclassed prior to instantiation on a project. That is, instances of architectural workers perform instances of specific subtypes of architectural work units to produce instances of architectural work products; they do not instantiate a generic abstract architectural work unit.

3. **Architectural worker:** a method component modeling something (e.g., person, team, or tool) that performs architectural work units to produce architectural work products.

    MFESA models architectural worker as an abstract class (in the object-oriented sense) that must be subclassed prior to instantiation on a project. That is, instances of architectural workers perform instances of specific subtypes of architectural work units to produce instances of architectural work products; they do not instantiate a generic abstract architectural worker.

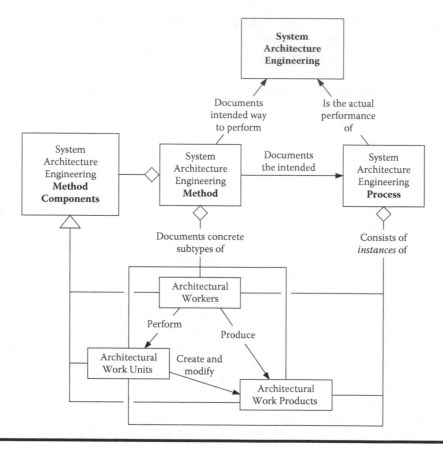

Figure 4.3 System architecture engineering methods and processes. This is a UML class diagram in which the diamonds represent composition (aggregation), the large white arrowhead represents subtyping (classification), and the lines with solid black arrowheads represent association (dependency).

- **System architecture engineering method:** a systematic documented way that system architecture engineering should be performed
- **System architecture engineering process:** the actual way that system architecture engineering is performed in practice.

As illustrated in Figure 4.3, a system architecture engineering method documents the intended way to perform system architecture engineering, whereas a system architecture engineering process is the actual way that system architecture engineering is performed. The system architecture engineering method is composed of method components that can be subtyped into *architectural work products* to be produced, *architectural work units* to be performed, and *architectural workers* that perform the work units to produce the work products. Note that a method consists of an integrated set of method components (classes of process components), whereas a process consists of instances of these method components.

4.3.3 Repository

As specified by the MFESA metamodel and illustrated in Figure 4.4, the MFESA repository contains three basic types of reusable architecture engineering method components: (1) *architecture*

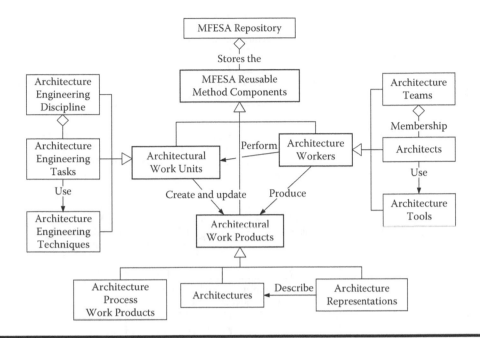

Figure 4.4 The primary contents of the MFESA repository.

work products, (2) *architecture work units* that create and update the architecture work products, and (3) *architecture workers* who perform the work units. The work products include system architectures and their representations. Architecture work units include the architecture engineering activity itself, its component tasks, and the techniques used to perform these tasks. The architecture workers include the architects, the architecture teams, and the architecture tools.

Remember that the MFESA repository stores method components (i.e., models of process components) and not the process components themselves. That is, the MFESA repository stores descriptions (classes) of process components rather than the actual process components. More specifically, the MFESA repository does not store actual architects, the actual performances of tasks, and actual architectures and their representations.

In addition to reusable method components, the MFESA repository can also be used to store reusable tailorable system architectural engineering methods that have been developed using the MFESA metamethod. However, this MFESA reference manual does not specify any such reusable methods, and development organizations producing a set of similar systems in similar ways are advised to create and store their own reusable methods.

4.3.3.1 Method Components: Tasks

As illustrated in Figure 4.5, the MFESA method consists of the following tasks, typically performed in a highly iterative, recursively incremental, parallel, and time-boxed manner. The numbering of and relationships between tasks illustrate the logical ordering and flow of work products between the tasks. Members of the architecture team(s):

1. **Plan and resource the architecture engineering effort.** Plan the overall architecture engineering effort and obtain the necessary resources to engineer the system architecture.

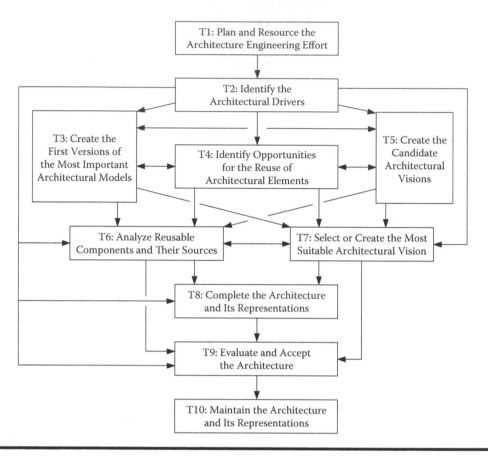

Figure 4.5   The logical ordering of MFESA tasks.

2. **Identify the architectural drivers.** Identify the architecturally significant product and process requirements that have been derived for and/or allocated to their system or subsystem and categorize them into a set of architectural concerns.

3. **Create the first versions of the most important architectural models.** This task typically consists of a consistent set of partial draft logical and physical static and dynamic models of the system or subsystem based on the architectural drivers, the associated architectural concerns, and the opportunities for reuse.

4. **Identify opportunities for the reuse of architectural elements.** Identify and initially analyze potential opportunities for the reuse of architectural decisions and inventions such as styles, patterns, and the relevant parts of existing architectural structures.

5. **Create the candidate architectural visions.** Use the initial architectural models to create a set of competing candidate architectural visions for their system or subsystem that supports meeting the derived and allocated architectural drivers and associated architectural concerns.

6. **Analyze reusable components and their sources.** Identify and evaluate potentially reusable physical architectural components and their sources for reuse within candidate architectural visions.

7. **Select or create the most suitable architectural vision.** Select the most suitable architectural vision for the system or subsystem from the competing candidate architectural visions. If combining consistent components from multiple competing candidate architectural

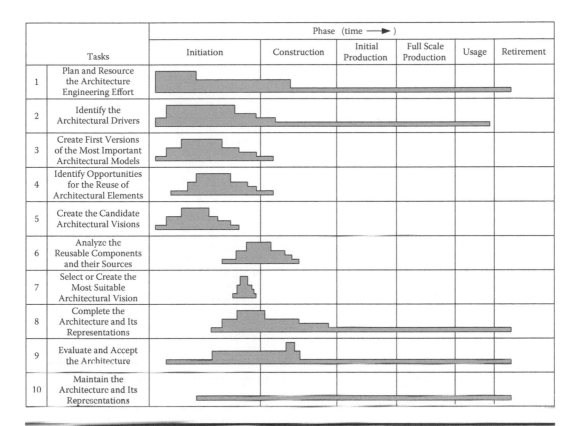

	Tasks	Phase (time ⟶)					
		Initiation	Construction	Initial Production	Full Scale Production	Usage	Retirement
1	Plan and Resource the Architecture Engineering Effort						
2	Identify the Architectural Drivers						
3	Create First Versions of the Most Important Architectural Models						
4	Identify Opportunities for the Reuse of Architectural Elements						
5	Create the Candidate Architectural Visions						
6	Analyze the Reusable Components and their Sources						
7	Select or Create the Most Suitable Architectural Vision						
8	Complete the Architecture and Its Representations						
9	Evaluate and Accept the Architecture						
10	Maintain the Architecture and Its Representations						

Figure 4.6   MFESA tasks by life-cycle phase.

visions would yield an even more suitable architectural vision, then do so to obtain the single new architectural vision.

8. **Complete the architecture and its representations.** Complete the system architecture and its representations based on the architectural vision selected or created.

9. **Evaluate and accept the architecture.** Evaluate the quality, completeness, and maturity of the system or subsystem architecture so that architectural risks can be managed, compliance with architecturally significant requirements can be determined, and the architecture can be accepted by its authoritative stakeholders.

10. **Maintain the architecture and its representations.** Maintain the architecture and its representations, while ensuring that the integrity of the system architecture and its representation do not degrade over time.

Note that Task 9 (Evaluate and Accept the Architecture) can be an ongoing task that overlaps *all* the other tasks. This is not completely indicated in Figure 4.5 because the extra lines would clutter up the diagram and make it less readable.

As illustrated in Figure 4.6, the MFESA tasks are performed in a largely concurrent manner. Therefore, their numbering is based on their logical rather than physical ordering. This illustration is very simplistic, is intended only to illustrate one general possibility, and neither time nor the level of effort for different tasks is necessarily drawn to scale. The actual effort will vary depending on whether:

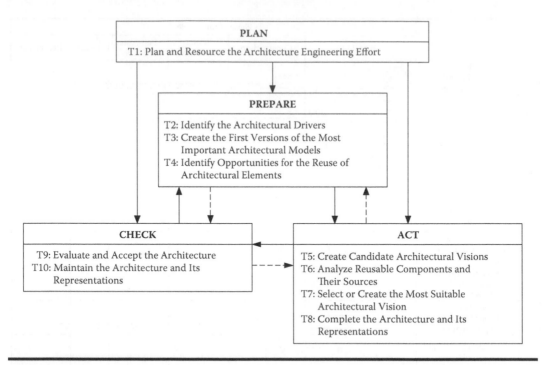

Figure 4.7 Plan, prepare, check, and act cycle for a single architectural element.

- System development is "greenfield" or for a new release of an existing system
- Development of the architecture is done all at once or incrementally on an individual subsystem-by-subsystem basis
- There are major updates to the architecturally significant requirements during the architecture engineering work

As illustrated in Figure 4.7, the tasks of the MFESA-instantiated method can be viewed as forming a plan, prepare, act, and check cycle.

As illustrated in Figure 4.8, Task 3 (Create first versions of most important architectural models), Task 4 (Identify opportunities for the reuse of architectural elements), and Task 5 (Create the candidate architectural visions) are performed concurrently and provide inputs to each other.

As illustrated in Figure 4.9, the MFESA method framework does not require that the architects immediately create the most suitable architecture in a big bang fashion. Instead, the initial architectural models are created during MFESA Task 5. These models are combined to produce several candidate competing architectural visions during MFESA Task 6. Then, during MFESA Task 7, either one of these candidate visions is selected as most suitable or a new architectural vision is created from the most appropriate consistent parts of the candidate architectural visions. Finally, during MFESA Task 8, the architecture is completed using the most suitable architectural vision as input.

As illustrated in Figure 4.10, MFESA has many stakeholders who provide inputs to MFESA and use the outputs of MFESA.

4.3.3.2 Method Components: Architectural Workers

As illustrated in Figures 4.3 and 4.4, the MFESA metamodel defines architectural workers and the MFESA repository stores the three subtypes of architectural workers: (1) architects, (2)

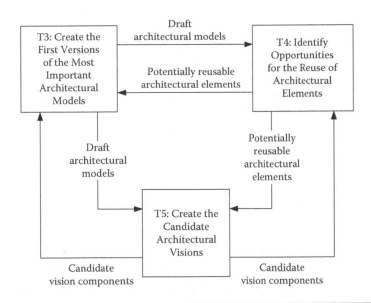

Figure 4.8   Interactions between concurrent Tasks 3, 4, and 5.

architectural teams, and (3) architectural tools. These architectural workers perform architectural work units to produce architectural work products. They will be documented in more detail in Chapter 16, "MFESA Method Components: Architectural Workers."

4.3.4 Metamethod

A metamethod is a method for constructing one or more methods. The fourth part of MFESA, the MFESA metamethod, is thus the method for constructing an MFESA method — that is, a method composed of MFESA method components. As a method itself, the MFESA metamethod can be viewed as being composed of its own component tasks and work products. During the second step of the first MFESA task, the relevant system architects collaborated closely with process engineers, technical leaders, experts in architectural engineering, and experts in MFESA to engineer one or more efficient and effective team-, project-, or organization-specific methods for engineering system architectures. As illustrated in Figure 4.11, the MFESA metamethod consists of the 13 tasks, which are documented in Chapter 17, "MFESA: The Metamethod for Creating Endeavor-Specific Methods."

4.4 Inputs

MFESA-compliant methods* typically use the following work products as inputs:

- **System request for proposal (RFP).** The system RFP is an invitation to system suppliers (e.g., system developers or integrators) to submit a proposal through a bidding process for

---

* Actually, it is the tasks of MFESA methods that actually use these inputs. Because different tasks use different inputs, this book does not have a chapter on inputs work products, but rather describes the inputs in the chapter on the task that first uses or produces the work products as inputs.

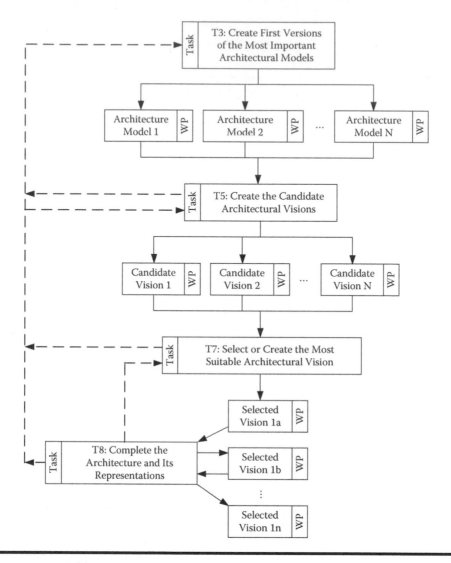

Figure 4.9   How architectural visions are created, selected, and iterated.

the system to be developed or extensively updated). The RFP provides sufficient information about the desired system (via a detailed technical specification, for example) to enable the bidders to propose an initial system architecture with associated cost, schedule, and other related information.

■ **System vision statement.** The system vision statement defines the system and documents its mission, the business problems it is to solve, the business opportunities it is to take advantage of, the major needs and desires of its stakeholders, and any major restrictions.

■ **System concept of operations (ConOps).** The system ConOps defines the major stakeholders' concept of how the system is to operate and often includes the top-level system use cases. Note that the ConOps may not be in its final form when provided as an input to the architecture engineering process; significant architectural engineering work and interaction with stakeholders may be required to complete it.

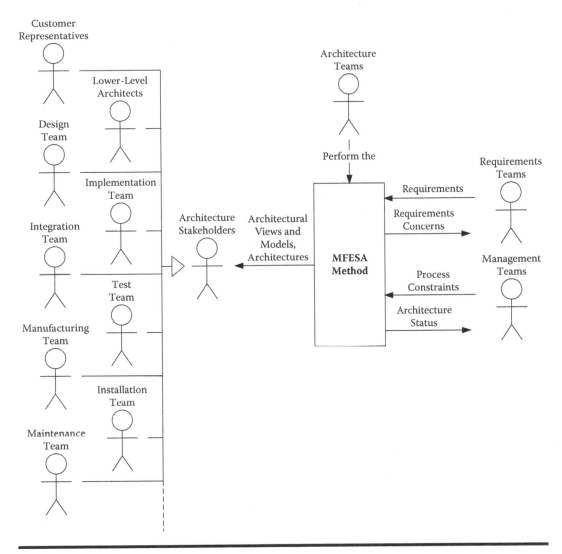

Figure 4.10　Primary MFESA inputs and outputs.

- **System requirements repository.** The system requirements repository is a database and the associated tools for storing and managing the system requirements (i.e., the functional, data, and quality requirements of the system as well as any mandatory constraints on the system architecture, design, implementation, and configuration). Due to the need to use an incremental development cycle to meet system delivery deadlines, the requirements repository will typically contain only the most important requirements when system architecture engineering begins. However, it is important that the requirements repository contain at least the most important architecturally significant requirements.

- **System requirements specification (SRS or SysRS).** A system requirements specification is a document that specifies the system requirements at a specific point in time during the development cycle. Due to the need to use an incremental development cycle, an SRS is typically only an initial partial draft when system architecture engineering begins. An SRS may (or may not) be automatically generated from the contents of the system requirements

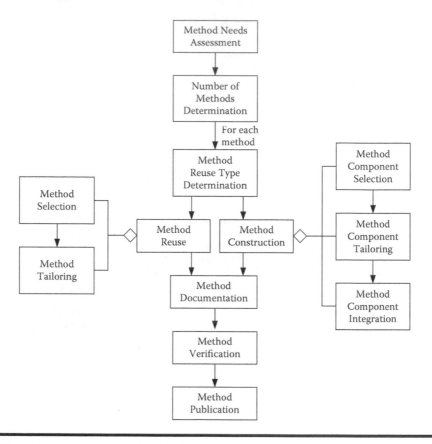

Figure 4.11   The MFESA metamethod tasks.

repository. There may (or may not) be different requirements specifications for different sub-systems and stakeholders.

■ **System product requirements.** These requirements constrain the behavior or characteristics of the system. They include architecturally significant business requirements (i.e., from the acquirer, customer, user, maintainer, etc.) and developer-derived architecturally significant requirements. They also include both system-level requirements as well as requirements that are derived for and allocated to lower-level architectural components.

– **Quality requirements.** System and subsystem quality requirements specify a minimum acceptable level of some quality characteristic or attribute as defined in the project quality model.

– **Functional requirements.** Functional requirements specify a mandatory behavior of the system or subsystem, especially functional requirements that when grouped into cohesive feature sets significantly drive the associated architecture.

– **Interface requirements.** Interface requirements specify mandatory interfaces, typically with existing external systems with which the system to be developed will interoperate.

– **Data requirements.** Data requirements specify mandatory characteristics of system or subsystem data such as data type and range.

– **Architecturally significant constraints:**

• **Architectural constraints.** Architectural constraints include architectural decisions and inventions that are mandated on the architecture as requirements. Architectural constraints are mandated by the requirements teams as opposed to being left to the architecture teams to decide.

- **Legacy architecture constraints.** Legacy architectural constraints are typically due to the requirement to update and enhance an existing legacy system rather than develop a new "greenfield" system and therefore to reuse as much of the existing architecture as is practical and appropriate. These constraints may be due to a mandate to use one or more system- or subsystem-level reference architectures when building a new variant system in a family of systems or a product line of systems. Finally, these constraints may be due to the need to conform to an enterprise architecture, which may greatly constrain architectural decisions, inventions, and trade-offs, especially when made to support interoperability between systems (such as the use of service-oriented architectures and standard protocols).
- **Laws, regulations, and policies constraints.** For example, these include organizational policies, and laws and regulations mandating levels of safety and security or the use of specific safety and security mechanisms.

■ **Process requirements.** These requirements constrain the development, operations, or retirement processes:

- **Budgetary requirements.** For example, budgetary requirements include the maximum acceptable budget to pay for research, development, initial production, full-scale production (e.g., manufacturing), sustainment (such as operations and maintenance), and retirement costs.
- **Facility requirements.** For example, facility requirements include the minimal acceptable number and capabilities of facilities for research, development, initial production, full-scale production (e.g., manufacturing), sustainment (including operations and maintenance), and retirement.
- **Schedule requirements.** For example, schedule requirements include the maximum acceptable deadlines associated with research, development, initial production, full-scale production (e.g., manufacturing), sustainment (including operations and maintenance), and retirement milestones.
- **Staffing requirements.** For example, staffing requirements include the minimal acceptable level of staffing that must be available to perform research, development, initial production, full-scale production, operations, and retirement tasks.
- **Conventions compliance requirements.** For example, conventions compliance requirements include required compliance with international or national formal process conventions (e.g., procedures, recommended practices, standards, and templates) such as [ANSI/EIA, 2003; DODAF, 2007a–c; IEEE, 1998; IEEE, 2000; ISO/IEC, 2001; and ISO/IEC, 2007]. It also includes industry *de facto* standards such as the INCOSE System Engineering Handbook [INCOSE, 2006] and the SEI Capability Maturity Model Integrated (CMMI) [Chrissis et al., 2007].

■ **Reference architecture.** A reference architecture is a reusable architecture with associated architectural representations that has been shown by experience to be appropriate for systems within a specific application domain or product line. The system under development may need to be compatible with a specified reference architecture. This is especially true when the system to be architected will be part of a product line of systems.

■ **Enterprise architecture.** An enterprise architecture is an organizational-level architecture with associated architectural representations that supports an organization's core goals and strategic direction and the current and/or future structure and behavior of the organization's business processes, organizational structure, and systems (especially information systems). The system under development may need to interoperate with other systems described by the organization's enterprise architecture.

4.5 Outputs

MFESA-compliant methods typically produce the following work products as outputs:*

- **Architectural work products:**
  - **Architectural models**
  - **Architectural documentation:**
    - Architecture analysis and simulation reports
    - Architecture documents
    - Architectural whitepapers
    - Architecture quality cases
    - Architecture training materials
  - **Architectural prototypes**
  - **Executable architectures**
  - **Architecture verification results:**
    - Architecture assessment outbriefs and reports
    - Architecture inspection results
    - Architecture model syntax checking results
    - Architecture quality metrics
    - Architecture test results
- **Requirements work products:**
  - **Metadata:** metadata identifying requirements and programmatic constraints as architecturally significant
  - **Architectural concerns:** cohesive collections of architecturally significant requirements

4.6 Assumptions

MFESA was not developed out of thin air. Its purpose is to address the challenges facing system architects that were described in Chapter 2. Furthermore, it is based on the system architecture engineering principles discussed in Chapter 3. Finally, MFESA is based on the following assumptions:

- **Responsibilities.** Requirements engineering is the responsibility of the requirements teams while architecture engineering is the responsibility of the architecture teams. The integration of requirements and architecture is typically the responsibility of the chief architect(s) collaborating with members of the relevant requirements teams and architecture teams.

---

* Actually, it is the tasks of MFESA methods that actually create these outputs. Because different tasks create different outputs, this book does not have a chapter on output work products, but rather describes the outputs in the chapter on the task that first produces the work products as outputs.

- **Multiple teams.** The system being architected is sufficiently large to require multiple requirements and architecture teams:
  - There will be an overall top-level system requirements team and system architecture team.
  - Individual subsystems will have their own requirements and architecture teams, although requirements engineers and architects may be members of more than one team and the same person may play the role of both requirements engineer and architect. These teams may also be logical subteams of subsystem integrated product teams (IPTs).
- **Method:**
  - To meet the specific needs of the project for which it is created, an architecture engineering method requires that:
    - Appropriate method components are selected from the MFESA repository of reusable method components.
    - Once these method components are integrated, they will still need tailoring (i.e., additions, modifications, and deletions).
  - MFESA methods can be used at all levels of the system hierarchy, and specifically at both the system and subsystem levels.

4.6.1 The Number and Timing of System Architecture Engineering Processes

Unless the system is quite small and simple, the mapping from system architecture engineering method to system architecture engineering process on a project is typically one too many. The method is typically followed many times by different teams on different parts of the system. As illustrated in Figure 4.12, systems can be quite large, consisting of many tiers of subsystems and lower-level architectural components. At the lowest levels, they also can consist of many different types of components, including hardware, software, data, manual procedures, facilities, etc. Note that the naming conventions for the different types of components at the different tiers in the system aggregation structure vary from organization to organization and are included here merely to emphasize the depth that such system structures can reach.

4.7 Relationships with Other Disciplines

Many steps of the MFESA tasks are closely related to other disciplines and activities other than architecture engineering. This includes:

- **Primary engineering disciplines:**
  - Requirements engineering
  - Design
  - Implementation
  - Integration
  - Testing
  - Quality assurance
  - Process engineering
  - Training
- **Management disciplines:**
  - Project and program management

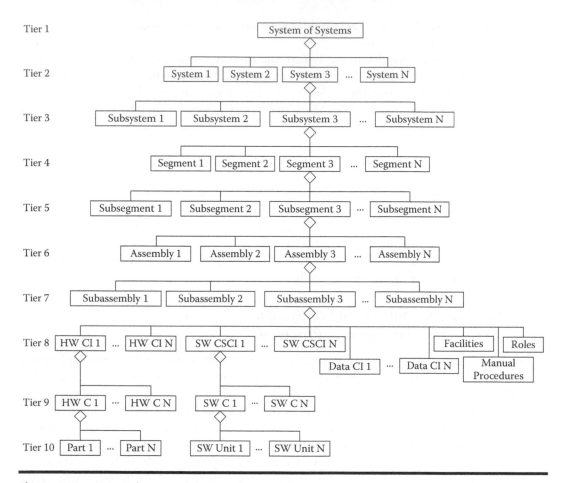

Figure 4.12 A generic system aggregation structure.

- Configuration management
- Risk management
- Measurement and metrics
■ **Specialty engineering disciplines:**
- Human factors engineering
- Reliability engineering
- Safety engineering
- Security engineering

The following subsections describe the relationships between system architecture engineering and the previously listed disciplines.

4.7.1 Requirements Engineering

Requirements engineering is the engineering discipline and activity within systems engineering consisting of the cohesive collection of all tasks that are primarily performed to produce and maintain the system requirements and other related requirements work products.

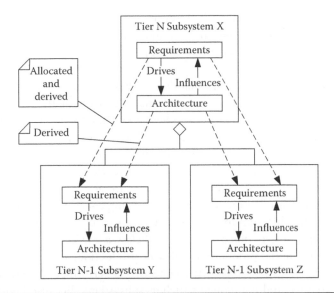

Figure 4.13   Interleaving of requirements engineering and architecture engineering.

The requirements team(s) have the following duties with regard to architecture engineering:

■ Engineering (such as identification, analysis, management, and specification) of the architecturally significant requirements including those derived for specific subsystems
■ Tracing the requirements to the architectural components (as well as to and from other work products)

Note, however, that the architects are responsible for allocating the architecturally significant requirements to the top-level architectural components and from the higher-level to lower-level architectural components. In essence, this implies some of the associated requirements tracing duties.

When integrated product teams (IPTs) are used to develop subsystems, the requirements teams and architecture teams may be virtual rather than official teams. Nevertheless, it is important for the virtual requirements team to include requirements engineers with sufficient expertise, training, and experience to perform requirements engineering and for the virtual architecture team to include architects with sufficient expertise, training, and experience to perform architecture engineering. Unfortunately, it is rare for a single person to be able to effectively and efficiently perform both requirements engineering and architecture engineering.

The system architecture engineering method is typically not followed in isolation. It is typically performed concurrently with other systems engineering activities during the course of a recursively iterative, recursively incremental, parallel, and time-boxed development cycle. As illustrated in Figure 4.13, requirements engineering and architecture engineering are typically performed in an incremental and concurrent manner for each architectural component at each tier in the system aggregation structure. Lower-level requirements are often engineered in one of three major ways:

1. The requirements of a higher-level architectural component can be allocated (i.e., passed down unchanged) to any of its architectural component parts.
2. New, more appropriate lower-level requirements can be directly derived for an architectural component from the requirements of the parent architectural component.

3. The architectural decisions, trade-offs, inventions, and assumptions made during the architectural engineering of the parent architectural component can be the sources of derived requirements for the child architectural components.

Many systems are so large and complex that they are best developed and released incrementally. Often, the development organization releases small increments internally (e.g., to the independent test team) as a means of managing risk. On the other hand, the development organization typically releases large increments having useful new capabilities (e.g., increases of functionality) externally to the acquisition organization or user community. As illustrated in Figure 4.14, the requirements and architecture are often simultaneously engineered for multiple releases, with the majority of the effort being spent in engineering the requirements and architecture of the next release, while some of the requirements and architecture of future releases are also being concurrently developed.

As illustrated in Figure 4.15 (not to scale), the architecture of a system is primarily engineered during the initiation and construction phases. Nevertheless, a low level of architecture engineering continues throughout the initial production, full-scale production, and usage phases.

As illustrated in Figure 4.13, architecture engineering is performed incrementally and concurrently on multiple architectural components such as subsystems. Because Figures 4.14 and 4.15 are only intended to clarify the concurrent nature of requirements engineering and architecture engineering, they are simplifications that do not show the iterative, recursively incremental, and concurrent (within architecture engineering) nature of architecture engineering.

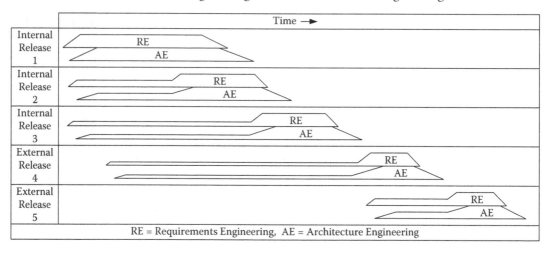

Figure 4.14 Incremental requirements and architecture engineering over multiple releases.

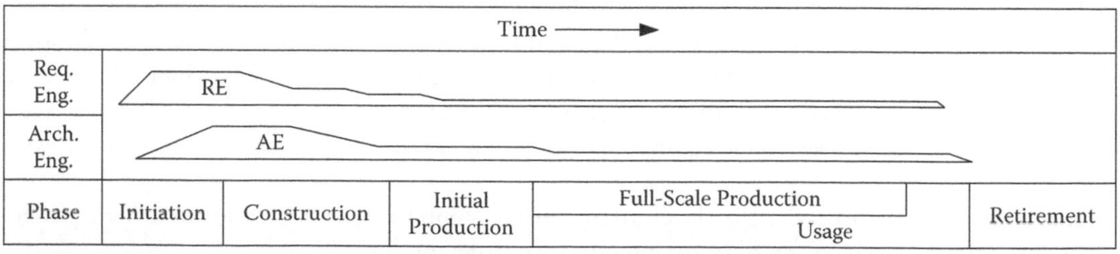

Figure 4.15 Architecture engineering effort as a function of phase.

4.7.2 Design

System design is the engineering discipline and activity within systems engineering consisting of the cohesive collection of all tasks involving all aspects of producing and maintaining the tactical design of the system and its components.

The design teams have the following duty with regard to architecture engineering:

■ Engineer the designs of the architectural components so that they are consistent with the component architectures.

4.7.3 Implementation

Implementation is the engineering discipline and activity within systems engineering consisting of the cohesive collection of all tasks involving the building of the architectural components of the system.

The implementation team(s) have the following duties with regard to architecture engineering:*

■ Fabricate the hardware so that it is consistent with the architecture and its representations.
■ Program the software so that it is consistent with the architecture and its representations.

4.7.4 Integration

Integration is the engineering discipline and activity within systems engineering consisting of the cohesive collection of all tasks involving the connecting of the architectural components to build a working system.

The integration team(s) have the following single duty with regard to architecture engineering:

■ Integrate the system's architectural components.

4.7.5 Testing

Testing is the engineering discipline and activity within systems engineering consisting of the cohesive collection of all tasks that are primarily performed to determine the quality of executable work products (such as models, software, and system) by attempting to cause them to fail under controlled conditions so that any existing and underlying defects may be identified, corrected, and avoided in the future.

The testing team(s) have the following duties with regard to architecture engineering:

■ Perform integration testing in a manner consistent with the architecture of the integrated components.
■ Test the architectural components to verify if they meet their allocated and derived architecturally significant requirements.
■ Test the overall system to verify that it meets the architecturally significant requirements.

---

* If the hardware and software are consistent with the design, and the design is consistent with the architecture, the hardware and software should largely be consistent with the architecture. However, the architecture may directly include implementation-level constraints such as the mandated use of a safe subset of C# or the mandated use of a given material.

4.7.6 Quality Engineering

Quality engineering (QE) is the engineering discipline and activity within systems engineering consisting of the cohesive collection of all tasks that are primarily performed to ensure and help continually improve the quality of an endeavor's process and work products. For example, quality engineering tasks include:

- Quality planning
- Quality modeling
- Quality assurance (QA)
- Quality control (QC)

The quality engineering team has the following duties with regard to architecture engineering:

- Collaborate with the requirements team to provide a quality model defining the quality characteristics underlying the architecturally significant quality requirements (quality modeling).
- Collaborate with the process team and architecture team to ensure and verify the quality of the architecture engineering methods (quality assurance).
- Collaborate with the architecture teams, the architecture evaluation teams, and other stakeholders to ensure and verify the quality of the architecture work products (quality control).

4.7.7 Process Engineering

Process engineering is the engineering discipline and activity within systems engineering consisting of the cohesive collection of all tasks directly involved in process improvement by the engineering of appropriate endeavor-specific and/or system-specific methods (i.e., process models) via the engineering and use of organizational method component frameworks (i.e., method metamodels), method component repositories or class libraries, reusable methods, and associated usage guidelines.

The process engineering team (also known as the system engineering process group, or SEPG) has the following duties with regard to architecture engineering:

- Collaborate with the architecture team to engineer an appropriate architecture engineering method (such as by selecting appropriate architecture engineering method components, tailoring these components, and integrating them to produce the actual method).
- Capture lessons learned from the architecture teams concerning the actual use of these methods.

Note that MFESA is useful even if there are no officially identified and dedicated process teams. Process engineer is a role that may or may not equate to a job title. Architects may well play the role of process engineers when it comes to creating a system architecture engineering method.

4.7.8 Training

Training is the engineering discipline and activity within systems engineering consisting of the cohesive collection of all tasks that are primarily performed to provide training to the members of the various teams and organizations involved in the endeavor. This training can be formal or informal. It can also be classroom-based or on-the-job training.

The training team has the following duties with regard to architecture engineering:

- Develop training materials regarding the architecture engineering method(s) and provide training to the architecture team and relevant stakeholders.
- Develop training materials regarding the architecture and its representations and provide training to the architecture team and relevant stakeholders.
- Develop training materials regarding the system's new concept of operations (including deployment, usage, operations, and maintenance) and provide training to the architecture team and relevant stakeholders.

4.7.9 Project/Program Management

Project/program management is the discipline and activity consisting of the cohesive collection of all tasks that are primarily performed to administer and lead an endeavor (e.g., project or program of related projects) to fulfill the endeavor's mission and to achieve its objectives.

The management team has the following duties with regard to architecture engineering:

- Determine and state some of the architecturally significant business drivers and quality goals
- Staff the architecture teams, including working with the chief architects to:
  - Establish requirements for architect recruitment, competency, and training
  - Interview, hire, and promote the architects
  - Assign via mandate appropriate responsibility, authority, and reporting channels to the architects
- Provide funding and cost oversight for the:
  - Performance of architectural engineering tasks
  - Acquisition of architecture engineering tools
- Monitor the:
  - Planning, performance, status, and effectiveness of architecture engineering tasks
  - Production of architecture work products
  - Achievement of architecture engineering schedule milestones
- Take corrective action as needed

4.7.10 Configuration Management

Configuration management is the discipline and activity within systems engineering consisting of the cohesive collection of all tasks performed to manage an endeavor's baselines of configuration items.

The configuration management team has the following duties with regard to architecture engineering:

- Configuration identification:
  - Assign configuration identifiers to architectural components identified as configuration items by the architecture teams or the project technical leaders.
  - Assign configuration identifiers to other architectural work products (e.g., architecture documents) identified by the architecture teams or the project technical leaders.
  - Identify baselines involving architectural components.

- Version control:
  - Manage multiple versions and variants of architectural components and other architectural work products.
- Configuration control:
  - Place architectural work products under configuration control.
  - Control changes to such architectural work products.
- Configuration status reporting:
  - Report the status of architectural baselines and configuration items.
- Configuration auditing:
  - Audit the identity, version, and completeness of architectural baselines.

4.7.11 Risk Management

Risk management is the discipline and activity within systems engineering consisting of the cohesive collection of tasks performed to lower project risks to acceptable levels. For example, risk management tasks include the following:

- Risk management planning
- Risk identification
- Risk analysis
- Risk control
- Risk monitoring

The risk management team has the following duties with regard to architecture engineering:

- Identify architectural risks.
- Analyze the probability and severity of architectural risks.
- Control architectural risks via mitigations and transfer of risks.
- Monitor the status of architectural risks and the effectiveness of their mitigations.

4.7.12 Measurements and Metrics

Measurement and metrics is the engineering discipline and activity within systems engineering consisting of the cohesive collection of all tasks involved with the measurement of project metrics. For example, risk management tasks include:

- Measurement needs assessment
- Measurement planning
- Metrics collection
- Metrics analysis
- Metrics reporting

The measurement and metrics team has the following duties with regard to architecture engineering:

- Collaborate with the architecture team(s) and the project/program management team(s) to determine the appropriate architecture metrics to collect.
- Collect these architecture metrics.

- Analyze these architecture metrics.
- Report these architecture metrics.

4.7.13 *Specialty Engineering Disciplines*

Systems architecture engineering has important relationships with general specialty engineering disciplines such as human factors engineering, reliability engineering, safety engineering, and security engineering as well as with application-domain-specific specialty engineering disciplines (e.g., chemical engineering, sensors, telecommunications, and weapons).

Human factors engineering is the engineering discipline and activity within systems engineering consisting of the cohesive collection of all tasks focusing on how people perform their tasks while interacting with the system.

The human factors team has the following duties with regard to architecture engineering:

- Influence architectural decisions to ensure that the architecture takes into account human (i.e., user, operator, and maintainer) limitations and capabilities.
- Verify that the architecture does not exceed human limitations and capabilities.

Reliability engineering is the engineering discipline and activity within systems engineering consisting of the cohesive collection of all tasks related to ensuring that a system will perform its intended function(s) when operated in a specified manner for a specified length of time under specified environmental conditions.

The reliability team has the following duties with regard to architecture engineering:

- Ensure that the architecture sufficiently supports the system meeting its reliability (and availability) requirements.
- Verify that the architecture sufficiently supports the system meeting its reliability (and availability) requirements.

Safety engineering is the engineering discipline and activity within systems engineering consisting of the cohesive collection of all tasks involved with preventing, reducing in probability or severity, detecting, and reacting to:

- *Accidental* harm to valuable assets (unauthorized* unintentional harm to valuable assets such as people, organizations, property, the environment, or services)
- Mishaps (the occurrence of one or more *events* that cause or could cause accidental harm to valuable assets, that is, accidents and safety incidents)
- Hazards (the existence of one or more *conditions* that enable or cause mishaps† to occur)
- Safety risks (expected accidental harm defined as the probability of accidental harm multiplied by maximum (or average) credible harm severity)

---

* Safety does not address authorized harm, such as the authorized harm that weapons systems cause to the enemy.
† Hazards often involve some combination of the existence of vulnerable assets that can be accidentally harmed, abusers who may accidentally cause the system to harm these assets, and system vulnerabilities that can cause accidents or enable the accidental abusers to cause the accidents. The accidental abusers may be people, systems, or aspects of the environment (e.g., earthquakes and storms).

The safety team has the following duties with regard to architecture engineering:

- Collaborate with the architecture teams to build safety patterns, mechanisms, and safeguards into the architecture.
- Evaluate the architecture and its representations for safety vulnerabilities and its support for meeting the safety requirements.

Security engineering is the engineering discipline and activity within systems engineering consisting of the cohesive collection of all tasks involved with preventing, reducing in probability or severity, detecting, and reacting to:

- *Malicious* harm to valuable assets (unauthorized* intentional harm to valuable assets such as people, property, the environment, or services)
- Misuses (the occurrence of one or more *events* that cause or could cause malicious harm to valuable assets, that is, attacks and security incidents such as probes)
- Threats (the existence of one or more *conditions* that enable or cause misuses† to occur)
- Security risks (expected malicious harm defined as the probability of malicious harm multiplied by maximum (or average) credible harm severity)

The security team has the following duties with regard to architecture engineering:

- Collaborate with the architecture teams to build security patterns, mechanisms, and countermeasures into the architecture.
- Evaluate the architecture and its representations for security vulnerabilities and its support for meeting the security requirements.

4.8 Guidelines

The following guidelines (and associated rationales) are useful with regard to the MFESA method framework:

- Remember that MFESA is not a system architecture engineering method.
- Remember that MFESA is more than a repository of reusable method components.
- Remember that MFESA has multiple uses.
- Remember that MFESA methods use both requirements and reusable architectural elements as inputs.
- Use MFESA tasks to organize and understand.
- Remember that MFESA affects more than just the architects.

These guidelines are discussed in more detail in the following:

---

* Security does not address authorized harm, such as the authorized harm that weapons systems cause to the enemy.
† Analogously to hazards, threats often involve some combination of the existence of vulnerable assets that can be maliciously harmed, abusers who may maliciously cause the system to harm these assets, and system vulnerabilities that can enable the malicious abusers to attack or probe the system or the valuable assets for which it is responsible.

■ **Remember that MFESA is not a system architecture engineering method.** Do not think of MFESA as a system architecture engineering method. Instead, think of it as a way to generate appropriate endeavor-specific system architecture engineering methods.

   Rationale: System architecture engineering methods document the intended way to perform architecture engineering. Thus, as-intended methods document and constrain the processes actually performed by architects. A major part of process improvement is method improvement. To improve methods, they must be made appropriate for the situation under which they will be used. Thus, methods must take into account the unique combination of characteristics of the system to be architected, the organizations and teams that will acquire and develop (and architect) the system, and properties of the project itself. Because no single method is appropriate for all situations, a method framework based on situational method engineering is needed to create appropriate situation-specific architecture engineering methods. MFESA is the first such framework.

■ **Remember that MFESA is more than a repository of reusable method components.** Do not consider MFESA as merely a repository of reusable method components from which to generate appropriate, endeavor-specific architecture engineering methods. Remember that MFESA consists of four main components: (1) an ontology of concepts and terminology, (2) the metamodel that specifies the types of reusable method components, (3) the repository of actual reusable method components, and (4) the metamethod for constructing endeavor-specific methods by selecting, tailoring, and integrating the appropriate method components.

   Rationale: Although critical to situational method engineering, a repository of reusable method components is insufficient *by itself* to create appropriate, situation-specific methods. These components need to be based on a solid foundation provided by an ontology defining concepts and terms as well as a metamodel defining the abstract types of the method components. Method engineering also requires a metamethod, an effective and efficient method for selecting, tailoring, and integrating method components into a usable method. MFESA derives its power from all four of its constituent parts: (1) the MFESA ontology, (2) the MFESA metamodel, (3) the MFESA repository of reusable MFESA method components and reusable methods, and (4) the MFESA metamethod for creating appropriate methods.

■ **Remember that MFESA has multiple uses.** Remember that MFESA is not just for constructing appropriate, endeavor-specific system architecture engineering methods.

   Rationale: The primary purpose of MFESA is to support the effective and efficient construction of appropriate architecture engineering methods. However, MFESA has other important uses. It provides a good structured introduction to system architecture engineering in terms of its work products; the work units that create, modify, and evaluate these work products; and the workers that perform the work units to engineer the work products. This helps clarify the great size and complexity of system architecture engineering, which is far too often over-simplified and under-resourced. MFESA uses situational method engineering to obtain the benefits of reuse of process engineering assets (i.e., the documentation of the method components). Finally, it provides a workable and attractive compromise for achieving the benefits of both flexibility and standardization.

■ **Remember that MFESA methods use both requirements and reusable architectural elements as inputs.** Remember that MFESA methods engineer system architectures based on both allocated and derived requirements as well as the availability of reusable architectural elements such as subsystems and architectural patterns.

   Rationale: The obvious inputs to any system architecture engineering method are the system requirements and related requirements work products such as the system vision statement and concept of operations. Because almost all systems contain some reusable components, it is therefore important to remember that the documentation of reusable components and their sources is an important input to almost all MFESA methods.

■ **Use MFESA tasks to organize and understand.** Use the MFESA tasks to organize the repository of reusable system architecture engineering method components and to understand the associated architectural work products and workers who produce these work products.

Rationale: The MFESA repository contains reusable method components that according to the MFESA metamodel are classified as architectural work products, work units, or workers. These method components could be documented in any order, and in fact, we intend to eventually document them in the form of an informational Web site with one or more Web pages per component and with hyperlinks connecting related components.* This will allow browsing and searching† in any order. However, in the meantime, we find it most useful to organize our thoughts and this book around the ten architecture engineering tasks and associate the work products with the tasks that create and use them.

■ **Remember that MFESA affects more than just the architects.** Remember that the system architecture engineering methods created using MFESA use outputs from and provide inputs to architectural stakeholders other than just the architects.

Rationale: Architects must interface with people performing other mainstream engineering activities (e.g., requirements engineering, design, implementation, integration, testing, and manufacturing); management activities (such as project management, configuration management, and risk management); and specialty engineering activities, including, but not limited to, human factors, reliability, safety, and security engineering. The architecture engineering tasks use inputs from these other disciplines and provide outputs to them. Thus, MFESA must enable the construction of architecture engineering methods that address these interfaces and flows. Other architecture stakeholders may therefore find MFESA valuable.

4.9 Summary

No single, general-purpose system architecture engineering method is appropriate for all projects, regardless of how tailorable it is, and this is because each situation is unique. The systems being architected vary greatly in terms of size, complexity, application domain, business criticality, functionality, and required levels of quality characteristics and attributes. Acquisition and development organizations (and their architecture teams) vary greatly in terms of size, experience, culture, degree of risk aversion, and geographic distribution. Projects vary greatly in terms of funding, schedule constraints, and contract type. Because no single method is appropriate for all situations, the need exists to be able to efficiently and effectively create appropriate project-specific architecture engineering methods. That is, a method framework supporting project-specific method engineering is needed.

---

* We welcome volunteers and sources of funding to transform this book's contents into an informational Web site with associated tools to enable architects to easily select, tailor, and integrate the MFESA method components. Please contact the authors if you are able and willing to help.

† Although paper books unfortunately do not support browsing and searching to the same extent that informational Web sites do, we have nevertheless provided a detailed table of contents and index that must serve the same uses until the Web site is completed.

4.9.1 MFESA Components

MFESA is not a system architecture engineering method. Instead, it is a method framework that is used to produce appropriate project-specific system architecture engineering methods. MFESA is composed of the following four components:

1. MFESA *ontology*, which clearly defines the concepts and terminology of system architecture engineering
2. MFESA *metamodel*, which defines the foundational abstract supertypes of MFESA reusable system architecture engineering method components and the relationships between them
3. MFESA *repository*, which stores the MFESA reusable method components
4. MFESA *metamethod*, which describes how to engineer project-specific system architecture engineering methods

4.9.2 Goal and Objectives

The primary goal of the MFESA method framework is to help architects effectively and efficiently engineer a consistent high-quality architecture for a single software-intensive system and its subsystems. The supporting objectives of MFESA are to:

■ Provide an organized introduction to system architecture engineering.
■ Clarify the large scope (size and complexity) of system architecture engineering.
■ Apply situational method engineering to system architecture engineering.
■ Provide the benefits of *flexibility* via method component selection and tailoring.
■ Provide the benefits of *standardization* via use of a unifying ontology, a repository of standard reusable method components based on a common metamodel, and restricting extension to subclassing of existing method components.

4.9.3 Inputs

The primary inputs to MFESA-compliant methods are the system's product and process requirements, which imply access to the project requirements repository where they are stored or the system requirements specifications in which they are specified and published. Other related inputs include the system request for proposal (RFP), system vision statement, and system concept of operations (ConOps). If they exist and are relevant, the organizational enterprise architecture and product line's reference architecture are also inputs. To support reuse, MFESA methods also use information about preexisting, potentially reusable architectural components and their sources such as product vendor marketing and technical literature.

4.9.4 Tasks

The MFESA repository of reusable method components contains ten system architecture engineering tasks. These tasks provide a convenient way to organize and explain both the reusable method components as well as generic MFESA-compliant methods:

1. Plan and resource the architecture engineering effort.
2. Identify the architectural drivers.

3. Create initial architectural models.
4. Identify opportunities for the reuse of architectural elements.
5. Create the candidate architectural visions.
6. Analyze reusable components and their sources.
7. Select or create the most suitable architectural vision.
8. Complete the architecture and its representations.
9. Evaluate and accept the architecture.
10. Maintain the architecture and its representations.

4.9.5 Outputs

The primary outputs of MFESA-compliant architecture engineering methods are the architectural work products consisting of the system architecture and its many representations. These representations include architectural models, architectural documents, architectural prototypes, and executable architectures. MFESA methods also produce requirements metadata and architectural concerns, which are cohesive collections of architecturally significant requirements.

4.9.6 Assumptions

MFESA is based on three primary assumptions. First, responsibilities should be clearly delineated between requirements engineers, architects, and the evaluators of architectural representations. Second, systems are typically so large that they require multiple architecture teams. Third and finally, method engineering is the best way to provide appropriate architecture engineering methods.

4.9.7 Other Disciplines

System architecture engineering methods (and thus, MFESA-compliant methods) must be consistent with the methods used by other disciplines because architecture engineering uses inputs produced by other disciplines and produces outputs used by other disciplines. This includes primary engineering disciplines such as requirements engineering, design, and testing. This also includes management disciplines such as project management, configuration management, and risk management. Finally, this also includes specialty engineering disciplines such as human factors, reliability, safety, and security engineering.

4.9.8 Guidelines

Use the following overall guidelines (and associated rationales) when using MFESA:

■ Remember that MFESA is not a system architecture engineering method.
■ Remember that MFESA is more than a repository of reusable method components.
■ Remember that MFESA has multiple uses.
■ Remember that MFESA methods use both requirements and reusable architectural elements as inputs.
■ Use MFESA tasks to organize and understand.
■ Remember that MFESA affects more than just the architects.

Chapter 5

MFESA: The Ontology of Concepts and Terminology

5.1 The Need for Mastering Concepts and Their Ramifications

System architecture engineering concepts and their ramifications must be mastered in order to understand the engineering of architectures for software-intensive systems. This chapter discusses and defines these terms, thereby providing the basis for mastering subsequent chapters. In fact, it goes beyond defining these concepts to providing an ontology capturing the relationships between the concepts.

5.2 Systems

The most fundamental concept underlying all system architectural engineering is that of a *system*. To illustrate the defining characteristics of a system, we start by describing two example systems: one natural and one artificial.

The natural world is filled with systems of all sizes, ranging from the microscopic to the size of the visible universe. Probably the best known and understood example of such a natural system is the human body, or the body of any mammal, for that matter. Considered from a scientific point of view (biology, medicine, and anatomy),* a body is a highly interconnected system of systems, some of the most important of which include the:

- Cardiovascular system:
  - *Functions*: transport gases (oxygen and carbon dioxide) between the lungs and the rest of the body; nutrients (such as glucose, fats, and amino acids) from the digestive system

---

* Note that different stakeholders of a system may view it as being composed of different components because of their different points of view. For example, the average person may decompose his or her body into a head, a torso, two arms, and two legs, whereas a philosopher might decompose a person into body, mind, and soul.

to the rest of the body; hormones from glands to the rest of the body; soluble cellular wastes from the rest of the body to the urinary system; and components of the immune system (white blood cells and antibodies) to sites of infection.

- *Components*: heart, blood, major blood vessels (arteries and veins), minor vessels (arterioles and venules), and microscopic vessels (capillaries).

■ Endocrine system:
- *Functions*: produce hormones.
- *Components*: glands and specialized cells, including the adrenal glands, heart, hypothalamus, intestines, pancreas, pituitary, stomach, thymus, thyroid, ovaries, and testes.

■ Gastrointestinal system:
- *Functions*: chop, chew, store, and digest food; store and eliminate solid wastes; and destroy swallowed germs.
- *Components*: mouth, throat, esophagus, stomach, pancreas, liver, gallbladder, small intestine, and large intestine.

■ Integumentary system:
- *Functions*: absorb physical shocks, insulate the body, protect the body from external sources of physical damage (e.g., germs and UV light), regulate body temperature (by sweating), and store energy.
- *Components*: skin, hair, nails, and the associated subcutaneous fat layer.

■ Lymphatic and immune system:
- *Functions*: provide protection against infectious diseases and certain internal malfunctions, distribute certain soluble nutrients and white blood cells, and collect certain soluble wastes.
- *Components*: antibodies; lymph fluid; lymph vessels, nodes, and ducts; spleen; tonsils and adenoids; thymus gland; and white blood cells.

■ Muscular system:
- *Functions*: provide pulling forces for bodily movement and control certain internal processes (blood distribution and digestions).
- *Components*: cardiac heart muscles, skeletal muscles connected to bones, smooth muscles within organs, and tendons connecting muscles to bones.

■ Nervous system:
- *Functions*: obtain sensory information, control other bodily systems, and provide mental (conscious) capabilities such as thoughts, emotions, and memories.
- *Components*: brain, spinal cord, peripheral nerves, and sense organs.

■ Reproductive system:
- *Functions*: reproduction.
- *Components*:
  • Female: ovaries, fallopian tubes, uterus, vagina, external genitalia, and breasts.
  • Male: testes, spermatic ducts, seminal vesicles, urethra, penis, and prostate and bulbouethral glands.

■ Respiratory system:
- *Functions*: primarily absorption of oxygen and removal of carbon dioxide, but secondarily vocalization.
- *Components*: nasal and other air passages in the skull, throat (pharynx), windpipe (trachea), lungs, major and minor lung airways (bronchi and bronchioles), diaphragm, and other respiratory muscles.

■ Skeletal system:
  - *Functions*: support the body, provide the muscular system with levers and anchor plates to support movement, protect the brain, produce blood for the cardiovascular system, and store minerals such as calcium that are needed for healthy functioning of the nervous system.
  - *Components*: axial skeleton (skull, spine, ribs, and breastbone), appendicular skeleton (limb bones, shoulders, and hips), and ligaments connecting bones.
■ Urinary system:
  - *Functions*: eliminate soluble wastes from the blood and help maintain a correct balance of water, fluids, salts, and minerals.
  - *Components*: kidneys, ureters, bladder, and urethra.

There are reasons why all the preceding 11 bodily systems were briefly summarized instead of just listing them or only summarizing one or two of them:

■ Systems and subsystems can often be characterized by the functions they perform and the components of which they are composed.

   Systems can be very complex, consisting of a large number of subsystems. Each of these systems may be further decomposed into smaller and smaller subsystems until one reaches a level of abstraction that can be considered atomic for the purpose at hand and therefore can be treated as a black box. For example, organs, tissues, cells, and organelles are four types of lower-level subsystems within the body.

   Note that the decomposition of a system into its component parts is potentially ambiguous and therefore partially based on convention. For example, one could argue that the blood producing bone marrow could be a part of the cardiovascular system instead of the skeletal system. Similarly, why is the spleen, which filters out old red blood cells and stores red blood cells, a part of the endocrine system instead of the cardiovascular system or the lymphatic and immune system (because people who have had their spleen removed are more susceptible to infection)? Also, are the white bloods cells a part of the lymphatic and immune system or the cardiovascular system?
■ Typically, each system or subsystem tends to be functionally cohesive, in that it performs a single or small number of highly related functions. However, some systems or subsystems are more functionally cohesive than others. Different systems within a system of systems perform different functions and consist of different architectural components, although there may be some overlap between them. However, some subsystems perform multiple functions; the skeletal system provides support, protection for the brain, and leverage for the muscles. On the other hand, some functions require the collaboration of multiple subsystems; the skin, digestive system, and immune system all provide protection from infection by germs.
■ There can be many types of interfaces between systems and subsystems, including the flow of information and control (nervous and endocrine systems) and the flow of materials (the cardiovascular, digestive, endocrine, respiratory, and urinary systems).

Because there are many more ways to look at the human body than from the viewpoint of an anatomist, the preceding static hierarchical decomposition of the human body into its component parts is not the only important architectural "structure." Hierarchical decomposition based on functional decomposition is *not* the only way to decompose a system, and there are typically many static and dynamic as well as logical and physical structures inherent in any non-trivial system, and the determination of which ones are important depends on the needs and interests of the system's stakeholders. For example, a modern doctor of Western medicine or a pharmaceutical researcher may be interested in a dynamic metabolic pathway, whereas a practitioner of Eastern alternative medicine may be interested in the flow of "Chi" through the body.

The systems comprising the human body are highly interrelated, connected by many different types of interfaces. Nerves connecting the brain, sense organs, muscles, and glands support the flow of data and control. The circulatory, lymphatic, and digestive systems are involved with the flow of materials, such as blood and lymph. Muscles, tendons, ligaments, and joints support the flow of mechanical energy, whereas chemical energy is also transported around the body in the form of glucose, fats, and ATP.* This existence of various interfaces transporting information, materials, and energy is also a characteristic of many artificial systems. The body and many artificial systems also have structural components and associated interfaces providing support (e.g., bones) and containment (e.g., skin).

Another interesting aspect of the human body is that the whole is much more than the sum of its component parts. At one end of the spectrum, a properly functioning body includes metabolism and is thus alive. At the other end of the spectrum, proper function of the human brain somehow produces consciousness (the human mind and all that it entails), including thoughts, emotions, and the ability to produce all the products of civilization (including artificial systems and this book). *It is through such emergent behavior and properties that a system can do and be things that its individual components cannot.*

Many other natural systems exist in size, ranging from the entire universe down to individual molecules. These systems vary greatly in complexity, composition, and how their components are interconnected and interact. Some natural systems naturally evolve due to the existence of the most fundamental of forces; for example, galaxies and stellar systems evolve because of gravitation. Other natural systems such as species naturally evolve via differential reproduction due to mutation, natural selection, and genetic drift. Finally, other complex natural systems such as ant nests and ecosystems naturally evolve due to other high-level types of interactions and collaborations.

Since the dawn of civilization, humanity has been producing ever-more complex artificial systems, including agricultural systems; civic systems from villages, towns, and cities; financial systems; power systems; manufacturing systems; and transportation systems. For the past 70 years, we have created software-intensive systems, which have a large part of their functionality and characteristics provided by the execution of computer software. Because engineering the architecture of such software-intensive artificial systems is the subject of this book, the word "system" when it occurs without any qualifier such as "artificial" will henceforth be restricted to software-intensive systems, and our next generic example will therefore be a software-intensive system.

One common type of such a system is a commercial passenger airliner. In fact, essentially all aircraft, whether commercial or military, are systems of systems. As illustrated in Figure 5.1, a generic modern commercial passenger airliner could be architected to consist of the aircraft itself and its associated supporting ground-based systems.

Figure 5.1 illustrates a highly incomplete hierarchical decomposition of an example *aircraft system of systems*. This static physical structure consists of the following top-level components:

■ **Aircraft system.** The aircraft system is the actual flying commercial passenger airliner, which is composed of its airframe, interiors, propulsion, mission, and vehicle segments.

---

* Adenotriphosphate (ATP) is the molecule that stores and transports energy within cells.

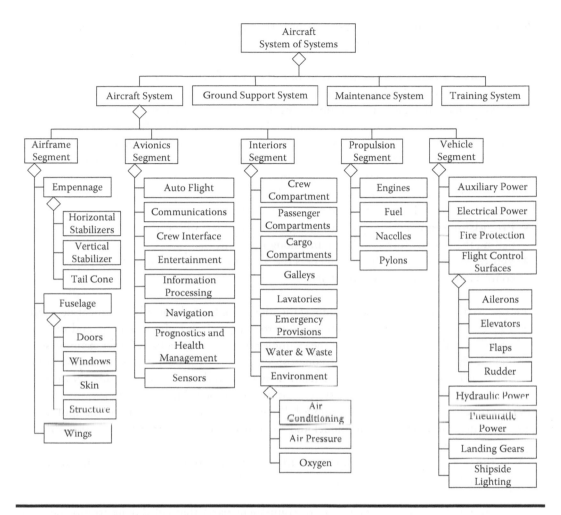

Figure 5.1 Example aircraft system of systems.

- **Ground support system.** The support system consists of the ground equipment that supports the aircraft when it is parked at the terminal. For example, this segment includes equipment that supplies ground-based electrical and hydraulic power.
- **Maintenance system.** The maintenance system consists of the maintenance facilities and equipment as well as the associated information system for managing the maintenance operation and supply chain management of aircraft parts.
- **Training system.** The training system provides training for the flight crew (pilot, copilot, and cabin crew) and aircraft maintainers. It consists of simulators, mock-ups, classroom and computer-based training materials, and a training management system for maintaining training plans, schedules, and records.

As illustrated in Figure 5.1, the example aircraft system consists of the following segments:

- **Airframe segment.** The airframe segment is the physical structural parts of the aircraft. The main functions of the airframe are to maintain structural integrity, sustain loads, and maintain aerodynamic profile and performance.

- **Avionics segment.** The avionics segment is the highly software-intensive, mission-oriented part of the aircraft consisting of all the electronic subsystems that manage and control the aircraft.
- **Interiors segment.** The interiors segment is the part of the aircraft consisting of the interior compartments within the fuselage as well as their support systems.

   Note that there are often multiple ways to decompose a system. For example, the interior segment could legitimately be considered a component of the fuselage as opposed to a component of the aircraft at the same level as the airframe, which contains the fuselage. In practice, the static physical hierarchical decomposition structure of a system is often organized more to optimize organizational support (e.g., domain expertise and subcontracting) and map to a functional decomposition of the requirements (e.g., the specification tree) than to represent actual physical inclusion.

- **Propulsion segment.** The propulsion segment is the part of the aircraft that concerns propelling the aircraft through the air and along the taxiways.
- **Vehicle segment.** The vehicle segment is the part of the aircraft consisting of physical flight control and various utility subsystems.

Notice that while being critically important, the preceding physical hierarchical decomposition aggregation structure is only one of numerous logical and physical static and dynamic structures that make up the majority of the example airliner's architecture. Other example structures include the airliner's functional decomposition structure, logical data structure, data flow structure, control flow structure, concurrency structure, and hardware structure. Notice that there are also numerous ways that the system can be decomposed into an aggregation hierarchy of architectural components; it is the responsibility of the architects to determine the appropriate types of architectural structures to model and to ensure that these architectural structures are properly engineered.

With the preceding concepts in mind, we are now in a position to define a system. We therefore provide the following definitions.

> **System:** any cohesive integrated set of component parts that collaborate to provide behavior and characteristics that the parts cannot provide individually.
> **Artificial system:** any human-produced system engineered to provide the behavior and characteristics needed to meet the needs and desires of legitimate* stakeholders.

In this book, we are interested in artificial systems that are engineered to meet the needs of their stakeholders. For the remainder of this book, we restrict the discussion to artificial systems and use the term "system" to mean "artificial system."

Systems rarely occur in isolation. System architects must almost always deal with systems of systems, either because they are architecting the overall system of systems or because they are architecting a system that is a part of a system of systems and that must properly interoperate with the other systems within the system of systems. We therefore provide the following definitions:

---

* Note that not all stakeholders are legitimate. If a stakeholder is defined as any person or organization that has an interest in the system, then some stakeholders are illegitimate in that they wish to harm the system or other valuable assets for which the system is responsible. These illegitimate stakeholders are called attackers or threat agents. Thus, to address security and survivability concerns, architects must address how the system will protect valuable assets from attack by attackers, detect when such attacks occur, and respond appropriately.

System-of-systems systems (SOS, a.k.a. network of systems): any single system, the primary component parts of which are themselves systems.

A system of systems exists when useful systems are integrated to form a larger system that delivers unique capabilities that are not directly achievable by its individual component systems. When the component systems are integrated, they tend to become interdependent.

Systems of systems form spectrums, based on the degree to which they interoperate and the degree to which they are developed as a group. At one end of the spectrum are systems of systems that from the very beginning were (or are being) architected and developed so that their component systems collaborate to provide a cohesive set of functionality. These systems of systems typically have a single overall acquisition organization and a single overall development organization that acts as the system (of systems) integrator and overall prime contractor. At the other end of the spectrum are systems of systems, the component systems of which were (or are being) developed totally independently of one another. These systems of systems do not have a single overall acquisition organization, do not have an overall development organization (i.e., no system integrator or prime contractor), and are often developed independently by competing development organizations. In the first case, there is typically an overall system of systems architecture team producing an overall system of systems architecture and its associated representations. In the latter case, there is no overall system of systems architecture team and thus no *cohesive* overall SOS architecture with its associated representations. Although all such systems of systems are technically systems, we will use the term "system of systems" to refer to one that is more toward the earlier end of the spectrum where the system of system is highly integrated and developed and maintained as a single overall system. We will reserve the use of the term "network of systems" to refer to one that is toward the later end of the spectrum where the component systems are developed and maintained independently of the overall system of systems.

A special case of the highly integrated system of systems occurs when a product line of highly similar systems is being developed and maintained. Such a family of systems is typically developed by a single development organization according to a single common reference architecture.

Family of systems (a.k.a. product line of systems): any set of systems of a single type sharing a common reference architecture.

We defined a system as a cohesive set of component parts. We now turn to defining and discussing the components of a system.

Architectural component: any part of the static physical *aggregation* structure* of a system.
Subsystem: any architectural component that itself is a system.

Almost all systems are sufficiently large and complex that their largest architectural components are heterogeneous subsystems. Thus, technically, almost all systems can be viewed as systems of systems. However, in practice, the terms "system of systems," "network of systems," and "family of systems" usually refer to aggregations or sets of smaller systems, which themselves are typically thought of as being somewhat logically independent of each other and individually useful in isolation. On the other hand, the term "subsystem" typically refers to a system, the primary use of

---

* Note that a system has static and dynamic as well as logical and physical structures. Because there are also multiple static physical structures, it is important for the definition to be specific about which structure is intended. Note also that software is considered a *physical* as opposed to a *logical* part of the system.

which is as a part of its larger containing system. For example, although an engine is a system, it is almost always considered to be a subsystem of some larger system for which it provides power.

According to the preceding definitions, a system is an aggregation structure of collaborating components. But what types of architectural components are there? A common misunderstanding is to think that artificial systems consist solely of combinations of hardware and software. It is true that software and hardware are often the two most important types of system components. It is commonly observed when engineering software-intensive systems that software cannot run without (computer) hardware, and that hardware cannot be controlled without software. With today's software-intensive systems, it is also true that quality characteristics such as availability, portability, reliability, safety, security, and usability are system characteristics and not achievable by either hardware or software alone. However, artificial systems (especially large and complex ones) are typically composed of many different types of architectural components in addition to hardware and software.

Although systems primarily consist of subsystems, the atomic, leaf node architectural components of a system can typically be:

- Consumable materials (such as ammunition, fuel, lubricants, reagents, and solvents)
- Data
- Documentation (both separate physical and built-in electronic documentation)
- Equipment (including maintenance, support, and training equipment)
- Facilities (such as maintenance, manufacturing, operations, support, training, and disposal facilities, including their component property, buildings, and furnishings)
- Firmware
- Hardware*
- Manual procedures
- Naturally occurring substances or entities (such as water, minerals, enzymes, and organisms†)
- Networks (for the flow of information, matter, and energy)
- Organizations
- Personnel‡
- Physical interfaces
- Roles played by people and organizations
- Software
- Tools

As mentioned previously, most of today's artificial systems contain very large amounts of software, which provide much of the systems' functionality and dynamic behavior. This software is used as an important glue to bind a system's subsystems together and allows them to intra-operate. Software has also become the primary technology for enabling systems to interoperate with each

---

* Note that there are many different types of hardware, such as actuators, computing devices, power supplies, sensors, and structural components.
† Note that specific strains of specific organisms may be critically important parts of systems that contain bioreactors. Such systems are used to produce medically important substances or to produce ethanol from sugar or other carbohydrates.
‡ Whether personnel are internal components of a system or external actors interacting with the system depends on the choice of where to place the system boundary and is an important decision that can have a significant impact on the requirements and architecture.

other and their human users, administrators, operators, and maintainers. For this reason, this book concentrates on software-intensive systems.

> **Software-intensive system:** any artificial system containing software components that provide a large and essential part (e.g., greater than 25 percent) of its behavior and characteristics.

Software partially or completely implements a large amount of a software-intensive system's requirements. Software-intensive systems also use software to implement a large amount of the key interactions among architectural elements.

In addition to the static aggregation structure of the system that decomposes it into a hierarchy of architectural components, there exist other important logical and dynamic structures, made up of other architectural elements and the relationships among them.

> **Architectural element:** any significant, cohesive *logical* or *physical* part of a system (i.e., a part of some system structure as seen from some viewpoint).

Note that an *architectural component* is a specialized subtype of *architectural element*. All architectural components are architectural elements, but there are architectural elements that are not architectural components. For example, in addition to the architectural components, architects must address the following typical types of logical system elements:*

- Data types
- Event types
- Functions
- Logical concurrent processes
- Logical control flows
- Logical data flows
- Logical failure, fault, and exception flows
- Logical interfaces
- Modes and states
- Services

Before leaving this discussion of systems, it would be useful to digress and summarize systems in terms of the following characteristics that they share, including:

- **Benefits.** Systems are engineered to provide benefits to their legitimate stakeholders. Meeting the needs and requirements of the stakeholders must be foremost in the minds of the architects of the system and drive their architectural decisions, inventions and trade-offs.
- **Legitimate stakeholders.** Systems can have both legitimate stakeholders and attackers (i.e., illegitimate stakeholders). Legitimate stakeholders have a legitimate stake in the proper behavior and characteristics of a system. On the other hand, illegitimate stakeholders (e.g., organizational, human, and malware attackers of the system) have a malicious, illegitimate stake in the proper behavior and characteristics of the system. Whereas legitimate stakeholders are a source of system goals and requirements, knowledge about attackers and malware

---

* Note that these example logical and physical system elements are of different types and may therefore belong to different views of different types of system structures and their associated different types of models.

may be used as sources of anti-requirements and to indirectly engineer security and survivability requirements and associated architectural countermeasures.

■ **Large decomposition structure.** Many modern software-intensive systems have some of the largest and most complex structures that humanity has ever built. Such systems can be viewed as an aggregation hierarchy consisting of a very large number of components of several different types. At the higher tiers in the aggregation tree, systems consist of lower-level systems (i.e., subsystems). However, the lowest leaves of the system aggregation hierarchy typically consist of hardware components, software components, data components, manual procedures, and roles played by people, facilities, equipment, and materials.

■ **Cohesive integrated set of system components.** To be a single system as opposed to a collection of unrelated systems, the system must be cohesive in the sense of supporting a single mission or a cohesive set of missions. Note that a system of systems, a network of systems, and a family (e.g., product line) of systems are also systems.

The architectural components of a system must be integrated into a single whole for the system to exist. For example, there is a big difference between a car parked in a garage (a system) and the collection of all the individual car parts lying on the floor of that garage (not a system).

■ **Aggregation structure.** Because a system consists of an integrated set of components, it has an important aggregation structure. Thus, every system, by definition, has at least one structure (i.e., its aggregation structure). However, all interesting systems have multiple structures. Because this aggregation structure may be hierarchical with many levels, systems can typically be decomposed into subsystems, which in turn can be decomposed into lower-level subsystems, whereby the subsystems are themselves systems.

Note that this aggregation structure is very important because people naturally tend to primarily think of systems in terms of their component parts. It is also important because of Conway's law, which states that the structure of a system tends to mirror the structure of the organization that produced it. Thus, systems are often developed by a largely hierarchical organization of cross-functional integrated product teams (IPTs) that are responsible for developing the individual subsystems and lower-level architectural components. Note that the mapping from system structure to organization structure is not exact because real organizations typically include additional teams such as the management team, the configuration management team, the quality engineering team, and several specialty engineering teams such as the safety team and the security team.

■ **External interactions with the environment.** Systems do not exist in isolation from their environments. To provide value, systems must interact with one or more of the following: some of their human stakeholders (e.g., users and operators), external artificial systems (e.g., as part of a system of systems or network of systems), and the natural environment (e.g., via sensors and actuators). Systems interact externally by exchanging *information* (e.g., data and control flows), *matter* (e.g., flows of materials and people), and *energy* (e.g., the flow of electrical, chemical, and mechanical energy).

■ **Collaborative internal interactions.** The different types of system components are not totally independent, and they are not just integrated statically (e.g., bolted together). Rather, they are related in multiple ways and interact in multiple ways. The system components must collaborate to provide the system's behavior and characteristics. These components interact with each other (and with the external environment of the system). System components do this by exchanging *information* (e.g., data and control flows), *matter* (e.g., materials and people), and *energy* (e.g., electrical, chemical, and mechanical).

- **Behavior and characteristics.** A system has both behavior (i.e., functionality) and characteristics (e.g., quality characteristics). When engineering the architecture of a system, it is critical to consider both its required functionality and its required attributes. The functionality of the system is largely provided by the internal processing and storage of the information, matter, and energy transmitted during the collaborative interactions between the system components.

- **Emergent behaviors and characteristics.** Often, the most important behaviors and characteristics of a system only emerge when the system's components are integrated together. Because of the integration and collaboration of its components, a system is able to do things that its components cannot do individually. For example, it is the proper integration of a vehicle's component subsystems that enables it to efficiently and safely transport people and goods from place to place. Similarly, many of the most important characteristics of a system (e.g., its quality) depend on the collaboration of the system's component parts. For example, quality characteristics such as availability, performance, reliability, safety, and security are primarily characteristics of the system as a whole rather than its component parts (e.g., its hardware and software), although the quality of individual system components often greatly influence the quality of the overall system.

- **Multiple structures.** The multiple types of components and component interactions imply that the systems have multiple structures, both logical and physical. Some of these structures may be relatively static whereas others are primarily dynamic. One such structure is the static physical aggregation (decomposition) structure capturing how the system is decomposed into subsystems; these subsystems are decomposed into lower-level subsystems, and so on down into appropriately small hardware, software, data, procedural, and other architectural components. Another important system structure is the logical static functional decomposition structure showing the functions performed by the system. Other structures include, but are not limited to, data flow structures, control flow structures, network structures, state transition structures, and allocation of software to hardware structures.

- **Multiple views and models.** Although some system structures can safely be seen as much more important than others (at least by a majority of stakeholders), it will be impossible to understand major aspects of a system without looking at it from multiple viewpoints based on the associated concerns of the system's various stakeholders.

  Because of the complexity of a typical system, it is impossible to understand the system without modeling it to abstract its essential behavior and characteristics from their diversionary details of its design and implementation. Because there are multiple structures of different types, it is impossible to use only a single type of model and a single associated modeling language to capture all important views of all important system structures.

  To understand systems and their structures, architects need to create multiple models providing multiple views. For example, architects can create a data view of the system by constructing a logical data model, a physical data model, and a data flow model.

- **Multiple focus areas.** Although models are abstractions and thus simplifications of reality, most systems are so complex that their models are also extremely complex. Instead of viewing an entire system by looking at an entire associated model, it is often important to focus on only the specific bits of the model that are relevant to the specific stakeholder concern at hand. For example, only a few of the parts of any single model capture the architectural decisions, inventions, and trade-offs that were made to support the achievement of any one type of system quality. To do that requires multiple parts of multiple models, and the collection of these relevant bits of the relevant models (and other related architecture representations), and this collection is called a focus area. For example, the two most important focus areas for one subsystem might focus on security and interoperability, whereas the three most important focus areas for another subsystem might focus on availability, reliability, and performance.

There are typically multiple ways to structurally model a system, and although vitally important, a physical aggregation structure is only one such way. In fact, a single system can typically be modeled as an aggregation structure in multiple ways because there usually exists more than one way to decompose a system into its architectural components (or compose a system out of existing architectural components). Thus, Figure 5.1 illustrates only one way to decompose an aircraft, and other ways would exhibit different branching with different branch and leaf nodes. Because decomposition is central to the definition of a system, one of the architect's more important responsibilities is to produce a relatively "optimal" aggregation structure and associated model. Note that this means that the identification of architectural components depends on many factors and is somewhat a matter of professional experience and judgment; it is not necessarily obvious when given an existing system.

5.3 System Architecture

The architecture of a system forms a top-level abstraction of the *to-be* or *as-is* system. The architecture is not a set of requirements of what the system must do or how well the system must do it. Rather, the architecture provides high-level strategic views of the system and how the system will meet its most important product and process requirements.

> **System architecture:** the set of all of the *most important, pervasive, higher-level, strategic decisions, inventions, engineering trade-offs, assumptions,* and their associated *rationales* concerning *how* the system meets its allocated and derived product and process requirements.

As illustrated in Figure 5.2, architects engineer the system architecture, which is an abstraction of the system and which consists of architectural decisions, inventions, trade-offs, assumptions, and rationales, whereby the trade-offs, assumptions, and rationales often drive the decisions and inventions.

This book deals with artificial software-intensive systems as opposed to natural systems such as solar systems and anthills, which evolve without the need of human architects. These technological systems were created by people who made architectural decisions, inventions, and engineering trade-offs during the course of the system's development and maintenance. Thus, all such systems have an architecture, even if that architecture only exists (or at one time existed) in peoples' heads and was never documented in the form of architectural representations (e.g., models and documents).

Note that our definition of system architecture differs significantly from the traditional definitions. Historically, system architecture has been explicitly defined in terms of "the structure" of the system. For example, the *International Council on Systems Engineering* (INCOSE) *Systems Engineering Handbook* defines system architecture as "the arrangement of elements and subsystems and the allocation of functions to them to meet system requirements." The *ANSI/IEEE Recommended Practice for Architectural Description of Software-Intensive Systems* (1471-2000) defines the system architecture as "the fundamental organization of a system embodied in its components, their relationships to each other and to the environment and the principles governing its

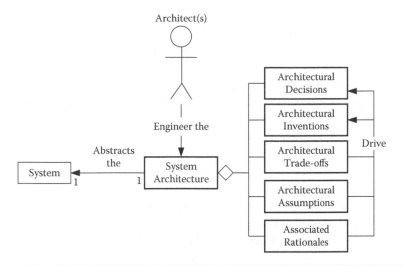

Figure 5.2 System architecture.

design and evolution." Although there are differences between software and system architectures, software architecture is often also defined in a traditional structural manner. For example, according to the Software Engineering Institute (SEI), *software* architecture* is defined as "the structure or structures of the system, which comprise software elements, the externally visible properties of those elements, and the relationships among them."

There are two key problems with these traditional definitions. As we shall see, there are many structures, and system architects cannot restrict themselves to the development and analysis of just one or two of them. More importantly, by restricting architecture to a structural viewpoint, the preceding definitions do not cover the entire breadth of a system architect's responsibilities. Remember that, in general, system architecture includes all of the most important, pervasive, higher-level, strategic decisions, inventions, engineering trade-offs, assumptions, and their associated rationales concerning how the system meets its allocated and derived product and process requirements. More specifically, system architecture includes the:

■ **Major system or subsystem structures.** The *most important* aspects of system architecture are typically the decomposition of the system into its major static and dynamic as well as logical and physical architectural elements, their associated black box characteristics and behavior, their relationships, and how the elements collaborate together to support the system's mission and architectural drivers.† As documented in section 5.4, "Architectural Structures," major system or subsystem structures may be logical or physical; they may also be static or dynamic.

---

* http://www.sei.cmu.edu/architecture/index.html.

† Note that a system architecture does not consist of only a single type of structure. The architecture of a system captures much more than "merely" how the system is physically decomposed recursively into lower and lower-level subsystems down to a level (e.g., configuration items) that no longer requires system architectural decisions but rather can be acquired, reused, or designed and implemented. Instead, there are numerous logical and physical, static and dynamic structures that can be modeled using associated modeling techniques to provide very useful views of the system architecture. Because the optimum type and number of structures depend on the system/subsystem being architected and many other factors, MFESA does *not* specify any specific set of architectural structures to model.

- **Other architectural decisions, inventions, trade-offs, assumptions, and rationales.** The architecture of a system also includes the:
  - Choice of major technologies and materials to incorporate into the architecture.
  - Application of architectural principles and heuristics.
  - Selection of architectural styles, patterns, and mechanisms, which are used to ensure that the system meets its architectural drivers (i.e., cohesive sets of its architecturally significant product and process requirements).
  - Architectural component *build* (directly or subcontract) versus *reuse* (COTS, GOTS, MOTS, open source, or internal reuse) decisions.
  - Approach to individual specialty engineering quality requirements (e.g., interoperability, performance, reliability, safety, and security). Note that to enable the rapid and easy location of the relevant parts of the representations (e.g., models and documents), they may well need to be identified in the representations via metadata.
  - Allocation of existing and derived product and process requirements to architectural elements.
  - Architectural analysis results (for trade-offs or rationales).
- **Strategic and pervasive *design*-level decisions.** For example, this includes system architects mandating the system-wide use of a specific design paradigm such as object orientation or common design patterns. Note that this is the responsibility of the system architects and should not be mandated by requirements engineers as constraints unless truly required.
- **Strategic and pervasive *implementation*-level decisions.** This includes system architects mandating the system-wide use of a specific language or language subset, such as Ada or a relatively safe subset of C or C++.

In practice, a major area of confusion is between *system architecture* and *system design*. Some people use them as synonyms. Others equate *architecture* with *preliminary design* and *design* with *detailed design*. These differences are illustrated in Table 5.1.

Note that these differences between the criteria used to differentiate system or subsystem architecture from system or subsystem design are to be interpreted relative to the scope of the system or

Table 5.1   Differences between Architecture and Design

Architecture	Design
Huge impact on quality, cost, and schedule	*Small impact* on quality, cost, and schedule
Pervasive—always affects multiple system components	*Local*—typically affects only individual components
Higher-levels of system aggregation hierarchy	*Lower-levels* of system aggregation hierarchy
Strategic decisions, inventions, and trade-offs	*Tactical* decisions, inventions, and trade-offs
Mirrors top-level development team *organization* (Conway's law)	*Little or no impact* on the top-level development team *organization*
Drives design and integration testing	*Drives* implementation and unit testing
Driven by requirements and even higher-level architecture	*Driven by* requirements, architecture, and higher-level design

subsystem being architected. Thus, a decision, invention, or trade-off may be considered to have a small impact on the overall system architecture (system design), but a huge impact on a lower-level subsystem (subsystem architecture). Similar scoping issues must be addressed in terms of having pervasive versus local effects, being at a higher versus lower level of the aggregation hierarchy, and being strategic versus tactical. Nevertheless, once the scope of discourse is clear, the difference between architecture and design within that scope should also be clear.

> **System design:** the entire set of all of the *secondary, local, lower-level, tactical decisions, inventions, engineering trade-offs, assumptions,* and their associated *rationales* concerning *how* the system meets its allocated and derived requirements.

A primary goal of the system architecture is to constrain the downstream development of the system including design, implementation, integration, testing, and deployment to:

- Ensure the achievement of the architecturally significant requirements
- Provide the benefits of consistency across the system

Note that a single view of a system architecture is insufficient to capture all of these inventions, decisions, trade-offs, and assumptions. Rather, the system and each of its subsystems include *multiple* architectural structures that should be documented using multiple architectural views. These multiple individual views must together provide a consistent overall view of the system.

5.4 Architectural Structures

As illustrated in Figure 5.3, a system architecture primarily consists of a number of logical and physical, static and dynamic architectural structures, each of which consists of a number of architectural elements connected by relationships between these elements. These architectural structures are abstractions of the system.

> **System architectural structure:** any cohesive set of architectural elements connected by associated relationships that captures a set of related architectural decisions, inventions, trade-offs, assumptions, and rationales.
> **Architectural element:** any part of an architectural structure.
> **Architectural relationship:** any significant relationship between two architectural elements.

This includes but is not limited to aggregation, association, collaboration, dependency, interaction, and specification/generalization.

System architectural structures can be either logical or physical. They can also be either static or dynamic. Thus, architectural structures tend to be logical and static, logical and dynamic, physical and static, or physical and dynamic.

> **Logical structure:** any architectural structure, the elements of which are logical or conceptual in nature.
> **Physical structure:** any architectural structure, the elements of which are physical or tangible in nature.

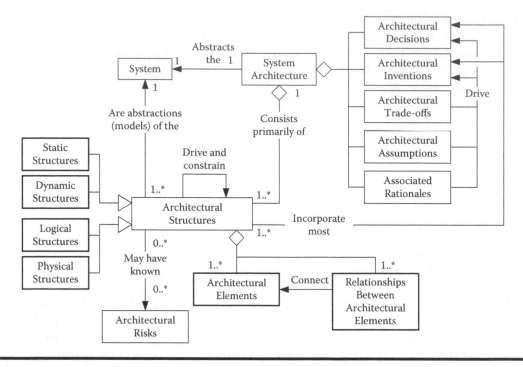

Figure 5.3 Architectural structures.

> **Static structure:** any architectural structure that captures the fixed (i.e., nonbehavior) aspects of the system.
> **Dynamic structure:** any architectural structure that captures the behavior of the system.

Once engineered, some system structures (such as the physical static aggregation decomposition structure) are so obvious that they seem to clearly and objectively exist in the system, regardless of whether they are modeled or documented via architecture representations. Each system has at least one such structure. However, some level of residual ambiguity resides in even such straightforward structures as the aggregation structure. For example, consider the aggregation structure of the aircraft system of systems illustrated in Figure 5.1. Although the lowest-level physical hardware components and interfaces seem physically and objectively built into the resulting aircraft, things are not necessarily so simple when it comes to the software, the organization of which varies depending on whether one is considering the source code, the object code, or the executing processes. Similarly, there is considerable flexibility as to how certain parts of the aircraft are decomposed. For example, do we use *functional* cohesion to group all aspects related to fuel into a single fuel system (as in Figure 5.1), or do we instead use *location* cohesion to allocate different fuel tanks, pumps, pipes, sensors, valves, etc., to different components of the airframe, depending on where they physically reside? Similar choices must be made concerning electrical power, information processing, prognostics and health management, and fire detection and suppression.

The architects are responsible for determining the appropriate architectural structures to model and document in the appropriate architectural representations. These choices depend on the importance of the associated architectural decisions, inventions, trade-offs, and assumptions. They also depend on the needs of the many architectural stakeholders, the architectural risks to be mitigated, and the uses to which the associated models and views will be put such as communication, analysis, and prototyping (e.g., to determine feasibility or verify assumptions).

Any system typically has a large number of potentially significant logical and physical static and dynamic structures that may need to be modeled. The following are typical examples of such structures that system architects may need to engineer as part of a system architecture:*

- **Logical static structures.** Logical static structures consist of logical architectural elements and the static relationships between them. Examples of such structures include:
  - **Allocation structures** (including static, configurable, or dynamic allocation) consisting of the allocation of something to functionality. This includes, but is not limited to, the allocation of:
    - Performance budgets to functionality
    - Maximum power consumption, weight, and volume to functionality
  - **Functional decomposition structure** consisting of the major functions performed by the system and their associated relationships (e.g., decomposition and invocation).
  - **Integration structures** consisting of groupings of logical structures that are suitable to demonstrate or verify that implementations of the logical structures conform to the architecture.
  - **Layering structure** consisting of the major layers of abstraction and their relationships (e.g., dependency).
  - **Logical class structure** consisting of the major classes of domain objects modeled, their relationships (e.g., association, aggregation, and generalization), and their behaviors.
  - **Logical data structures** consisting of logical data types that are visible at the boundaries of logical architectural elements (e.g., functions) and the relationships between them.†
  - **Logical data management structures** consisting of distinct categories of data that are managed similarly. For example, data of different lifetimes may provide a set of categories: ephemeral data that is conveyed across interfaces, longer-lived context data associated with a runtime mode, persistent data that is stored in a database or similar structure, and static data such as configuration parameters stored in firmware.
  - **Logical management structures** consisting of the orchestration and control of the logical structures; this is also known as the logical *infrastructure*.
  - **Service structure** consisting of the major services provided by the system
- **Logical dynamic structures.** Logical dynamic structures consist of logical architectural elements and their dynamic relationships capturing their dynamic behavior. Examples of such architectural structures include:
  - **Logical flow structures** consisting of a cohesive set of system functions and the significant flows between them:
    - *Logical control flow structures* consisting of a cohesive set of system functions and the significant control flows between them
    - *Logical data flow structures* consisting of a cohesive set of system functions and the significant flows of data (types) between them
    - *Logical energy flow structures* consisting of a cohesive set of system functions and the significant flows of energy (e.g., electrical, mechanical, rotational) between them

---

* Although relatively complete when compared to architectural engineering standards and methods, it is impractical to provide a "complete" list. Architects may need to add additional project-specific structures and models to meet project- or system-specific needs.

† Note the difference between a logical data structure and a physical data structure that is captured in the difference between a logical data model and a physical data model.

- *Logical failure flow structures* consisting of a cohesive set of system functions and the significant potential failures and exceptions that flow between them (e.g., showing where they are raised, how they flow through the system, and where they are handled)
    - *Logical material flow structures* consisting of a cohesive set of system functions and the significant flows of energy (e.g., chemicals, fuel, or products) between them
  - **Mode and state structures** consisting of the modes and states of major logical elements and their relationships (e.g., state transitions)
  - **Mode allocation structures** consisting of the allocation of functionality to system mode and state
  - **Process structures** consisting of the major logical parallel threads of execution
  - **Temporal structures** consisting of the major events and their temporal relationships (e.g., initiation and synchronization)
- **Physical static structures.** Physical static structures consist of physical architectural elements and their static relationships. Examples of such structures include:
  - **Allocation structures** (including static, configurable, or dynamic allocation) consisting of the allocation of one item to another:
    - *Allocation of air conditioning* to hardware components
    - *Allocation of data* (e.g., databases or files) to platforms (i.e., OS, middleware, and computational hardware), mass storage devices, and/or networks
    - Allocation of functionality to:
      - Architectural components such as subsystems, equipment, facilities, hardware, software, and personnel (i.e., manual procedures)
      - Dedicated or shared resources (e.g., processors, memory, buses, and I/O ports)
    - *Allocation of memory and performance budgets* to hardware components, processes, and software
    - *Allocation of power* to hardware components
    - *Allocation of processes* to platforms
    - *Allocation of software* to platforms
    - *Allocation of weight* to hardware components
  - **Computational structures** consisting of architectural components that perform computations and the communication channels between them.
  - **Physical configuration structure** consisting of the major physical components (such as subsystems, assemblies, configuration items, and line replaceable units) and their relationships (e.g., decomposition and interaction).

    Note that the physical configuration structure is the basis for Conway's law, which states that the structures of a system tend to mirror the structure of the development organization that produces it. Note also that Conway's law is based primarily on the integrated product teams responsible for the subsystems and typically does not include management or specialty engineering groups.
  - **Connection structures** consisting of significant architectural components and the physical connections (e.g., electrical, hydraulic, laser, mechanical, microwave, pneumatic, and radio) between them.
  - **Data structures** consisting of major data types and their relationships (e.g., attributes, pointers, and references).
  - **Hardware structure** consisting of the major types of hardware and their relationships (e.g., wired and wireless communication, physical connection, and physical location).

- **Interface structures** consisting of the major physical interfaces and their relationships (e.g., media, protocols, and connections to physical components).
- **Network structure** consisting of the major types of network hardware (e.g., computational and network devices) and their relationships (e.g., network connections).
- **Role structures** consisting of groupings of functional behavior that may be delegated to actor/agent structures or humans.
- **Deployment structure** consisting of the allocation of software and data components to hardware components.
- **Technology allocation structure** consisting of the allocation of technologies to physical architectural elements.

■ **Physical dynamic structures.** Physical dynamic structures consist of physical architectural elements and their dynamic relationships capturing their dynamic behavior. Examples of such structures include:

- **Actor/agent structures** consisting of the active elements that can initiate/execute behavior defined by role structures, usually employing a process structure to achieve concurrent/parallel execution (the role structure may be invariant or may dynamically alter based on mode/state or context-dependent division of labor between the agent structure and one or more human actors)
- **Concurrency structure** consisting of the major logical parallel threads of execution and their relationships (e.g., initiation, communication, synchronization, and termination)
- **Logical to physical allocation structure** consisting of the allocation of logical structures to hardware components, including the deployment of logical configuration structure to physical configuration structures as well as the delegation of logical structures to hardware components (e.g., execution of a given behavior in an analog circuit)
- **Performance allocation structure** consisting of the allocation of performance budgets to physical hardware, software, and process architectural elements
- **Physical flow structures** consisting of a cohesive set of architectural components and the significant flows between them
  - *Physical control flow structures* consisting of a cohesive set of architectural components and the significant control flows between them
  - *Physical data flow structures* consisting of a cohesive set of architectural components and the significant flows of data (types) between them
  - *Physical energy flow structures* consisting of a cohesive set of architectural components and the significant flows of energy (e.g., electrical, chemical, hydraulic, mechanical, and thermal) between them
  - *Physical failure flow structures* consisting of a cohesive set of architectural components and the significant potential failures and exceptions that flow between them (e.g., showing where they are raised, how they flow through the system, and where they are handled)
  - *Physical material flow structures* consisting of a cohesive set of architectural components and the significant flows of energy (e.g., gases, liquids, solids, and plasmas such as chemicals, fuels, or products) between them
  - *Physical message flow structures* consisting of a cohesive set of architectural components and the significant messages between them
- **Process structures** consisting of the major physical processes and threads and their relationships

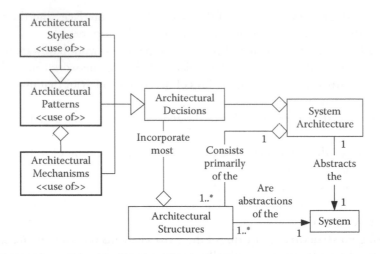

Figure 5.4   Architectural styles, patterns, and mechanisms.

5.5 Architectural Styles, Patterns, and Mechanisms

When architecting and modeling a system's architectural structures, architects strive to reuse previous architectural decisions, inventions, and engineering trade-offs that have been successfully used in the past. Major reusable architectural decisions are often identified and documented in the form of architectural styles and patterns. These styles and patterns often incorporate specific architectural mechanisms as elements.

As illustrated in Figure 5.4, the use of architectural styles, patterns, and mechanisms are architectural decisions and therefore part of the system architecture. These design decisions are incorporated into the architectural structures, which are where most of the architectural decisions are made.*

> **Architectural pattern:** any well-documented reusable solution to a commonly occurring architectural problem within the context of a given set of existing architectural decisions, inventions, engineering trade-offs, and assumptions.

This use of the concept of pattern initially came from the architectural (building) community [Alexander, 1979]. The software community adapted and applied it to the documentation of reusable software designs. Since then, a thriving software patterns community with its own books, journals, and conferences has developed. Software patterns have been widely applied to many disciplines, including, but not limited to, project management, requirements, architecture, implementation (where they are called idioms), and testing.

---

* Note here the important difference between the system and its architectural structures. Some people are strongly tempted to equate the system with its physical static hierarchical decomposition structure into physical components, such as hardware and software. This seems, after all, to be the system that one sees when looking at the "system." However, there are three major reasons why this is not identical to the system. First, one is looking at only the leaf nodes of the system rather than at the entire hierarchical decomposition tree. Second, there are multiple ways to decompose the system hierarchically, which yield multiple trees, possibly with different leaves. Third, this is only one of the many logical and physical, static and dynamic kinds of structures that a system can have. Thus, the relationship from architectural structure to system is many-to-one.

A similar situation is developing within the systems engineering community. Although patterns exist in system engineering, extremely few have been explicitly identified and documented as such. This also applies to system architecture engineering, where the primary patterns identified are essentially those of large software applications.

Examples of system and software architectural patterns include:

- Automatic failover (hot, warm, and cold)
- Automatic garbage collection
- Automatic restart
- Built-in testing (BIT)
- Bus, star, ring, tree, and mesh network topology patterns
- Communication via proxies and wrappers
- Cryptography
- Data verification
- Firewalls
- Heartbeat
- Load balancing
- Peer-to-peer, shared data, central bus, and virtual enterprise (grid computing) communication patterns between subsystems
- Physical barriers
- Physical separation
- Redundancy/replication (i.e., replicated subsystems, hardware, and/or software to support availability, capacity, reliability, and robustness)
- Reflection at hardware/software boundary
- Roll-back or roll-forward
- Voting

The largest grained architectural patterns are often called architectural styles.

Architectural style: any top-level architectural pattern that provides an overall context in which lower-level architectural patterns exist.

Traditionally, one of the most common system architectural styles has been *functional allocation*, in which the architects allocate disjoint functionally cohesive sets of capabilities to individual subsystems. The functional allocation pattern thus attempts to create a one-to-one mapping between the static functional decomposition logical architecture and the static system decomposition physical architecture. This style works well with some systems, such as when aircraft propulsion is allocated to the jet engine subsystem. However, it no longer tends to work well when allocating functionality implemented by software to hardware platforms. For example, although modern aircraft have numerous embedded computers, the amount of functionality is so great that it must be allocated across multiple platforms (i.e., a single processing box may collaborate with several other boxes to implement a certain set of functionality).

Other example architectural styles include:

- Agents
- Blackboard communication
- Client/server

- Event-driven scheduling
- Federated architecture with individual platforms performing functionally cohesive tasks
- Grid computing
- Layered software architecture
- Model view controller (MVC)
- Object orientation (OO)
- Pipe-and-filter software architecture
- Publish and subscribe communication
- Service-oriented architecture (SOA)

> **Architectural mechanism:** a major architectural decision or invention, often an element of an architectural pattern.

For example, the architects might decide to incorporate the following architectural mechanisms to implement architecturally significant security requirements:

- Access control, including:
  - Identification
    - User identifiers
  - Authentication
    - Biometrics
  - Authorization
    - Biometrics or else user identification and pass phrases for identification and authentication
- Digital signatures and timestamps (to support confidentiality, integrity, and non-repudiation)
- Encryption/decryption (to support confidentiality and non-repudiation)
- Firewalls (to support confidentiality and integrity)
- Hash codes (to support integrity)
- Virus detection and suppression (to support integrity)
- Intrusion detection and suppression
- Tamper proofing (to support confidentiality and integrity)

5.6 Architectural Drivers and Concerns

As illustrated in Figure 5.5, architectural drivers are either architecturally significant product requirements or process requirements that therefore drive the engineering of the system architecture. Architectural concerns are cohesive collections of architectural drivers, whereby some of the most important architectural concerns are the architectural focus areas that concentrate on quality characteristics, specialty engineering disciplines, or areas of programmatic interest (e.g., cost, schedule, and available resources).

> **Architectural driver:** any architecturally significant product or process requirement that drives the engineering of the system architecture.

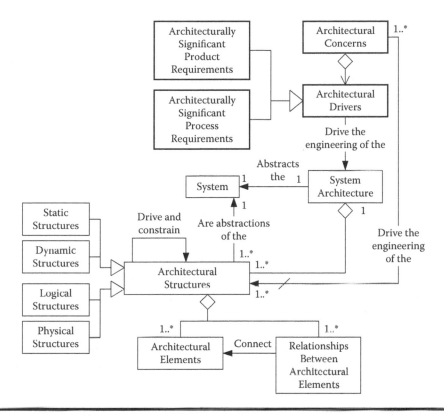

Figure 5.5 Architectural concerns and drivers.

Example *types* of architectural drivers include:

- **Product requirements:** architecturally significant product requirements such as:
  - Quality requirements
  - Functional requirements
  - Interface requirements
  - Data requirements
  - Architectural constraints such as mandated use of specific technologies
- **Process requirements:** architecturally significant process requirements such as:
  - Cost requirements, including limitations on development, maintenance, operation, and retirement costs
  - Schedule requirements, including mandatory development and delivery milestones
  - Resource requirements, including ability to be developed by existing staff and manufactured using existing manufacturing facilities
  - Strategic organizational constraints, including partnering agreements
  - Policies, laws, and regulations

Architectural concern: *cohesive* collection of architectural drivers.

Some example *architectural concerns* include:

■ **Quality characteristic and quality attribute**[*] **concerns:** cohesive collections of quality requirements associated with either quality characteristics or quality attributes such as availability, capacity, interoperability, performance (including its quality attributes such as jitter, latency, response time, schedulability, and throughput), portability, reliability, safety, security, stability, testability, and usability

Note that quality characteristics are often in direct conflict with each other because increasing one quality characteristic or attribute typically decreases one or more others. Stakeholders and requirements engineers must therefore be careful when engineering quality requirements to ensure that they are both individually and collectively feasible and practical. Similarly, architects must take care when making engineering trade-offs between conflicting architectural concerns.

■ **Functional feature set concerns:** cohesive collections of functional requirements such as those specifying the reading of sensors, the control of actuators, the calculation of data, and communication of messages.

■ **Prototype development cost concerns:** development requirements limiting the costs to develop the various prototype versions of the system prior to small-scale production.

■ **Schedule concerns:** development requirements specifying deadlines such as for "first flight" of a new aircraft.

As cohesive collections of requirements, architectural concerns should not be confused with the architectural mechanisms used to implement them. For example, the following are security architectural concerns and are therefore not security architectural mechanisms:

■ **Security problem quality attributes:**
  - Malicious harm to valuable assets:
    • Corruption (loss of integrity) of data, messages, software, hardware
    • Infection (software corruption)
    • Loss of availability due to attack
    • Loss of confidentiality (i.e., privacy and anonymity)
    • Repudiation of transaction (service)
  - Attacks and security incidents
  - Security vulnerabilities
  - Attackers
  - Security threats
  - Security risks
■ **Security solution quality attributes:**
  - Prevention of:
    • Malicious harm to valuable assets
    • Successful attacks
    • Security vulnerabilities
    • Attacker opportunity
    • Security threats
    • Security risks

---

[*] See Chapter 18 ("Architecture and Quality") for a more in-depth discussion of quality characteristics and quality attributes.

- Detection of:
  - Vulnerabilities
  - Attacks
  - Intrusion
- Reaction:
  - Identification of attackers and intruders
  - Containment of attacks and intrusions
  - Active countermeasures against attackers and intruders
- Adaptation
  - Isolation of attackers/intruders
  - Denial of resources to attackers/intruders

The following are some representative example architectural concerns and associated requirements:

- **Quality concerns:**
  - **Availability:** the set of all architecturally significant subsystem requirements that specify the availability of subsystem functionality or components:
    - R145-34: during normal mode of operation, subsystem X shall provide sensor data with an availability of at least 99.99 percent over the expected service life of the system.
    - R145-42: during normal mode of operation, subsystem X shall provide function F_1 with an availability of at least 99.9 percent over the expected service life of the system.
    - R145-54: during normal mode of operation, subsystem X shall provide function F_2 with an availability of at least 99.99 percent over the expected service life of the system.
  - **Performance:** the set of all architecturally significant subsystem requirements that specify the performance of subsystem functionality or components:
    - Response time:
      - R145-21: during normal mode of operation, subsystem X shall provide services F_3, F_4, and F_5 with a response time of Z milliseconds.
    - Schedulability:
      - R145-28: during normal mode of operation, subsystem X shall provide function F_5 every 5 minutes.
      - R145-30: during normal mode of operation, subsystem X shall provide function F_6 every day at 6 PM EST.
    - Throughput:
      - R145-25: during normal mode of operation, subsystem X shall provide services F_3, F_7, F_8, and F_9 with a throughput of Z transactions per second.
- **Constraints concern:**
  - **Architectural constraints:** the set of all architecturally significant system or subsystem requirements that specify architectural constraints:
    - R1-14: the system shall use a COTS real-time operating system.
    - R1-23: the system shall use COTS computing and networking hardware.
    - R1-31: the system shall reuse GOTS system X.
    - R1-44: the system shall be implemented using industry-standard protocols for at least 95 percent of its interfaces with external systems.
    - R1-52: the system shall incorporate a service-oriented architecture (SOA).
    - R1-74: the system shall comply with CJCSI 6212.01D *Interoperability and Supportability of Information Technology and National Security Systems.*

- R1-83: the system shall use user names and pass phrases for user identification and authentication.*
- R1-93: the system shall use a COTS system for encryption and decryption.
- R1-123: the system shall be implemented in a safe subset of C++.

■ **Functional concern:**
- **Subsystem X sensors and sensor fusion:** the set of all architecturally significant subsystem X requirements that specify functional requirements related to sensors and sensor fusion — R5-13, R5-14, R5-15, R5-18, R5-21, and R5-23.

■ **Process concern:**
- **Cost:** the set of all architecturally significant system or subsystem process requirements that specify maximum costs:
  - Development cost:
    - R1-104: the cost to develop the system shall not exceed $20 million USD.
  - Manufacturing cost:
    - R1-105: the mean unit cost per system shall not exceed $50,000 USD during initial small scale development.
    - R1-106: the mean unit cost per system shall not exceed $8000 USD during full-scale development.
  - Sustainment cost:
    - R1-107: the mean unit maintenance cost per system shall not exceed $500 USD per year.
    - R1-108: the mean unit operations cost per system shall not exceed $2500 USD per year.
  - Retirement cost:
    - R1-109: the mean disposal unit cost per system shall not exceed $250 USD in July 1, 2008 dollars.

5.7 Architectural Representations

As illustrated in Figure 5.6, MFESA clearly distinguishes a system's architecture, which is conceptual, from particular representations of the system's architecture, which are concrete work products describing one or more facets of the architecture. Whereas an existing system always has an architecture,† representations of that architecture may or may not exist; and if architectural representations do exist, they may or may not conform to the system's architecture (e.g., they may incorporate defects or may be obsolete, having not been properly maintained once developed).

> **Architectural representation:** any cohesive collection of information that documents a system architecture.

---

* Note that because these are probably the weakest mechanisms for identifying externals and authenticating these identifications, there needs to be a good reason for specifying this constraint. The requirements should probably let the architect have the freedom to use other means such as identification cards or biometrics.

† Note that although a system always has an architecture, at least in theory, it is practically impossible to know the complete contents of the architecture after the fact due to the impossibility of producing "complete" documentation. The architecture is thus a matter of differing professional opinions and debate.

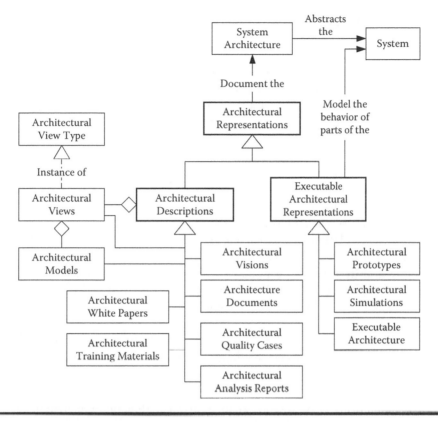

Figure 5.6 Architectural representations.

Representations of the architecture enable the architects to:

■ Communicate (and demonstrate the appropriateness of) their architectural decisions, inventions, trade-offs, and assumption to the system's stakeholders
■ Constrain the downstream activities of design, implementation, integration, and testing
■ Analyze and simulate the system's behavior and characteristics

> **Architectural description:** any architectural representation that is not executable (e.g., documents and most models).

Architectural descriptions are essentially documents containing text, sketches, drawings, diagrams, charts, tables, photographs, and videos.* They usually consist primarily of architectural views consisting of one or more architectural models. Whereas architectural descriptions are essentially static, executable representations are dynamic executable architectural prototypes, models, or simulations. For example, there is often enough information in state models and AADL models to execute them to learn or verify information about the future behavior of the system.

---

* The formality of the medium in which architectural descriptions are made will vary depending on the maturity of the architecture, the criticality of the documentation, the audiences of the documentation, and the uses to which it will be put.

> **Executable architectural representation:** any architectural representation with sufficient dynamic semantics to be executable.

An executable architectural representation implements a subset of the architecture and is used to verify (e.g., via test or demonstration) that the implemented subset of the architecture correctly fulfills some subset of the architecturally significant requirements. When presented to stakeholders, an executable architecture can also be used to validate these architecturally significant requirements.

> **Architectural prototype:** any architectural representation that (1) prototypes an aspect of the system with sufficient dynamic semantics to be executable and (2) is used to verify or validate the quality or appropriateness of one or more architectural decisions, inventions, trade-offs, or assumptions.
>
> **Architectural simulation:** any executable architectural representation (e.g., dynamic model) that is used to simulate the behavior of the system.
>
> **Executable architecture:** any executable architectural representation that is a very early partial implementation of the system that (1) is used to verify aspects of the architecture and (2) may evolve into the completed system.

As noted in the Recommended Practice 1471-2000 of the Institute of Electrical and Electronics Engineers (IEEE), architectural representations have many uses including the following [IEEE, 2000]:

- **Analysis.** Architectural representations can be used to support the static and dynamic technical analysis of different parts of the architecture, including input to analysis tools.
- **Automatic software generation.** Architectural representations can provide information needed by tools to automatically generate system software.
- **Certification and accreditation.** Architectural representations contain information needed to certify compliance of the system with its architecturally significant requirements and to accredit the system as ready for use.
- **Communication.** Architectural representations are used to communicate the architecture to the architecture's many stakeholders, including communication among organizations, teams, and individuals.
- **Contractual documentation.** Architectural representations provide input to requests for proposals, proposals, contracts, and statements of work. For example, proposals typically include initial architectural representations of the proposed system, thereby helping the acquisition organization determine which development organization will win the contract to develop the system.
- **Contract negotiations.** Architectural representations contain architectural information needed by the acquisition organization and development organization to negotiate major changes to the system.
- **Development driver.** Architectural representations provide information that drives downstream development activities such as design, implementation, integration, testing, and manufacturing.
- **Estimation.** Architectural representations can provide information needed to estimate the development costs and schedules.
- **Evaluation.** Architectural representations can provide information needed to evaluate (e.g., inspect, assess, review, and test) the system quality, maturity, status, and fitness for purpose.

- **Feasibility.** Architectural representations may provide information needed to estimate the feasibility of constructing or manufacturing the system. Note that this includes affordability as well as resource (such as the availability of adequate staffing as well as development, manufacturing, and maintenance facilities), schedule, and technological feasibility.
- **Integration.** Architectural representations provide information about architectural components needed to integrate those components into the system.
- **Maintenance, repair, and sustainment documentation.** Architectural representations can provide information needed to maintain, repair, and sustain the system.
- **Manufacturing documentation.** Architectural representations can provide information needed to manufacture multiple instances or variants of the system.
- **Operational documentation.** Architectural representations can provide information needed to operate the system.
- **Quality.** Architectural representations (especially architectural quality cases) can provide information needed to demonstrate the quality of the architecture and estimate the quality of the eventual system.
- **Reference architecture documentation.** Architectural representations can provide information needed to specify common architectural decisions, inventions, and trade-offs across a family (e.g., product line) of systems.
- **Replacement documentation.** Architectural representations can provide information needed to support future replacement of the system.
- **Requirements verification.** Architectural representations can provide information needed to verify system support for its architecturally significant requirements.
- **Retirement documentation.** Architectural representations can provide information needed to properly retire the system when it is no longer needed. This includes information useful for decommissioning as well as the sale or safe disposal of component parts.
- **Reuse documentation.** Architectural representations can provide information needed to determine if architectural components are suitable for reuse as parts of other systems.
- **Risk.** Architectural representations can provide information needed to identify, avoid, and mitigate system risks, especially those associated with architectural concerns such as sufficient support for system qualities.
- **Selection.** Architectural representations can provide information needed to select from among multiple competing architectural visions, architectural components, or systems.
- **Transition planning.** Architectural representations can provide information needed to plan the transition from one or more legacy systems to the new system.

5.8 Architectural Models, Views, and Focus Areas

Perhaps the largest part of most architects' work is the production of architectural models. Architects produce these models to reduce and manage the extreme complexity of today's systems. They also use models to increase the understandability of the architecture by abstracting away diversionary details and restricting information to a single stakeholder viewpoint.

> **Architectural model:** any architectural representation that models a single system structure in terms of the structure's architectural elements and the relationships between them.
> **Modeling language:** any formal or semiformal language that is used to consistently document models.

Architectural models can also be implemented using some combination of both graphical and textual modeling languages and associated frameworks such as:

- AADL — Architecture Analysis and Design Language [Feiler et al., 2006]
- BPMN — Business Process Modeling Notation [BPMN, 2004]
- DODAF — United States Department of Defense Architecture Framework [DODAF, 2007a–c]
- E2A — Institute for Enterprise Architecture Development's (IFEAD's) Extended Enterprise Architecture (E2A) Framework [IFEAD, 2006]
- FEAF — Federal Enterprise Architecture Framework [CIOC, 1999]
- MODAF — United Kingdom Ministry of Defense Architecture Framework [MOD, 2007]
- NAF — NATO System Architecture Framework [NATO, 2005]
- SysML — System Modeling Language [OMG, 2007]
- TOGAF — The Open Group Architecture Framework [TOG, 2007]
- UML — Unified Modeling Language* [ISO/IEC, 2005]
- Zachman — Zachman Framework for Enterprise Architecture [Zachman, 1987, 2008]

Graphical architecture modeling languages are very popular and are often used to produce a large number of different types of architectural views and associated architectural diagrams that are appropriate to describe the architecture of the system as a whole, the architecture of individual subsystems, as well as the architecture of pure data, hardware, and software architectural elements. These graphical views are typically supported by associated ancillary textual information that further describes the nodes and arcs on the diagrams.

Architectural models are typically used to model the various types of architectural structures. Because there are often multiple ways to model the same structures, an important part of any system architecture engineering method will be the selection of a consistent set of the most appropriate modeling languages† to use to create the appropriate models of the relevant structures.

The following is a relatively complete listing of potentially useful logical and physical, static and dynamic architectural models. When using MFESA to instantiate an appropriate, endeavor-specific system architecture engineering method, it is important for the architecture teams (in collaboration with the process engineering team) to select the relevant models that are appropriate (e.g., useful and cost-effective) to model the important architectural structures. Note also that every model is not appropriate for every system, and different subsystems within the same system may well need to be modeled differently.

- **Logical static models.** Logical static models represent logical architectural elements and their static relationships. Examples include the following:
  - *Functional decomposition models.* These are models of functional decomposition structures, typically documenting the major functions of the system and the decomposition and/or invocation relationships between them. Examples include:
    - Component diagrams (UML 2.x)

---

* Note that UML was originally developed as a language for documenting low-level, object-oriented software designs in order to support automated forward and reverse engineering of source code. Because it was not developed to document system architectures, architecture teams have had to tailor, extend, or reinterpret some of UML for use at the system level. SysML was developed to avoid these limitations of UML, and UML2 provides better support than UML1.

† Note that multiple modeling languages are typically needed because no single language supports all of the models and views that might be useful and cost-effective to produce, use, and maintain.

- Composite structure diagrams
- Data flow diagrams (DFDs) and their associated textual function and data specifications
- Functional decomposition diagrams (i.e., class diagrams with classes representing functions and aggregation relationships)
- Operational activity model (DODAF Operational View OV-5)* and associated integrated dictionary (DODAF All View AV-2)
- Package diagrams (UML)

- *Layering models.* These are models of logical layering structures, typically documenting the major layers of abstraction and their relationships (e.g., dependency).

  Layer models documented in the form of two- or three-dimensional layer diagrams are commonly used in the software community to document layering structures. The horizontal software layers (from top-most to lowest) are typically the application layer, middleware layer (sometimes divided into system-specific over COTS subordinate layers), operating system, and (sometimes) computer hardware. The application layer is sometimes decomposed into labeled vertical partitions representing logical functional feature sets. Examples include:
  - Layer diagrams

- *Logical allocation models.* These are models of logical static structures, typically documenting the allocation of something to functionality. Examples include:
  - Performance to function allocation tables
  - Weight to function allocation tables

- *Logical class models.* These are models of logical class structures, typically documenting major domain concepts and their associated relationships (e.g., aggregation, association, and generalization). Examples include:
  - Class diagrams
  - High-level operational concept graphic (DODAF Operational View OV-1) and associated integrated dictionary (DODAF All View AV-2)

- *Logical data models.* These are models of logical data structures, typically documenting data type and their associated relationships (e.g., decomposition and invocation). Examples include:
  - Entity relationship diagrams (ERDs)
  - Integration definition for information modeling (IDEF1X) models [FIPS, 1993]
  - Class diagrams (without associated methods)
  - Logical data model (DODAF Operational View OV-7) and associated integrated dictionary (DODAF All View AV-2)
  - Data dictionaries
  - Tables defining logical data types

- *Logical service models.* These are models of logical service structures, typically documenting the major services provided by the system. Examples include:
  - Service diagrams and textual specifications of the services

---

* One can easily argue that the DODAF operational views are at the enterprise architecture level rather than at the system architecture level. Accepting that argument, then only the DODAF system views should be used to model the system architecture.

■ **Logical dynamic models.** Logical dynamic models represent logical dynamic structures and consist of logical architectural elements and their dynamic relationships capturing their dynamic behavior. Examples of such architectural models include:

– *Logical flow models.* These are models of logical flow structures, typically documenting a cohesive set of system functions and the significant flows between them. Examples of such models include:

• *Logical control flow models.* These are models of logical control flow structures, typically documenting a cohesive set of system functions and the significant control flows between them. Examples include:

■ Control flow context diagrams and their associated textual terminator and control specifications

■ Control flow diagrams (CFDs) and their associated textual function and control specifications (possibly including state charts and decision tables)

• *Logical data flow models.* These are models of logical data flow structures, typically documenting a cohesive set of system functions and the significant data flows between them. Examples include:

■ Data flow context diagrams and their associated textual terminator and data specifications

■ Data flow diagrams (DFDs) and their associated textual function and data specifications

■ Operational node connectivity description (DODAF Operational View OV-2) with operational nodes and associated activities replacing functions and associated integrated dictionary (DODAF All View AV-2)

■ Organizational information exchange matrix (DODAF Operational View OV-3) and associated integrated dictionary (DODAF All View AV-2)

• *Logical energy flow models.* These are models of logical energy flow structures, typically documenting a cohesive set of system functions and the significant flows of energy (e.g., chemical, electrical, hydraulic, mechanical, and pneumatic) between them. Examples include:

■ Energy flow context diagrams and their associated textual terminator and energy specifications

■ Energy flow diagrams and their associated textual function and energy specifications

• *Logical failure flow models.* These are models of logical failure flow structures, typically documenting a cohesive set of system functions and the significant flows of failures, faults, and exceptions between them. Examples include:

■ Failure flow context diagrams and their associated textual terminator and failure, fault, and exception specifications (e.g., showing where they are raised, how they flow through the system, and where they are handled)

■ Failure flow diagrams and their associated textual function and failure, fault, and exception specifications

• *Logical material flow models.* These are models of logical material flow structures, typically documenting a cohesive set of system functions and the significant flows of materials (e.g., chemicals, fuel, and products) between them. Examples include:

■ Material flow context diagrams and their associated textual terminator and material specifications

■ Material flow diagrams and their associated textual function and material specifications

- *Logical process models.* These are models of process structures, typically documenting the major logical parallel threads of execution and their interactions (e.g., initiation, communication, synchronization, and termination). Examples include:
  - Activity diagrams (restricted to logical processes or threads)
  - Business process diagrams (BPMN) [BPMN, 2004]
  - Data flow diagrams (with logical processes represented by functions)
  - Swim lane diagrams (restricted to logical processes or threads)
- *Mode allocation models.* These are models of mode allocation structures, typically documenting the allocation of functionality to system modes. Examples include:
  - Function to model tables
- *Mode and state models.* These are models of mode and state structures, typically documenting the modes and states of a system or subsystem, the transitions between them, and associated events and guards. Examples include:
  - State transition diagrams and their associated textual state/mode specifications
  - State charts and their associated textual state/mode specifications
  - State transition tables
  - Operational state transition description (DODAF Operational View OV-6b) and associated integrated dictionary (DODAF All View AV-2)
  - Systems state transition description (DODAF Operational View SV-10b) and associated integrated dictionary (DODAF All View AV-2)
- *Temporal models.* These are models of temporal structures, typically documenting the major events and their temporal relationships (e.g., initiation, synchronization, and termination). Examples include:
  - Interaction diagrams, especially timing diagrams (UML)
  - Operational rules model (DODAF Operational View OV-6a) and associated integrated dictionary (DODAF All View AV-2)
  - Operational event-trace description (DODAF Operational View OV-6c) and associated integrated dictionary (DODAF All View AV-2)

■ **Physical static structures.** Physical static models represent physical architectural elements and their static relationships. Examples of such models include:
- *Allocation models.* These are models of allocation structures, typically documenting the allocation of something to architectural components. Examples include:
  - *Functionality allocation models.* These are models of functionality allocation structures, typically documenting the allocation of functionality to architectural components such as subsystems, equipment, facilities, hardware, software, and personnel (i.e., manual procedures) as well as the allocation of functionality to dedicated or shared resources (e.g., processors, memory, buses, and I/O ports). Examples include:
    - Function allocation tables
    - Requirements traceability tables
    - UML deployment diagrams showing allocation of functionality
  - *Hardware component allocation models.* These are models of hardware component allocation structures, typically documenting the allocation of something to hardware components. Examples include allocation of air conditioning, data (e.g., databases or files), power, processes, software, volume, and weight to hardware components

such as mass storage devices, networks, and platforms (i.e., OS, middleware, and computational hardware). Examples include:

- Hardware allocation tables
- UML deployment diagrams

- *Logical to physical allocation models.* These are models of logical to physical allocation structures, typically documenting the allocation of logical structures to hardware components, including the deployment of logical configuration structure to physical configuration structures as well as the delegation of logical structures to hardware components (e.g., execution of a given behavior in an analog circuit). Examples include:
  - Logical to physical allocation tables
- *Memory allocation models.* These are models of memory budget allocation structures, typically documenting the allocation of memory budgets to hardware components, processes, and software. Examples include:
  - Memory budget tables
- *Performance allocation models.* These are models of performance allocation structures, typically documenting the allocation of performance budgets to physical hardware, software, and process architectural elements. Examples include:
  - Performance budget tables
  - Systems performance parameters matrix (DODAF Systems and Services View SV-7) and associated integrated dictionary (DODAF All View AV-2)
  - Services performance parameters matrix (DODAF Systems and Services View SV-7) and associated integrated dictionary (DODAF All View AV-2)
- *Process allocation models.* These are models of process allocation structures, typically documenting the allocation of processes to platforms. Examples include:
  - UML deployment diagrams
- *Technology allocation models.* These are models of technology allocation structures, typically documenting the allocation of technologies to physical architectural elements. Examples include:
  - Technical standards profile (DODAF Technical View TV-1) and associated integrated dictionary (DODAF All View AV-2)

- *Connection models.* These are models of connection structures, typically documenting the significant architectural components and the physical connections (e.g., electrical, hydraulic, LASAR, mechanical, microwave, pneumatic, and radio) between them. Examples include:
  - Architecture interconnect context diagrams and their associated textual architecture interconnect specifications
  - Architecture interconnect diagrams and their associated textual architecture interconnect specifications
  - Connection diagrams and their textual connector specifications
  - Block diagrams

    Figure 5.7 is an example block diagram for an example system that creates electricity and fresh water from sunlight and sea water. Note that this incomplete diagram only shows physical interfaces (such as flows of water and steam); it does not show the many electrical and informational flows involving and auxiliary power subsystem and control subsystem (such as data and control flows involving the large number of sensors and actuators).
  - UML deployment diagrams

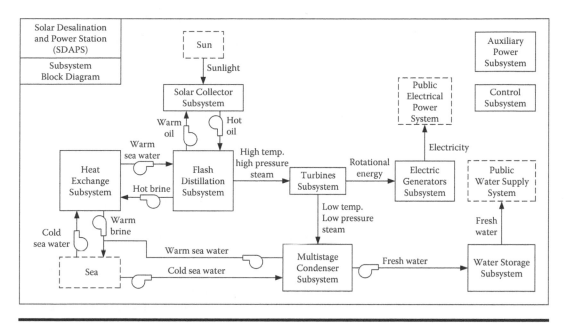

Figure 5.7   Example block diagram.

- *Data models.* These are models of physical data structures, typically documenting the major data types and their relationships (e.g., attributes, pointers, and references). Examples include:
  - Entity relationship diagrams (ERDs) and associated textual entity and relationship specifications
  - Class diagrams and associated textual class and relationship specifications
  - Tables defining physical data types
  - Physical schema (DODAF Systems and Services View SV-11) and associated integrated dictionary (DODAF All View AV-2)
  - Systems data exchange matrix (DODAF Systems and Services View SV-6) and associated integrated dictionary (DODAF All View AV-2)
  - Services data exchange matrix (DODAF Systems and Services View SV-6) and associated integrated dictionary (DODAF All View AV-2)
- *Deployment models.* These are models of deployment structures, typically documenting the allocation of software and data components to hardware components. Examples include:
  - Installation diagrams
  - UML deployment diagrams
- *Hardware models.* These are models of hardware structures, typically documenting the major types of hardware and their relationships (e.g., data communication protocols, physical connection, wired and wireless communication, and physical location). Examples include:
  - Block diagrams, possibly with relevant subsystems indicated
      Figure 5.7 illustrates an example block diagram for a system that produces electricity and drinkable water from sunlight and sea water.
  - Hardware schematics

- Two-dimensional perspectives or renderings, which show what a system, physical subsystem, or hardware component would look like when viewed from above, the back, or the side
- Three-dimensional perspectives or renderings, which show what a system, physical subsystem, or hardware component would look like when viewed from an angle
- Placement diagrams, which show the physical location of physical subsystems or hardware components within the overall system or higher-level subsystem, and which may be two- or three-dimensional perspectives possibly with all or part of one or more surfaces cut away to show the location and orientation of relevant physical subsystems or hardware components (that is, cut-away diagrams)
- Parts explosion diagrams, which are typically special three-dimensional perspectives or renderings of the system or physical subsystem decomposed with its relevant physical subsystems or hardware components separated along lines showing physical interfaces between them
- System size compatibility diagrams, which are typically 2D perspectives showing how [an instance of] a system or subsystem will fit and maneuver within the dimensional constraints of its physical environment (e.g., showing how one or more airliners can be parked at the gates of an airport terminal, and showing how one or more military aircraft can be positioned and taxi on the deck of an aircraft carrier)
- *Interface models.* These are models of interface structures, typically documenting the major physical interfaces and their relationships (e.g., media, protocols, and connections to physical components). Examples include:
  - Interface diagrams and their associated textual interface specifications
- *Integration models.* These are models of integration structures, typically documenting the groupings of architectural components that are intended to be connected together. Examples include:
  - Aggregation diagrams documenting the order of integration of architectural components
  - Arrangement diagrams showing the location, size, shape, and type of internal architectural components

    Note that there are also specialized aggregation diagrams such as structural arrangement diagrams that show the location, size, shape, and type of internal structural components, such as frames, keels, ribs, and spars, as well as seam arrangement diagrams that show the location, size, shape, and type of seams between external surface coverings (such as the skin of airplanes and rockets and the hull plates of ships and armored vehicles).
  - Manufacturing sequence flow diagram documenting the order in which system components are manufactured and integrated
- *Network models.* These are models of network structures, typically documenting the major types of hardware components (e.g., computational and network devices) and the networks (e.g., buses) that connect them. Examples include:
  - Block diagrams
  - Network diagrams (ERDs) and their associated textual descriptions of the hardware components and network connections
  - Wiring diagrams

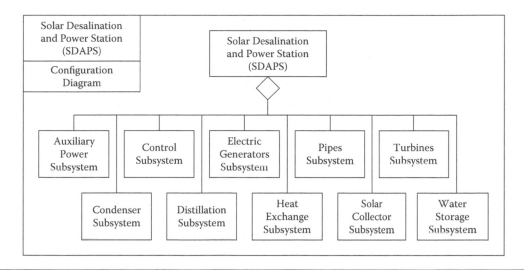

Figure 5.8   Example configuration diagram.

- *Physical configuration models.* These are models of physical configuration structures, typically documenting the major physical components (e.g., subsystems, assemblies, and configuration items) and the aggregation relationships between them. Examples include:
  - Class diagrams with classes representing an aggregation hierarchy of architectural components
  - Component diagrams
  - Configuration diagrams showing the decomposition of the system or a subsystem into its component subsystem
    Because of Conway's law, this diagram also tends to show the decomposition of larger integrated product teams (IPTs) into their component IPTs. Figure 5.8 illustrates a top-level configuration diagram for an example system that creates electricity and fresh water from sunlight and sea water.
- *Physical flow models.* These are models of physical flow structures, typically documenting a cohesive set of architectural components and the significant flows between them. Examples of such models include:
  - *Physical data flow models.* These are models of physical data flow structures, typically documenting a cohesive set of architectural components and the significant flows of data (types) between them. Examples include:
    - Physical data flow diagrams with associated textual architectural component and data specifications
    - Systems functionality description (DODAF Systems and Services View SV-4a) and associated integrated dictionary (DODAF All View AV-2)
    - Services functionality description (DODAF Systems and Services View SV-4b) and associated integrated dictionary (DODAF All View AV-2)
    - Confidential data flow diagrams, which show the flow of confidential (e.g., financially sensitive or classified) data and show relevant security information (such as where the data may be encrypted and decrypted)
      Note that the data flows may be annotated using color, for example, black for secure data (such as data that is encrypted and has associated hash code and

digital signature) and red for insecure data (such as data that is unencrypted and is without hash code and digital signature).

- *Physical energy flow models.* These are models of physical energy flow structures, typically documenting a cohesive set of architectural components and the significant flows of energy (e.g., electrical, chemical, hydraulic, mechanical, and thermal) between them. Examples include:
  - Block diagrams with energy flows annotated on connections
  - Class diagrams with directional associations representing energy flows
  - Connection diagrams with connections annotated with energy flows
  - Network diagrams with networks representing connections and energy flows annotated
  - Physical energy flow diagrams with associated textual architectural component and energy specifications
  - Sequence diagrams with messages representing energy flows
  - Wiring diagrams
- *Physical failure flow models.* These are models of physical failure flow structures, typically documenting a cohesive set of architectural components and the significant potential failures and exceptions that flow between them (e.g., showing where they are raised, how they flow through the system, and where they are handled). Examples include:
  - Fault propagation diagrams annotated with the allocation of fault tolerance to architectural components
  - Failure flow diagrams annotated with the allocation of failure tolerance to architectural components
- *Physical material flow models.* These are models of physical material flow structures, typically documenting a cohesive set of architectural components and the significant flows of materials (e.g., gases, liquids, solids, and plasmas such as chemicals, fuels, or products) between them. Examples include:
  - Block diagrams with material flows annotated on connections
  - Class diagrams with directional associations representing material flows
  - Connection diagrams with connections annotated with materials flows
  - Network diagrams with networks representing connections and material flows annotated
  - Physical material flow diagrams with associated textual architectural component and material specifications
  - Sequence diagrams with messages representing material flows
- *Physical message flow models.* These are models of physical control flow structures, typically documenting a cohesive set of architectural components and the significant messages that flow between them. Examples include:
  - Architecture message context diagrams and their associated textual terminator and message specifications
  - Architecture message diagrams and their associated textual message specifications
  - Class diagrams with classes representing architectural components
  - Physical control flow diagrams with associated textual architectural component specifications
  - Sequence diagrams with classes representing architectural components
  - Swim lane diagrams with classes representing architectural components

- *Role models.* These are models of role structures, typically documenting groupings of functional behavior that may be delegated to actor/agent structures or humans. An example of such models includes:
  - Function delegation to role table
- **Physical dynamic models.** These are models of physical dynamic structures, typically documenting physical architectural components and their dynamic relationships capturing their dynamic behavior. Examples of such models include:
  - *Actor/agent models.* These are models of actor/agent structures, typically documenting the active elements that can initiate/execute behavior defined by role structures, usually employing a process structure to achieve concurrent/parallel execution. Note that the role structure may be invariant or may dynamically alter based on mode/state or context-dependent division of labor between the agent structure and one or more human actors. An example includes:
    - Class diagram with classes representing actors or agents
  - *Migration models.* These are models that show the active migration of software between hardware components.
  - *Physical process models (a.k.a. concurrency models).* These are models of process structures, typically documenting the major physical processes, threads (i.e., virtual processes), and their interactions (e.g., initiation, communication, synchronization, and termination). Examples include:
    - Activity diagrams (restricted to processes or threads)
    - Swim lane diagrams (restricted to processes or threads)

As illustrated in Figures 5.9, 5.11, and 5.12, architects use multiple architectural views to document and communicate the various architectural structures that form the primary components of a system architecture to their many stakeholders. These stakeholders include (but are not limited to) acquirers, managers, requirements engineers, other architects,* integrators, testers, manufacturers, certifiers, users, operators, and administrators. Because these different stakeholders have very different needs and backgrounds, no single view or model of the architecture is sufficient.

For each of the four types of architectural structures, there is a corresponding type of architectural view. Each architectural view is largely composed of one or more corresponding architectural models, and each architectural view is an instance of an architectural viewpoint that is appropriate for the architectural structure being modeled and documented. An architectural viewpoint specifies the purpose of its views, who their stakeholders are, and how to create, use, and analyze them.

> **Architectural view:** any architectural representation describing a single architectural structure that consists of one or more related models of that structure.

Although architectural views typically represent the entire system from the perspective of a related set of stakeholder concerns, they can be restricted to a single architectural component such as a subsystem.

---

* Architects also use these views to communicate with themselves, whereby the views become memory aids for future reference.

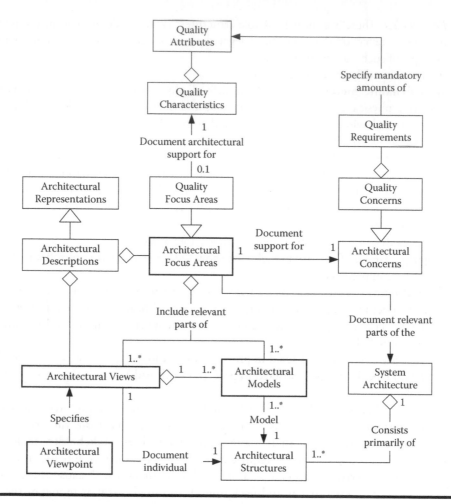

Figure 5.9 Views versus models versus structures versus focus areas.

> **Architectural viewpoint:** a specification of the conventions (e.g., standards, procedures, guidelines, and templates) for constructing, analyzing, and documenting a type of architectural view.

A viewpoint may also provide analysis and evaluation techniques that may be used to develop model completeness and consistency tests. A viewpoint may also include patterns, guidelines, and heuristics that help in the development of the associated view.

Note that each architectural view is an instance of an architectural viewpoint. Examples of architectural viewpoints include logical static view, logical dynamic view, physical static view, and physical dynamic view.

Although architectural views are very popular and even necessary to properly document architectural models and the corresponding architectural structures, they have one serious drawback. An architectural view provides a complete view of a single type of logical or physical, static or dynamic model, either that of the entire system or of an entire subsystem. Thus, a class view might show all architectural classes and their relationships. Unfortunately, the architectural decisions, inventions, and trade-offs associated with the architecture's support for a single architectural concern will be scattered across multiple models and views. Therefore, to understand how the architecture achieves a mandatory level of a specific quality characteristic (e.g., interoperability, reliability, and security) requires focusing on multiple parts of multiple views and models. This is the rationale for also including architectural focus areas.

> **Architectural focus area:** the documented part of an architecture that is related to a specific architectural concern, regardless of the architectural structures in which the architectural decisions, inventions, trade-offs, assumptions, and rationales are found or the architectural views and models in which they are documented.
> **Quality focus area:** any architectural focus area restricted to a single quality characteristic or quality attribute.*

Although the quality focus areas are often the most important architectural focus areas, other focus areas can also be quite important. These focus areas related to architectural constraints or process requirements and include affordability (development, sustainment, or life-cycle cost); schedule (ability to deliver the system on schedule); and manufacturability.

> **Quality characteristic:** any high-level property of a system or architectural component that characterizes an aspect of its quality.
> **Quality attribute:** any major component or part (aggregation) of a quality characteristic or of another quality attribute.

Note that quality attributes are not subtypes or subclasses of quality characteristics, but rather parts of quality characteristics. That is, a quality attribute is not a specialized (but complete) abstraction of the quality characteristic, but rather only captures a part of the abstraction of the quality characteristic. One way to tell the difference between a subtype of a quality characteristic and one of its quality attributes is whether or not it makes sense to append the name of the quality characteristic to the name of the subtype quality characteristic or quality attribute. For example, protocol interoperability is a subtype of interoperability and thus a lower-level quality characteristic. On the other hand, jitter is a part of performance (i.e., a quality attribute) and it does not sound right to say jitter performance.

As illustrated in Figure 5.10, a potentially very large number of quality characteristics exist that may be used to categorize architecturally significant quality-related requirements into quality focus areas. Chapter 18, "Architecture and Quality," discusses quality characteristics and attributes in considerably more detail.

As illustrated in Figure 5.11, there tends to be a natural flow from the identification of architectural concerns to the selection of architecture tools. To meet the architectural concerns, the architects must typically incorporate certain architectural structures. These architectural concerns also result in the identification of corresponding architectural focus areas (e.g., performance, reliability, safety, and security). The architectural structures and focus areas then tend to drive the selection of appropriate architectural views of these structures. The architectural views and focus areas then tend to drive the selection of appropriate architectural models of the architectural structures viewed by means of the architectural views. Once the architects understand what models they need to create, they can then select appropriate modeling languages with which to implement the

---

* A quality focus area can be effectively and efficiently documented using an associated architectural quality case.

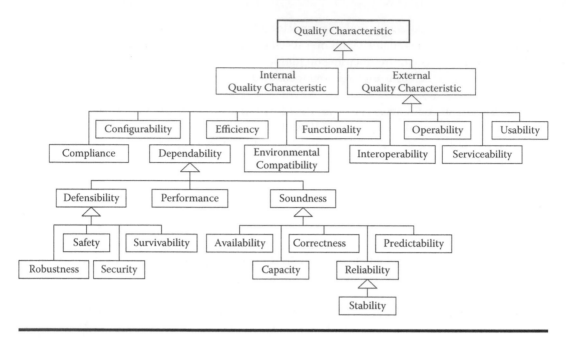

Figure 5.10 Some example quality characteristics.

models. Finally, the architects can select appropriate architecture tools that they can use to create the architectural models comprising the architectural views of the architectural structures.

The reader should note that this is a much better ordering of architectural work than for architects to let a tool vendor sell them a modeling tool before they even know which models they need to create or even why the models are appropriate for the system or architectural component being architected. Unfortunately, this happens all too often.

Note too unlike some methods and standards, MFESA does *not* use terms such as "logical architecture," "physical architecture," "technical architecture," and "security architecture." MFESA uses the terms "view" and "focus area" instead of "architecture" when referring to a system-wide view or concern-specific focus area.

As illustrated in Figure 5.12, each large and complex system has a single similarly large and complex multifaceted architecture, whether or not the architecture has been properly documented in some set of appropriate architectural representations. This system architecture consists largely of a significant number of structures, each of which can be either static or dynamic as well as either logical or physical. To understand, analyze, and document these different structures, architects must create a set of associated views, each of which consists largely of one or more associated architecture models. For simplicity's sake, Figure 5.12 only illustrates a small number of these views and models. In addition to creating these views and models, the architects must work to ensure that they are and remain consistent with one another. Another responsibility of the architects is to incrementally develop these models and ensure their integrity over time.

5.9 Architecture Work Products

Every system has an architecture in the sense that its architects consciously and unconsciously made a large number of architectural decisions, inventions, trade-offs, and assumptions. They also

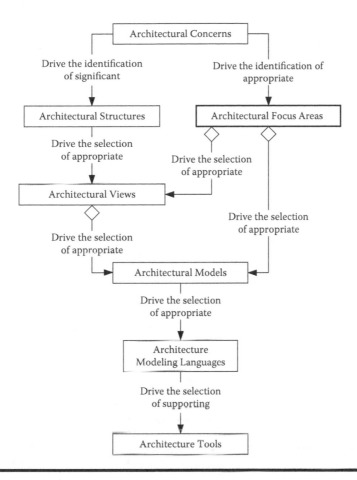

Figure 5.11   The natural flow from architectural concerns to architecture tools.

have explicit or implicit rationales for what they did. However, unless the architecture is properly documented in the form of one or more interdependent, mutually consistent representations, it cannot be easily communicated to its stakeholders (especially once the architects have moved on to new systems and new companies) and its value rapidly decreases over time. The architecture must be captured in a more permanent medium than the architect's memories. The architectural structures (especially the logical and dynamic ones) are, for the most part, not obvious to others. The many models tend to be very large and complex. Thus, unlike properly designed and commented software, system architectures are, for the most part, not self-documenting.

A system or subsystem architecture can be documented and implemented in the following forms:

- ■ **Architectural Descriptions:**
    - **Architectural views, focus areas, and models.** Focus areas document architectural support for individual architectural concerns. Architectural views provide views of architectural structures and consist of architectural models specified using architectural modeling languages.
    - **Architectural analysis reports.** Reports describing the results of various types of architectural analysis, including cost, feasibility, performance, reliability, risk, safety, security, schedule, trade-off, etc.

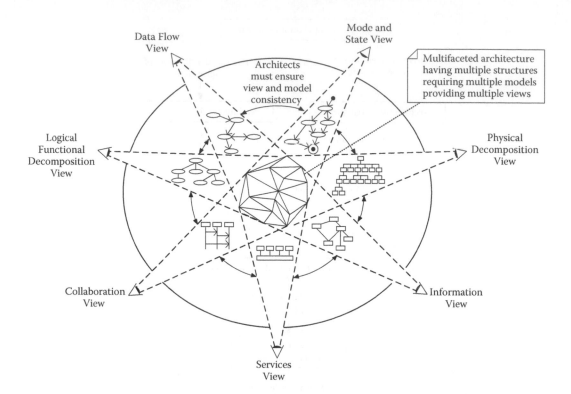

Figure 5.12 Multiple views of multiple structures of a single multifaceted architecture.

- **Architecture documents.** Architecture documents describe the system architecture or the architecture of one of its components, including views and models of its structures as well as other architectural decisions, inventions, trade-offs, assumptions, and their associated rationales.
- **Architectural white papers.** White papers are often used to document a single architectural focus area, such as architectural support for individual architectural concerns (e.g., performance, reliability, robustness, safety, and security).
- **Architecture quality cases.** An architectural quality case makes the architects' case that their architecture has sufficient quality [Firesmith et al., 2006]. As illustrated in Figure 5.13, an architectural quality case documents architectural *claims* (e.g., the architecture sufficiently supports meeting the architecturally significant requirements); clear and compelling *arguments* (e.g., architectural inventions and decisions, associated engineering trade-offs and assumptions, and their rationales) justifying belief in these claims; and official *evidence* supporting these arguments.
- **Architecture training materials.** Training materials are sometimes produced to train new members of the development team as well as other stakeholders of the architecture.
■ **Architectural prototypes.** An architectural prototype is any executable representation of the architecture that faithfully models one or more behaviors or characteristics of the system architecture. It may be as simple as a "fit check" model of a gimbaled optical platform produced via stereolithography, or a scale model that actually executes some subset of the behavior of the system.

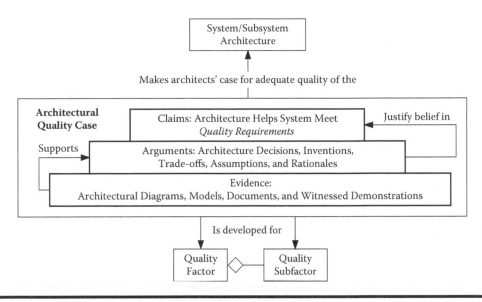

Figure 5.13 Structure of architecture quality cases.

- **Executable architecture.** An executable architecture is a partially implemented architecture intended to evolve into the system or subsystem. It is used to obtain early verification of architectural decisions and assumptions as well as to support the iterative and incremental development of the system or subsystem.

5.10 Architectural Visions and Vision Components

To understand the architecture of a system or subsystem, it helps to start with the architectural vision that was expanded and completed to form the architecture's complete set of representations. As an overview, an architectural vision is an incomplete, top-level summary of the architecture. An architectural vision is a mental image, representing in broad brush strokes how one or more architects imagine, conceive, and foresee the eventual architecture.

> **Architectural vision (a.k.a. architectural concept):** a conceptual overview of the architecture of the overall system or one of its architectural elements consisting of a cohesive and consistent set of architectural vision components.
>
> **Architectural vision component:** documentation of one of the more important actual or potential architectural decisions, inventions, or trade-offs addressing one or more architectural concerns.

Although they are related concepts and some architects confuse an architectural *vision* with the set of the first *versions* of the most important architectural models, there are several differences between them:

- An architectural vision is composed of architectural vision components, which are very important individual architectural decisions, inventions, trade-offs, or assumptions. There are some vision components that are not easily documented using models (i.e., they are

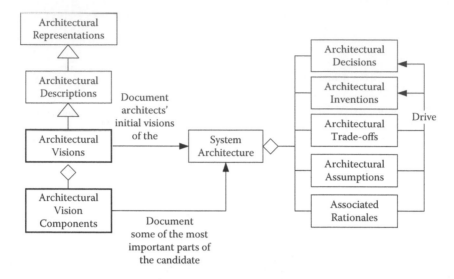

Figure 5.14 Architecture visions composed of architectural vision components.

better documented in architectural documents or white papers), and architectural models (even the first versions of the most important models) contain information that is not sufficiently important to be called a vision component. Thus, architectural visions and the first versions of the most important architectural models greatly overlap, but each contains information that is missing from the other.

■ Architectural vision components tend to be at a higher level of abstraction than most parts of individual architectural models.

■ One typically should develop multiple competing architectural visions but one is somewhat less likely to develop multiple competing versions of architectural models.

As illustrated in Figure 5.14, architectural visions are types of architectural descriptions composed of architectural vision components. Architectural vision components document some of the most important parts of the candidate architecture.

5.11 Guidelines

Use the following guidelines (and associated rationales) associated with the MFESA ontology of system engineering concepts and terminology:

■ Remember that systems typically contain more than software and hardware.
■ Assumptions and rationales are important parts of the architecture.
■ Engineer the entire system architecture, not just the system structures.
■ Architect all important types of system structures.
■ Create other system architectural representations in addition to graphical models and views.
■ Address architectural focus areas that cut across multiple models and views.
■ Strive to keep the different types of models, views, and focus areas consistent.
■ While useful, architectural patterns and styles are insufficient by themselves.

- Architectural concerns should center on architecturally significant requirements.
- Architectural quality cases are critical.
- Create multiple candidate architectural visions.
- Do not forget executable architectural representations.

These guidelines are discussed in more detail as follows:

- **Remember that systems typically contain more than software and hardware.** When decomposing the system into its component parts, consider components consisting of data, manual procedures, persons (or the roles they play), facilities, materials, equipment, etc.

  Rationale: Systems are typically composed of an aggregation hierarchy of subsystems, and the lowest subsystems are largely composed of hardware and/or software. However, systems typically contain more than just hardware and software. Systems can also be composed of data, manual procedures, persons (or the roles they play), facilities, materials, equipment, and other types of components. Whereas hardware engineers concentrate on hardware and software engineers concentrate on software, systems engineers (and system architects) must consider all the different types of system architectural components.

- **Assumptions and rationales are important parts of the architecture.** When engineering a system's architecture, do not stop with the architectural decisions, inventions, and trade-offs. Also address the architectural assumptions and the rationales for making the architectural decisions, inventions, trade-offs, and assumptions. Document the assumptions and rationales in the architectural representations.

  Rationale: Assumptions that may be obvious to the architects may not be so obvious to the other stakeholders of the architecture. These assumptions may also be false or may not hold as the system and its architecture evolve over time. Rationales improve the understandability of the architecture and are very useful when evaluating the architectural impact of changing requirements.

- **Engineer the entire system architecture, not just the system structures.** When engineering the system architecture, consider *all* the most important, pervasive, higher-level, strategic decisions, inventions, trade-offs, assumptions, and rationales regarding how the system will meet its requirements. For example, address the choice of technologies and materials, the application of architectural principles and heuristics, the selection of architectural patterns, styles, and mechanisms, the build-versus-reuse decisions, and the approach to meeting specialty engineering quality requirements. Also address the strategic and pervasive design-level decisions (e.g., choice of design paradigm) and implementation-level decisions (e.g., choice of programming language or programming language subset).

  Rationale: Many architects believe that a system's architecture is nothing more than the system's structures (i.e., its components, their relationships, and how they collaborate to meet the system's requirements). Although these are the most obvious, and typically the most important, parts of the system architecture, system architects must also make many other architectural decisions, inventions, trade-offs, and assumptions in addition to those involving system structuring. In fact, a system architecture should be defined as the most important, pervasive, higher-level, strategic decisions, inventions, trade-offs, assumptions, and rationales regarding how the system will meet its requirements. Although the system architecture is primarily incorporated in architectural structures, there are many other architectural decisions, inventions, and trade-offs that are not easily documented in the form of architectural models of system structures. These other parts of the architecture are often found in documents such as system architecture documents and architecture white papers. Therefore, engineer all of the system architecture, not just the system's structures.

■ **Architect all important types of system structures.** The architecture team should address all the important system structures. The architecture team should not just address the logical static functional decomposition structure and the physical static hierarchical aggregation decomposition structure of the system into its subsystems.

Rationale: A system can have a great number of different types of system structures. These structures can be logical or physical. They can also be static or dynamic. Architects should not automatically create the same models of the same architectural structures that they did on previous projects or blindly follow some standard or industry guideline that mandates or recommends a generic set of models. Given the system's architecturally significant requirements, it is more important for architects to determine which of the many possible system structures are relevant and worth architecting given limited funding, schedule, and staffing. Only then can the architects determine the appropriate set of system structures to model and represent.

■ **Create other system architectural representations in addition to graphical models and views.** The architects should create textual documents and other architectural representations that capture the architectural decisions, inventions, trade-offs, assumptions, and rationales that either are not easily captured in graphical models or are difficult to identify because they are scattered among a potentially huge amount of modeling information (i.e., a few relevant trees hidden in plain sight within massive forests).

Rationale: Traditionally, the only real architectural representation was the system architecture document, which contained mostly text with a few supporting cartoons and block diagrams. The trend now is to replace such informal documents with architectural models, often generated using a graphical modeling language such as UML, SysML, and DODAF [DODAF, 2007a–c]. While this has made the architectural representations less ambiguous and more objective, the move to graphical models has not come without its price. Although pictures can be worth a thousand words, very large and complex graphical models can easily overwhelm readers. It is not uncommon to see diagrams of huge UML models that cover an entire wall of a conference room. Similarly, when one wishes to see such graphics on computer monitors or in printed documents, one is often left with the choice of either seeing only a tiny part of the diagram or else a picture consisting of a large number of tiny boxes and lines, labeled in text that is far too small to read. Thus, views consisting of huge graphical models are not always easy to understand. They also do not communicate all the architects' decisions, inventions, trade-offs, assumptions, and rationales.

Unfortunately, it is now no longer uncommon for architects and their plans to state that "the models are the architecture,"* only to later augment their models with architecture documents and white papers containing textual descriptions and other forms of architectural information when they discover that many stakeholders cannot use the models to meet their needs and help them perform their tasks. It is best to use graphical models for what they are good for, and to use textual models, free-form text, tables, and informal diagrams for what they are best at.

■ **Address architectural focus areas that cut across multiple models and views.** The architects should identify and address the important architectural focus areas that cut across multiple architectural models and views. The architects should document their related

---

* Note that this statement also confuses the architecture with models that are representations of the architecture.

architectural decisions, inventions, trade-offs, assumptions, and rationales in architectural quality cases (e.g., interoperability or reliability quality cases), architectural white papers (e.g., whitepapers on performance or throughput), or both. Architects should explicitly include architectural focus areas in their plans and methods. The architects should also address architectural focus areas as a natural part of their work.

Rationale: Typical models abstract an entire architecture from a specific viewpoint in terms of one of the system's structures. Thus, for example, architects might create models that portray a system's architecture in terms of its allocation of software to hardware, data flows, control flows, hardware components, layers, networks, and states. While highly useful for certain purposes, such models and views are of little value if the architect wishes to clearly convey how the architecture supports the achievement of individual architecture concerns by supporting the achievement of individual types of quality requirements (i.e., requirements specifying minimum acceptable amounts of individual quality characteristics or quality attributes). Architectural decisions, inventions, and trade-offs related to individual quality characteristics are typically scattered across many models (and other architectural representations). It is like looking at a forest containing hundreds of trees of many different species and trying to see only the handful of oaks; one cannot see the relevant trees for the forest.

To clearly document what the architects have done regarding specific architectural concerns, the architects need to develop focus areas in addition to models. Although a few focus areas address process requirements such as cost and schedule, by far the largest number (and often the most important) focus areas address quality characteristics and possibly even quality attributes. This is the reason why most focus areas are best documented using architectural quality cases, whitepapers on architectural support for individual quality characteristics and attributes, or both. An intrinsic part of any complete architectural engineering method should therefore be the production of architectural quality cases (e.g., interoperability or reliability quality cases), architectural whitepapers (e.g., whitepapers on performance or throughput), or both as a natural part of the architects' work.

■ **Strive to keep the different types of models, views, and focus areas consistent.** The architects should strive to maintain consistency when adding information to architectural models, views, and focus areas. They should regularly verify the consistency of the existing models, views, and focus areas.

Rationale: Given the large number of architectural structures that may need to be developed and communicated and therefore the large number of models and views to be engineered, it should be of little surprise that it becomes difficult to keep these different models and views consistent with one another. Unfortunately, there is currently insufficient tool support to ensure that all the models created are mutually consistent. Significant effort needs to be expended to ensure (and maintain) the consistency of the many logical and physical, static and dynamic models and views.

■ **While useful, architectural patterns and styles are insufficient by themselves.** The architects should not concentrate solely on the identification, usage, and verification of architectural patterns and styles. When scheduling their work and estimating the amount of time needed to perform their work, the architects should carefully consider all of their tasks. The architects should not forget those tasks that occur after the initial architecture is developed.

Rationale: Architectural patterns, styles, and mechanisms are proven reusable solutions to common architectural problems. Although they are very useful when developing architectural models, only a limited number of *system* architectural patterns have been published. There is far more to the ten MFESA architectural tasks than merely identifying and applying appropriate architectural patterns, and architecture engineering methods need to provide architects with far more guidance than "Use the appropriate architectural patterns."

- **Architectural concerns should center on architecturally significant requirements.** The architects should not get lost in an analysis paralysis by treating each potentially relevant requirement equally. When the architects are identifying architecturally significant requirements and grouping them into architectural concerns, they should concentrate on the quality requirements and other requirements that strongly influence the architecture. Note that the architecturally significant requirements need to have unambiguous and feasible thresholds so that the architects can make engineering trade-offs between them, know when their architecture is sufficient, and thus know when to stop iterating the architecture.

  Rationale: On large systems, there are often a very large number of requirements, many of which do not have significant architectural ramifications. The architects will have insufficient time in which to treat all requirements equally. They therefore need a means to quickly identify the architecturally significant requirements.

  Architects must obviously produce a system architecture that enables the system to meet its functional requirements. However, in many ways, that is the easy job. Far more central to the architecture is the degree to which it supports the system's ability to meet its quality requirements, which are usually its most important architecturally significant requirements. Thus, the primary architectural concerns correspond to cohesive sets of requirements grouped by quality characteristic or quality attribute. For example, good architects spend a lot of their time and effort ensuring that their architectures enable the system to achieve sufficient levels of availability, capacity, interoperability, maintainability, performance, portability, reliability, safety, security, and usability.

- **Architectural quality cases are critical.** The architects should produce architectural quality cases for all of the quality characteristics for which they have quality requirements or other quality-related requirements. The detail and formality of these quality cases should be commensurate with the criticality of the associated quality characteristic. The production and verification of quality cases should be explicitly included in the architects' schedules, budgets, plans, and methods (e.g., standards, procedures, and templates).

  Rationale: Given the central importance of quality to architectural engineering, architects should carefully consider creating architectural quality cases as a natural part of their work. By (1) cohesively documenting their claims that the architecture meets a cohesive set of quality requirements; (2) their clear and compelling arguments justifying belief in their claims; and (3) official and relevant evidence supporting these arguments, architectural quality cases document quality focus areas and support both architectural assessments as well as associated certification and accreditation activities. By making the production of architectural quality cases an official part of the architectural engineering method, chief architects and process engineers can help ensure that the system and subsystem architectures properly meet the related quality concerns and requirements. Note that MFESA *architectural* quality cases are taken from the Quality Assessment of System Architecture and their Requirements (QUASAR) method for assessing system requirements and architecture [Firesmith et al., 2006].

- **Create multiple candidate architectural visions.** When producing the architecture and its representations, the architects should create several candidate architectural visions from which to choose the most suitable one for further development into the completed architecture. Note that "most suitable" merely means the best of the candidate architectural visions and not the absolute best architectural vision theoretically possible.

Rationale: There are typically many ways to architect a system or subsystem. Some of these ways will be significantly better than others, whereas there may not be a lot of difference between others. Even highly experienced and proficient architects rarely envision an optimal architecture at first attempt. Typically, it is not until several competing architectural visions have been identified and compared that an architectural approach that adequately addresses the often competing quality concerns is identified.

■ **Do not forget executable architectural representations.** When creating the architectural representations, the architects should not concentrate solely on architectural documents and static models. To the extent practical and cost-effective, the architects should produce executable architectural representations (e.g., executable models, simulations, prototypes, and executable architectures). The architects should use these executable representations to evaluate the appropriateness of their decisions, inventions, and trade-offs.

Rationale: Because most architectural representations consist of text and relatively simplistic graphics, they are inherently static in nature and intended for communication among the architects and other stakeholders. This is even true of models portraying dynamic information, such as activity diagrams, control flow diagrams, and state charts.* However, architects should remember that some architectural representations are executable, including models for simulations, architectural prototypes, and executable architectures. These more formal architectural representations can be used to answer important questions that the more static representations cannot. Therefore, architects should remember to consider engineering executable architectural representations as well as the more traditional and static representations.

5.12 Pitfalls

Avoid the following common pitfalls associated with the MFESA ontology of system engineering concepts and terminology and mitigate their negative consequences when they occur:

■ Architectures are confused with designs and architecture engineering is confused with designing.
■ Architectures are confused with structures or models of structures.
■ Trade-offs, assumptions, and rationales are ignored.
■ Models and views are engineered, but not focus areas.
■ Architectural quality cases are not developed or are only developed long after the fact for assessments.

The following express these pitfalls in more detail, describing their negative consequences and the steps the architects can use to mitigate them:

■ **Architectures are confused with designs and architecture engineering is confused with designing.**
  – **Pitfall.** Architects and other architecture stakeholders do not recognize the difference between *system architecture* and *system design,* and therefore confuse the two. Similarly, they also confuse the *system architecture engineering* and *designing* disciplines and

---

* State models are the foundation for some simujlation tools but are often incomplete, therefore ambiguous, and insufficient for execution.

activities. As a result, the system architecture becomes difficult to differentiate from the system design, and the system architecture representations (especially architecture documents and models) are often full of design-level information.

Given the critical nature of system architecture engineering in the development cycle and the corresponding importance of the role of the architect, it is quite strange to frequently meet developers whose official job title is architect but who neither can define the term "architecture" nor differentiate it from the term "design." It should therefore not be unexpected that many managers and process engineers also have the same difficulty.

- **Negative consequences.** This pitfall often causes the following negative consequences:
  - Turf battles occur between the architecture and design teams when architects inappropriately make design decisions.
  - Important architectural decisions, inventions, and trade-offs are not made.
  - System architecture representations are incomplete, missing higher-level architectural information.
  - Important parts of the architecture are not communicated to the stakeholders of the architecture.
  - The system may not meet its architecturally significant requirements.
- **Mitigations.** To avoid or mitigate the negative consequences of this pitfall:
  - The project glossary should clearly define the term architecture and differentiate it from the term design. This mitigation is built into the MFESA ontology of concepts and terminology.
  - The project-specific system architecture engineering method should clearly differentiate the system architecture engineering discipline and activity from system designing. This mitigation is built into the MFESA repository of reusable system architecture engineering method components.
  - Architects, designers, managers, and process engineers should be trained to recognize the difference between architecture and design.
  - Architectural representations should be evaluated for inappropriate inclusion of design materials.
  - Models containing both architecture and design should have the architectural decisions, inventions, trade-offs, assumptions, and rationales identified as such.

■ **Architectures are confused with structures or models of structures.**
  - **Pitfall.** Architects and other architecture stakeholders confuse the architecture with either architectural structures or models of these structures. The entire architecture engineering process is restricted to the creation of architectural models, which become the only representations of the architecture. One may hear architects say that "the models are the architecture."

    Architects do far more than just model system structures. Their primary work is in the making and documentation of strategic, high-level, pervasive decisions, inventions, trade-offs, and assumptions regarding how the system will meet its requirements.
  - **Negative consequences.** This pitfall often causes the following negative consequences:
    - The resulting architectural representations are incomplete, missing many of the decisions, inventions, trade-offs, assumptions, and rationales that are not easy to capture in graphical models.
    - Because their relevant information is scattered across multiple models, it is difficult to locate the contents of focus areas (especially areas focusing on individual quality characteristics) and thus they are often ignored.

- There is inadequate architectural support for some of the quality characteristics and associated quality requirements.
- It is difficult to verify adequate support for the system meeting its quality requirements.
- The system is much less likely to meet all of its architecturally significant requirements.
- Because architectural models can often be difficult for non-architects to read and understand, architecture evaluators and other architecture stakeholders have difficulty assessing, verifying, and validating the architectural models.

- **Mitigations.** To avoid or mitigate the negative consequences of this pitfall:
  - The project-specific system architecture engineering method should clearly define the terms "system architecture" and "system structure," and clearly state the relationship between them. This mitigation is built into the MFESA ontology of concepts and terminology.
  - The system architecture engineering method should include not just system structural modeling, but also all the other important and relevant architectural tasks and steps. This mitigation is built into the MFESA repository of reusable system architecture engineering method components.

■ **Trade-offs, assumptions, and rationales are ignored.**
  - **Pitfall.** Architects record their architectural decisions and inventions but ignore the rest of the architecture. System architecture documents, architectural models, and other architecture representations do not document (1) trade-offs made between conflicting quality concerns; (2) assumptions on which the architectural decisions, inventions, and trade-offs are based; and (3) rationales for making specific architectural decisions, inventions, and trade-offs.
  - **Negative consequences.** This pitfall often causes the following negative consequences:
    - Because of the lack of emphasis on engineering trade-offs, this pitfall can result in architectures that do not adequately support all of the potentially conflicting quality requirements.
    - The architects cannot respond rapidly to changing requirements or assess the impact of these changing requirements.
    - It is more difficult to identify inconsistencies between architectural components, resulting in an increased risk of defects not being visible until integration and integration testing.
    - It is much more difficult to correctly remember what was done and why it was done months or years after the fact than it is to document it as you go. Therefore, not documenting trade-offs, assumptions, and rationales in an attempt to save time, effort, and cost is often counterproductive.
    - By not making assumptions explicit, the resulting incomplete architectural representations may cause miscommunication between the architects and subject matter experts.
    - By ignoring rationales, the architecture and its architectural representations may become less maintainable and the integrity of the architecture is more likely to be lost over time.
  - **Mitigation.** To avoid or mitigate the negative consequences of this pitfall:
    - The project-specific system architecture engineering method needs to explicitly incorporate trade-offs, assumptions, and rationales into the relevant tasks and work products. These mitigations are built into the MFESA definition of architecture and

incorporated into the repository of reusable system architecture engineering method components, both in tasks (e.g., MFESA Task 3: *Create the first versions of the most important architecture models* and Task 5: *Create the candidate architectural visions*) as well as several architectural work products, including architecture documents and architectural quality cases.

- ◼ **Models and views are engineered, but not focus areas**.
  - – **Pitfall.** Architects concentrate on constructing architectural views and their component models, but ignore focus areas such as those documenting support for quality-specific architectural concerns.
  - – **Negative consequences.** This pitfall often causes the following negative consequences:
    - • Incomplete architectural representations are produced.
    - • There is inadequate architectural support for some of the quality characteristics and associated quality requirements.
    - • Architecture evaluators and other architecture stakeholders have difficulty assessing and validating the degree to which the architectural models because architectural models can often be difficult for non-architects to read and understand as well as locate information that is relevant to them when it is scattered across multiple models support architectural concerns.
  - – **Mitigations.** To avoid or mitigate the negative consequences of this pitfall:
    - • The project-specific system architecture engineering method should explicitly include focus areas as well as views and their models. These mitigations are built into the MFESA ontology of concepts and terminology as well as into the MFESA repository of reusable system architecture engineering method components.
    - • Architects should also be taught the value of focus areas to address quality-related architectural concerns.
- ◼ **Architectural quality cases are not developed or are only developed long after the fact for assessments.**
  - – **Pitfall.** Architects either (1) do not develop architectural quality cases at all or (2) only develop the architectural quality cases immediately prior to architecture quality assessments such as Quality Assessments of System Architectures and their Requirements (QUASAR) [Firesmith et al., 2006]). In the latter case, the architects produce incomplete and poor-quality architectural quality cases. The resulting architectural quality cases often contain irrelevant evidence (e.g., architecture plans, architectural engineering procedures, and architectural engineering schedules) merely because such documentation happens to exist.

    This happens because architectural quality cases are often not a part of the system architecture engineering method and are not mandated by contract on the development organization.
  - – **Negative consequences.** This pitfall often causes the following negative consequences:
    - • Because the architectural quality cases are not built until the last minute, they are not able to positively influence the architecture's support for the associated quality concern (e.g., by being a known hoop to be jumped through).
    - • The system architecture is less likely to sufficiently support the associated quality requirements.
    - • It is more difficult to assess, verify, and validate that the system architecture sufficiently supports the architecturally significant requirements.

- Architectural defects are more likely to be difficult and expensive to correct in later builds and releases.
  - **Mitigation.** To avoid or mitigate the negative consequences of this pitfall:
    - The project-specific system architecture engineering method should explicitly incorporate the development of architectural quality cases as a natural ongoing part of the development and documentation of the architecture. This mitigation is built into the MFESA Task 3: *Create initial architectural models*.

5.13   Summary

This chapter defined the foundational system architectural concepts and their relationships with emphasis on the architectural work products. Figure 5.15 illustrates the complete ontology of these concepts and terminology defined in the previous paragraphs. This chapter also provided a useful set of guidelines as well as a set of pitfalls to avoid or mitigate.

The following are guidelines associated with the MFESA ontology of system engineering concepts and terminology:

- Remember that systems typically contain more than software and hardware.
- Assumptions and rationales are important parts of the architecture.
- Engineer the entire system architecture, not just the system structures.
- Architect all important types of system structures.
- Create other system architectural representations in addition to graphical models and views.
- Address architectural focus areas that cut across multiple models and views.
- Strive to keep the different types of models, views, and focus areas consistent.
- While useful, architectural patterns and styles are insufficient by themselves.
- Architectural concerns should center on architecturally significant requirements.
- Architectural quality cases are critical.
- Create multiple candidate architectural visions.
- Do not forget executable architectural representations.

The following are common pitfalls associated with the MFESA ontology:

- Architectures are confused with designs and architecture engineering is confused with designing.
- Architectures are confused with structures or models of structures.
- Trade-offs, assumptions, and rationales are ignored.
- Models and views are engineered, but not focus areas.
- Architectural quality cases are not developed or are only developed long after the fact for assessments.

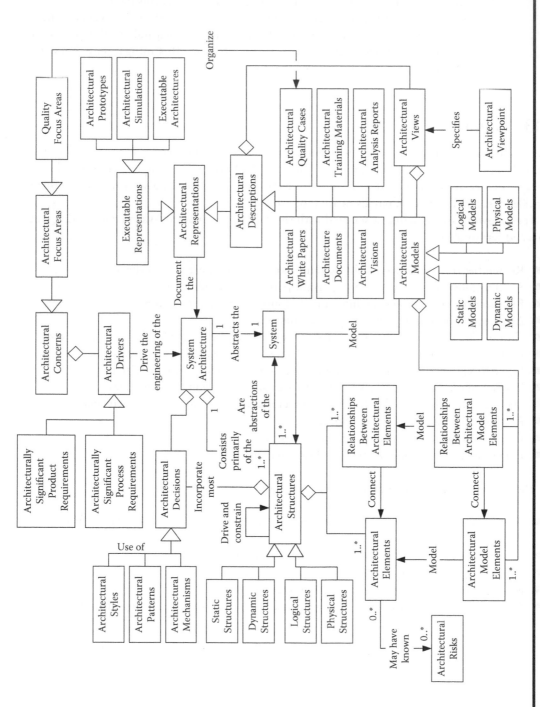

Figure 5.15 Complete ontology of architectural work product concepts and terminology.

Chapter 6

Task 1: Plan and Resource the Architecture Engineering Effort

6.1 Introduction

During this MFESA task, the system architecture team plans the system architecture engineering effort and obtains the resources necessary to engineer the system architecture and its representations. Figure 6.1 summarizes this task by illustrating its typical inputs, steps, and outputs.

This task is a reusable MFESA method component and as such is intended to be tailorable to meet the specific needs of the endeavor. Therefore, it is to be expected that at least some of the task's objectives, preconditions, inputs, steps, postconditions, work products, guidelines, and pitfalls will be tailored during the creation of the endeavor-specific system architecture engineering method. Tailoring of a method component includes adding missing content, modifying existing content, or removing existing content that is inappropriate, unnecessary, or not cost-effective to perform.

6.2 Goal and Objectives

The *Plan and Resource the Architecture Engineering Effort* task has the following overall goal:

- Prepare the system engineering team(s) to engineer the system architecture.

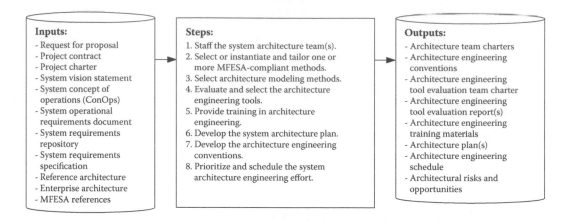

Inputs:
- Request for proposal
- Project contract
- Project charter
- System vision statement
- System concept of operations (ConOps)
- System operational requirements document
- System requirements repository
- System requirements specification
- Reference architecture
- Enterprise architecture
- MFESA references

Steps:
1. Staff the system architecture team(s).
2. Select or instantiate and tailor one or more MFESA-compliant methods.
3. Select architecture modeling methods.
4. Evaluate and select the architecture engineering tools.
5. Provide training in architecture engineering.
6. Develop the system architecture plan.
7. Develop the architecture engineering conventions.
8. Prioritize and schedule the system architecture engineering effort.

Outputs:
- Architecture team charters
- Architecture engineering conventions
- Architecture engineering tool evaluation team charter
- Architecture engineering tool evaluation report(s)
- Architecture engineering training materials
- Architecture plan(s)
- Architecture engineering schedule
- Architectural risks and opportunities

Figure 6.1 Summary of Task 1 inputs, steps, and outputs.

The typical subordinate objectives of this task are to:

- Staff the various system architecture teams.
- Create or reuse one or more appropriate project-specific system architecture engineering methods, including associated engineering standards, procedures, and templates.
- Select appropriate architecture tools to help the architects perform the system architecture engineering tasks.
- Train the various system architecture teams in how to perform system architecture engineering using the project-specific method(s) and supporting architectural tools.
- Develop one or more plan(s) documenting how the teams intend to perform the system architecture engineering tasks and produce the associated architectural work products.
- Schedule the architecture teams' work.

6.3 Preconditions

The *Plan and Resource the Architecture Engineering Effort* task can typically begin when the following preconditions have been met:

- **The project is initially funded.** Sufficient funding is available to begin staffing the top-level system architecture team(s) and to begin planning the architecture engineering work.
- **Requirements engineering has started.** A sufficient amount of systems requirements engineering has been performed to get an initial partial understanding of the:
  - Scope of the system architecture engineering effort
  - Architectural concerns

6.4 Inputs

The *Plan and Resource the Architecture Engineering Effort* task typically has the following inputs:*

---

* Remember to tailor out those inputs that are not appropriate when you are instantiating a project-specific architecture engineering method. For example, there will probably not be a system RFP when the system is being developed in-house for use in-house.

- **System request for proposal (RFP).** The system RFP is an invitation to system suppliers (e.g., system developers, prime contractor, or system integrators) to submit a proposal through a bidding process for the system to be developed or extensively updated. The RFP provides sufficient information about the desired system (such as via a detailed technical specification) to enable the bidders to propose an initial system architecture with associated cost, schedule, and other related information.
- **Project contract.** The project contract is a contract between the system acquirer and the system developer (or between the system prime contractor and a system subcontractor) legally documenting the agreement between the two parties regarding the development of a new system or a new version of an existing system. The project contract may include the system's customer requirements that were or are being used to engineer the system's derived technical requirements.
- **Project charter.** The project charter is a document that officially documents the scope, objectives, and participants in a system development project. The project charter also documents the roles, responsibilities, and authority of the major project participants. It also identifies the project's main stakeholders.
- **System vision statement.** The system vision statement defines the system and documents its mission, the business problems it is to solve, the business opportunities it is to take advantage of, the major needs and desires of its stakeholders, and any major restrictions.*
- **System concept of operations (ConOps).** The system ConOps defines the major stakeholders' concept of how the system is to operate and often includes the top-level system use cases.
- **System operational requirements document (ORD).** The system ORD defines the users' requirements for the system in terms of its performance and operational parameters.
- **System requirements repository.** The system requirements repository is a database and associated tools for storing and managing the system requirements (i.e., the functional, data, and quality requirements of the system as well as any mandatory constraints on the system architecture, design, implementation, and configuration). Due to the need to use an incremental development cycle to meet system delivery deadlines, the requirements repository will typically contain only the most important requirements when system architecture engineering begins. However, it is important that the requirements repository contain at least the most important architecturally significant requirements.
- **System requirements specification (SRS or SysRS).** A system requirements specification is a document that specifies the system requirements at specific points of time during the development cycle. Due to the need to use an incremental development cycle, an SRS is typically only an initial partial draft when system architecture engineering begins. An SRS may (or may not) be automatically generated from the contents of the system requirements repository. There may (or may not) be different requirements specifications for different stakeholders.
- **Reference architecture.** A reference architecture is a reusable architecture with associated architectural representations that has been shown by experience to be appropriate for systems within a specific application domain or product line. The system under development may need to be compatible with a specified reference architecture. This is especially true when the system to be architected will be part of a product line of systems.
- **Enterprise architecture.** An enterprise architecture is an organizational level architecture with associated architectural representations that supports an organization's core goals and

---

* Note that this is a requirements work product and should not be confined with an architectural vision.

strategic direction and the current and/or future structure and behavior of the organization's business processes, organizational structure, and systems (especially information systems). The system under development may need to interoperate with other systems described by the organization's enterprise architecture.

■ **MFESA references.** The MFESA references document the Method Framework for Engineering System Architectures (MFESA), including the reusable method components that can be used to create situation-specific system architecture engineering methods. These references include this book as well as any associated conventions (e.g., standards, guidelines, and templates).

6.5 Steps

During the *Plan and Resource the Architecture Engineering Effort* task, the architecture teams collaborate with other teams (such as the management team, process engineering team, and training team) to perform the following steps, typically in an incremental, iterative, concurrent, and time-boxed manner:

1. **Staff the system architecture team(s).** The management team staffs the overall system architecture team by selecting the endeavor's chief system architects. The management team also staffs the architecture teams for any top-level subsystems (or other architectural components) that have already been identified (e.g., during the proposal production process). The number of architects staffed will depend on such factors as the expected size of the system, the number of subsystems envisioned, and the number of specialty engineering areas where specialized domain expertise is needed. Over time as new architectural components are identified, the management team typically identifies and staffs their associated architecture teams. Members of the architecture teams collaborate to produce team charters as well as collaborate with the management team to update the project organizational chart to show the architecture teams.

2. **Select or instantiate and tailor one or more MFESA-compliant methods.*** The architects collaborate with process engineers, technical leaders, experts in architectural engineering, and experts in MFESA to either select one or more existing MFESA-compliant architecture engineering methods or to use the MFESA metamethod and repository of reusable method components to instantiate one or more MFESA-compliant architecture engineering methods to meet the specific needs of the endeavor and its associated organizations, especially the system development organization.

3. **Select the architecture modeling methods.** The architects collaborate with process engineers, technical leaders, and experts in architectural engineering to select provisional standard modeling methods and modeling languages for creating the appropriate architectural models that make up the views and focus areas of the system's architecture representations.

   Note that the final selection of modeling methods depends upon the identification of the architectural concerns that occurs during Task 2: *Identify the Architectural Drivers*, which will determine the appropriate viewpoints, views, focus areas, and therefore the appropriate models to produce.

---

* The metamethod for performing this step is documented in detail in Chapter 17: "MFESA: The Metamethod for Creating Endeavor-Specific Methods." A method is MFESA-compliant if it is primarily composed of reusable MFESA method components and if any additional method components are consistent with the MFESA ontology and metamodel.

4. **Evaluate and select the architecture engineering tools.** The architecture engineering organization forms one or more architecture engineering tool evaluation teams. Based on the tailored architecture engineering methods, including their associated modeling methods and modeling languages, these teams create evaluation criteria for evaluating, comparing, and selecting potential architecture engineering tools and their sources (such as tool vendors or open source organizations). The teams identify and obtain evaluation copies of potential architecture engineering tools. The teams evaluate the tools, document the results in one or more tool evaluation reports, and make their recommendations to management regarding tool acquisition. The management team collaborates with the system architecture teams and process engineering team to select and acquire appropriate architecture engineering tools to support the use of the endeavor architecture engineering methods.

5. **Provide training in architecture engineering.** The training team collaborates with the chief system architecture team to develop or obtain training materials for architecture engineering (in general) and MFESA. The training team develops or obtains training materials for the architecture modeling methods instantiated from MFESA and the architecture engineering tools. The training team incrementally trains the architects in MFESA, the instantiated MFESA methods, the architecture modeling methods, and the proper use of the architecture engineering tools. The training team also provides overview training to relevant stakeholders who need to understand how the architecture is being produced, such as relevant members of other teams who will interact with the architects (e.g., requirements engineers, managers, and acquisition representatives).

6. **Develop the system architecture plan(s).** The architecture teams develop one or more system or subsystem architecture plans documenting their plans for performing the architecture engineering tasks and producing the system architectural work products. The top-level architecture team also collaborates with the system chief engineer to develop the system architecture engineering section of the project system engineering management plan (SEMP). The architecture teams obtain management approval of the architecture plans. The development or sustainment organization obtains acquisition organization approval of or concurrence with these plans. The configuration management team baselines the plans and places them under configuration control.

7. **Develop the architecture engineering conventions.** Based on the tailored architecture engineering method(s), the selected architecture modeling methods and languages, and the selected architecture engineering tools, the architecture teams collaborate with the process team to develop appropriate standards, procedures, guidelines, tool manuals, and documentation templates for performing system architecture engineering and producing the associated architectural work products. These conventions are evaluated, approved, and baselined.

8. **Prioritize and schedule the system architecture engineering effort.** The architecture teams collaborate with the requirements teams and management team to prioritize, order, and schedule the performance of the architecture engineering work units. This includes adding architectural tasks and architecture-related development or life-cycle milestones to the endeavor work breakdown structure, master schedule, and subsystem schedules.

6.6 Postconditions

The *Plan and Resource the Architecture Engineering Effort* task is typically complete for the current subsystem (or other architectural component) when the following postconditions hold:

- **Architecture teams are staffed.** Incrementally, as architectural components are identified, the architecture teams are properly staffed with sufficient members having appropriate expertise and experience.
- **MFESA-compliant system architecture engineering method(s) exist.** MFESA has been used to produce one or more compliant system architecture engineering methods that meet the specific needs of the endeavor.
- **Architecture tools are selected.** The tools for engineering the system architecture have been identified, evaluated, selected, acquired, installed, configured, and verified.

  Note that this task overlaps several other MFESA tasks, such as Task 3 *Create the First Versions of the Most Important Architectural Models*, and inputs from these tasks will influence the architectural tools selected.
- **Architecture teams are trained.** Incrementally, as architecture teams are identified and staffed, the architecture teams are properly staffed and trained in the tailored MFESA method, selected architecture modeling methods, and associated tools.
- **Architecture engineering plans are approved.** The architecture engineering plans and architecture engineering section of the system engineering management plan are developed or obtained, approved, and baselined.
- **Architecture engineering conventions are approved.** The architecture engineering standards, procedures, guidelines, tool manuals, and documentation templates are developed or obtained, approved, and baselined.
- **Architecture engineering schedules are approved.** The endeavor work breakdown structure (WBS), master endeavor schedule, and subsystem schedules have been updated with system architecture engineering work units and milestones and have also been approved.

6.7 Work Products

The *Plan and Resource the Architecture Engineering Effort* task typically produces the following work products:

- **Architecture team charters:** a team charter for each architecture team, including such items as:
  - Team name.
  - Team goals and objectives.
  - Architectural tasks to perform.
  - Architectural work products to produce.
  - Team profile including roles, responsibilities, required skills and expertise, and membership.
  - Authority and management support. (Note that this may need to explicitly delineate the relationships and boundaries of scope with other project roles and teams such as the chief engineer, chief scientist, requirements engineering team, and systems engineering integration team [SEIT].)
  - Stakeholders including associated communication and management reporting.
  - Ground rules, including conflict resolution and elevation.
  - Success criteria.
  - Schedule and milestones.
  - Required resources, including staffing, budget, schedule (time), and tools.

- **Architecture plan(s):** the system/subsystem architecture plans (and parts of other plans such as system engineering management plans) documenting the:
  - Goal and scope of architecture engineering
  - Organization of the architecture engineering teams, including roles and responsibilities
  - Summary of the architecture engineering method, including modeling methods and associated tools
  - How architecture engineering fits into the system development/life cycle
- **Architecture engineering conventions:** architecture engineering content and format standards, procedures, guidelines, tool manuals, and documentation templates documenting the proper, effective, and efficient:
  - Performance of the:
    - Tailored endeavor-specific MFESA method(s)
    - Architecture modeling methods
  - Development of:
    - Architecture models
    - Architecture documents
    - Executable architecture and architectural prototypes
  - Use of:
    - Architecture modeling languages
    - Architecture engineering tools
- **Architecture engineering tool evaluation team charter:**
  - Team name
  - Team goals and objectives
  - Method for performing tool evaluation
  - Tool evaluation work products to produce
  - Team profile, including roles, responsibilities, required skills and expertise, and membership
  - Authority and management support
  - Stakeholders, including communication and management reporting
  - Ground rules, including conflict resolution and elevation
  - Success criteria
  - Schedule and milestones
  - Required resources, including staffing, budget, schedule (time), and tools
- **Architecture engineering tool evaluation report(s):** report documenting the evaluation and selection of architectural engineering tools including the:
  - Evaluation goals
  - Evaluation team membership, roles, responsibilities, and experience and expertise
  - Evaluation resources, including budget and schedule
  - Evaluation method
  - Facilities
  - Evaluation criteria
  - Evaluation tools and tool vendors evaluated
  - Evaluation results
  - Recommendations
- **Architecture engineering training materials:** training materials for the tailored endeavor-specific MFESA method(s), the architecture modeling methods, and architecture engineering tools.

■ **Architecture engineering schedule:** the endeavor work breakdown structure (WBS), master endeavor schedule, and subsystem schedules, including architectural tasks and milestones.

6.8 Guidelines

Use the following guidelines (and associated rationales) when performing the *Plan and Resource the Architecture Engineering Effort* task:

■ Properly staff the top-level architecture team(s).
■ Properly plan the architecture engineering effort.
■ Produce and maintain a proper and sufficient schedule.
■ Create or reuse appropriate MFESA method(s).
■ Select appropriate architecture modeling method(s).
■ Select appropriate architecture engineering tools.
■ Provide appropriate training.

The following descriptions express these guidelines in more detail and provide their associated rationales:

■ **Properly staff the top-level architecture team(s).** It is important for management to properly staff the initial top-level architecture team(s) at the very beginning of the endeavor. Ensure that the members of this team have the necessary architecture engineering experience and expertise to perform their tasks. Ensure that they can quickly earn the necessary respect of those with whom they must collaborate.

Rationale: Architecture engineering starts at the beginning of the endeavor and runs concurrently with requirements engineering. Also, time is required for the top-level system architecture team to determine the standard endeavor-specific method and generate its associated conventions (such as standards, procedures, guidelines, templates, and tool manuals) that will be used by the lower-level architecture teams. Without the proper experience and expertise, they will not be able to properly perform their tasks and earn the respect of those with whom they must collaborate. Without the respect of others, they will not be able to ensure that the architectural integrity is not lost as lower-level architects make unilateral changes.

■ **Properly plan the architecture engineering effort.** The management team should ensure that the architecture team properly produces one or more architecture engineering plans to guide the architecture engineering effort. If multiple plans are produced, then produce a top-level, overall system architecture engineering plan that takes precedence over the others. The management, architecture, and quality teams should ensure that all system architecture engineering plans are consistent with other project plans (such as the system engineering management plans and configuration management plans) as well as with each other.

Rationale: Plans are needed to ensure that (1) there are sufficient resources, (2) the architectural tasks are properly coordinated, and (3) the architectural work products are available when they are needed. Depending on such factors as the size of the system, the size of the architecture engineering organization, and the existence of contractually separate or geographically separate architecture teams (such as the prime and subcontractor architecture teams), there may be more than one architecture plan. Nevertheless, the overarching system architecture plan should take precedence over the lower-level subsystem architecture plans, which should not needlessly differ from it.

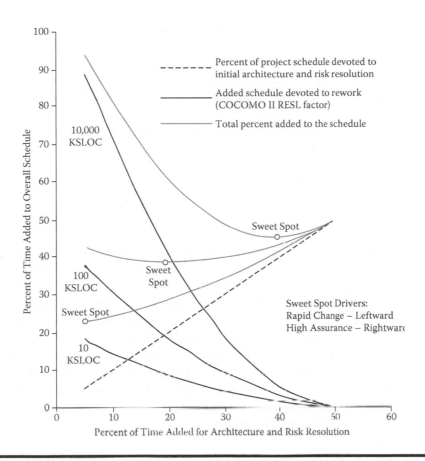

Figure 6.2 The optimum amount of architecture engineering. (*Source*: From Boehm et al., 2003.)

- **Produce and maintain a proper and sufficient schedule.** The architecture team should collaborate closely with the management team to provide an optimal schedule for the architecture engineering effort. When developing software-intensive systems, use a tool to estimate the optimal amount of architecture engineering in order to minimize the combination of architecture engineering and rework effort. Note that as illustrated in Figure 6.2, the optimum amount of architecture engineering (i.e., the sweet spot) increases with the amount of software [Boehm et al., 2003].

    Rationale: Architecture engineering requires a major amount of work in order to be properly and completely performed. Insufficient architecture engineering results in excessive rework.

- **Create or reuse appropriate MFESA method(s).** The architecture team should collaborate with the process team to use the MFESA repository and the MFESA metamethod documented in Chapter 17 to either identify an appropriate existing reusable MFESA-compliant architecture engineering method or to instantiate one or more new endeavor-specific MFESA methods that meet the specific needs of the endeavor. When selecting and tailoring reusable methods or method components, take into account the characteristics of the system, the organizations involved with developing the system, and the type of project being used (including the characteristics of the contract, if any).

    Rationale: The method components within the MFESA repository capture a set of industry best practices for engineering the architecture of large and complex systems. Not all of these practices may be cost-effective for all endeavors, especially those developing smaller and simpler systems. The MFESA repository of system architecture engineering method components includes tasks and work products covering the entire life cycle of systems, and some of these may be significantly simplified if the scope of the endeavor is limited.

■ **Select appropriate architecture modeling method(s).** The top-level system architecture team should carefully select a consistent set of modeling methods and associated modeling languages that capture all the important views of the architecture.

Rationale: System architectures typically include a large number of static and dynamic, logical and physical structures. To properly model these structures typically requires multiple modeling methods producing multiple types of architectural models, often documented using multiple modeling languages. Some of these models are complementary, whereas other models are largely redundant or have contradictory underlying philosophies. For example, functional decomposition and object-oriented methods decompose the system in very different ways. A consistent set of modeling methods and associated modeling languages that capture all important views of the architecture is needed to properly represent the structures of a large and complex architecture.

■ **Select appropriate architecture engineering tools.** The architecture team should collaborate with the tool evaluation team to carefully evaluate and select the architecture engineering tool(s) to be used to produce the architectural representations. Perform this evaluation and selection only after the MFESA method(s) have been determined and after the architecture modeling methods (and architecture modeling languages) have been selected.

Rationale: The tools should support the needed architecture engineering process, not the other way around. Tool vendor representatives may attempt to sell their tools as the solution to all architecture engineering problems, regardless of the project-specific challenges that the architects face.

■ **Provide appropriate training.** In collaboration with the chief architects and architecture subject matter experts, the training team should train the initially staffed architects very early in the architecture engineering process. Incrementally train the members of additional subsystem architecture teams as these new teams are staffed. Finally, train new architects as they join existing architecture teams. Provide adequate training to other stakeholders who will need to understand how the system will be architected.

Rationale: Some architects will require some level of training in system architecture engineering. A higher level of training in MFESA will typically be needed by those architects who will help select and tailor an appropriate reusable architecture engineering method or else instantiate a new method from the method components within the MFESA repository. The architects who will use these endeavor-specific MFESA-compliant methods will typically benefit from training in these methods. A minimum level of training will also benefit the stakeholders who will need to understand the architecture, its representations, and how the architects are engineering them.

6.9 Pitfalls

Avoid the following common pitfalls associated with performing the *Plan and Resource the Architecture Engineering Effort* task and mitigate their negative consequences when they occur:

- Architects produce incomplete architecture plans and conventions.
- Management provides inadequate resources.
- Management provides inadequate staff and stakeholder training.
- Architects lack authority.
- Architects instantiate the entire MFESA repository without tailoring.
- Tool vendors drive architecture engineering and modeling methods.
- Planning and resourcing are unsynchronized.
- Planning and resourcing are only done once up front.

The following express these pitfalls in more detail, describing their negative consequences and the steps the architects can use to mitigate them:

- **Architects produce incomplete architecture plans and conventions.**
  - **Pitfall.** The architecture plans and conventions are incomplete. The system architecture engineering method is not adequately documented in the system engineering management plan, the system architecture plan, and relevant conventions (such as standards, procedures, guidelines, and tool manuals).
  - **Negative consequences.** This pitfall often causes the following negative consequences:
    - Architects, process engineers, quality engineers, and managers may have different understandings of how the system architecture is to be engineered.
    - Different architecture teams may use different versions of the official project-specific architecture engineering method, resulting in some teams unnecessarily performing certain tasks, not performing certain necessary tasks, and developing inconsistent architecture work products.
    - Time will be wasted arguing as to what the appropriate method should be.
  - **Mitigations.** Do the following to avoid or mitigate the negative consequences of this pitfall:
    - Perform the MFESA metamethod to create the appropriate project-specific method(s) very early in the development process, before different architecture teams have time to inconsistently perform their tasks.
    - Clearly document the official project-specific method in the appropriate project documents.
    - Verify the completeness and consistency of different architecture plans (such as the system architecture plan and lower-level subsystem architecture plans).
- **Management provides inadequate resources.**
  - **Pitfall.** Management does not provide the architects and architecture teams with sufficient resources to effectively and efficiently perform their system architecture engineering tasks and produce high-quality architecture work products. This can include:
    - **Inadequate architecture team staffing.** The architecture teams are not properly staffed with sufficient members having sufficient training and expertise to fulfill their responsibilities. This is especially prevalent on large programs when is it impossible to hire or assign sufficiently trained and experienced system architects.
    - **Inadequate funding.** The architects and architecture teams are not provided sufficient funding to perform their architecture engineering tasks and to purchase the tools needed to perform their tasks.
    - **Inadequate schedule.** The architects and architecture teams are not given sufficient time in which to adequately perform their tasks.

- **Negative consequences.** This pitfall often causes the following negative consequences:
  - Without adequate resources, the architects and architecture teams may create an architecture and associated architecture representations that are incomplete, of inadequate quality, inadequately usable by the architecture's stakeholders, and that do not adequately enable the system to meet its architecturally significant requirements.
  - If understaffed architecture teams fall behind and are unable to address issues as they arise (such as new or changed architecturally significant requirements), then they can easily fall into catch-up mode that at best follows behind the rest of the project and fixes lost architectural integrity rather than leading the design, implementation, and integration work.
- **Mitigations.** Do the following to avoid or mitigate the negative consequences of this pitfall:
  - The chief system architecture team and process team should collaborate to:
    - Determine the necessary resources
    - Clearly communicate to management the specific resources needed and the negative consequences of not having them

■ **Management provides inadequate staff and stakeholder training.**
- **Pitfall.** The architects, other members of the architecture teams, and the associated stakeholders are not provided with sufficient training to effectively and efficiently perform their system architecture engineering tasks and produce high-quality architecture work products.
- **Negative consequences.** This pitfall often causes the following negative consequences:
  - Without adequate training, the architects and architecture teams may create an architecture and associated architecture representations that are of inadequate quality, inadequately usable by the architecture's stakeholders, and that do not adequately enable the system to meet its architecturally significant requirements.
  - The time required to train and grow people into capable architects may cause the architecture team to miss critical early milestones for incorporating architectural representations into the associated baselines. The resulting loss of architecture team credibility may require years from which to recover.
- **Mitigation.** Do the following to avoid or mitigate the negative consequences of this pitfall:
  - Have the training team (or chief architects) provide members of the architecture teams and related stakeholders with sufficient training in system architecture engineering, the corresponding modeling method(s), and the associated tools.

■ **Architects lack authority.**
- **Pitfall.** Although management gives the architects and architecture teams the *responsibility* to perform their architecture tasks and produce the architecture work products, management does not give them the associated authority needed to successfully do their work. Lower-level architecture teams may ignore common architectural decisions, inventions, and trade-offs made by higher-level architecture teams. Similarly, designers and implementers may ignore architectural decisions, inventions, and trade-offs.
- **Negative consequences.** This pitfall often causes the following negative consequences:
  - The integrity and consistency of the architecture can be lost.
  - The architectural representations may not be consistent with the design and implementation leading to a system that is difficult to maintain and extended once developed.

- Endeavors staffed by engineers who are unfamiliar with the authority and responsibilities of a properly functioning architecture team may at best fail to fully engage with the team and at worst feel threatened by and resist what they may perceive as encroachment on their "turf."
  - **Mitigations.** Do the following to avoid or mitigate the negative consequences of this pitfall:
    - Management must give the architects the authority to ensure the integrity of the architecture and the consistency of the architecture, design, and implementation.
    - Management may assign the chief system architect the role of project technical leader.
    - There should be a chief architect and top-level architecture team that oversees the lower-level architects and architecture teams.
    - The top-level architecture team should include representatives from lower-level architecture teams.
    - There should be a clear procedure for the architects to use to escalate conflicts between architecture teams, design teams, and implementation teams.
    - Subsystem architecture, design, and implementation teams may be virtual and combined into integrated product teams (IPTs).
    - Other teams should receive at least minimal training in architecture engineering that properly incorporates when and how they should collaborate with the architects.
- **Architects instantiate the entire MFESA repository without tailoring.**
  - **Pitfall.** The chief architecture team, process team, and/or management team create the system architecture engineering method by selecting *all* of the system architecture engineering method components from the entire MFESA repository. They also do not tailor any of the method components to meet the specific needs of the project.
  - **Negative consequences.** This pitfall often causes the following negative consequences:
    - The resulting system architecture engineering method is:
      - Generic and thus not project specific
      - Too large and complex for the needs of the project
      - Overkill, being neither cost-effective nor efficient
    - This decreases the productivity of the architecture teams and increases project costs and schedule due to the performance of unnecessary tasks and the production of unnecessary architectural work products.
  - **Mitigations.** Do the following to avoid or mitigate the negative consequences of this pitfall:
    - Use the MFESA metamethod, including the *Method Needs Assessment* task as well as the steps involving the selection and tailoring of appropriate method components of the *Method Construction* task. Look for this pitfall when performing the *Method Verification* task. (See Chapter 17.)
    - Train the process engineers and chief architects in method engineering.
    - If needed, hire a consultant who is familiar with method engineering to help engineer the project-specific system architecture engineering method.
- **Tool vendors drive architecture engineering and modeling methods.**
  - **Pitfall.** The vendors of architecture (or design) computer-aided system engineering tools may market their tool(s) as the silver-bullet solution to the system architecture engineering problem. However, the reality is that the architecture modeling market is still maturing along with (or behind) the state of the industry. Once a tool has successfully

supported one or a few projects, its vendor may seek to establish the associated supported method as a *de facto* industry standard so that their tool will fare well in a tool competition. This approach is backwards. The MFESA metamethod's *Method Needs Assessment* task should drive the system architecture engineering method, and the method should drive architecture tool selection.

- **Negative consequences.** This pitfall often causes the following negative consequences:
  - Tool selection determines the system architecture engineering and modeling methods.
  - The tool(s) tend to support generic method(s) that may not be appropriate for engineering the system.
  - The tools may not support the development of the appropriate models.
  - The completeness and relevance of the architectural representations may suffer.
- **Mitigations.** Do the following to avoid or mitigate the negative consequences of this pitfall:
  - Perform the MFESA metamethod to create one or more appropriate project-specific system architecture engineering methods.
  - Perform a trade study to evaluate existing architecture tools, including their support for the selected architecture engineering method as well as support for chosen architecture modeling languages.

■ **Planning and resourcing are unsynchronized.**
- **Pitfall.** The architecture planning and resourcing performed by the acquisition and development organizations get out of synchronization. Similarly, the planning and resourcing performed by the system architecture team and the lower-level subsystem architecture teams become inconsistent.
- **Negative consequence.** This pitfall often causes the following negative consequence:
  - Different architecture teams end up with inconsistent system architecture engineering methods, schedules, and funding.
- **Mitigations.** Do the following to avoid or mitigate the negative consequences of this pitfall:
  - Clearly document the official project-specific method* in the appropriate project documents.
  - Verify the completeness and consistency of different architecture plans (such as the system architecture plan and lower-level subsystem architecture plans).
  - Evaluate the architecture-related items in the subsystem schedules for their consistency with the project master plan.
  - Evaluate the architecture-related subsystem funding for its consistency with the overall project funding.
  - Report and properly remove inconsistencies.

■ **Planning and resourcing are only done once up front.**
- **Pitfall.** Planning and resourcing of the system architecture engineering effort is only performed once, early in the system architecture engineering effort even though:
  - Changes to the architecturally significant requirements or the project schedule may invalidate the original effort estimate.

---

* This assumes a single project-wide architecture engineering method. There may be valid technical and managerial reasons for using different methods on different parts of the project.

- Early planning and resourcing may have only addressed the initial engineering of the architecture and its representations and ignored downstream architectural tasks related to architecture maintenance and the preservation of architectural integrity in spite of changes to the requirements, design, and implementation.
- **Negative consequences.** This pitfall often causes the following negative consequences:
  - Important architectural tasks and steps are not completed.
  - The integrity of the architecture is lost as architectural representations get out of sync with the architecture as the architecture evolves.
  - The architectural representations become less useful over time.
- **Mitigations.** Do the following to avoid or mitigate the negative consequences of this pitfall:
  - Schedule architectural tasks and the creation and delivery of architectural work products at appropriate project milestones.
  - Treat the architectural representations as valuable project and organizational assets.
  - Maintain the architectural representations and place them under configuration control.
  - Incrementally and iteratively repeat system architecture engineering multiple times for different architectural components and also for the same component as requirements and the architecture evolve during the life of the system.

6.10 Summary

During the *Plan and Resource the Architecture Engineering Effort* task, the system engineering team prepares to engineer the system architecture.

6.10.1 Steps

When instantiating the *Plan and Resource the Architecture Engineering Effort* task, the following steps may be included, modified, or tailored out:

1. Staff the system architecture team(s).
2. Select or instantiate and tailor one or more MFESA-compliant methods.
3. Select architecture modeling methods.
4. Evaluate and select the architecture engineering tools.
5. Provide training in architecture engineering.
6. Develop the system architecture plan(s).
7. Develop the architecture engineering conventions.
8. Prioritize and schedule the system architecture engineering effort.

6.10.2 Work Products

When instantiating the *Plan and Resource the Architecture Engineering Effort* task, the following work products may be included, modified, or tailored out:

- Architecture team charters
- Architecture plan(s)

- Architecture engineering conventions
- Architecture engineering tool evaluation team charter
- Architecture engineering tool evaluation report(s)
- Architecture engineering training materials
- Architecture engineering schedule
- Architectural risks and opportunities

6.10.3 Guidelines

Use the following guidelines (and associated rationales) when performing the *Plan and Resource the Architecture Engineering Effort* task:

- Properly staff the top-level architecture team(s).
- Properly plan the architecture engineering effort.
- Produce and maintain a proper and sufficient schedule.
- Create or reuse appropriate MFESA method(s).
- Select appropriate architecture modeling method(s).
- Select appropriate architecture engineering tools.
- Provide appropriate training.

6.10.4 Pitfalls

Avoid the following common pitfalls associated with performing the *Plan and Resource the Architecture Engineering Effort* task and mitigate their negative consequences when they occur:

- Architects produce incomplete architecture plans and conventions.
- Management provides inadequate resources.
- Management provides inadequate staff and stakeholder training.
- Architects lack authority.
- Architects instantiate the entire MFESA repository without tailoring.
- Tool vendors drive architecture engineering and modeling methods.
- Planning and resourcing are unsynchronized.
- Planning and resourcing are only done once up front.

Chapter 7

Task 2: Identify the Architectural Drivers

7.1 Introduction

During this MFESA task, the system or subsystem architecture team identifies the architecturally significant requirements that have been derived for or allocated to their system or architectural component and categorize them into a set of architectural concerns.

The architecture team begins by obtaining and comprehending the potentially relevant requirements. The architecture team then verifies their completeness, maturity, and feasibility to ensure their adequacy to drive the engineering of the architecture. The architecture team collaborates with the requirements team to fix any defects identified. The architecture team then identifies the architecturally significant drivers and groups them to identify and form cohesive architectural concerns. Finally, the architecture team verifies the quality of the architectural concerns.

This task is a reusable MFESA method component and as such is intended to be tailorable to meet the specific needs of the endeavor. Therefore, it is to be expected that at least some of the task's objectives, preconditions, inputs, steps, postconditions, work products, guidelines, and pitfalls will be tailored during the creation of the endeavor-specific system architecture engineering method. Tailoring of a method component includes adding missing content, modifying existing content, or removing existing content that is inappropriate, unnecessary, or not cost-effective to perform.

7.2 Goal and Objectives

The *Identify the Architectural Drivers* task has the following overall goal:

- Identify the architecturally significant requirements that drive the development of the architecture so that the architects can base their architecture and architectural representations on them.

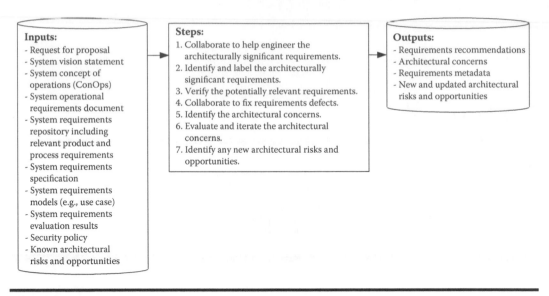

Figure 7.1   Summary of Task 2 inputs, steps, and outputs.

The typical subordinate objectives of this task are to:

- Understand and verify the completeness, maturity, and feasibility of the requirements that have been allocated to the system or architectural component being architected.
- Categorize the architecturally significant architectural requirements into cohesive concerns that will subsequently drive the performance of the "downstream" MFESA tasks.

7.3 Preconditions

Because of the large potential degree of overlap between the *Identify the Architectural Drivers* task and the requirements engineering tasks, no hard-and-fast preconditions actually exist for this task. However, the degree of effectiveness and efficiency of this task increases with the degree to which the relevant architecturally significant requirements have been properly engineered (i.e., exist with sufficient quality and maturity).*

The *Identify the Architectural Drivers* task can typically begin when the following preconditions† have been met:

- **Most important potentially relevant requirements exist.** A *majority* of the most important architecturally significant product and process requirements have been derived and allocated to the system or architectural component being architected.

---

* On the other hand, the requirements can also drive the requirements. This includes, but is not limited to, lower-level architectural components, modification of requirements to enable the use of OTS architectural components, and certain safety requirements based on hazards related to the architecture.

† Note that these preconditions have been made conditional so as to allow iterative, incremental, and concurrent development and not mandate a strict waterfall development cycle. This task does *not* require that *all* the architecturally significant requirements exist and be perfectly complete and mature.

- **Architecturally significant requirements are sufficiently complete and mature.** The relevant architecturally significant product and process requirements are *sufficiently* complete and mature for the current point in the development schedule to drive the engineering of the system or architectural component architectures:
  - Sufficient product and process requirements have been identified, analyzed, and specified.
  - Each of these requirements has sufficient quality (i.e., have the appropriate characteristics such as being complete, concise, consistent, mandatory, verifiable, validated, and unambiguous).
- **The architecture team has access to the requirements team.** If the most important potentially relevant requirements do not exist or are not sufficiently mature, then the architecture team must be able to collaborate with the requirements team to engineer the architecturally significant requirements.

7.4 Inputs

The *Identify the Architectural Drivers* task typically uses the following inputs:

- **Request for proposal (RFP).** The system RFP is an invitation to system suppliers (such as system developers or integrators) to submit a proposal through a bidding process for the system to be developed or extensively updated. The RFP provides sufficient information about the desired system (via a detailed technical specification, for example) to enable the bidders to propose an initial system architecture with associated cost, schedule, and other related information.
- **System vision statement.** The system vision statement defines the system and documents its mission, the business problems it is to solve, the business opportunities it is to take advantage of, the major needs and desires of its stakeholders, and any major restrictions.
- **System concept of operations (ConOps).** The system ConOps defines the major stakeholders' concept of how the system is to operate and often includes the top-level system use cases or usage scenarios.

  Note that many architects would count themselves very lucky to have a customer that provides them with a well-formed ConOps. If such a ConOps is not available or it is incomplete or inconsistent, then the architecture team may need to collaborate with the requirements team (and stakeholder community including customer representatives) to create a suitable ConOps.
- **System operational requirements document (ORD).** The system ORD defines the users' requirements for the system in terms of its performance and operational parameters.
- **Relevant product requirements:**
  - Functional requirements
  - Quality requirements
  - Data requirements
  - Interface requirements
  - Architectural constraints
    - Acquisition constraints
    - Budgetary constraints
    - Facility constraints

- Schedule constraints
- Staffing constraints

■ **Relevant process requirements:**

■ **System requirements repository.** The system requirements repository is a database and the associated tools for storing and managing the system product and process requirements. Due to the need to use a recursively incremental development cycle to the large size and complexity or most systems, the requirements repository will typically contain only the most important requirements when system architecture engineering begins. However, it is important that the requirements repository contain at least the most important architecturally significant requirements.

■ **System requirements specification (SRS or SysRS).** A system requirements specification is a document that specifies the system requirements at specific points in time during the development cycle. Due to the need to use a recursively incremental development cycle, an SRS is typically only an initial partial draft when system architecture engineering begins. An SRS may (or may not) be automatically generated from the contents of the system requirements repository. There may (or may not) be different requirements specifications for different stakeholders.

■ **System requirements models.** System requirements models are models developed during requirements engineering to help identify and analyze the system requirements. For example, they may include use case models for functional requirements and misuse case models and attack trees for security requirements.

■ **System requirements evaluation results.** These may be the results of peer reviews, inspections, formal milestone reviews, and quality audits of the system or the architectural component being architected.

■ **Security policy.** A security policy is a document that specifies the policies for ensuring that security goals are met. Security policies typically include policies implemented procedurally by people (such as administrators and users) but may also include policies implemented by the system.

Although security policies should have been used by the requirements team to engineer security requirements, this may not have taken place and the architecture team needs to collaborate with the security team to ensure that the system (including its associated procedures) will meet the security goals of its stakeholders.

■ **Known architectural risks and opportunities:** Known architectural risks in the endeavor risk repository, as well as known architectural opportunities.

7.5 Steps

During the *Identify the Architectural Drivers* task, members of the relevant architecture team typically collaborate with the requirements teams, requirements evaluation teams, and quality assurance teams to perform the following steps in a *highly* iterative, recursively incremental, concurrent, and time-boxed manner:

1. **Collaborate to help engineer the architecturally significant requirements.** If the requirements team has not previously engineered the architecturally significant requirements, then the architecture team collaborates with the corresponding requirements team to engineer the architecturally significant requirements. This includes identifying, analyzing, managing, and specifying these requirements.

Note: Engineering requirements (even the architecturally significant ones) is ultimately a requirements engineering task, not an architecture engineering task. However, the requirements team may (or may not) have already engineered these requirements when the architecture team begins performing this MFESA task.

2. **Identify and label the architecturally significant requirements.** The members of the architecture team determine which of the existing requirements are relevant and architecturally significant. Members of the architecture team:
   - Obtain access to the potentially relevant requirements, including all the product and process requirements derived for and allocated to the architectural component (e.g., system or subsystem) being architected.
   - Read and comprehend these requirements, obtaining input from the requirements engineering team, stakeholders, and subject matter experts as needed.
   - Identify candidate architecturally significant product requirements based on:
     • Their knowledge, skills, and experience
     • Relevant program or program office documentation, interviews with program office leadership, and subject matter experts
   - Confirm that these candidate architecturally significant requirements actually have significant architectural ramifications.

     Remember that an architecturally significant requirement is any *individual* requirement that has a significant impact on the architecture. Although the cohesive set of all functional requirements in a feature set collectively influence the architecture, most do not individually affect the architecture. Instead, it is often the individual quality requirements that have the largest effect on the architecture. Ultimately, this identification of architectural significance is typically based on the professional experience and expertise of the architects with input from subject matter experts as needed.
   - Update the requirements repository with metadata labeling the architecturally significant requirements so that these relevant requirements can be easily identified as such in the future:*
     • **Origin:** the authoritative source for the driver. If found in a governing artifact, provide a reference.
     • **Owner:** the entity that is the controlling stakeholder for a given driver. This is the entity one would be required to approach in order to secure a deviation or an alteration to the driver in question.
     • **Derivation:** the parent requirement from which the architectural driver has been derived and the method of derivation. For example, a timing budget may derive throughput requirements for lower-level architectural elements based on an allocation of the overall maximum time given by a higher-level throughput requirement.
     • **Criticality:** the degree of importance of the requirement. Not all drivers are created equal, and it is important to know their relative importance when trade-offs among competing drivers must be entertained and when prioritizing requirements when scheduling their implementation.
     • **Probable impact:** the degree to which the architectural driver or concern impacts the architecture. For example, how big would the impact be on the architecture if the driver or concern were to be significantly changed?

---

* Note that the first four of the following types of metadata are not architecture-specific and usually identified by requirements engineers. The architects are typically responsible for the last two types of metadata.

- **Competing or contradictory drivers:** a listing of other architectural drivers whose satisfaction conflicts with the driver in question. This can be particularly important where the drivers for a particular system will be bounded by threshold or objective values, because there will be trade-offs and even adjustment in one or more of these drivers based on trade studies and related analyses.

3. **Verify the potentially relevant requirements.** If the members of the architecture team have not already verified the quality of these requirements as members of requirements evaluation teams, then they verify whether the potentially relevant requirements:
   - Are sufficiently complete and mature for the current point in the development schedule to drive the engineering of the architectures
   - Are consistent with existing architectural representations such as models and analyses
   - Are architecturally feasible*
   - Have the proper characteristics (e.g., are cohesive, consistent, complete, unambiguous, and verifiable)
   - Do not *unnecessarily* constrain the architecture and prohibit the appropriate reuse of existing architectural components (such as COTS components)
   - Do not unnecessarily increase architectural risks†

   One way to do this is to compare the potentially relevant requirements against the existing architectural representation (such as models and analyses) for consistency. The architecture team may also verify that the quality requirements are unambiguous with specific conditions, quality criteria, and clear thresholds on appropriate measurement scales.

4. **Collaborate to fix requirements defects.** The members of the architecture team notify the requirements team of all architecturally significant requirements that are insufficiently engineered and thus unsuitable for driving the architecture/engineering effort. Members of the architecture team collaborate closely with the members of the requirements team to fix any defects identified in the potentially relevant requirements. This will likely mean both deriving new requirements or modifying existing requirements. It will also likely mean negotiation of the architecturally significant requirements with their stakeholders to ensure that the requirements are feasible and not unnecessarily restrictive.

5. **Identify the architectural concerns.** The members of the architecture team identify architectural concerns by categorizing the architecturally significant system or subsystem requirements into cohesive sets. Architects give each concern a meaningful name, a brief description, and a list (or reference) of each of the concern's component requirements. Architects identify the following types of architectural concerns:
   - **Quality concerns.** Architects identify quality concerns by categorizing architecturally significant quality requirements, first by quality characteristic and then possibly by quality attributes using the same project quality model‡ that was used to engineer the quality requirements.

---

* A requirement may not be feasible for many (such as technical, physical, financial, and schedule) reasons.
† It may become necessary to renegotiate requirements that cause excessive architectural risks.
‡ If the project did not use a quality model that defined the different types of quality characteristics (see Chapter 18, "Architecture and Quality"), then the requirements team and architecture team should at least use the same definitions for the relevant quality characteristics and attributes.

- **Constraint concerns.** Architects identify constraint concerns by categorizing architecturally significant constraints such as architectural, design, implementation, and testing constraints.
- **Data concerns.** Architects identify data concerns by categorizing architecturally significant data requirements by major cohesive sets of data or data types.
- **Functional concerns.** Architects identify functional concerns by categorizing architecturally significant functional requirements into cohesive feature sets.
- **Interface concerns.** Architects identify interface concerns by categorizing architecturally significant interface requirements by major cohesive sets of interfaces.
- **Process concerns.** Architects identify process concerns by categorizing architecturally significant process requirements such as development costs, production costs, sustainment (i.e., operational and maintenance) costs, and schedule constraints (such as required delivery dates).

6. **Evaluate and iterate the architectural concerns.** The members of the architecture team evaluate the architectural concerns for cohesion, completeness, and usability. They then fix any defects identified during the evaluation.

7. **Identify any new architectural risks and opportunities.** Concurrently with the performance of the previous steps, architects identify any new architectural risks and opportunities associated with the architectural concerns that the new architecture should mitigate.

7.6 Postconditions

The *Identify the Architectural Drivers* task is complete for the current system or subsystem when the following postconditions hold:

- **Architecturally significant requirements are identified.** The architecture team has identified all the architecturally significant product and process requirements.
- **Architecturally significant requirements exist and are sufficiently mature.** The architecture team has confirmed that the architecturally significant requirements are sufficiently complete, mature, feasible, consistent, and of sufficient quality to drive the development of the architecture (and subsequent assessment of the architecture against the requirements).
- **Requirements metadata is updated.** The architecture team has updated the requirements metadata in the requirements repository to identify the architecturally significant requirements.
- **Architecture concerns are identified and evaluated.** The architecture team has identified and evaluated a set of associated architectural concerns (i.e., cohesive collections of architecturally significant requirements).

7.7 Work Products

The *Identify the Architectural Drivers* task produces the following work products:

- **Requirements recommendations:** recommended changes to the requirements to make them more suitable for driving and assessing the architecture.

- **Architectural concerns:** the documented set of all architectural concerns, whereby each concern includes its name, brief description, and list of (or references to) its component architecturally significant requirements.

  These concerns may affect the ConOps and result in recommended changes that must be negotiated with the requirements engineers and validated with the stakeholders.

- **Requirements metadata:** metadata (e.g., in the requirements repository) that identifies the relevant architecturally significant product and process requirements as being architecturally significant.

- **Architectural risks and opportunities:** any new or modified architectural risks documented in the endeavor risk repository as a part of endeavor risk management activity. This also includes an updated list of architectural opportunities that the architecture teams may wish to take advantage of in the future.

7.8 Guidelines

Use the following guidelines (and associated rationales) when performing the *Identify the Architectural Drivers* task:

- Collaborate closely with the requirements team.
- Notify the requirements team(s) of relevant requirements defects.
- Challenge difficult requirements.
- Consider the impact of the architecture on the requirements.
- Respect team boundaries and responsibilities.
- If necessary, clarify relevant requirements with the stakeholders.
- Concentrate on the architecturally significant requirements.
- Remember that quality attributes can be architectural concerns too.
- Formally manage architectural risks.

The following descriptions express these guidelines in more detail and provide their associated rationales:

- **Collaborate closely with the requirements team.** The architecture team should closely collaborate with the requirements team to ensure that the relevant architecturally significant requirements exist and have sufficient quality and maturity to drive the development of the architecture. This includes verifying the requirements (e.g., using an architecture-oriented requirements assessment method such as QUASAR) as well as working with the requirements team to fix the requirements.

  Rationale: The requirements team is responsible for engineering the requirements, not the architecture team. The members of the requirements team should have the training and expertise to produce all the requirements, including the architecturally significant requirements. Because the members of the architecture team typically do not have the expertise necessary to properly engineer requirements, they need to collaborate with the requirements team to ensure that the architecturally significant requirements are engineered correctly. Note, however, that this does not prohibit a single person from being a member of both a requirements team and an architecture team if that person has the proper training and expertise.

In practice, stakeholders are poor at providing architecturally significant requirements and many practicing requirements engineers have been poor at engineering them, especially the nonfunctional requirements. For example, requirements repositories and specifications often include quality "requirements" such as "the system shall be interoperable" without specifying (1) with what the system shall be interoperable, (2) the type of interoperability, and (3) to what degree the system shall be interoperable. This is especially true of quality requirements because current popular requirements engineering methods (e.g., use case modeling) tend to emphasize only the functional requirements.

■ **Notify the requirements team(s) of relevant requirements defects.** Notify the requirements team of all architecturally significant requirements that are insufficiently engineered and thus unsuitable for driving the architectural effort. For example, ensure that the quality requirements are verifiable with specific conditions, quality criteria, and measurable thresholds.

Rationale: It is the requirements team's responsibility to make changes to the requirements. The architecture team may also not have write access to the requirements repository. The requirements may be baselined or the requirements specifications may be under configuration control and only modifiable by requirements engineers following requirements engineering procedures.

■ **Challenge difficult requirements.** When verifying the requirements, challenge the architecturally significant requirements that are very difficult to fulfill. Identify the requirements that are high cost, make it difficult to meet schedule deadlines, or have large associated technical risks.

Rationale: The stakeholders often do not understand the negative ramifications of their requirements on feasibility, cost, schedule, and risk. Once they understand the negative effect of such requirements, the stakeholders are often willing to either waive such requirements or change their thresholds to make it easier to engineer an acceptable architecture.

■ **Consider the impact of the architecture on the requirements.** The architecture team should determine the effects that their architecture decisions, inventions, trade-offs, and assumptions could have on the requirements. As stakeholders of the requirements, the members of the architecture teams should provide input to drive the engineering of new requirements as well as change requests and trouble reports to drive the modifications of existing requirements. The architects should also provide major input into prioritizing the requirements for scheduling of requirements implementation.

Rationale: Not only do the requirements drive the architecture, but the architecture will also often drive the requirements. The architecture will also drive the allocation of the requirements to architectural components and the derivation of new requirements that are appropriate for the components. The reuse of architectural components (such as OTS components) can be inappropriately limited if the requirements are unnecessarily restrictive. Architectural prototypes can also be used to help determine appropriate thresholds on quality requirements.

When customer requirements are given in the form of missions (such as business operations or military missions) to be supported, then it is the responsibility of the development organization to derive the system technical requirements. These requirements, especially those requirements derived for and allocated to lower levels in the system static physical decomposition hierarchy, are strongly influenced by the structure of that hierarchy (i.e., by the related architectural decisions, trade-offs, and assumptions).

Note that because systems often have multiple stakeholders with inconsistent needs and desires, it is often useful to use the concept of "satisficing" stakeholders as opposed to satisfying requirements. Satisficing the stakeholders means that every stakeholder does not get everything they want, but every stakeholder gets something with which they can be satisfied.

■ **Respect team boundaries and responsibilities.** Architecture teams should not engineer the architecturally significant requirements, and requirements teams should not engineer unnecessary architectural constraints.

Rationale: The requirements teams are responsible for engineering the requirements, including the architecturally significant requirements. It is not the responsibility of the architecture teams to engineer requirements. The primary responsibilities of the architecture team are to:

- Ensure that the requirements teams completely and properly engineer the architecturally significant requirements, for example, via close collaboration and by taking part in requirements evaluations (such as quality control inspections, walk-throughs, and reviews).
- Label the architecturally significant requirements engineered by the requirements team as such (e.g., by means of metadata in the requirements repository).

■ **If necessary, clarify relevant requirements with the stakeholders.** If necessary, architects should discuss the potentially relevant requirements with appropriate stakeholders, such as members of the relevant requirements and management teams to obtain clarification.

Rationale: For example, the architecture team often needs clarification to verify whether some of the relevant architecturally significant requirements are truly required and not unnecessary architecture constraints.

■ **Concentrate on the architecturally significant requirements.** When reviewing the requirements, the architecture team should concentrate on the architecturally significant requirements such as quality requirements, architectural constraints, and major functional feature sets or capabilities.

Rationale: Not all requirements are architecturally significant. Given their heavy workload, the architects do not have time to devote extra effort to comprehend and evaluate the requirements that do not have architectural ramifications.

■ **Remember that quality attributes can be architectural concerns too.** Consider quality attributes when grouping architecturally significant requirements into architectural concerns.

Rationale: Although quality characteristics (e.g., performance) are often architectural concerns, so can their quality attributes (such as event schedulability, jitter, latency, response time, and throughput) if they are sufficiently important to the architecture and sufficient attribute-specific quality requirements exist.

■ **Formally manage architectural risks.** As part of the risk management activity, identify architectural risks during the performance of this task. Add any architectural risks identified to the official endeavor risk management database.

Rationale: Architectural risks that are not identified or that are not added to the project risk management database when identified are much less likely to be effectively managed.

Note that risk is a "four letter word" to some managers, making it very difficult for the architects to identify architectural risks as actual risks. This fear of risk management is itself a risk, and a symptom of this dysfunction is the architects being explicitly or implicitly forced to speak of architectural issues or concerns when they obviously mean risks.

7.9 Pitfalls

Avoid the following common pitfalls associated with performing the *Identify the Architectural Drivers* task and mitigate their negative consequences when they occur:

- All requirements are architecturally significant.
- Well-engineered architecturally significant requirements are lacking.
- Architects rely excessively on functional requirements.
- The architects ignore the architecturally significant functional and process requirements.
- Specialty engineering requirements are misplaced.
- Unnecessary constraints are imposed on the architecture.
- Architects engineer architecturally significant requirements.
- Requirements lack relevant metadata.
- Architects fail to clarify architectural drivers.

The following express these pitfalls in more detail, describing their negative consequences and the steps the architects can use to mitigate them:

- **All requirements are architecturally significant.**
  - **Pitfall.** Although only a relatively small portion of the product and process requirements have a significant impact on the architecture, junior architects may mistakenly assume that all requirements are architecturally significant.
  - **Negative consequences.** This pitfall often causes the following negative consequences:
    - Architects easily become overwhelmed when they try to take all the requirements into account when making architectural decisions and trade-offs.
    - Architects waste time trying to determine the specific impact of requirements on the architecture when these requirements are not architecturally significant.
    - Architects do not take into account some of the architecturally significant requirements that get overlooked in the mass of requirements that are not architecturally significant.
    - Architectural engineering takes longer than necessary and exceeds its budget and schedule milestones.
    - The resulting architecture does not address all architectural concerns and does not meet all the architecturally significant requirements.
    - Architectural trade-offs between different quality characteristics and attributes are suboptimal.
  - **Mitigations.** Do the following to avoid or mitigate the negative consequences of this pitfall:
    - Provide the architects (especially the junior architects) with training in how to recognize architecturally significant requirements (such as quality requirements derived for and allocated to the part of the system architecture for which the architects are responsible).
    - Provide the architects with sample architectural quality cases that map standard architectural decisions (such as patterns and mechanisms) that support their achievement to architectural concerns (e.g., cohesive sets of quality requirements) [Firesmith et al., 2006].
    - Provide junior architects with mentoring and oversight by senior architects.
- **Well-engineered architecturally significant requirements are lacking.**
  - **Pitfall.** The requirements teams have not properly engineered the architecturally significant requirements, especially quality requirements and those derived requirements allocated to individual architectural components.

- **Negative consequences.** This pitfall often causes the following negative consequences:
  - It is very difficult to identify and understand the architectural drivers and associated architectural concerns.
  - There are no objective criteria against which to assess the architecture against its support for sufficient levels of quality characteristics and quality attributes. If performed anyway, such assessments tend to devolve into arguments between the assessors and architects as to whether the architecture is adequate. The architects may well feel that it is unfair to assess their architectures against criteria that are not officially requirements and did not drive the development of the architecture.
  - It is unlikely that the architecture will adequately meet the quality needs of the system stakeholders and therefore be acceptable to them.
- **Mitigations.** Do the following to avoid or mitigate the negative consequences of this pitfall:
  - Ensure that the proper engineering of the architecturally significant requirements is an explicit part of the requirements engineering method.
  - Ensure that the architecture teams closely collaborate with the requirements teams.
  - Ensure that an architect is part of each requirements engineering team to ensure that all the architecturally significant requirements are properly engineered.
  - Ensure that the members of the architecture teams properly verify the requirements.
  - Perform quality assessments of system architectures and their requirements (QUASAR) [Firesmith et al., 2006; Firesmith and Capell, 2007].

◼ **Architects rely excessively on functional requirements.**
- **Pitfall.** The architects incorrectly assume that the requirements engineers will adequately engineer the architecturally significant requirements. Both traditional (e.g., structured analysis) and more modern (e.g., use case modeling) system requirements engineering methods tend to emphasize functional requirements. Similarly, many traditional system architecture engineering methods and patterns tend to overemphasize the impact of the functional requirements on the system architecture, especially on the static system hierarchical aggregation structure, which traditionally is often decomposed in functionally cohesive subsystems.
- **Negative consequences.** This pitfall often causes the following negative consequences:
  - Because the system architecture should largely be driven by the associated quality requirements and concerns and because the acceptability and quality of the architecture depends on it fulfilling the quality requirements, this tendency to concentrate on functional feature sets can cause the architects to overlook significant quality characteristics and attributes.
  - Quality characteristics and attributes tend to be quasi-orthogonal to functional features, in that it is often a set of subfunctions or subsystems that must collaborate to implement the corresponding quality concerns (i.e., cohesive sets of quality requirements).
- **Mitigations.** Do the following to avoid or mitigate the negative consequences of this pitfall:
  - Incorporate techniques for engineering quality requirements in the overall requirements engineering method.
  - Train the requirements engineers in how to engineer quality requirements and the critical importance of these requirements to the architecture and acceptability of the system.
  - Include a system architect as a member of the requirements engineering team.

- Include system architects as members of the requirements verification (review) team.
- Evaluate the completeness and quality of the requirements with regard to quality characteristics and attributes. Have the requirements team produce requirements-level quality cases and present them to an assessment team, including at least one system architect [Firesmith et al., 2006; Firesmith and Capell, 2007].

■ **The architects ignore the architecturally significant functional and process requirements.**
- **Pitfall.** The architects ignore the architecturally significant functional* and process requirements when identifying architecturally significant requirements. For example, process requirements (such as those concerning cost, schedule, and use of existing manufacturing facilities) can greatly influence architectural decisions.
- **Negative consequences.** This pitfall often causes the following negative consequences:
  - The architecture may cause the system to fail to meet its associated functional or process requirements, thereby making the system unacceptable to its stakeholders.
- **Mitigations.** Do the following to avoid or mitigate the negative consequences of this pitfall:
  - Look for the cohesive feature sets that can be usefully implemented as an architectural component.
  - Ensure that the architecturally significant process requirements are inputs to the architecture engineering method.

■ **Specialty engineering requirements are misplaced.**
- **Pitfall.** Many specialty engineering requirements (i.e., quality requirements such as availability, reliability, safety, and security requirements) are not stored in the requirements repository and published in the requirements specifications. Instead, they are published in specialty engineering documents (such as security policies) that typically are not provided as direct input into the system architecture engineering effort. The architects may base their architecture on the requirements in the requirements repository and specifications, and ignore (or be unaware of) the specialty engineering requirements stored in specialty engineering documents. They may only become aware of these requirements after the architecture is largely completed and significant design and implementation have been based on it.
- **Negative consequences.** This pitfall often causes the following negative consequences:
  - These requirements are often overlooked as architectural drivers and do not influence the resulting architecture.
  - The architecture fails to adequately support the associated quality characteristics and quality attributes.
  - The system fails to meet these architecturally significant requirements.
  - The system is not acceptable to its stakeholders.
  - Major changes must be made to the architecture (and any associated design and implementation).
- **Mitigations.** Do the following to avoid or mitigate the negative consequences of this pitfall:
  - Ensure the early and close cooperation between the requirements, architecture, and specialty engineering teams.

---

* To an extent, this is the opposite situation as the preceding pitfall: architects rely excessively on functional requirements.

- Ensure that the specialty engineering teams store their architecturally significant requirements in the project requirements repository and specify them in the system requirements specifications. If this is not possible, then ensure that the architects are aware of, and have access to, whatever means are used to capture the architecturally significant specialty engineering requirements.

■ **Unnecessary constraints are imposed on the architecture.**
  - **Pitfall.** Although requirements engineers and other stakeholders are typically not qualified to make architectural decisions, they nevertheless unnecessarily specify architectural decisions as requirements (constraints).
  - **Negative consequences.** This pitfall often causes the following negative consequences:
    - Unnecessary architectural constraints inappropriately tie the architects' hands.
    - The architecture team produces suboptimal architectures because of the inappropriate engineering trade-offs these constraints force the architects to make.
    - Time is wasted in unproductive arguments and turf battles between the requirements engineers and architects.
  - **Mitigations.** Do the following to avoid or mitigate the negative consequences of this pitfall:
    - Train the requirements engineers in how to recognize and avoid unnecessary architectural constraints.
    - Ensure close collaboration between the requirements and architecture teams.
    - Include an architect as a member of the requirements engineering team.
    - Include architects as members of the requirements verification (e.g., review) team.
    - Evaluate the requirements with regard to unnecessary quality constraints. Have the requirements team produce requirements-level quality cases and present them to an assessment team, including at least one architect.

■ **Architects engineer architecturally significant requirements.**
  - **Pitfall.** The architects try to engineer architecturally significant requirements as a way of addressing the common lack of well-engineered architecturally significant requirements. Although architects are responsible for making architectural decisions, inventions, and trade-offs, they are not qualified to engineer requirements. They also do not have adequate access to the stakeholders.
  - **Negative consequences.** This pitfall often causes the following negative consequences:
    - The quality of these requirements created by the architecture teams is poor because the requirements do not have all the characteristics of good requirements. For example, the requirements may be ambiguous, incomplete, incorrect, and unverifiable.
    - These requirements may unnecessarily constrain future architects when the requirements are reused on other systems.
    - The requirements add unnecessary costs and schedule delays due to the cost of allocation, tracing, and verification (such as via testing), which would be considerably less if they had been documented as architectural decisions, inventions, and trade-offs
    - The thresholds on quality measurement scales of quality requirements may not match the needs of the stakeholders because the architects may not have access to the stakeholders and the architects have not been trained in the development of proper quality requirements.*

---

* Unfortunately, the same can also be said about far too many requirements engineers.

- **Mitigations.** Do the following to avoid or mitigate the negative consequences of this pitfall:
  - Have the architects collaborate closely with the requirements engineers to properly divide up their duties (i.e., the proper separation of requirements engineering tasks and architecture engineering tasks).
  - Verify the quality of all identified architecturally significant requirements and not just the functional requirements. For example, hold QUASAR assessments of the requirements quality cases produced by the requirements engineers to ensure that the architecturally significant requirements are produced early enough to drive the engineering of the architecture and provide objective criteria against which to assess the architecture [Firesmith and Capell, 2007].
  - Have architects be members of the requirements verification teams to ensure that the requirements are complete (i.e., quality requirements with feasible and correct thresholds on appropriate associated quality measurement scales).
- **Requirements lack relevant metadata.**
  - **Pitfall.** Architects, process engineers, technical leaders, and managers incorrectly assume that the architecturally significant requirements are properly identified as such. Most requirements engineers are not qualified to determine the architectural ramifications of their requirements and thus do not tend to label the architecturally significant requirements as such within the requirements repositories. It is typically up to the system architects to provide requirements metadata to label the various types of architecturally significant requirements and to group these requirements into architectural concerns.
  - **Negative consequences.** This pitfall often causes the following negative consequences:
    - When relevant requirements metadata does not exist, it is often difficult or impossible to query a requirements database for all architecturally significant requirements or all requirements related to specific quality characteristics.
    - Architectural concerns are incomplete, missing some of the associated requirements.
    - The architects cannot ensure that their architecture sufficiently supports all the architecturally significant requirements.
    - The architects cannot take all relevant requirements into account when making architectural trade-offs between competing quality characteristics.
    - Unless properly documented, the trade-offs, assumptions, and rationales may be lost when architects leave the project.
  - **Mitigations.** Do the following to avoid or mitigate the negative consequences of this pitfall:
    - Incorporate the identification of architecturally significant requirements as an integral part of the architecture engineering method and as an important specific responsibility of the architecture teams.
    - Make the evaluation of the identification of architecturally significant requirements as metadata within the requirements repository an explicit part of the verification of the completeness and correctness of requirements.
    - Ensure that the architects have write access to the appropriate metadata in the requirements repository.
    - Provide junior architects with mentoring and oversight by senior architects.
- **Architects fail to clarify architectural drivers.**
  - **Pitfall.** The system architects do not properly collaborate with specialty engineering subject matter experts, requirements engineers, and other stakeholders when they lack

adequate training and expertise to understand the architectural ramifications of the architectural drivers and concerns.

- **Negative consequences.** This pitfall often causes the following negative consequences:
  - The inexperienced architects make poor architectural decisions, inventions, trade-offs, and assumptions.
  - The inexperienced architects engineer incomplete architectural representations.
  - The inexperienced architects lose credibility with management, requirements engineers, designers, implementers, and testers.
  - The resulting architecture may cause the system to fail to meet its architecturally significant requirements.
  - Project budget and schedule constraints are not met.
  - The system is not acceptable to its stakeholders.
- **Mitigations.** Do the following to avoid or mitigate the negative consequences of this pitfall:
  - Hire qualified architects.
  - Provide architects with training in architecture engineering.
  - Apprentice junior architects to master architects.

7.10 Summary

During the *Identify the Architectural Drivers* task, the architecture teams identify the architecturally significant requirements that have been derived for and/or allocated to their system or architectural component and categorize them into a set of architectural concerns.

7.10.1 Steps

When instantiating the *Identify the Architectural Drivers* task, the following steps may be included, modified, or tailored out:

1. Collaborate to help engineer the architecturally significant requirements.
2. Identify and label the architecturally significant requirements.
3. Verify the potentially relevant requirements.
4. Collaborate to fix requirements defects.
5. Identify the architectural concerns.
6. Evaluate and iterate the architectural concerns.
7. Identify any new architectural risks and opportunities.

7.10.2 Work Products

When instantiating the *Identify the Architectural Drivers* task, the following work products may be included, modified, or tailored out:

- Requirements recommendations
- Architectural concerns
- Requirements metadata
- Architectural risks and opportunities

7.10.3 Guidelines

Use the following guidelines (and associated rationales) when performing the *Identify the Architectural Drivers* task:

- Collaborate closely with the requirements team.
- Notify the requirements team(s) of relevant requirements defects.
- Challenge difficult requirements.
- Consider the impact of the architecture on the requirements.
- Respect team boundaries and responsibilities.
- If necessary, clarify relevant requirements with the stakeholders.
- Concentrate on the architecturally significant requirements.
- Remember that quality attributes can be architectural concerns too.
- Formally manage architectural risks.

7.10.4 Pitfalls

Avoid the following common pitfalls associated with performing the *Identify the Architectural Drivers* task and mitigate their negative consequences when they occur:

- All requirements are architecturally significant.
- Well-engineered architecturally significant requirements are lacking.
- Architects rely excessively on functional requirements.
- The architects ignore the architecturally significant functional and process requirements.
- Specialty engineering requirements are misplaced.
- Unnecessary constraints are imposed on the architecture.
- Architects engineer architecturally significant requirements.
- Requirements lack relevant metadata.
- Architects fail to clarify architectural drivers.

Chapter 8

Chapter 8

Task 3: Create the First Versions of the Most Important Architectural Models

8.1 Introduction

During this MFESA task, the system or subsystem architecture team creates the first versions of the most important architectural models. Typically this task primarily creates a consistent set of partial draft logical and physical, static and dynamic models of the system or subsystem based on the architectural drivers, the associated architectural concerns, and the opportunities for reuse (Figure 8.1).

When taken together, these initial partial draft architectural models:

- Provide the most important views of the architecture
- Capture the most important candidate architectural decisions, inventions, and trade-offs
- Adequately address the architectural concerns

The architectural team begins by selecting the types of logical and physical architectural views to create for the architectural element. For each of these architectural views, the team starts creating one or more initial partial draft architectural models. The team selects a set of potential architectural styles, patterns, and mechanisms for meeting the derived and allocated architectural drivers and concerns. The team captures these decisions within the appropriate models by selecting candidate model elements and their associated relationships. While producing models for physical

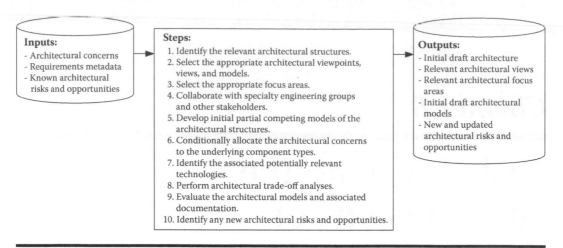

Figure 8.1   Summary of Task 3 inputs, steps, and outputs.

views, the team also selects potentially reusable architectural components, evaluates them within the context of the models, and then determines potential ways to address any associated limitations or incompatibilities. The team identifies, evaluates, and selects potential technologies for use within the architectural element. Finally, the team evaluates the models and associated documentation to ensure that, when taken together, the resulting set of potential architectural decisions, inventions, and trade-offs sufficiently support the architectural drivers and concerns derived for and allocated to the architectural element being architected.

Note that the required staffing, budget, and schedule cannot be accurately estimated until a minimum amount of this task is performed, although this estimation occurs as part of Task 1, *Plan and Resource the Architecture Engineering Effort.* This is not a contradiction because MFESA does not assume a waterfall sequencing between its tasks. Instead, the MFESA tasks are typically performed in a recursively incremental, iterative, concurrent, and time-boxed manner. As illustrated in Figure 4.6, there is also significant overlap between the tasks for any one incremental build or release.

As illustrated from top to bottom on the left side of Figure 8.2, architecturally significant requirements are grouped into architectural concerns, which are partially implemented in architectural structures and represented by architectural focus areas. Each architectural structure to be modeled is represented by an appropriate corresponding logical or physical and static or dynamic architectural viewpoint. One or more views are instantiated for each viewpoint, and one or more models are created for each view. Architectural focus areas include the relevant parts of the relevant models that implement the associated architectural concern. The right side of Figure 8.2 mirrors the left side using confidentiality as a specific example concern, which consists of a set of confidentiality security requirements. Confidentiality is partially implemented by three structures and represented by the confidentiality focus area. Each of these confidentiality-related system structures has its own viewpoint, view, and model implemented using appropriate diagrams.

This task is a reusable MFESA method component and as such is intended to be tailorable to meet the specific needs of the endeavor. Therefore, it is to be expected that at least some of the task's objectives, preconditions, inputs, steps, postconditions, work products, guidelines, and pitfalls will be tailored during the creation of the endeavor-specific system architecture engineering method.

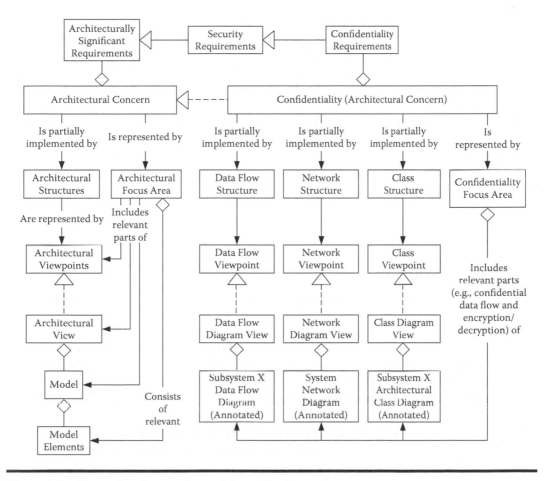

Figure 8.2   General and example model creation from concerns.

Tailoring of a method component includes adding missing content, modifying existing content, or removing existing content that is inappropriate, unnecessary, or not cost-effective to perform.

As illustrated in Figure 4.8, this task is typically performed concurrently with the next two tasks: Task 4, *Identify Opportunities for the Reuse of Architectural Elements,* and Task 5, *Create the Candidate Architectural Visions.* This task produces input that they use and they produce inputs that this task uses.

8.2 Goal and Objectives

The *Create the First Versions of the Most Important Architectural Models* task has the following overall goal:

- Create an initial set of partial draft architectural models of the system architecture.

The typical subordinate objectives of this task are to:

- Perform adequate architectural modeling* to sufficiently support:
  - Identification of opportunities for the reuse of architectural elements (MFESA Task 4). That is, the models should be sufficiently complete to identify such opportunities.
  - Creation of candidate architectural visions (MFESA Task 5). That is, sufficient variants of these models should exist to support the creation of a sufficient number of architectural visions.
- Capture the most important candidate system architectural decisions, inventions, and trade-offs.
- Provide the most important views and focus areas of the system architecture.
- Ensure that the candidate architectural decisions sufficiently support the relevant architectural concerns.
- Provide a foundation of architectural models from which to create a set of competing candidate architectural visions.

8.3 Preconditions

The *Create the First Versions of the Most Important Architectural Models* task can typically begin when the following preconditions have been met:

- **Architecturally significant requirements are identified.** The architecture team has identified a significant number of the most important of the relevant architecturally significant product and process requirements.
- **Architectural concerns are determined.** The architecture team has grouped these architecturally significant requirements into architectural concerns.
- **Requirements metadata identifies architecturally significant requirements.** The requirements repository has been updated with metadata to identify those requirements that are architecturally significant.

8.4 Inputs

The *Create the First Versions of the Most Important Architectural Models* task typically uses the following inputs:

- **Architectural concerns:** a set of architectural concerns for the system or the architectural component being architected that includes all relevant architecturally significant product and process requirements.
- **Relevant requirements metadata:** all the metadata describing the relevant architecturally significant product and process requirements.

---

* This objective is a major reason why this task should be performed iteratively, recursively incrementally, and concurrently with MFESA Tasks 4 and 5. Note that a fine line exists between performing sufficient architectural modeling to support the next two tasks without wasting too much time producing models that will be not be used because they are associated with architectural visions that will not be selected. Because this determination cannot be made perfectly before the competing architectural visions exist, the architects must merely do their best based on their training and experience, knowing that they will often perform somewhat too little or too much modeling during this task.

■ **Known architectural risks and opportunities:** known architectural risks in the endeavor risk repository, as well as known architectural opportunities.

8.5 Steps

During the *Create the First Versions of the Most Important Architectural Models* task, members of the relevant architecture team collaborates to perform the following steps in an incremental, iterative, concurrent, and time-boxed manner:

1. **Identify the relevant architectural structures.** Identify the types of architectural structures comprising the system or subsystem architecture based on the architectural concerns as well as the opportunities for architectural reuse. This includes identifying all appropriate logical and physical as well as static and dynamic structures that it will be important, cost-effective, and practical to model.

2. **Select the appropriate architectural viewpoints, views, and models.** For each architectural structure to be modeled, select the appropriate corresponding logical or physical and static or dynamic architectural viewpoint. Select one or more views for each viewpoint, and select one or more models for each view.

3. **Select the appropriate focus areas.** For each architectural concern, determine if an associated architectural focus area should be produced.

4. **Collaborate with specialty engineering groups and other stakeholders.** Collaborate with specialty engineering groups such as performance, reliability, safety, and security, as well as other stakeholders, to ensure that the initial architectural decisions, inventions, and trade-offs properly address their concerns.

5. **Develop initial partial competing models of the architectural structures.** For each architectural view of each logical and physical static and dynamic architectural structure, develop one or more initial partial draft competing architectural models. Determine the primary component element (types) of the model and the relationships between these model elements. Note that multiple competing versions of all or part of the same model may be created for inclusion in competing architectural visions.

6. **Conditionally allocate the architectural concerns to the underlying component types.** Conditionally determine the gross allocation of the architectural concerns to the underlying types of components such as data, facilities, firmware, hardware, manual processes, software, and subsystems. Determine the ramifications of this conditional allocation. Because of its major effect on the architectural concerns (fulfillment of the architecturally significant requirements), determine the effects of the use of software on the architectural decisions, inventions, trade-offs, and assumptions.

7. **Identify the associated potentially relevant technologies.** Identify the potentially relevant technologies associated with the underlying types of components. For example, this could include:
   - **Interoperability technologies:**
     - Mechanical linkages versus hydraulic linkages versus fly-by-wire
   - **Hardware technologies:**
     - Materials technologies such as metals versus carbon composites based on their ramifications on such characteristics as strength, weight, reliability, and cost

- Capabilities such as computer processor speed and memory, and network bandwidth
- **Software technologies:**
  - Software design paradigm such as object orientation (OO)
  - Types of databases such as relational or object-oriented
  - Type(s) of operating system* such as real-time OS
  - Middleware technologies such as CORBA, .NET, and J2EE
  - Programming language(s) such as C++, C#, and Java†

8. **Perform architectural trade-off analyses.** Use the architectural models and architectural concerns as inputs to perform architectural trade-off analyses between the different competing architectural concerns to determine how well the architectural models (and associated structures) support the architectural concerns. Update the models (and structures) based on the results of the trade-off analysis. Notify the requirements team if the competing architectural concerns are determined to contain inconsistent or infeasible requirements.

   Note that this step is actually performed in an ongoing concurrent manner, both with other steps within this task and with other tasks.

9. **Evaluate the architecture models and associated documentation.** The architecture team evaluates the architectural models and other relevant documentation to ensure that, when taken together, the resulting set of potential architectural decisions, inventions, and trade-offs sufficiently support the architectural drivers and concerns derived for and allocated to the architectural element being architected. Evaluate the reuse decisions made, the architectural trade-offs made, and the appropriateness of the technologies selected.

10. **Identify any new architectural risks and opportunities.** Concurrently with the performance of the previous steps, identify any new architectural risks and opportunities associated with the candidate reusable architectural elements that the new architecture should mitigate or take advantage of. As appropriate, update any previously known architectural risks and opportunities.

8.6 Postconditions

The *Create the First Versions of the Most Important Architectural Models* task is complete for the current subsystem (or architectural element) when the following postconditions hold:

- **Candidate potential architectural vision components are selected.** The subsystem architecture team has identified candidate potential architectural vision components that credibly support one or more architecturally significant relevant product or process requirements comprising each relevant architectural concern.
- **Representations of the architectural vision components exist.** A set of draft architectural representations exist that properly describe the selected candidate potential architectural vision components.

---

* The selection of the actual OS can be postponed until the *Complete the Architecture and Its Representations* task.

† This may include the determination of whether to use the entire programming language or only a safe or secure subset of the language.

- **Architectural risks and opportunities are managed.** Any newly identified credible architectural risks have been analyzed and documented in the endeavor risk management repository. Any newly identified architectural opportunities have also been analyzed and documented.

8.7 Work Products

The *Create the First Versions of the Most Important Architectural Models* task produces the following work products:

- **Initial draft architecture:** a list of potentially key architectural decisions, inventions, trade-offs, and assumptions with associated rationales.
- **Relevant architectural views:** a list of relevant architectural views of the logical and physical static and dynamic architectural structures.
- **Relevant architectural focus areas:** a list of relevant focus areas addressing specific architectural concerns.
- **Initial draft architectural models:** a set of the initial partial draft architectural models of the architectural structures.
- **New and updated architectural risks and opportunities:** any new or modified architectural risks documented in the endeavor risk repository as a part of endeavor risk management activity. This also includes an updated list of architectural opportunities that the architecture teams may wish to take advantage of in the future.

8.8 Guidelines

Use the following guidelines (and associated rationales) when performing the *Create the First Versions of the Most Important Architectural Models* task:

- Perform architectural trade-off analysis.
- Reuse architectural principles, heuristics, styles, patterns, vision components, and metaphors.
- Use a recursively incremental, iterative, parallel, and time-boxed development cycle.
- Begin developing logical models before beginning to develop physical models.
- Do not overemphasize the physical decomposition hierarchy.
- Use explicitly documented system-partitioning criteria.
- Concentrate on the interfaces.
- Model concurrency.
- Consider the impact of hardware decisions on usability and software.
- Consider human limitations when allocating system functionality to manual procedures.
- Do not start from scratch.
- Formally manage architectural risks.

The following descriptions express these guidelines in more detail and provide their associated rationales:

- **Perform architectural trade-off analysis.** It is critical that the architects perform architectural trade-off analysis between the different architectural concerns. This step actually is performed in an ongoing parallel manner, both with other steps within this task and with other tasks.

Rationale: The architecture cannot be locally optimized for each competing architectural concern. Rather, the architecture must be globally optimized to sufficiently support all of the architectural concerns.

■ **Reuse architectural principles, heuristics, styles, patterns, vision components, and metaphors.** When creating architectural models to solve the problems associated with an architectural concern, reuse architectural:

- *Principles* such as abstraction,* low coupling, high cohesion, modularity, and separation of concerns
- *Heuristics* such as:
  - Controlling the number of elements addressed in a particular model representation to that which is easily understood by a reader, such as the "7±2" heuristic
  - Emphasizing one particular physical factor in a model representation, such as vibration environment
- *Styles, patterns, mechanisms,* and *metaphors*
- *Vision components* successfully used in the past on similar systems†
- *Inventions*‡

Rationale: Reuse of architectural principles, heuristics, styles, patterns, vision components, and metaphors has proven effective in efficiently producing effective models.

Note that reusing *vision* components (i.e., critical design decisions, inventions, trade-offs, assumptions, and rationales) is not the same thing as reusing *architectural* components (i.e., actual subsystems, hardware components, and software components). Because they are at a higher level of abstraction, vision components are inherently more reusable than the less generic architectural components, which may not have been developed with reuse in mind.

■ **Use a recursively incremental, iterative, parallel, and time-boxed development cycle.** Perform this task iteratively, incrementally, and in parallel with the next two tasks:

- Identify candidate architectural visions.
- Select the architectural vision.

Rationale: Most systems have architectures that are so large that they are best developed in a recursively incremental manner. Most systems have architectures that are so large and complex that they must be developed incrementally in order to identify and remove architectural defects. The output of this task can be useful as inputs to the other tasks, and their outputs can be useful as inputs to this task. Performing these tasks iteratively, incrementally, and in parallel helps keep their work products consistent. Using time boxes to specify architectural milestones helps ensure that architects do not become trapped in an architectural

---

* Although logical system architectures have historically relied heavily on functional abstraction to produce a functionally decomposed logical architecture, other forms of abstraction (such as object abstraction, data abstraction, agent abstraction, and process abstraction) should also be considered in order to decrease coupling, increase cohesion, provide better modularity, better separate concerns, and the like.

† Care should be taken to avoid simply reusing the architectures of previous systems because the older architectures may not scale up to the size, complexity, increased functionality, and new technology that can be used by the current system.

‡ Developing the architecture of new, larger, more complex systems incorporating increased functionality and new technology often requires innovation and invention to better solve the architectural concerns. Older architectural vision components may not scale up.

analysis paralysis and that sufficient architectural representations exist in a timely manner to support system design, implementation, integration, and testing.

■ **Begin developing logical models before beginning to develop physical models.** While making architectural decisions, inventions, trade-offs, and assumptions, the architects should typically *begin* engineering the logical models of the system architecture's logical structures before they *begin* engineering the physical models of the architecture's physical structures. However, the architects should not wait until all logical models are completed to begin building physical models. The architects should develop static and dynamic models concurrently.

Rationale: Several architecture engineering standards and methods refer to logical and physical "architectures." Conversely, MFESA takes the position that a system (or subsystem) has only a single architecture that typically has multiple logical and physical structures that will be modeled using appropriate logical and physical architectural models. Similarly, a system typically has both static and dynamic structures requiring both static and dynamic models. Thus, a system tends to have logical static structures, logical dynamic structures, physical static structures, and physical dynamic structures.

Especially during the "greenfield" development of new systems, the development of logical architectural structures tends to drive the development of physical structures. Depending on the modeling techniques used, dynamic modeling of dynamic structures either precedes (functional decomposition) or follows (object orientation) static models of static structures. The structures associated with subsystems at higher tiers in the system aggregation hierarchy tend to drive and constrain the structures primarily associated with lower-tier subsystems. Finally, many of the logical and physical static and dynamic structures cut across subsystem boundaries.

■ **Do not overemphasize the physical decomposition hierarchy.** It is unwise to only consider the physical configuration structure of a system hierarchically decomposed into subsystems, regardless of the importance of Conway's law.

Rationale: There are different kinds of views of the multiple structures in a system's architecture. Because the architectural elements of the different structures do not map one-to-one, there will need to be multiple types of architecture "teams," although the same architect may be a member of more than one team.

■ **Use explicitly documented system-partitioning criteria**. Develop a consistent set of appropriate criteria for partitioning the system into its architectural components.* Document these criteria and their rationale. Examine these criteria to ferret out implicit assumptions that have been made by the systems engineering team regarding the requirements or architecture. Treat all such assumptions critically. The criteria employed at any given point in the architecture development process should be selected from a more complete set of potential criteria, such as:

  – **Abstraction.** The system is decomposed into a hierarchy of architectural components, each of which captures or represents a single concept:

    • Functional decomposition partitions the system into lower-level architectural components, each of which captures a single system function or functionally cohesive set of subfunctions.

---

* A significant challenge facing system architects is that there exists a large number of useful partitioning criteria, several of which are incompatible with each other. System architects must use their training, experience, and expertise to determine when to use the different partitioning criteria.

- Data decomposition partitions the system into lower-level architectural components, each of which creates, reads, updates, and deletes a cohesive set or type of data.
- Object-oriented decomposition partitions the system into lower-level architectural components, each of which represents a class of objects.
- **Information hiding.** The system is decomposed into a hierarchy of architectural components, each of which hides design and implementation details behind a well-defined interface.
- **Coupling.** The system is decomposed into a hierarchy of architectural components, each of which is loosely coupled to the others (i.e., internal coupling is greater than external coupling). (See "Information hiding" above.)
- **Cohesion.** The system is decomposed into a hierarchy of architectural components, each of which is highly cohesive. (See "Abstraction" above.)
- **Complexity.** The system is decomposed into a hierarchy of architectural components that minimizes the complexity of the individual architectural components and the complexity of the interfaces between the architectural components.
- **Quality requirements.** The system is decomposed into a hierarchy of architectural components that supports the allocation, derivation, and achievement of the associated quality requirements.
- **Reuse of OTS components.** The system is decomposed into a hierarchy of architectural components so that individual architectural components can be reused. This includes reusable components from organizational reuse repositories, COTS components obtained from vendors, MOTS components obtained from the military, GOTS components obtained from the government, and open-source components.
- **Software feasibility.** The system is decomposed into a hierarchy of architectural components so that it is feasible (i.e., technically as well as within budget and schedule constraints) for software to enable architectural component intra-operability. While technical and programmatic feasibility is critical for all types of interfaces (including data, energy, and material), ensuring that software can be used to achieve interoperability and *intra*-operability is critical because software is often the primary glue used to connect software-intensive architectural units. Note that this involves more than simple message passing and the defining of message formats (syntax) and protocols. There are many types of intra-operability. The interfaces must deal with semantics and support for quality and functional requirements, which typically involve cutting across multiple architectural interfaces of multiple components.
- **Expertise.** The system is decomposed according to Conway's law into a hierarchy of architectural components so that the individual components can be developed by separate teams or organizations having the relevant experience and expertise in the associated application domain and technologies. For example, a weapons subsystem is assigned to a team having weapons expertise, while a communications subsystem is assigned to a team having communications expertise.
- **Contractual boundaries.** The system is decomposed into a hierarchy of architectural components so that individual architectural components can be acquired via contract from external organizations. For example, the acquisition organization can acquire a network of systems from various development organizations and a prime contractor (system integrator) can acquire architectural components from subcontractors.
- **Maintenance.** The system is decomposed into a hierarchy of architectural components so that separate components (such as line-replaceable units) can be independently replaced during maintenance.

- **Physical separation.** A physically distributed system is decomposed into a hierarchy of architectural components so that separate components can be physically separated.
- **Time separation.** The system is decomposed into a hierarchy of architectural components so that components segregate behavior that will operate at different rates, degrees of urgency, or different latency requirements. This is primarily an issue with the software components of real-time systems, and it often involves the use of a cyclic scheduler that slices time into different "binary progression rate groups" operating, for example, at 1 Hz, 2 Hz, 4 Hz, 8 Hz, 16 Hz, 32 Hz, and so forth.
- **Safety.** The system is decomposed into a hierarchy of architectural components in a way that separates components of high safety criticality from components of low safety criticality and minimizes the size of the high safety criticality components. By limiting the size of safety-critical components, safety certification is simplified and the cost of developing these components to a higher standard is lowered.
- **Security.** The system is decomposed into a hierarchy of architectural components in a way that separates components of high security criticality from components of low security criticality and minimizes the size of the high security criticality components. For example, the decomposition of the system separates architectural components that process data having different levels of security classifications. By limiting the size of security-critical components, security certification is simplified and the cost of developing these components to a higher standard is lowered.
- **Certification and accreditation.** The system is decomposed into a hierarchy of architectural components in a way that separates components that require certification or accreditation by external agencies from those that do not.
  Rationale: When used appropriately, these criteria have proven very effective at producing good system decompositions in a cost-effective manner.
■ **Concentrate on the interfaces.** When modeling systems, architects should ensure that the architecture adequately supports both interoperability and intra-operability requirements. Architects should properly engineer both the external *and* the internal interfaces.
  Rationale: Because every system interfaces with people and other systems within its environment (i.e., the world outside the system boundary), every system can be viewed as a part of a larger system of systems. Every system can also be viewed as a system of systems because, by definition, it can be decomposed into its architectural components (i.e., its subsystems). The system's emergent properties and behaviors largely result from the interactions between the systems architectural components. Because they must solve global problems, architects are responsible for cross-component integrity and solving problems that local designers and integrated product teams (IPTs) are less able and less likely to address. Many architectural defects and risks occur at the interfaces and are not identified until the architectural components are integrated, undergo integration testing, or the system is fielded and begins interoperating with other systems. Concentrating on interfaces early helps avoid problems later on when they are more difficult and costly to correct.
■ **Model concurrency.** The architects should model system concurrency. This should include modeling system start-up and shut-down.
  Rationale: Modeling system concurrency helps avoid concurrency failures such as deadlock, live lock, race conditions, and priority inversion.
  Because concurrency failures commonly occur during system start-up and shut-down, modeling concurrency at start-up and shut-down helps to ensure that executable architectural components start up, interact, and shut down in the proper order.

For example, communicating executable architectural components (such as processors) may start up in the wrong order (i.e., a client component before a server component on which it depends). When this happens, the client component cannot communicate with the server component and both early data transfers and service requests may be lost unless the client component buffers the data and postpones the service requests until the server component has successfully started up.

■ **Consider the impact of hardware decisions on usability and software.** When making hardware decisions, consider their potential negative impact on the usability of architectural components as well as how the hardware will affect software. Consider how the hardware will affect both software allocated to the hardware component as well as software allocated to any interfacing architectural components.

Rationale: Decomposing a system into subsystems to minimize hardware development and manufacturing costs tends to produce architectural components having incompatible human interfaces and software infrastructures. This, in turn, tends toward the production of poorer software architectures, decreased usability, and significantly increased development, operations, and maintenance costs.

■ **Consider human limitations when allocating system functionality to manual procedures.** Use human factors engineering when *provisionally* partitioning the system and allocating system functionality to manual procedures.* Because alternative versions of the same models may be allocated to alternative architectural visions, such partitioning may vary from one architectural vision to another. As a minimum, consider:
  - Limitations on human abilities such as likely human error rates and response times
  - Safety requirements, safety analysis, human actions with safety ramifications, and the potential consequences of human failures
  - How human–machine interactions will influence the achievement of quality requirements such as performance, availability, reliability, and the like

Rationale: Compared with hardware and software, humans are extremely slow, which can prevent the system from meeting its performance requirements. Because humans can be very unreliable, their errors and failures can also have very negative impacts on a system's ability to meet its reliability and availability requirements.

■ **Do not start from scratch.** Do not automatically assume "greenfield" development and start developing the architectural models and candidate visions from scratch.

Rationale: It is important to avoid wasting time and effort reinventing architectures from scratch, especially when few architects have the luxury of being a part of "greenfield" development. There are typically ample sources of reusable architectural information to consider.

■ **Formally manage architectural risks.** As part of the risk management activity, identify architectural risks during the performance of this task. Add any architectural risks identified to the official endeavor risk management database.

Rationale: Architectural risks that are not identified or that are not added to the project risk management database when identified are much less likely to be effectively managed.

Note that management fear of risk management is itself a risk, and a symptom of this dysfunction is the architects being explicitly or implicitly forced to speak of architectural issues or concerns when they obviously mean risks.

---

* Note that this allocation of system functionality is provisional during this task. The allocation of functionality to hardware, software, and manual procedures will need to be revisited during the *Complete the Architecture and Its Representations* task.

8.9 Pitfalls

Avoid the following common pitfalls associated with performing the *Create the First Versions of the Most Important Architectural Models* task and mitigate their negative consequences when they occur:

- The architects succumb to analysis paralysis.
- The architects engineer too few architectural models.
- The architects engineer inappropriate models and views.
- The architects construct views but no focus areas.
- Some stakeholders believe that the models are the architecture.
- Inconsistencies exist between models, views, and focus areas.
- The architects use inappropriate architectural patterns.
- System decomposition is performed by the acquisition organization.

The following express these pitfalls in more detail, describing their negative consequences and the steps the architects can use to mitigate them:

- **The architects succumb to analysis paralysis.**
  - **Pitfall.** In their zeal to replace vague textual documents and architectural "cartoons" with more rigorous architectural models, the architecture teams suffer from analysis paralysis by modeling everything in ever-increasing detail, eventually evolving the architectural models into design and implementation models.
  - **Negative consequences.** This pitfall often causes the following negative consequences:
    - The architects confuse architectural models with the architecture.
    - A turf battle occurs between the architects and the designers.
    - Development costs increase and schedule delays occur.
  - **Mitigations.** Do the following to avoid or mitigate the negative consequences of this pitfall:
    - Incorporate model completion criteria into the system architecture engineering method.
    - Time-box the architecture engineering effort.
    - Provide training in modeling the architecture.
    - Collaborate with experienced architects or architecture consultants.
- **The architects engineer too few architectural models.**
  - **Pitfall.** The architecture teams engineer too few architectural models to sufficiently support:
    - Identification of opportunities for the reuse of architectural elements (Task 4). That is, the models are insufficiently complete to identify such opportunities.
    - Creation of candidate architectural visions (Task 5). That is, insufficient variants of these models exist to support the creation of a sufficient number of architectural visions.

    Remember that the goal of this task is *not* to create a complete set of completed models. Rather, it is to create only the first versions of the most important models so that the architects can identify opportunities for reuse (Task 4) and create candidate architectural visions (Task 5). Because of the goal of this task, it is relatively likely for this pitfall to occur.

- **Negative consequences.** This pitfall often causes the following negative consequences:
  - Some important models of the architecture are missing or are too incomplete.
  - Some important views of some architectural structures are missing or are too incomplete.
  - The architects will not make some important architectural decisions, inventions, trade-offs, and assumptions.
  - The architects will not engineer an adequate set of architectural representations.
  - The probability increases that the architecture will not support the achievement of all architecturally significant requirements.
  - The probability increases that the system will not meet all of its architecturally significant requirements.
  - The probability increases that the architecture will not be acceptable to all of its stakeholders.
- **Mitigations.** Do the following to avoid or mitigate the negative consequences of this pitfall:
  - Do not tailor the first two steps out of this task in the project system architecture engineering method.
  - Continue modeling until the objectives of this task are achieved.
  - Perform this task concurrently with Tasks 4 and 5.
  - Provide training and mentoring in the selection of appropriate architectural views, viewpoints, and models.

■ **The architects engineer inappropriate models and views**.
- **Pitfall.** The architecture teams engineer inappropriate models and views, possibly because they are popular, expected, or demanded by stakeholders rather than engineering those models and views that are most appropriate, useful, and cost-effective.
- **Negative consequences.** This pitfall often causes the following negative consequences:
  - Important models and views are not engineered.
  - Excessive unnecessary effort producing unnecessary models and views increases development cost and creates schedule delays.
- **Mitigations.** Do the following to avoid or mitigate the negative consequences of this pitfall:
  - Incorporate the appropriate models and views into the system architecture engineering method. Note that this may prove difficult as the appropriate models and views may vary, depending on the architectural component being modeled, and may not be known at the beginning of the architecture engineering effort when the appropriate models and views are first being selected.
  - Incorporate model-specific completion criteria into the system architecture engineering method.
  - Provide training in modeling the architecture.
  - Collaborate with experienced architects or architecture consultants.

■ **The architects construct views but no focus areas.**
- **Pitfall.** The architects only produce system- or subsystem-wide views during modeling and thereby fail to capture important architecture-concern-specific focus areas. For example, it is difficult to determine, given a series of views, exactly which parts of which views are relevant to achieve the necessary threshold specified for any one quality characteristic or quality attribute. Thus, the evidentiary part of any architectural quality case may involve only a few specific nodes and arcs on specific model

diagrams as well as specific parts of white papers and other architectural representations [Firesmith et al., 2006].
- **Negative consequence.** This pitfall often causes the following negative consequence:
  • The resulting architectural representations may not be adequate for all stakeholders.
- **Mitigations.** Do the following to avoid or mitigate the negative consequences of this pitfall:
  • Ensure that architecture modeling includes both views and focus areas by including both in the project system architecture engineering method.
  • Verify the existence of both during model evaluations.

■ **Some stakeholders believe that the models are the architecture.**
- **Pitfall.** The stakeholders (and even the architects themselves) equate models with the architecture. This is incorrect in two ways. First, it confuses the architecture with representations of the architecture. Second, it also ignores the fact that there are numerous architectural representations beyond the models, and many important architectural decisions, inventions, and trade-offs are better captured in native language textual descriptions and informal architectural "cartoons" rather than in more formally defined graphical and textual models.

  Because the goal of this task is to construct models, this task can reinforce the mistaken equating of models with the architecture. It is important for the architects to remember to also develop the aspects of the architecture that are either not easy to capture in architectural models or that can be very difficult to locate within extremely large and complex models.
- **Negative consequences.** This pitfall often causes the following negative consequences:
  • The architectural representations are incomplete.
  • Important architectural decisions, inventions, trade-offs, and assumptions are not made.
  • Rationales are not documented, making maintenance more difficult.
  • Architectural representations are more difficult for the less technically trained stakeholders to understand and validate.
- **Mitigations.** Do the following to avoid or mitigate the negative consequences of this pitfall:
  • Provide training that covers the ontology of concepts and terminology of architecture engineering, including both the distinction between architecture and architectural representations and the different kinds of architectural representations.
  • Emphasize the need to produce all appropriate architectural representations that describe the architecture in the system architecture engineering method.
  • Require the creation of architectural quality cases to capture in a convenient manner many of the critical architectural decisions, inventions, trade-offs, assumptions, and rationales associated with meeting quality requirements and supporting the system's ability to achieve sufficient levels of important quality characteristics and attributes.

■ **Inconsistencies exist between models, views, and focus areas.***
- **Pitfall.** The architects (and especially their management) think that it is easy to create and maintain a consistent set of models, views, and focus areas. Inconsistencies develop

---

* This is restricted to the consistent set of models to be associated with a single architectural vision. Variants of models to be associated with multiple inconsistent architectural visions often are (and should be) inconsistent.

within and between different architectural representations. Consistency is especially difficult, given the fact that tools provide very little support for consistency between the different models, and essentially no support for consistency between the models and other textual information in the form of documents.

- **Negative consequences.** This pitfall often causes the following negative consequences:
  - The architecture is ambiguous.
  - It is difficult or impossible to ensure architectural integrity.
- **Mitigations.** Do the following to avoid or mitigate the negative consequences of this pitfall:
  - Incorporate model-consistency verification in the system architecture engineering method.
  - Include time for consistency checking in the schedule.
  - Understand the schema underpinning any modeling tool used. The architects may use schema diagrams to become aware of how changes made on one model might ripple to other models, causing unintended consequences.
  - Acquire or develop consistency checking tools.

■ **The architects use inappropriate traditional architectural patterns.**
- **Pitfall.** The architects use inappropriate architectural patterns. For example, the architects automatically assume that just because a system architecture pattern was successfully used on previous systems that it will continue to be appropriate on the current system.

  For example, when relatively little functionality was allocated to software on traditional military aircraft, a popular architecture pattern was to allocate functionally cohesive sets of software to individual computers. This pattern of one-feature-set-to-one-computer was successful because the aircraft could support a small number of computers and because it fit well with the traditional hardware functional decomposition mindset. However, as more and more functionality was allocated to software, the corresponding increasing number of individual computers that this pattern demanded became impractical because of severe limitations on weight, volume, electricity, and air conditioning. A new pattern is now developing where functionally diverse sets of software are allocated to individual collaborating computers. The new pattern uses partitions and real-time operating systems to separate the different software applications or feature sets.

- **Negative consequences.** This pitfall often causes the following negative consequences:
  - The architects do not make the most suitable architectural decisions, inventions, and trade-offs.
  - The probability of architectural weaknesses and risks increases.
  - The architects miss opportunities for significant improvement of the architecture over that of previous systems.
- **Mitigations.** Do the following to avoid or mitigate the negative consequences of this pitfall:
  - Incorporate the production of architectural *inventions* as well as decisions in the system architecture engineering method.
  - The trainers should emphasize the value of creativity and invention during architecture engineering.
  - Experienced architects should provide independent consulting and review of the architectural models.

■ **System decomposition is performed by the acquisition organization.**
- **Pitfall.** Representatives of the acquisition organization decompose the system of systems into individual systems (or the system into architectural components) for acquisition purposes.

  This typically happens before the development organizations are involved and before the requirements have been properly engineered. The acquisition representatives are typically not professional system architects, are often inadequately trained and experienced, and are therefore rarely qualified to make such critical architectural decisions. Instead, they decompose the current system to match the decompositions of previous systems, even though the functionality, size, and complexity of the new system may be quite a bit larger than that of legacy systems.
- **Negative consequences.** This pitfall often causes the following negative consequences:
  • The resulting architecture may not be the most suitable for the system, given its specific architecturally significant requirements.
  • Such early decomposition made by people who are not professional architects may result in excessive complication in the architectural components, the interfaces between the components, or both.
- **Mitigations.** This pitfall is very difficult to avoid and deal with because many of the people who are most likely to stumble into this pitfall are high-level managers who are unlikely to read a thick book on system architecture engineering. It is therefore up to the chief architects to help their management to not fall into this pit.

  To avoid or mitigate the negative consequences of this pitfall:
  • Postpone the top-level architectural decision until (1) a significant amount of the requirements (especially the quality requirements) has been engineered, (2) a significant amount of architectural modeling has been done, (3) several architectural visions have been developed.
  • Analyze the pros and cons of several decompositions before deciding on the most suitable approach.
  • Assign the top-level decomposition to an experienced system integrator (prime contractor) that is then responsible for the highest-level architectural decisions, inventions, and trade-offs.
  • Obtain the consulting services of an experienced architect to help engineer the top-level architecture.

8.10 Summary

During the *Create the First Versions of the Most Important Architectural Models* task, the architecture teams create an initial set of partial draft architectural models of the system architecture.

As illustrated in Figure 4.8, this task is typically performed concurrently with the next two tasks: Task 4, *Identify Opportunities for the Reuse of Architectural Elements,* and Task 5, *Create the Candidate Architectural Visions.* This task produces input that they use and they produce inputs that it uses.

8.10.1 Steps

When instantiating the *Create the First Versions of the Most Important Architectural Models* task, the following steps may be included, modified, or tailored out:

1. Identify the relevant architectural structures.
2. Select the appropriate architectural viewpoints, views, and models.
3. Select the appropriate focus areas.
4. Collaborate with specialty engineering groups and other stakeholders.
5. Develop initial partial competing models of the architectural structures.
6. Conditionally allocate the architectural concerns to the underlying component types.
7. Identify the associated potentially relevant technologies.
8. Perform architectural trade-off analyses.
9. Evaluate the architecture models and associated documentation.
10. Identify any architectural risks and opportunities.

8.10.2 Work Products

When instantiating the *Create the First Versions of the Most Important Architectural Models* task, the following work products may be included, modified, or tailored out:

- Initial draft architecture
- Relevant architectural views
- Relevant architectural focus areas
- Initial draft architectural models
- New and updated architectural risks and opportunities

8.10.3 Guidelines

Use the following guidelines (and associated rationales) when performing the *Create the First Versions of the Most Important Architectural Models* task:

- Perform architectural trade-off analysis.
- Reuse architectural principles, heuristics, styles, patterns, vision components, and metaphors.
- Use a recursively incremental, iterative, parallel, and time-boxed development cycle.
- Begin developing logical models before beginning to develop physical models.
- Do not overemphasize the physical decomposition hierarchy.
- Use explicitly documented system-partitioning criteria.
- Concentrate on the interface.
- Model concurrency.
- Consider the impact of hardware decisions on usability and software.
- Consider human limitations when allocating system functionality to manual procedures.
- Do not start from scratch.
- Formally manage architectural risks.

8.10.4 Pitfalls

Avoid the following common pitfalls associated with performing the *Create the First Versions of the Most Important Architectural Models* task and mitigate their negative consequences when they occur:

- The architects succumb to analysis paralysis.
- The architects engineer too few architectural models.
- The architects engineer inappropriate models and views.
- The architects construct views but no focus areas.
- Some stakeholders believe that the models are the architecture.
- Inconsistencies exist between models, views, and focus areas.
- The architects use inappropriate architectural patterns.
- System decomposition is performed by the acquisition organization.

Chapter 9

Task 4: Identify Opportunities for the Reuse of Architectural Elements

9.1 Introduction

During this MFESA task, the system or subsystem architecture team identifies and initially analyzes opportunities for architectural reuse of architectural decisions and inventions such as styles, patterns, and the relevant elements of existing architectural structures (Figure 9.1).

A key reason for performing this task is to avoid reinventing "architectural wheels" by leveraging knowledge and experience gained from prior architecture engineering efforts. Where practical and appropriate, it is important to reuse what has been found to be "good" or "suitable," and to avoid pitfalls identified from experience with prior architectures. Note that the architectural sources being considered for reuse "mining" should not necessarily be limited to architectural work products from specific internal legacy systems. Rather, a broader scope of architectural sources should be considered, such as architectural knowledge that is generally available within the industry and even knowledge about the architecture of competitive systems.

This task is a reusable MFESA method component and as such is intended to be tailorable to meet the specific needs of the endeavor. Therefore, it is expected that at least some of the task's objectives, preconditions, inputs, steps, postconditions, work products, guidelines, and pitfalls will be tailored during the creation of the endeavor-specific system architecture engineering method. Tailoring of a method component includes adding missing content, modifying existing content, or removing existing content that is inappropriate, unnecessary, or not cost-effective to perform.

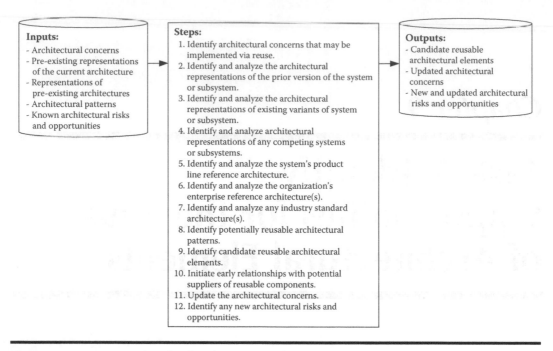

Inputs:
- Architectural concerns
- Pre-existing representations of the current architecture
- Representations of pre-existing architectures
- Architectural patterns
- Known architectural risks and opportunities

Steps:
1. Identify architectural concerns that may be implemented via reuse.
2. Identify and analyze the architectural representations of the prior version of the system or subsystem.
3. Identify and analyze the architectural representations of existing variants of system or subsystem.
4. Identify and analyze architectural representations of any competing systems or subsystems.
5. Identify and analyze the system's product line reference architecture.
6. Identify and analyze the organization's enterprise reference architecture(s).
7. Identify and analyze any industry standard architecture(s).
8. Identify potentially reusable architectural patterns.
9. Identify candidate reusable architectural elements.
10. Initiate early relationships with potential suppliers of reusable components.
11. Update the architectural concerns.
12. Identify any new architectural risks and opportunities.

Outputs:
- Candidate reusable architectural elements
- Updated architectural concerns
- New and updated architectural risks and opportunities

Figure 9.1   Summary of Task 4 inputs, steps, and outputs.

9.2 Goal and Objectives

The *Identify Opportunities for the Reuse of Architectural Elements* task has the following overall goal:

■ Identify any opportunities to reuse existing architectural elements as part of the architecture of the system or architectural component being developed. Any opportunities so identified become candidate reusable architectural elements.

The typical subordinate objectives of this task are to:

■ Understand the relevant legacy or existing architectures sufficiently well to identify potentially reusable architectural elements.
■ Identify the architectural risks and opportunities associated with any aspects of the architectures of relevant existing systems should they be selected for reuse and incorporation within the target environment.
■ Identify any additional architectural concerns due to the constraints associated with reusing or enhancing legacy or existing architectures.
■ Provide a set of reusable architectural element candidates to influence (and possibly include in) a set of initial draft architectural models.

9.3 Preconditions

The *Identify Opportunities for the Reuse of Architectural Elements* task can typically begin when the following preconditions have been met:

- **Architectural concerns exist.** Sufficient architectural concerns exist to begin to identify associated opportunities for the reuse of architectural elements.
- **Initial draft architecture exists.** Potentially key architectural decisions, inventions, trade-offs, and assumptions have been documented in a list.
- **Relevant architectural views exist.** A set of relevant architectural views of the logical and physical, static and dynamic architectural structures exists for the system or architectural component being architected.
- **Initial draft architectural models exist.** A set of initial partial draft architectural models of the architectural structures exists for the system or architectural component being architected.
- **Relevant architectural focus areas exist.** A set of relevant focus areas exists for the system or architectural component being architected.

9.4 Inputs

The *Identify Opportunities for the Reuse of Architectural Elements* task typically has the following inputs:

- **Architectural concerns:** cohesive sets of architecturally significant requirements.
- **Representations of current architecture:** the draft initial representations of the current architecture, including:
  - List of key architectural decisions, inventions, trade-offs, and assumptions
  - Relevant architectural views
  - Relevant architectural focus areas
  - Initial draft architectural models
- **Representations of preexisting architectures:** the representations of relevant preexisting architectures, including:
  - Prior versions of system
  - Existing variants of system
  - Competitors' systems
  - Reference architecture of product line of systems
  - Enterprise architecture
  - Industry standard architectures
- **Known architecture patterns:** architecture patterns, including architectural styles.
- **Known architectural risks and opportunities:** known architectural risks in the endeavor risk repository for the system or the architectural component being architected, known risks associated with preexisting architectures, and known architectural opportunities.

9.5 Steps

As illustrated in Figure 4.8, the *Identify Opportunities for the Reuse of Architectural Elements* task is performed largely in parallel with Task 3, *Create the First Versions of the Most Important Architectural Models.* During this task, members of the relevant architecture team collaborate to perform the following steps in an incremental, iterative, concurrent, and time-boxed manner:

1. **Identify the architectural concerns that may be implemented via reuse.** Identify those architectural concerns that could credibly be implemented by reusing existing architectural

components, regardless of the source of these components. Because architectural concerns are generally related to specialty engineering disciplines or quality focus areas (such as availability, capacity, interoperability, maintainability, performance, reliability, safety, security, and stealth) and programmatic process requirements (such as cost, schedule, and use of available resources), this step focuses on identifying which architectural concerns are promoted by existing architectural components.

2. **Identify and analyze the architectural representations of the prior version of the system or subsystem.** If there exists a previous version of the system or subsystem that is currently being architected, then obtain or reverse-engineer its architectural representations to identify and initially analyze any relevant, potentially reusable architectural elements contained within them.

3. **Identify and analyze the architectural representations of existing variants of the system or subsystem.** If there exist any variants of the system or subsystem that is currently being architected, then if practical obtain or reverse engineer their architectural representations to identify and initially analyze any relevant, potentially reusable architectural elements contained within them.

4. **Identify and analyze the architectural representations of any competing systems or subsystems.** If there exist any competing systems of the system or subsystem that is currently being architected, then if practical obtain or reverse-engineer their architectural representations to identify and initially analyze any relevant, potentially reusable architectural elements contained within them.

5. **Identify and analyze the system's product line reference architecture.** If the system or subsystem that is currently being architected is part of a product line or family of systems, then obtain and initially analyze the representations of the product line's reference architecture to identify any relevant, potentially reusable architectural elements contained within it.

6. **Identify and analyze the organization's enterprise architecture(s).** If the system or subsystem that is currently being architected must conform to one or more organizational enterprise architectures, then obtain and initially analyze the representations of these enterprise architectures to identify any relevant, potentially reusable architectural elements contained within them.

7. **Identify and analyze any industry standard architecture(s).** If the system or subsystem that is currently being architected should conform to one or more industry standard architectures, then obtain and initially analyze the representations of these standard architectures to identify any relevant, potentially reusable architectural elements contained within them.

8. **Identify potentially reusable architectural patterns.** Identify and initially analyze potentially reusable architectural patterns and select appropriate architectural mechanisms.

   - **Select potential architectural styles.** Identify the architectural styles that are potentially appropriate for the system or subsystem being architected. Classify each such architectural style as an individual candidate architectural vision component and document it using an architectural vision component description. Update the relevant initial partial models to include the identified potential architectural styles.

   - **Identify potential architectural patterns.** Identify system, hardware, and software architectural patterns that are potentially appropriate for the system or subsystem being architected. Consider the patterns that are appropriate to address the quality architectural concerns.* Classify each such architectural pattern as a single candidate

---

* There are relatively standard architectural patterns supporting availability, interoperability, portability, reliability, safety, and security.

architectural element and document it using an architectural vision component description. Update the relevant initial partial models to include the identified potential architectural patterns.

- **Select potential architectural mechanisms.** Identify the associated architectural mechanisms that are potentially appropriate for the system or subsystem being architected. Classify each such architectural mechanism as a single candidate architectural vision component and document it using an architectural vision component description. Update the relevant initial partial models to include the identified potential architectural mechanisms.
- **Address reuse incompatibilities.** Decide how to address inconsistencies between potentially reusable elements and the architectural concerns they address. Update the relevant architectural models to address these inconsistencies.

9. **Identify candidate reusable architectural elements.** Using the architecturally significant requirements, associated architectural concerns, known architectural risks, architectural heuristics, and the training and experience of the architects as a sieve, select candidate reusable architectural elements from the potentially reusable architectural elements identified and initially analyzed during the preceding steps of this task.

10. **Initiate early relationships with potential suppliers of reusable components.** Where appropriate, instigate initial relationships with the key vendors and suppliers of reusable architectural components that are to be incorporated in one or more of the competing architectural visions.

11. **Update the architectural concerns.** Concurrently with the performance of the previous steps, update existing architectural concerns and identify new architectural concerns based on the initial analysis of the existing, potentially reusable architectures and architectural elements. That is, work where appropriate to renegotiate the architecturally significant requirements to enable the reuse of architectural elements that may have been prohibited by excessively restrictive requirements.

12. **Identify any new architectural risks and opportunities.** Concurrently with the performance of the previous steps, identify any new architectural risks and opportunities associated with the candidate reusable architectural elements that the new architecture should mitigate or take advantage of. As appropriate, update any previously known architectural risks and opportunities.

Figure 9.2 summarizes how the steps of this task use the potential sources of architectural reuse to identify candidate reusable architectural elements. The top classification hierarchy summaries Steps 2 through 7 of this task. The small classification hierarchy in the middle left part of the diagram results from the first three parts of Step 8. The identification of candidate reusable architectural elements using the sieve produced by addressing architectural risks, architectural heuristics ("rules of thumb"), the architects' experience and training, and the requirements comprising the architectural concerns is addressed by Step 9.

9.6 Postconditions

The *Identify Opportunities for the Reuse of Architectural Elements* task is typically complete for the current subsystem when the following postconditions hold:

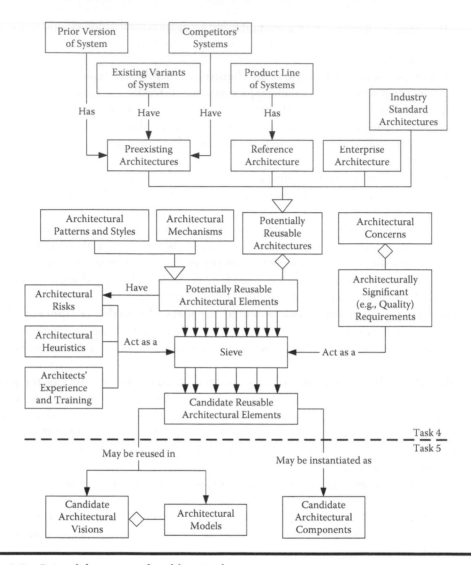

Figure 9.2 Potential sources of architectural reuse.

- **Architectural concerns are updated.** The architectural concerns and their associated architecturally significant product and process requirements have been updated where appropriate.
- **Architectural patterns are identified and analyzed.** A sufficient number of potentially reusable architectural patterns have been identified and initially analyzed for suitability.
- **Reusable architectures are identified and analyzed.** A sufficient number of potentially reusable architectural representations have been identified and initially analyzed for reuse suitability.
- **Reusable architectural elements are identified and analyzed.** A sufficient number of potentially reusable architectural elements have been identified and initially analyzed for suitability as reusable architectural element candidates.
- **Relationships with vendors and suppliers are initiated.** Relationships have been initiated with key vendors and suppliers of reusable architectural components that are to be

incorporated in one or more of the competing architectural visions to provide needed insights into components and their sources.
- **Architectural risks and opportunities are managed.** Any newly identified credible architectural risks have been analyzed and documented in the endeavor risk management repository. Any newly identified architectural opportunities have also been analyzed and documented.

9.7 Work Products

The *Identify Opportunities for the Reuse of Architectural Elements* task typically produces the following work products:

- **Candidate reusable architectural elements:** a list of candidate reusable architectural elements for the system or architectural component being architected.
- **Updated architectural concerns:** any updated architectural concerns that include changes made in the architecturally significant requirements to support reuse.
- **New and updated architectural risks and opportunities:** any new or modified architectural risks documented in the endeavor risk repository as a part of endeavor risk management activity. This also includes an updated list of architectural opportunities that the architecture teams may wish to take advantage of in the future.

9.8 Guidelines

Use the following guidelines (and associated rationales) when performing the *Identify Opportunities for the Reuse of Architectural Elements* task:

- Do not start from scratch.
- Do not be excessively constrained by the past.
- Conform to the enterprise architecture.
- Conform to the product line reference architecture.
- Consider system architecture patterns.
- Support modeling.
- Formally manage architectural risks.

The following descriptions express these guidelines in more detail and provide their associated rationales:

- **Do not start from scratch.** Do not automatically assume "greenfield" development and start developing the architectural models and candidate visions from scratch. Instead, reuse architectural elements where appropriate.
  Rationale: It is important to avoid wasting time and effort recreating architectures from scratch, especially when few architects have the luxury of being a part of truly "greenfield" development.
- **Do not be excessively constrained by the past.** Although the architects should carefully consider the existing architecture of the previous version of the system, they should not feel themselves to be excessively constrained by the existing architecture.

Rationale: Existing architectures may not adequately support the achievement of the requirements of the new version of the system. Starting with past architectures may replicate their defects and weaknesses. There may be good reasons for replacing parts of past architectures. Although it is tempting to take the easier road and just reuse the existing architecture, it may prove far too constricting to position the existing architecture as the "incumbent" that can only be displaced with major justification.

■ **Conform to the enterprise architecture.** Where practical, the system architecture should be consistent with any enterprise architecture defining how it should interoperate with other systems within the enterprise. Values of the enterprise such as the target price that it seeks to service in the market or the willingness of the enterprise to invest in training and job satisfaction of the users will tend to affect the suitability of various architectural alternatives. For example, a portion of the legacy system architecture that may be ripe for revision at first assessment may end up being retained in the new architecture due to its particular alignment with the enterprise architecture.

Rationale: Deviation from the enterprise architecture typically decreases interoperability while increasing development and maintenance costs.

■ **Conform to the product line reference architecture.** If engineering a system that is part of a product line of systems, then the architectural models and candidate architectural visions should conform to any relevant reference architecture.

Rationale: Deviation from a reference architecture typically causes the loss of the benefits of standardization (e.g., it increases development and maintenance costs). If the system is part of a product line family of systems, the architect will require very good reasons to deviate from the reference architecture of the product line.

■ **Consider system architecture patterns.** Although fewer system-level architecture patterns exist than software architecture patterns, it is nevertheless important to consider them when identifying probably reusable architectural elements for the system.

Rationale: System architecture patterns capture proven solutions to common problems within specific contexts. If the problem and context apply, then the pattern's solution will also likely apply. Reuse of system architecture patterns reduces risk and increases both productivity and understandability.

■ **Support modeling.** Use this task to identify reusable architectural element candidates so that architectural models can be engineered.

Rationale: Reuse will typically have a major impact on the architectural models.

Warning: Determining whether these candidate architectural elements will actually be reused does not happen until the candidate architectural visions are engineered and the optimal architectural vision is selected and iterated.

■ **Formally manage architectural risks.** As part of the risk management activity, identify architectural risks during the performance of this task. Add any architectural risks identified to the official endeavor risk management database.

Rationale: Architectural risks that are not identified or that are not added to the project risk management database when identified are much less likely to be effectively managed.

9.9 Pitfalls

Avoid the following common pitfalls associated with performing the *Identify Opportunities for the Reuse of Architectural Elements* task and mitigate their negative consequences when they occur:

- The architects start from scratch.
- The architects ignore past lessons learned.
- The architects overly rely on previous architectures.
- The architects select specific OTS components too early.
- The architects assume the reusability of immature architectural components.
- The architects assume the reusability of immature technologies.
- Inadequate information exists to determine reusability.

The following express these pitfalls in more detail, describing their negative consequences and the steps that architects can use to mitigate them:

- **The architects start from scratch.**
  - **Pitfall.** Architects assume "greenfield" development in which the system architecture is developed largely or completely from scratch.

    Most systems are variations or upgrades of existing systems, and almost all system architectures can usefully reuse something from previous systems, even if it is only the reuse of architectural patterns and mechanisms.

    Even the architectures of new "greenfield" systems are often constrained by the need to conform to enterprise, reference, or industry standard architectures.
  - **Negative consequences.** This pitfall often causes the following negative consequences:
    - The new architecture is inappropriately inconsistent with existing architecture.
    - Productivity is lost as large parts of the architecture are unnecessarily reengineered.
    - The return on investment (ROI) made in engineering the previous architecture (as well as producing the associated design, implementation, and tests) is lost.
    - The schedule is greatly extended and deadlines will be missed.
    - The probability increases that the new architecture and architectural representations will contain defects and weaknesses.
    - Project (as well as architectural) risks are increased.
  - **Mitigations.** Do the following to avoid or mitigate the negative consequences of this pitfall:
    - Include this task in the project-specific architecture engineering method.
    - Verify that the steps of this task are actually being performed.
- **The architects ignore past lessons learned.**
  - **Pitfall.** Architects ignore past lessons learned while trying to engineer a totally new and better architecture.

    Note that it is extremely difficult to invent good new architectural styles, patterns, and mechanisms. It is much easier (and more likely to be successful) to base one's architectural decisions on those that experience has shown to be successful in the past. Failure to consider what has been tried in the past but was ultimately unsuccessful increases the likelihood of failing again.
  - **Negative consequences.** This pitfall often causes the following negative consequences:
    - Productivity is lost as work is unnecessarily repeated.
    - The return on investment (ROI) made in engineering the previous architectures is lost.
    - The schedule is extended and deadlines may be missed as work is repeated.
    - The probability increases that the new architecture and architectural representations will contain defects and weaknesses.
    - Architectural risks are increased.

- **Mitigations.** Do the following to avoid or mitigate the negative consequences of this pitfall:
  - Include this task in the project-specific architecture engineering method.
  - Verify that the steps of this task are actually being performed.
  - Provide training in the benefits (and risks) of architectural reuse.

  The following are examples of such previously learned lessons:
  - The use of stealth airframe pattern X caused difficulties in meeting Stealth Specification Y when total aircraft surface area exceeded approximately Z ft^2, which resulted in additional costs and schedule time to empirically fine-tune the stealth airframe pattern.
  - Use of an n-tier architecture pattern within an overall Enterprise Java Bean (EJB) architecture in conjunction with an aggregate network interconnect bandwidth to the database servers limited to 44 Mbps caused unacceptable transaction response times (greater than 2 seconds) when the number of simultaneous active users exceeded about 500.
  - Distribution of the critical event processing software across multiple processors that were arranged in a "cross point" switching fabric architecture performed an order of magnitude better as compared to an architecture in which processors were arranged in "port modular" fashion, which requires a multilevel switching fabric. However, the cross point architecture cannot be physically packaged as densely as the port modular architecture for the switching fabric. Therefore, the cross point architecture could not be scaled beyond X processors while staying within physical size constraints.

■ **The architects overly rely on previous architectures.**
  - **Pitfall.** Either the architects rely too much on their experience engineering previous architectures or they rely too much on the architecture of the previous version or variant of the system being architected.

    Note that this pitfall is the opposite of the previous pitfall. A moderate position between these two extremes is often the best approach to take.
  - **Negative consequences.** This pitfall often causes the following negative consequences:
    - Such an over-reliance will tend to limit the benefits of the new system relative to previous systems by preventing the incorporation of major new technologies and architectural inventions.
    - Such an over-reliance may in fact cause the system to fail to meet its architecturally significant requirements if the size, complexity, and context of the new system differ sufficiently from those of previous systems.
    - The potential for accidents may increase if inadequate analysis and testing are performed. For example, many accidents (such as the Ariane 5 rocket) have resulted from the reuse of a component that worked fine as part of a previous system but which failed when incorporated into a new system with different requirements.
  - **Mitigations.** Do the following to avoid or mitigate the negative consequences of this pitfall:
    - Actively identify any weaknesses, limitations, lost opportunities, or risks associated with the previous architectures.
    - Actively brainstorm the use of new architectural patterns, mechanisms, and technologies.
    - Carefully consider any changes regarding architecturally significant requirements between the new system and the previous system, especially changes in quality requirements.

- Provide training in the risks (and benefits) of architectural reuse.
- Ensure that the architects are able to defend the correctness and appropriateness of any elements they identify as candidates for reuse.

■ **The architects select specific OTS components too early.**
  - **Pitfall.** Architects or other stakeholders in positions of authority (e.g., project managers and chief system engineers) select specific OTS components before proper trade studies and engineering trade-offs have been made.
  - **Negative consequences.** This pitfall often causes the following negative consequences:
    - The wrong OTS component is selected for inclusion in the system architecture.
    - The architecture includes defects, weaknesses, or missed opportunities.
  - **Mitigations.** Do the following to avoid or mitigate the negative consequences of this pitfall:
    - Ensure that potential OTS components are only identified during this task by postponing the selection of OTS components until the *Analyze Reusable Components and Their Sources* task.
    - Verify the proper performance of this task.
    - Include trade studies in the system architecture engineering method.
    - Mandate the performance of engineering trade-offs between candidate OTS components before selecting the appropriate ones.

■ **The architects assume the reusability of immature architectural components.**
  - **Pitfall.** The architects or other stakeholders in positions of authority assume that they can rely on the reuse of architectural components (such as planned COTS components) that either do not yet exist or are not sufficiently mature for reuse.
  - **Negative consequences.** This pitfall often causes the following negative consequences:
    - When such vaporware ends up being late for delivery, the entire project schedule will be delayed.
    - When such vaporware ends up not having the necessary functionality or needed levels of quality characteristics or quality attributes, significant changes to the architecture may be required.
  - **Mitigations.** Do the following to avoid or mitigate the negative consequences of this pitfall:
    - The architects should add this pitfall as a risk in the project risk management database.
    - The architects should not rely solely on the availability of such components.
    - The architects should produce alternative architectures as backups if the architectural components are not available in time.
    - If this must be assumed, then consider it an architectural risk and manage it accordingly.

■ **The architects assume the reusability of immature technologies.**
  - **Pitfall.** Architects or other stakeholders in positions of authority rely on the reuse of architectural components that are based on new and immature technologies.
  - **Negative consequences.** This pitfall often causes the following negative consequences:
    - Architectural components incorporating such technology may not achieve the required levels of quality characteristics such as availability, performance, and reliability.
  - **Mitigations.** Do the following to avoid or mitigate the negative consequences of this pitfall:

- Perform technology readiness assessments to ensure that the technology is sufficiently mature to be incorporated into the architecture.
- When immature technology must be used, then note that as a risk in the endeavor risk repository.

■ **Inadequate information exists to determine reusability.**
  - **Pitfall.** The architectural representations of pre-existing architectural elements do not provide the architects with sufficient information to determine whether they are reusable in the current system. Missing information may include
    - **Requirements.** The requirements are completely missing, ambiguous, emphasize only functional capabilities, or ignore quality characteristics (i.e., no quality requirements).
    - **Architectural representations.** There are few or no architectural models, the models are at a very high level of abstraction, the architectural element is treated as a black box, there is little or no information about how the architecture addresses quality characteristics (i.e., no quality cases), and there is little or no rationale for architectural decisions, inventions, trade-offs, and assumptions.
  - **Negative consequences.** This pitfall often causes the following negative consequences:
    - The degree to which the potentially reusable architectural elements support the required level of relevant quality characteristics and quality attributes is unknown.
    - The major architectural decisions, inventions, trade-offs, assumptions, and rationales documented in the available architectural representations of the preexisting architectural elements are unclear.
    - Without the necessary information, architects must (1) assume a large risk in accepting elements for reuse, (2) potentially miss an architectural opportunity by rejecting a suitable reusable element due to lack of information, or (3) incur a substantial increase in cost or delay in schedule while attempting to gain sufficient information.
  - **Mitigations.** Do the following to avoid or mitigate the negative consequences of this pitfall:
    - Where practical, the architects should reverse-engineer the requirements of the potentially reusable architectural elements.
    - The architects should determine the degree to which the potentially reusable architectural elements support the relevant quality characteristics and quality attributes.
    - Where practical, the architects should reverse-engineer the major architectural decisions, inventions, trade-offs, assumptions, and rationales from the available architectural representations of the preexisting architectural elements.
    - The architects should add extra time and effort to the schedule estimates.

9.10 Summary

During the *Identify Opportunities for the Reuse of Architectural Elements* task, the architecture teams identify any opportunities to reuse existing architectural work products as part of the architecture of the target system or subsystem being developed.

Remember that this task is typically performed concurrently with its neighboring two tasks: Task 3, *Create the First Versions of the Most Important Architectural Models,* and Task 5, *Create the Candidate Architectural Visions.* This task produces input that they use, and they produce inputs that it uses.

9.10.1 Steps

When instantiating the *Identify Opportunities for the Reuse of Architectural Elements* task, the following steps may be included, modified, or tailored out:

1. Identify architectural concerns that may be implemented via reuse.
2. Identify and analyze the architectural representations of the prior version of the system or subsystem.
3. Identify and analyze the architectural representations of existing variants of the system or subsystem.
4. Identify and analyze the architectural representations of any competing systems or subsystems.
5. Identify and analyze the system's product line reference architecture.
6, Identify and analyze the organization's enterprise reference architecture(s).
7. Identify and analyze any industry standard architecture(s).
8. Identify potentially reusable architectural patterns.
9. Identify candidate reusable architectural elements.
10. Initiate early relationships with potential suppliers of reusable components.
11. Update the architectural concerns.
12. Identify any new architectural risks and opportunities.

9.10.2 Work Products

When instantiating the *Identify Opportunities for the Reuse of Architectural Elements* task, the following work products may be included, modified, or tailored out:

- Candidate reusable architectural elements
- Updated architectural concerns
- New and updated architectural risks and opportunities

9.10.3 Guidelines

Use the following guidelines (and associated rationales) when performing the *Identify Opportunities for the Reuse of Architectural Elements* task:

- Do not start from scratch.
- Do not be excessively constrained by the past.
- Conform to the enterprise architecture.
- Conform to the product line reference architecture.
- Consider system architecture patterns.
- Support modeling.
- Formally manage architectural risks.

9.10.4 Pitfalls

Avoid the following common pitfalls associated with performing the *Identify Opportunities for the Reuse of Architectural Elements* task and mitigate their negative consequences when they occur:

- The architects start from scratch.
- The architects ignore past lessons learned.
- The architects overly rely on previous architectures.
- The architects select specific OTS components too early.
- The architects assume the reusability of immature architectural components.
- The architects assume the reusability of immature technologies.
- Inadequate information exists to determine reusability.

Chapter 10

Task 5: Create the Candidate Architectural Visions

10.1 Introduction

During this MFESA task, the system or subsystem architecture team uses the architectural drivers, initial architectural models, and potentially reusable architecture elements to create a set of competing candidate architectural visions for their system or subsystem that support meeting the derived and allocated architectural drivers and associated architectural concerns.

That is, the architecture team identifies potential architectural visions that sufficiently fulfill the architecturally significant requirements of the system or subsystem (Figure 10.1). For the architecture team to begin this task, a reasonable number of potential architectural visions components must have been identified. The architecture team begins by grouping the identified architectural vision components into a documented set of competing architectural visions that adequately address all the architectural concerns. Then, the architecture team verifies the quality (such as adequacy, cohesiveness, consistency, and feasibility) of the candidate architectural visions.

This task is a reusable MFESA method component and as such is intended to be tailorable to meet the specific needs of the endeavor. Therefore, it is expected that at least some of the task's objectives, preconditions, inputs, steps, postconditions, work products, guidelines, and pitfalls will be tailored during the creation of the endeavor-specific system architecture engineering method. Tailoring of a method component includes adding missing content, modifying existing content, or removing existing content that is inappropriate, unnecessary, or not cost-effective to perform.

As illustrated in Figure 4.8, this task is largely performed concurrently with Task 3, *Create the First Versions of the Most Important Architectural Models,* and Task 4, *Identify Opportunities for the Reuse of Architectural Elements.*

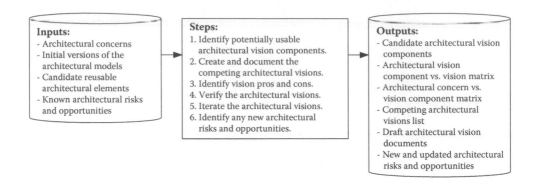

Figure 10.1   Summary of Task 5 inputs, steps, and outputs.

10.2 Goal and Objectives

The *Create the Candidate Architectural Visions* task has the following overall goal:

- Create multiple candidate architectural visions of the system architecture.

  The typical subordinate objectives of this task are to:

- Verify that the candidate subsystem architectural visions sufficiently support the relevant architecture concerns and their associated architecturally significant requirements.
- Provide a sufficiently large and appropriate set of competing candidate architectural visions from which a single vision may be selected as most suitable.

10.3 Preconditions

The *Create the Candidate Architectural Visions* task can typically begin when the following preconditions have been met:

- **Architectural vision components are identified.** The architecture team has identified an initial, reasonably complete set of architecture vision components.

10.4 Inputs

The *Create the Candidate Architectural Visions* task typically uses the following inputs:

- **Architectural concerns:** the architectural concerns driving the system or architectural component being architected.
- **Initial versions of the architectural models:** the first partial versions of the most important architectural models.
- **Candidate reusable architectural elements:** the candidate reusable architectural elements for the system or architectural component being architected.
- **Known architectural risks and opportunities:** known architectural risks in the endeavor risk repository for the system or the architectural component being architected and known architectural opportunities.

10.5 Steps

The *Create the Candidate Architectural Visions* task is typically performed concurrently with Task 3, *Create First Versions of the Most Important Architectural Models,* and Task 4, *Identify Opportunities for the Reuse of Architectural Elements.* During this task, members of the relevant architecture team collaborate to perform the following steps in an incremental, iterative, concurrent, and time-boxed manner:

1. **Identify potentially usable architectural vision components.** Obtain access to, read, and understand:
   - The architectural concerns driving the architecture
   - Any existing first versions of the most important architectural models
   - The potentially reusable architectural elements
   - Known architectural risks and opportunities

   Based on the preceding information, identify the potentially usable architectural vision components (i.e., the most critical architectural decisions, inventions, trade-offs, and assumptions).

2. **Create and document the competing architectural visions.** Based on their support for meeting the architectural concerns, group consistent sets of these identified potential architectural vision components to produce set integrated, internally consistent competing architectural visions that adequately address the architectural concerns. Document these visions in:
   - An architectural vision component versus vision matrix
   - An architectural concern versus architectural component matrix
   - Draft architecture vision documents

   Techniques: Architects can use a joint application development (JAD) workshop with subject matter experts to brainstorm the candidate visions and their components. Architects can create a matrix that maps architectural concerns to architectural vision components to help ensure that architectural visions adequately address the architectural concerns. Architects can create a vision versus vision components matrix to identify the vision components and which vision components are part of which visions.

3. **Identify vision pros and cons.** Identify the respective advantages, disadvantages, and risks associated with the individual architectural visions. Document them in the draft architectural vision documents.

4. **Verify the architectural visions.** Verify, both internally and with external stakeholders, the competing candidate architectural visions to determine if the components within each vision:
   - Are internally consistent
   - Exhibit appropriate levels of engineering principles (e.g., high cohesion and low coupling)
   - Reuse architectural elements in a way that makes sense within the context of the candidate vision
   - Credibly support the architecturally significant product and process requirements comprising each relevant architectural concern (e.g., by repeating the architectural trade-off analysis)
   - Are properly and adequately documented

5. **Iterate the architectural visions.** To the extent practical, correct any defects, weaknesses, and risks identified during verification. Update the individual candidate architectural vision documents.

Note that the purpose of this step is not to select the final vision but rather to ensure that the candidate architectural visions are adequate as input into the selection process.

6. **Identify any new architectural risks and opportunities.** Concurrently with the performance of the previous steps, identify any new architectural risks and opportunities associated with the candidate architectural visions that the new architecture should mitigate or take advantage of. As appropriate, update any previously known architectural risks and opportunities.

10.6 Postconditions

The *Create the Candidate Architectural Visions* task is complete for the current subsystem when the following postconditions hold:

- **Candidate architectural visions exist.** Properly documented candidate competing internally consistent architectural visions exist that credibly support the architecturally significant subsystem requirements comprising each of the relevant architectural concerns.
- **Architectural risks and opportunities are managed.** Any newly identified credible architectural risks have been analyzed and documented in the endeavor risk management repository. Any newly identified architectural opportunities have also been analyzed and documented.

10.7 Work Products

The *Create the Candidate Architectural Visions* task produces the following work products:

- **Candidate architectural vision components:** a list of potentially relevant architectural vision components for the system or architectural component being architected.
- **Architectural vision component versus vision matrix:** a matrix mapping the architectural vision components to their architectural visions.

  As illustrated in Table 10.1, the entries in the matrix cells merely show whether the candidate architectural vision contains the corresponding architectural vision component. Thus, for example, all four architectural visions include Architectural Vision Components 1 and 8, whereas only Architectural Vision 3 includes Architectural Vision Component 11.

- **Architectural concern versus vision component matrix:** a matrix mapping the architectural concerns to the architectural vision components that collaborate to implement them.

  As illustrated in Table 10.2, the entries in the matrix cells represent the degree to which the architectural vision components enable the architecture to satisfy the concern or make it more difficult to satisfy the concern. Two plus signs (++) signify that the architectural component strongly supports the architecture meeting the requirements of the architectural concern, one plus sign (+) signifies weak support for the concern, a zero (0) signifies no significant impact on the concern, one minus sign (–) signifies that the use of the architectural component makes it somewhat more difficult to meet the concern, and two minus signs (– –) signify that use of the component makes it very difficult (if not impossible) to meet the concern.

  Note that although many architectural concerns will be product requirements for quality characteristics and attributes, others will be process requirements, such as for cost and schedule.

- **Competing architectural visions list:** a list of the competing architectural visions.

Table 10.1 Architectural Vision Component versus Vision Matrix

Architectural Vision Component vs. Architectural Vision Matrix		Candidate Architectural Visions			
		Architectural Vision 1	*Architectural Vision 2*	*Architectural Vision 3*	*Architectural Vision 4*
Candidate Architectural Vision Components	Component 1	X	X	X	X
	Component 2	X		X	X
	Component 3	X	X		
	Component 4	X	X	X	
	Component 5	X		X	X
	Component 6	X	X	X	
	Component 7	X		X	X
	Component 8	X	X	X	X
	Component 9		X		X
	Component 10		X	X	X
	Component 11			X	
	Component 12			X	X
	Component 13			X	X

- **Draft architectural vision documents:** a set of draft architectural vision documents, each of which contains text and graphical models that document the:
  - Architectural vision components comprising the architectural vision (i.e., components in a single column of the architectural vision component versus vision matrix)
  - Architectural concerns addressed by the architectural vision
  - Strengths and weaknesses (including risks) of the architectural vision
- **New and updated architectural risks and opportunities:** any new or modified architectural risks documented in the endeavor risk repository as a part of endeavor risk management activity. This also includes an updated list of architectural opportunities that the architecture teams may wish to take advantage of in the future.

10.8 Guidelines

Use the following guidelines (and associated rationales) when performing the *Create the Candidate Architectural Visions* task:

- Identify an appropriate number of candidate architectural visions.
- Complete candidate architectural visions to the appropriate level of detail.
- Prepare architectural components for OTS incorporation.
- Formally manage architectural risks.

Table 10.2   Example Partial Architectural Concern versus Architectural Component Matrix

Architectural Concern vs. Vision Component Matrix		Architectural Vision Components												
		User Identifier and Pass Phrase	COTS Security Smart Card	COTS Biometrics Reader	Locked Rooms with Keyed Access	Guards with Security Cameras	Dedicated Encryption/Decryption HW Server	COTS Encryption/Decryption Software	Encrypted Messages	Encrypted Database Records	COTS Intrusion Detection Software	COTS Antivirus Software	Digital Signature	Single Sign-On
Architectural Concerns	Access control	++	++	++	++	++	0	0	0	0	0	0	+	++
	Confidentiality	+	+	+	+	+	++	++	++	++	0	0	0	0
	Integrity (message)	+	+	+	+	+	0	0	+	+	+	++	+	0
	Integrity (software)	+	+	+	+	+	0	0	0	0	+	++	+	0
	Integrity (data)	+	+	+	+	+	0	0	0	+	+	++	+	0
	Non-repudiation	0	0	0	0	0	0	0	0	+	+	+	++	0
	Availability	–	–	0	0	0	0	0	0	–	+	+	0	0
	Cost	++	–	– –	0	0	– –	– –	– –	0	–	+	–	–
	Performance	–	–	0	0	0	+	–	–	– –	–	–	–	+
	Usability	–	–	0	0	0	0	0	0	0	0	0	–	++

The following descriptions express these guidelines in more detail and provide their associated rationales:

■ **Identify an appropriate number of candidate architectural visions.** The goal of this task is not to produce the final architectural vision for the system or subsystem, but rather to identify a reasonable set of potential architectural visions from which to make the best available choice. Thus, a subgoal of this task is to identify a set of potentially relevant vision components that can then be combined into a set of consistent architectural visions from which to choose.

Emphasize brainstorming and creative, out-of-the-box thinking that may result in the most suitable architectural vision.

Rationale: Identify multiple candidate visions because the most suitable architectural vision is rarely the first one identified. Identify multiple vision components because the vision components are needed from which to construct the candidate architectural visions.

■ **Complete candidate architectural visions to the appropriate level of detail.** Develop the competing architectural visions to a level of detail that is consistent with the architecturally significant requirements, that is consistent with any associated known architectural risks, that enables the candidate visions to be compared, and that enables the most suitable candidate vision to be selected.

Rationale: The work expended to produce any additional detail on a candidate vision that is not selected as most suitable will be wasted.

■ **Prepare architectural components for OTS incorporation.** Where appropriate, prepare relevant architectural components for the incorporation of off-the-shelf components (such as by adding additional architectural or design elements [such as wrappers, proxies, and bridges]) to support interoperability of the OTS component as illustrated in Figure 10.2. Ensure that these additional elements support all relevant types of interoperability, such as physical, energy, material, and information (such as syntax, semantics, and protocol) interoperability.*

Rationale: An OTS component will rarely be exactly what the architecture needs, and the interfaces of an OTS component will rarely be consistent with the interfaces of the other components with which the OTS component must interoperate.

■ **Formally manage architectural risks.** As part of the risk management activity, identify architectural risks during the performance of this task. Add any architectural risks identified to the official endeavor risk management database.

Rationale: Architectural risks that are not identified or that are not added to the project risk management database when identified are much less likely to be effectively managed.

10.9 Pitfalls

Avoid the following common pitfalls associated with performing the *Create the Candidate Architectural Visions* task and mitigate their negative consequences when they occur:

■ The architects engineer only one architectural vision.
■ Management provides insufficient resources.

---

* See Interoperability in Chapter 18, Section 4 *External Quality Characteristics* for a more detailed discussion of these types of interoperability.

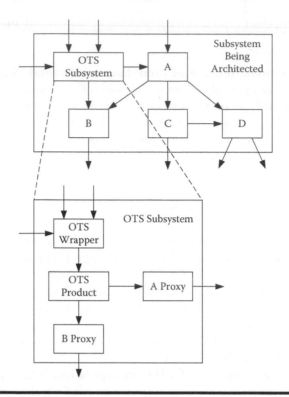

Figure 10.2 Architecting OTS subsystems.

- ■ Management confuses the architectural vision with the completed architecture.
- ■ Management does not permit architects to make mistakes.
- ■ The architects compare the architectural visions prematurely.
- ■ The architects do not compare the pros and cons of the candidate visions.

The following express these pitfalls in more detail, describing their negative consequences and the steps the architects can use to mitigate them:

- ■ **The architects engineer only one architectural vision.**
  - – **Pitfall.** Architects ignore this architecture engineering task by basing the system architecture on the very first architectural vision developed.
  - – **Negative consequences.** This pitfall often causes the following negative consequences:
    - • A poor-quality architectural vision. Unfortunately, the first such architectural vision is rarely the best one on which to found one's architecture. This is especially true when engineering architectures for new systems or when the architecture teams are largely staffed by less-experienced architects. It typically takes several significant iterations before even experienced architects are satisfied that they should move forward. This is especially true if there is only a single system architect because the multiple members of architecture teams are typically able to more easily brainstorm the creation of multiple alternative candidate architectural visions.
    - • A substandard architecture is produced that fails to meet all of its architecturally significant requirements.

- The resulting architecture may unnecessarily include weaknesses and fail to take advantage of opportunities that would have been available in other candidate architectural visions.
- By attempting to save a little time and effort during the performance of this task, the architecture team (and other downstream development teams) may end up having to expend significantly more time and effort later in the development cycle to correct architectural defects and weaknesses incorporated in the initial architectural vision.
  - **Mitigations.** To avoid or mitigate the negative consequences of this pitfall:
    - Ensure that the project-specific system architecture engineering method explicitly includes Task 5, *Create the Candidate Architectural Visions,* and Task 7, *Select or Create the Most Suitable Architectural Vision* for use as a foundation for constructing the actual architecture.
    - Verify that this task is properly performed.
    - Train the architects and other stakeholders in the importance of creating multiple competing architectural visions.
    - Ensure that this task is included in the architecture plan, project schedule, and project work breakdown.

■ **Management provides insufficient resources.**
  - **Pitfall.** Management does not provide sufficient resources (such as time, staff, effort, and funding) to generate multiple competing candidate architectural visions.

    Note that it takes a nontrivial amount of the architecture team's valuable time and effort to properly generate, document, and compare multiple candidate architectural visions.
  - **Negative consequences.** This pitfall often causes the following negative consequences:
    - With limited resources, there will be a strong tendency for the architecture team to go with the first architectural vision they come up with and this may not be the most suitable architectural vision they could have created had they had more time. Note that the task is not to select the most suitable architectural vision in absolute terms, so much as to select the most suitable vision within the programmatic constraints of schedule and budget.
    - The resulting architecture may unnecessarily include weaknesses and fail to take advantage of opportunities that would have been available in other architectural visions.
    - By attempting to save a little time and effort during the performance of this task, the architecture team (and other downstream development teams) may end up having to expend significantly more time and effort later in the development cycle to correct architectural defects and weaknesses incorporated in the initial architectural vision.
  - **Mitigations.** To avoid or mitigate the negative consequences of this pitfall:
    - Include this task in the project system architecture engineering method.
    - Verify that this task is properly performed within budget and schedule constraints.
    - Train the architects and other stakeholders in the importance of performing this task.
    - Ensure that this task is included in the architecture plan, project schedule, and project work breakdown.

■ **Management confuses the architectural vision with the completed architecture.**
  - **Pitfall.** Management and other stakeholders confuse a draft architectural vision with the actual architecture that needs to be built based on the eventually selected vision.
  - **Negative consequences.** This pitfall often causes the following negative consequences:

- Management does not realize that there is significant work to be performed in completing, verifying, and maintaining the architectural vision because a vision is just a high-level overview or summary of an architecture, with some of the most important architectural decisions and inventions made but with many more yet to be made.
- Management may incorrectly consider the architecture engineering work to be done.
- Management may therefore withdraw the resources necessary to finish the architecture.
- The architecture is not sufficiently completed and mature for the associated point in the development cycle.
- Downstream activities based on the vision are negatively affected by the incompleteness of the vision.
- **Mitigations.** To avoid or mitigate the negative consequences of this pitfall:
  - Train architecture stakeholders (including architects and management) in the value of producing multiple competing candidate visions such as an improved architecture and the better meeting of architecturally significant requirements.
  - Clearly identify the subsequent tasks that must yet be accomplished in the project plan, work breakdown schedule (WBS), and other documents.
  - Use a project glossary that clearly defines the architecture, an architectural vision, and the differences between them.
  - Explicitly budget the major sub-steps of architecture engineering as opposed to only having a single, monolithic architecture budget item.

- **Management does not permit architects to make mistakes.**
  - **Pitfall.** The management culture in the development or acquisition organization does not tolerate mistakes.
  - **Negative consequences.** This pitfall often causes the following negative consequences:
    - Managers in such organizations may consider the development of multiple competing candidate architectural visions a waste of time because it involves developing visions that will ultimately be neither selected nor used.
    - Managers may consider it a sign of incompetence if the architects do not come up with the correct vision on the first try.
  - **Mitigations.** To avoid or mitigate the negative consequences of this pitfall:
    - Train and counsel managers to understand the value and necessity of iterating the architecture.
    - Train and counsel managers to realize that it is entirely normal for the best architectural vision to be created as a result of comparing multiple candidate architectural visions and either selecting the best one or combining the best aspects of multiple architectural visions.

- **The architects compare the architectural visions prematurely.**
  - **Pitfall.** The architects spend a lot of time comparing the architectural visions as they are being developed and before they are sufficiently complete to provide a basis for comparison.

    Although it is only natural to have some overlap between this task and the *Select or Create the Most Suitable Architectural Vision* task, within budget and schedule constraints, it is important to allow the members of the architecture team to freely brainstorm the development of multiple architectural visions, without feeling that their raw ideas are being criticized.

    Note also that although the vision components within individual architectural visions need to be self-consistent, it is okay for different architectural visions to be inconsistent with each other. Thus, consistency across multiple architectural visions is not a restriction.

- **Negative consequences.** This pitfall often causes the following negative consequences:
  - Certain valid architectural visions may be ruled out based on incomplete and inadequate information.
  - The most suitable architectural vision may not be selected or created because another architectural vision may be selected prematurely.
  - The premature comparison of the candidate architecture visions may underemphasize the reuse of architectural components during the following *Analyze Reusable Components and Their Sources* task.
  - The premature selection may inappropriately limit the creation and further development of competing architectural visions.
- **Mitigations.** To avoid or mitigate the negative consequences of this pitfall:
  - The competing architectural visions need to incorporate sufficient architectural decisions, inventions, trade-offs, and assumptions to support a fair comparison.
  - Emphasize that it is okay (and in fact preferable) to produce multiple competing architectural visions.

  Note that the opposite extreme is to thoroughly study all candidate architectural visions to identify every last virtue, defect, and weakness even if these pros and cons are insufficiently significant to drive the selection process. This is especially likely to happen when one or more people have a pet candidate vision and seek to prove its superiority by identifying and documenting its every possible strength.

■ **The architects do not compare the pros and cons of the candidate visions.**
- **Pitfall.** The architects do not sufficiently identify, adequately document, and compare the absolute and relative advantages and disadvantages of the competing candidate architectural visions.
- **Negative consequences.** This pitfall often causes the following negative consequences:
  - It is difficult to *verify* the correctness and appropriateness of the candidate architectural visions.
  - It is difficult to identify, document, and analyze the risks associated with the different competing architectural visions.
  - It is difficult for architects to *communicate* the relative pros and cons and risks of each candidate architectural vision.
  - It is difficult for architects to *select* the most suitable architectural vision.
- **Mitigations.** To avoid or mitigate the negative consequences of this pitfall:
  - Determine the degree to which each architectural vision supports the associated architectural concerns.
  - Document the respective advantages and disadvantages of each candidate architectural vision.
  - Document the architectural risks and opportunities associated with each candidate architectural vision.

10.10 Summary

During the *Create the Candidate Architectural Visions* task, the architecture teams create multiple candidate architectural visions of the system architecture.

Remember that this task is typically performed concurrently with its previous two tasks: Task 3, *Create the First Versions of the Most Important Architectural Models,* and Task 4, *Identify Opportunities for the Reuse of Architectural Elements.* This task produces input that the other tasks use, and they produce inputs that it uses. For example, the candidate architectural visions include vision components derived from the first versions of the most important architectural models as well as any potentially reusable architectural elements.

10.10.1 Steps

When instantiating the *Create the Candidate Architectural Visions* task, the following steps may be included, modified, or tailored out:

1. Identify potentially usable architectural vision components.
2. Create and document the competing architectural visions.
3. Identify vision pros and cons.
4. Verify the architectural visions.
5. Iterate the architectural visions.
6. Identify any new architectural risks and opportunities.

10.10.2 Work Products

When instantiating the *Create the Candidate Architectural Visions* task, the following work products may be included, modified, or tailored out:

- Candidate architectural vision components
- Architectural vision component versus vision matrix
- Architectural concern versus vision component matrix
- Competing architectural visions list
- Draft architectural vision documents
- New and updated architectural risks and opportunities

10.10.3 Guidelines

Use the following guidelines (and associated rationales) when performing the *Create the Candidate Architectural Visions* task:

- Identify an appropriate number of candidate architectural visions.
- Complete candidate architectural visions to the appropriate level of detail.
- Prepare architectural components for OTS incorporation.
- Formally manage architectural risks.

10.10.4 Pitfalls

Avoid the following common pitfalls associated with performing the *Create the Candidate Architectural Visions* task and mitigate their negative consequences when they occur:

- The architects engineer only one architectural vision.
- Management provides insufficient resources.
- Management confuses the architectural vision with the completed architecture.
- Management does not permit architects to make mistakes.
- The architects compare the architectural visions prematurely.
- The architects do not compare the pros and cons of the candidate visions.

Chapter 11

Task 6: Analyze Reusable Components and Their Sources

11.1 Introduction

During this MFESA task, the system or subsystem architecture team identifies and evaluates potentially reusable components and their sources (Figure 11.1).

Many of the most important architectural decisions typically revolve around how to decompose the system or subsystem being architected into lower-level architectural elements. While some of these lower-level architectural elements will probably have to be developed from scratch, several of them should probably be reused from various sources:

- Commercial-off-the-shelf (COTS) components from product and component vendors
- Government-off-the-shelf (GOTS) components from civilian governmental agencies
- Military-off-the-shelf (MOTS) components from military acquisition agencies
- Reusable components from the development organization's:
  - Reuse repositories or inventory
  - Legacy systems
  - Enterprise or product line reference architecture
- Open-source software components from open-source organizational repositories
- Freeware software components from freeware sources

These potentially reusable components can be:

- Reusable at the system or subsystem level
- Consumable materials or products, data, equipment, facilities, firmware, hardware, manual procedures, or software components

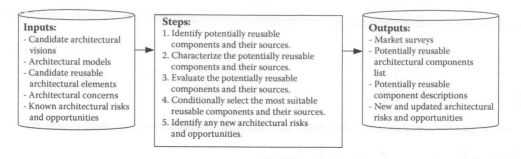

Figure 11.1   Summary of Task 6 inputs, steps, and outputs.

- Services and service infrastructures
- Small-scale components, general components such as the operating system (OS), middleware, and common hardware
- Domain-specific components

This task is a reusable MFESA method component and as such is intended to be tailorable to meet the specific needs of the endeavor. Therefore, it is expected that at least some of the task's objectives, preconditions, inputs, steps, postconditions, work products, guidelines, and pitfalls will be tailored during the creation of the endeavor-specific system architecture engineering method. Tailoring of a method component includes adding missing content, modifying existing content, or removing existing content that is inappropriate, unnecessary, or not cost-effective to perform.

11.2 Goal and Objectives

The *Analyze Reusable Components and Their Sources* task has the following overall goal:

- Determine if any existing architectural components are potentially reusable as part of the architecture of the current system or subsystem.

The typical subordinate objectives of this task are to:

- Identify any existing components that are potentially reusable as part of the architecture of the current system or subsystem.
- Evaluate these components for suitability (e.g., functionality, performance, and reliability).
- Evaluate the sources of these components for suitability (e.g., market share and financial stability).
- Provide a set of potentially reusable components to influence (and possibly include in) a set of initial draft architectural models.

11.3 Preconditions

The *Analyze Reusable Components and Their Sources* task can typically begin when the following preconditions have been met:

- **Architectural concerns exist.** The primary architectural concerns for the system or the architectural component being architected have been identified and documented (Task 2).
- **First versions of the architectural models exist.** The first versions of the most important architectural models exist (Task 3).
- **Opportunities for architectural reuse exist.** The primary opportunities for architectural reuse, including reusable architectural elements, have been identified and initially analyzed for probable reuse (Task 4).
- **Candidate architectural visions exist.** Draft candidate architectural visions meeting these concerns, incorporating these models, and exploiting these reuse opportunities exist (Task 5).

11.4 Inputs

The *Analyze Reusable Components and Their Sources* task typically uses the following inputs:

- **Architectural models:** the first versions of the most important architectural models.
- **Architectural concerns:** the architectural concerns of the system or affecting the architectural component being architected.
- **Candidate reusable architectural elements:** the candidate reusable architectural elements for the system or architectural component being architected.
- **Reuse documentation:** documentation concerning potentially reusable components and their sources (such as the component vendor technical and marketing literature)
- **Candidate architectural visions:** potential architectural visions meeting the architectural concerns
- **Known architectural risks and opportunities:** a list of known architectural risks and opportunities for the system or the architectural component being architected.

11.5 Steps

As illustrated in Figure 4.8, the *Analyze Reusable Components and Their Sources* task is performed largely in parallel with Task 7, *Select or Create the Most Suitable Architectural Vision*. During this task, members of the relevant architecture team collaborate to perform the following steps in an incremental, iterative, concurrent, and time-boxed manner:

1. **Identify potentially reusable components and their sources.** Identify the potentially reusable components and their sources based on the architectural concerns that have been identified as credibly being implementable by reusing existing architectural components. The candidate architectural visions, the associated models, and the candidate reusable architectural elements and their documentations identify the potentially usable components and their sources. Obtain reuse documentation of the components and their sources. Consider all sources of such reusable components such as commercial vendors, governmental agencies, reuse repositories, and open-source organizations. Consider reusable component types such as subsystems, hardware, software, data, facilities, and others.
2. **Characterize potentially reusable components and their sources.** Characterize the potentially reusable components in terms of such characteristics as type (such as COTS, GOTS,

MOTS, and open source), cost, availability, flexibility, documentation, market share, and licensing. Characterize the sources of these components in terms of market share, financial stability, and the like.

3. **Evaluate the potentially reusable components and their sources.** Technically evaluate the selected potentially reusable components for quality, maturity, stability, and how well they address the associated architectural concerns. Determine the technical and programmatic risks associated with reusing these components.

   – **Determine the evaluation resources.** Determine the appropriate amount of limited project resources (e.g., cost, schedule, and staffing) to invest in the evaluation of the selected potential reusable components and their sources.

   – **Determine the evaluation approach.** Review the potential methods that may be used to evaluate the selected potentially reusable components and their sources (see "Appendix D: *Decision-Making Techniques*" for guidance). Determine the appropriate approach to use to evaluate the selected potentially reusable components and their sources. Identify the associated evaluation criteria and measurement scales and determine the associated decision rule.

   – **Evaluate the selected potentially reusable components and their sources.** Use the determined evaluation approach to:

     • Ensure that the potentially reusable components and their sources have the required characteristics (such as functionality and levels of quality characteristics and quality attributes).

     • Perform an engineering trade-off analysis of the selected potentially reusable components and their sources.

     Document the analysis results in a manner that makes the relative pros and cons of each potentially reusable component obvious.

4. **Conditionally select the most suitable reusable components and their sources.** Use the decision rule on the results of the preceding evaluation to select the most suitable of the evaluated potentially reusable components and their sources. Perform a quick informal reevaluation of the results of the evaluation of the selected potentially reusable components and their sources to double check the reasonableness and logical consistency of the selection results.

   Note that the final selection of the reusable components and their sources occurs during Task 8, *Complete the Architecture and Its Representations*, Step 6, *Address Remaining Architectural Reuse Issues*.

5. **Identify any new architectural risks and opportunities.** Concurrently with the performance of the previous steps, identify and formally document any new architectural risks and opportunities that were discovered.

11.6 Postconditions

The *Analyze Reusable Components and Their Sources* task is typically complete for the current subsystem when the following postconditions hold:

■ **Potentially reusable components are identified, evaluated, and conditionally selected.** A sufficient set of potentially reusable components has been identified. These potentially reusable components have been evaluated (analyzed and tested) for suitability and conditionally selected to provide a basis for the selection of the most suitable architectural vision (Task 7).

■ **Sources are evaluated for suitability.** The sources of these components have also been evaluated for suitability (e.g., marketshare, financial stability, and track record).

■ **Architectural risks and opportunities are managed.** Any newly identified credible architectural risks have been analyzed and documented in the endeavor risk management repository. Any newly identified architectural opportunities have also been analyzed and documented.

11.7 Work Products

The *Analyze Reusable Components and Their Sources* task typically produces the following work products:

■ **Market surveys:** market surveys of potentially reusable components and their sources, each including:
  - Executive overview
  - Survey objectives
  - Survey method, including evaluation criteria
  - Survey results
■ **Potentially reusable architectural components list:** a short list of the potentially reusable components that have been conditionally selected for incorporation into the system.
■ **Potentially reusable architectural component descriptions:** descriptions of the potentially reusable components, including:
  - Description
  - Support for architectural concerns, including any mismatches
  - Required update to business process
  - Source
  - Estimated development and life-cycle costs
  - Descriptions of associated license agreements, contracts, and procurement agreements
■ **New and updated architectural risks and opportunities:** any new or modified architectural risks documented in the endeavor risk repository as a part of endeavor risk management activity. This also includes an updated list of architectural opportunities that the architecture teams may wish to take advantage of in the future.

11.8 Guidelines

Use the following guidelines (and associated rationales) when performing the *Analyze Reusable Components and Their Sources* task:

■ Use appropriate decision techniques.
■ Perform task concurrently.
■ Formally manage architectural risks.

The following descriptions express these guidelines in more detail and provide their associated rationales:

■ **Use appropriate decision techniques.** Use the most appropriate decision technique from Appendix D to select the most suitable reusable components.

Rationale: Inappropriate decision techniques are less likely to result in the right decision and may even take more time and effort than is necessary.

■ **Perform tasks concurrently.** Begin this task during Task 6, *Identify Opportunities for the Reuse of Architectural Elements*. Perform this task concurrently with Task 7, *Select or Create the Most Suitable Architectural Vision*.

Rationale: The reuse of existing architectural components should pervade architecture engineering methods. Although the potentially reusable architectural components are identified during this task, the analysis and determination of their actual suitability is performed in an ongoing manner as models and candidate architectural visions are created and selected. Note that the final selection of reusable components does not occur until Task 8, *Complete the Architecture and Its Representations*.

■ **Formally manage architectural risks.** As part of the risk management activity, identify architectural risks during the performance of this task. Add any architectural risks identified to the official endeavor risk management database.

Rationale: Architectural risks that are not identified or that are not added to the project risk management database when identified are much less likely to be effectively managed.

11.9 Pitfalls

Avoid the following common pitfalls associated with performing the *Analyze Reusable Components and Their Sources* task and mitigate their negative consequences when they occur:

■ Authoritative stakeholders assume that reuse will improve cost and schedule.
■ Insufficient information exists for evaluation and reuse.
■ Stakeholders have an unrealistic expectation of "exact fit."
■ Developers have little or no control over future changes.
■ The source organization (such as a vendor) fails to adequately maintain a reusable architectural component.
■ Legal rights are unacceptable.
■ Incompatibilities with underlying technologies exist.

The following express these pitfalls in more detail, describing their negative consequences and the steps that architects can use to mitigate them:

■ **Authoritative stakeholders assume that reuse will improve cost and schedule.**
 – **Pitfall.** Acquisition personnel, management, and even some architects automatically assume that reuse of existing architectural components will significantly improve system cost and the development schedule.

 The predicted savings in budget and schedule due to reusing architectural components within the system architecture are often highly over-optimistic. Stated another way, the estimates of the amount of effort required to incorporate potentially reusable components within new systems have been typically far too low.

 • The specialization and integration of potentially reusable components tends to be unpredictably complex.

- The costs associated with locating, understanding, and evaluating potentially reusable components may be a significant portion of the cost of creating similar components from scratch.
- For software, the productivity for creating a line of "wrapper" code or for modifying potentially reusable software components is likely to be *much* lower than the productivity of producing new software. Thus, costs per line of code for wrapping or modifying potentially reusable software components are likely to be much higher than for producing new software. Thus, the ratio of the amount of reused code to wrapper code must be quite substantial for the cost of code development to justify reuse.
- The effort and costs associated with integration and post-deployment support for systems that incorporate multiple reusable components can scale in a dramatically nonlinear fashion with the number of different reusable components that are integrated. The *aggregate* component-based system is much more difficult to develop and support than is a system that incorporates only a single reusable component.
- Reusable components may not have been developed for inclusion in the current type of system.

Note that while the above problems are typically true of major software-intensive subsystems, this is often not the case with regard to the reuse of *highly generic* software components such as operating systems, middleware, databases, and programming language class libraries.

- **Negative consequences.** This pitfall often causes the following negative consequences:
  - Because the underlying levels of quality characteristics provided by a reusable architectural component may not be sufficient to meet quality concerns and requirements, significantly more work may need to be done within other components or within wrappers to achieve these mandatory levels of quality.
  - The architecture budget and the overall project budget are exceeded.
  - The architecture schedule and project schedule deadlines are not met.
  - The system architecture needs to be reengineered (e.g., to replace the reusable components with components developed from scratch).
- **Mitigations.** To avoid or mitigate the negative consequences of this pitfall:
  - Carefully estimate the costs and schedules associated with reuse of one or more architectural components, basing these estimates on real data.
  - Include cost and schedule underestimation as a risk in the project risk management database, and then develop effective risk mitigation approaches.

■ **Insufficient information exists for evaluation and reuse.**
- **Pitfall.** Potentially reusable architectural components do not come with sufficient information to be objectively evaluated and easily reused.

  Sometimes this is because the potentially reusable components were not created with reuse in mind. Examples include government off-the-shelf (GOTS) components and components that were internally developed for previous systems. Sometimes this is because the potentially reusable components contain commercially valuable proprietary information that would be exposed if made available for evaluation and reuse purposes.

  For example, the potentially reusable components:
  - Do not come with well-engineered requirements (e.g., the requirements are completely missing, emphasize only functional capabilities, and ignore quality characteristics)
  - Do not come with well-engineered architectural representations (e.g., few or no architectural models, models are at a very high level of abstraction, the component

is treated as a black box, and there is little or no information about how the architecture addresses quality characteristics)
  - Are only supplied with marketing material
- **Negative consequences.** This pitfall often causes the following negative consequences:
  - An inappropriate selection may be made and may not be discovered until development or integration, which will drive up costs and lengthen schedules due to rework (or extra work to compensate).
  - The architects do not know what quality concerns drove the development of the components, and therefore do not know how well the components might meet the requirements of the current system.
  - The system architects do not know the component's architecture in terms of its major architectural decisions, inventions, trade-offs, assumptions, and rationales.
  - The architects do not know if the component's internals implement patterns and mechanisms that are consistent with the rest of the system architecture.
  - The architects must consider that the components will be more difficult to integrate and test.
  - The architects will have to develop (primarily software) wrapper components to connect the reusable components with the rest of the architectural components without sufficient information to effectively and efficiently do so.
  - It will be difficult for the architects to estimate the size of the architectural integration effort (i.e., the time and effort needed to include reuse in modeling the overall architecture, creating architectural visions, and completing the architecture).
- **Mitigations.** To avoid or mitigate the negative consequences of this pitfall:
  - Treat the potentially reusable components as black boxes that do not contain clearly identifiable subcomponents.
  - Include extra model analysis and simulation to ensure that both the architectural visions and the architecture meet the architecturally significant requirements.
  - Add extra time and effort to the schedule estimates.
  - Negotiate with the source (such as vendor) for access to additional information, typically with a nondisclosure agreement and considerable license constraints.

■ **Stakeholders have an unrealistic expectation of "exact fit."**
- **Pitfall.** Architects and their management ignore the fact that potentially reusable components will rarely exactly meet the requirements of the system.
  - The potentially reusable components will probably not provide the exact set of functions required.
  - The potentially reusable components are even less likely to have the required level of nonfunctional requirements such as quality characteristics (e.g., availability, interoperability, performance, reliability, robustness, safety, security, and survivability). This is especially true of COTS software components evaluated for potential reuse for many military and governmental systems because they often have insufficient performance, reliability, and security.
  - The security model and mechanisms of the potentially reusable components are unlikely to be consistent with each other and may not be sufficient for the overall system architecture.
  - The granularity of the potentially reusable components is likely too coarse or too fine for the rest of the system architecture.
- **Negative consequences.** This pitfall often causes the following negative consequences:

- The primary consequence of these unrealistic expectations is unrealistic plans, budgets, and schedules.
- Because the requirements of the potentially reusable components are unlikely to be changed to enable them to be integrated into the system architecture, it is likely that the system's requirements will need modification to be consistent with those of the selected potentially reusable components. Stated differently, systems that consist primarily of newly developed components typically have corresponding requirements that at some point in time during development are "frozen" into a baseline, and therefore these baselined requirements will drive the downstream architecture engineering. However, when reusable components are incorporated into the architecture, this is no longer likely and the requirements must often be modified to suit the potentially reusable components. This is particularly true for selected reusable components that provide human interface logic.
- Significant collaboration, iteration, and compromise between requirements engineering and architecture engineering are probably necessary.
  - **Mitigations.** To avoid or mitigate the negative consequences of this pitfall:
    - Include the influence of architectural reuse on the requirements and the consequential negotiation of requirements in the overall system development method (and the requirements engineering and architecture engineering methods).
    - Carefully evaluate and document the impact of potentially reusable components on the architecturally significant requirements.
    - Carefully evaluate and document the fit of the potentially reusable components within the current system architecture.
    - Identify any necessary architectural components or architectural patterns (e.g., software wrappers and proxies) needed to incorporate the potentially reusable architectural components.
- ■ **Developers have little or no control over future changes.**
  - **Pitfall.** Reusable components are changed by their sources in ways that are inconsistent with the system's architecturally significant requirements or existing architecture.

    The system acquisition, development, and sustainment organizations typically have little or no control over future changes made to potentially reusable components by the component development organizations. Market conditions unrelated to the existing system development effort are likely to drive the futures of the potentially reusable components, and these futures are likely to be driven by additional or changing requirements that have little to do with the requirements of the system being architected.
  - **Negative consequences.** This pitfall often causes the following negative consequences:
    - The frequency and kinds of updates available for potentially reusable components may be under- or overestimated.
    - Worse, there needs to be a planned effort to track and monitor these updates throughout the project life cycle in order to determine their impact.
    - The sources' schedules for updating their reusable components does not match the project schedule for new builds or releases of the system, which increases the difficulty of performing and scheduling architecture engineering, integration, and testing tasks. This is particularly true for COTS software components because it is possible that multiple updates per year will be available from the vendor, depending upon the specific product.

- The system being architected may need to be rapidly or frequently updated to remain maintainable and current with the updated versions of the reusable components. Note that the severity of this consequence increases rapidly with the number of components used.
- The possible existence of "unused code" in potentially reusable software components may violate the system's security constraints.
- Different versions of the architecture containing different versions of the reusable components may need to be worked on simultaneously, and this may include the engineering of different versions of the architecture.

  – **Mitigations.** To avoid or mitigate the negative consequences of this pitfall:
    - Create a highly modular architecture with low coupling between the reused architectural components and the rest of the system so as to limit the impact of any changes to the reused components.
    - Develop a project master schedule that includes regular internal builds and external releases, in which the time between builds and releases is manageable (i.e., neither too little nor too much time between the builds and releases).
    - Regardless of the sources' schedule of reusable components update, incorporate new versions of the reusable components into the system architecture based on the master project schedule.
    - If necessary, develop multiple versions of the system and its architecture, whereby different versions of the system contain different versions of the reusable components.

- **The source organization (such as a vendor) fails to adequately maintain a reusable architectural component.**
  – **Pitfall.** The business or organization that supplies a reused architectural component fails to adequately maintain or support it because:
    - The health of the source organization fails during the life of the system.
    - The reused architectural component becomes obsolete (e.g., because it is based on obsolete technology), thereby causing the component to be abandoned.
    - Business conditions dictate the development and release of a significantly updated architectural component that is not backward-compatible.

    Suppliers may be vendors of COTS components or organizations supplying free open-source components. It is difficult to predict the longevity and financial viability of suppliers of reusable architectural components. Markets also come and go all the time. The continuing interest of a supplier to serve similar markets or technologies over time should not be assumed as guaranteed. The stability over many years of good business relationships with component suppliers may be difficult or impossible to maintain.

  – **Negative consequences.** This pitfall often causes the following negative consequences:
    - The architectural component is not truly reusable within the system.
    - Architectural visions and architectures based on the reuse of the component are not feasible.
    - Scarce architecture engineering resources are wasted in producing architectural visions based on the reuse of these components.
    - It becomes impossible to integrate and maintain the architectural component within schedule and budget constraints.
    - The architecture may lock the system into an obsolete technology.
    - Maintenance costs for the component may increase dramatically as the system developer takes over maintenance responsibilities for the no-longer-supported component.

- **Mitigations.** To avoid or mitigate the negative consequences of this pitfall:
  - Identify this as a risk, incorporate it into the project risk management system, and then develop an appropriate mitigation strategy.
  - Include the *Evaluate Potentially Reusable Components and Their Sources* step in the system architecture engineering method.
  - Carefully evaluate support for potentially reusable architectural components during the *Evaluate Potentially Reusable Components and Their Sources* step of the system architecture engineering method.
  - Include support for the potentially reusable components in the decision criteria used during the *Conditionally Select the Most Suitable Reusable Components and Their Sources* step of this task.
  - Carefully evaluate source viability and future plans when evaluating the sources of potentially reusable components.
  - Carefully evaluate technology trends that may obsolete the technology used in the architectural component.
  - Negotiate license agreements to include the archiving of source code for reusable software components so that the developer may take over maintenance if the vendors no longer maintain the components.
- **Legal rights are unacceptable.**
  - **Pitfall.** Potentially reusable architectural components (especially software components) have inappropriate or excessively restrictive licenses, data rights, and access rights.

    Note that the suppliers of COTS components can dramatically change licensing provisions over time with correspondingly dramatic increases in licensing costs, thus necessitating the assessment of alternatives should chosen reusable components become unaffordable or unacceptably restricted over time.

    There are two possibilities:
    - The components have unacceptable licensing at the time of selection. In this case, the component should be removed from consideration.
    - The component licensing becomes unacceptable at some later date. In this case, the component is reusable at the time of selection but becomes non-reusable at some later time.
  - **Negative consequences.** This pitfall often causes the following negative consequences:
    - The components are not truly reusable within the system.
    - Architectural visions and architectures based on the reuse of these components are not feasible.
    - Scarce architecture engineering resources are wasted in producing architectural visions based on the reuse of these components.
  - **Mitigations.** To avoid or mitigate the negative consequences of this pitfall, ensure:
    - Careful evaluation of the licenses, data rights, and access rights during the *Evaluate Potentially Reusable Components and Their Sources* step of the system architecture engineering method.
    - Identification of an alternate architectural component or architectural strategy as a backup to deal with future licensing problems.
- **Incompatibilities with underlying technologies exist.**
  - **Pitfall.** Incompatibilities exist between the technologies used within the potentially reusable components and those used within the remainder of the system:

- The program languages, middleware, and interface protocols used within potentially reusable components may be incompatible either among themselves or with those used in the remainder of the system.
- Existing project team members may have no prior experience with the technologies (e.g., program languages, middleware, interface protocols, and materials) that are incorporated within the potentially reusable components.
- Version control for potentially reusable components may be much more complex than for the remainder of the system, especially when multiple external sources of bug fixing or updates exist.
- Very often, the training and learning curve effort needed to bring existing program team members "up to speed" on the internals of potentially reusable components is seriously underestimated.

Not having access to the original creators of the potentially reusable components is a serious drawback. Coupled with inadequate reuse documentation or data rights, this can be a knockout factor for potentially reusable components.

- **Negative consequences.** This pitfall often causes the following negative consequences:
  - The resulting architecture may have unnecessary defects, weaknesses, and risks due to the incompatibilities.
  - The budget and schedule may be inadequate due to the late discovery of these unnecessary architectural defects, weaknesses, and risks.
  - The resulting complete architecture may not meet the architecturally significant requirements.
  - The resulting system may not be acceptable to its stakeholders.
- **Mitigations.** To avoid or mitigate the negative consequences of this pitfall:
  - Incorporate the *Evaluate Potentially Reusable Components and Their Sources* step into the system architecture engineering method.
  - During this step, evaluate the potentially reusable components for incompatible technologies, including (1) incompatibilities between technologies, and (2) incompatibilities with requirements, especially the quality requirements.

The preceding list is by no means exhaustive. Instead, it is intended to illustrate some representative examples of pitfalls and risks that should be examined when architects plan to incorporate and integrate reusable architectural components within the system being architected. The list is not intended to imply that integration of such components is either better or worse than developing all components from scratch. Rather, it is intended to illustrate that the risk considerations are different from the all-new development approach, especially with regard to software components.

11.10 Summary

During the *Analyze Reusable Components and Their Sources* task, the architecture teams determine if any existing architectural components are potentially reusable as part of the architecture of the current system or subsystem.

11.10.1 Steps

When instantiating the *Analyze Reusable Components and Their Sources* task, the following steps may be included, modified, or tailored out:

1. Identify potentially reusable components and their sources.
2. Characterize the potentially reusable components and their sources.
3. Evaluate the potentially reusable components and their sources.
4. Conditionally select the most suitable reusable components and their sources.
5. Identify any new architectural risks and opportunities.

11.10.2 Work Products

When instantiating the *Analyze Reusable Components and Their Sources* task, the following work products may be included, modified, or tailored out:

- Market surveys
- Potentially reusable architectural components list
- Potentially reusable architectural component descriptions
- New and updated architectural risks and opportunities

11.10.3 Guidelines

Use the following guidelines (and associated rationales) when performing the *Analyze Reusable Components and Their Sources* task:

- Use appropriate decision techniques.
- Perform task concurrently.
- Formally manage architectural risks.

11.10.4 Pitfalls

Avoid the following common pitfalls associated with performing the *Analyze Reusable Components and Their Sources* task and mitigate their negative consequences when they occur:

- Authoritative stakeholders assume that reuse will improve cost and schedule.
- Insufficient information exists for evaluation and reuse.
- Stakeholders have an unrealistic expectation of "exact fit."
- Developers have little or no control over future changes.
- The source organization (such as a vendor) fails to adequately maintain a reusable architectural component.
- Legal rights are unacceptable.
- Incompatibilities with underlying technologies exist.

Task 7: Select or Create the Most Suitable Architectural Vision

12.1 Introduction

During this MFESA task, the system or subsystem architecture team selects the most suitable architectural vision for their system or subsystem from the competing candidate architectural visions (Figure 12.1). If combining consistent components from multiple competing candidate architectural visions would yield an even more suitable architectural vision, then the team does so to obtain the single new architectural vision.

The essence of the selection task is to select a single vision from the available competing candidate architectural visions based upon evaluation against a set of evaluation criteria. Because architectural engineering in practice always involves difficult and untidy trade-offs between competing concerns, the selected architectural vision will typically not be the *best* choice but rather the vision that has been judged most suitable based upon the evaluation and associated decision technique used.

This task is a reusable MFESA method component and as such is intended to be tailorable to meet the specific needs of the endeavor. Therefore, it is expected that at least some of the task's objectives, preconditions, inputs, steps, postconditions, work products, guidelines, and pitfalls will be tailored during the creation of the endeavor-specific system architecture engineering method. Tailoring a method component includes adding missing content, modifying existing content, or removing existing content that is inappropriate, unnecessary, or not cost-effective to perform.

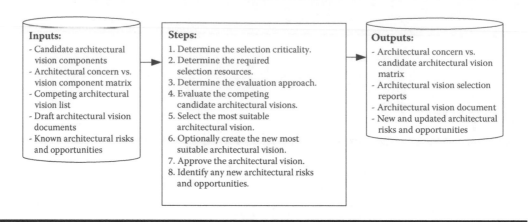

Inputs:
- Candidate architectural vision components
- Architectural concern vs. vision component matrix
- Competing architectural vision list
- Draft architectural vision documents
- Known architectural risks and opportunities

Steps:
1. Determine the selection criticality.
2. Determine the required selection resources.
3. Determine the evaluation approach.
4. Evaluate the competing candidate architectural visions.
5. Select the most suitable architectural vision.
6. Optionally create the new most suitable architectural vision.
7. Approve the architectural vision.
8. Identify any new architectural risks and opportunities.

Outputs:
- Architectural concern vs. candidate architectural vision matrix
- Architectural vision selection reports
- Architectural vision document
- New and updated architectural risks and opportunities

Figure 12.1   Summary of Task 7 inputs, steps, and outputs.

12.2 Goal and Objectives

The *Select or Create the Most Suitable Architectural Vision* task has the following overall goal:

■ Obtain a single architectural vision for the system or subsystem from the competing candidate architectural visions.

The typical subordinate objectives of this task are to:

■ Ensure that the selected architectural vision has been properly judged to be most suitable for the system or subsystem.
■ Provide a proper foundation on which to complete the engineering of the system or subsystem architecture.

12.3 Preconditions

The *Select or Create the Most Suitable Architectural Vision* task can typically begin when the following preconditions have been met:

■ **Architectural concerns are mature.** The architectural concerns are relatively complete, stable, and authoritative.
■ **Candidate competing architectural visions exist.** The architecture team has identified and documented multiple internally consistent candidate competing architectural visions.
■ **Candidate architectural visions meet requirements.** The candidate competing architectural visions credibly support the architecturally significant subsystem requirements comprising each relevant architectural concern.

12.4 Inputs

The *Select or Create the Most Suitable Architectural Vision* task typically uses the following inputs:

■ **Candidate architectural vision components:** a list of potentially relevant architectural vision components for the system or architectural component being architected.

■ **Architectural concern versus vision component matrix:** a matrix that maps the architectural concerns to the architectural vision components.

■ **Competing architectural visions list:** a list of the competing architectural visions.

■ **Draft architectural vision documents:** a set of draft architectural vision documents, each containing text and graphical models that document:
  – Architectural vision components comprising each architectural vision
  – Architectural concerns addressed by the architectural vision
  – Strengths and weaknesses (including risks) of the architectural vision

■ **Known architectural risks and opportunities:** a list of associated architectural risks and opportunities.

12.5 Steps

During the *Select or Create the Most Suitable Architectural Vision* task, members of the relevant architecture team collaborate to perform the following steps in an incremental, iterative, concurrent, and time-boxed manner:

1. **Determine the selection criticality.** Determine how critical it is to select the most suitable competing candidate architectural vision. Take into account the importance of the architecturally significant requirements allocated to and derived for the system or subsystem.

   Note that selecting the most suitable architectural vision from the competing set of candidate architectural visions is *always* a critically important decision to the success of the system. However, the architects must nevertheless decide on the amount of time and effort they should invest in this decision-making process and thus which decision-making technique to use. There are therefore differing degrees of criticality from which to choose.

2. **Determine the required selection resources.** Based on the previously determined criticality of the selection, determine the appropriate amount of limited project resources (such as the funding, schedule, and staffing) appropriate to invest in the vision selection process.

   Note that Steps 1 and 2 may be reversed. If the architects know how much time and effort they can afford to spend in making this decision, they have implicitly determined the selection criticality and can move directly to determining the evaluation approach to use.

3. **Determine the evaluation approach.** Review the potential methods that may be used to select the most suitable of the competing candidate architectural visions (see Appendix D: *Decision-Making Techniques* for guidance). Based on the criticality of the selection and the resources available for selecting the most suitable architectural vision, determine the appropriate approach to use to evaluate the competing candidate architectural visions. Identify the associated evaluation criteria and measurement scales and determine the associated decision rule.

4. **Evaluate the competing candidate architectural visions.** Evaluate the competing candidate architectural visions against their support for the architectural concerns and suitability of reusable architectural components.
   – **Perform an interaction analysis.** Evaluate the competing candidate architectural visions in terms of the interactions between their component architectural elements. Also evaluate them in terms of their interactions with external systems and humans (such as the users, operators, and maintainers).

- **Perform a cost/benefit analysis.** Evaluate the competing candidate architectural visions in terms of their adverse and beneficial characteristics. Note that the candidate architectural vision with the most beneficial characteristics (in number, individual importance, or overall weight) may well not be the same as the candidate architectural vision with the fewest or least important adverse characteristics. Document the analysis results in a manner that makes the relative pros and cons of each candidate vision obvious.
- **Perform a risk analysis.** Evaluate the competing candidate architectural visions in terms of their associated risks, as well as both the cost and effectiveness of their associated risk mitigations.
- **Perform an engineering trade-off analysis.** Use the determined evaluation approach to perform an engineering trade-off analysis of the competing candidate architectural visions against the architectural concerns (i.e., cohesive sets of architecturally significant requirements).
- **Test reusable component suitability within relevant competing candidate architectural visions.** If a candidate architectural vision incorporates one or more reusable architectural components, obtain evaluation copies of these components, use them to create one or more executable architectural prototypes, and test the prototypes to determine if the reusable architectural components adequately support meeting the architectural concerns within the context of the candidate architectural vision.

  Informally review the results of the evaluation of the competing architectural visions (and vision creation, if appropriate) to determine their reasonableness and the logical consistency of the results.

5. **Select the most suitable architectural vision.** Use the decision rule on the evaluation results to select the most suitable of the evaluated competing candidate architectural visions.

   The success of this step is limited by the architects' previous success in creating candidate architectural visions. Given time and effort limitations as well as the architects' natural human limitations, the *optimum* most suitable architectural vision may not be among the candidate architectural visions. In fact, it is likely that there is no objectively most suitable architectural vision, and it is up to the architects to determine if the architectural vision selected using the decision rule on the evaluation results is sufficient to meet the system's architecturally significant requirements.

6. **Optionally create the new most suitable architectural vision.** If combining consistent vision components from multiple competing candidate architectural visions would yield an even more suitable architectural vision, then do so to create the single new vision.

   Repeat Step 4 to evaluate the suitability of the newly created architectural vision.

   The success of this step is limited by the architects' previous success in creating optimal vision components. Given time and effort limitations as well as the architects' natural human limitations, the optimal vision components may not have been previously determined. There is therefore no guarantee that the architects can create the *absolute* most suitable architectural vision from these vision components.

7. **Approve the architectural vision.** Stakeholders review and approve the selected vision, thereby authorizing progress to the next task.

   Note that this step may be tailored out of the system architecture engineering methods used on some endeavors if the stakeholders do not officially approve of the architects' selection and the architects merely need to notify the appropriate stakeholders of their decision.

8. **Identify any new architectural risks and opportunities.** Concurrently with the performance of the previous steps, identify and formally document any new architectural risks and opportunities that were discovered.

12.6 Postconditions

The *Select or Create the Most Suitable Architectural Vision* task is typically complete for the current subsystem when the following postconditions hold:

■ **The most suitable architectural vision exists.** The most suitable architectural vision for the system or subsystem has either been selected or created.
■ **Results are documented.** The results of the selection or creation process have been properly documented.
■ **Architectural risks and opportunities are managed.** Any newly identified credible architectural risks have been analyzed and documented in the endeavor risk management repository. Any newly identified architectural opportunities have also been analyzed and documented.

12.7 Work Products

The *Select or Create the Most Suitable Architectural Vision* task typically produces the following work products:

■ **Architectural concern versus candidate architectural vision matrix:** as illustrated in Table 12.1, a matrix that documents how well a set of competing candidate architectural visions support the relevant architectural concerns.

The cell contents have the following meanings: ++ means very strong support, + means significant support, 0 means neither support nor hindrance, – means significant hindrance,

Table 12.1 Example Architectural Concern versus Candidate Architectural Vision Matrix

Architectural Concern vs. Architectural Visions Matrix		*Candidate Competing Architectural Visions*			
		Architectural Vision 1	*Architectural Vision 2*	*Architectural Vision 3*	*Architectural Vision 4*
Architectural Concerns	Availability	+	+	+ +	+
	Development cost	0	+ +	– –	0
	Development schedule	+	+ +	– –	–
	Interoperability	+	–	+	+
	Performance	+	– –	+	+ +
	Portability	0	+	0	+
	Reliability	+	–	+ +	–
	Safety	–	–	+ +	0
	Security	–	+ +	–	0
	Usability	–	0	–	+

and – – means strong hindrance (i.e., the architectural vision makes it very difficult for the system to achieve its associated requirements).

- **Architectural vision selection report:** a report describing the chosen evaluation and selection approach, including evaluation criteria, measurement scales, decision rule, and rationale, as well as the evaluation results (e.g., including an architectural concern versus candidate architectural vision matrix).
- **Architectural vision document:** a document describing the selected or created most suitable architectural vision.*
- **New and updated architectural risks and opportunities:** any new or modified architectural risks documented in the endeavor risk repository as part of endeavor risk management activity. This also includes an updated list of architectural opportunities that the architecture teams may wish to take advantage of in the future.

  Because any risks that were associated with the candidate architectural visions that were not selected are no longer relevant at this point, they should be retired and removed from the endeavor risk repository.

12.8 Guidelines

Use the following guidelines (and associated rationales) when performing the *Select or Create the Most Suitable Architectural Vision* task:

- Ensure a commensurate approach.
- Ensure a consistent evaluation approach.
- Ensure complete evaluation criteria.
- Avoid unwarranted assumptions.
- Use common sense when using decision methods to select the most suitable candidate architectural vision.
- Take reuse into account.
- Test reusable architectural component suitability.
- Maintain the architectural vision.
- Formally manage architectural risks.

The following descriptions express these guidelines in more detail and provide their associated rationales:

- **Ensure a commensurate approach.** Ensure that the evaluation approach is consistent with the criticality of the selection and the resources that are available to make the selection.

  Rationale: There are limited resources (such as the schedule and availability of architects) with which to decide on the most suitable architectural vision. Invest too little and the wrong architectural vision may be chosen, whereas invest too much and some other work will remain undone.
- **Ensure a consistent evaluation approach.** Ensure that the evaluation criteria, measurement scales, and decision-making technique used are consistent.

---

* Note that this information may be incorporated into a system-subsystem architecture document as opposed to forming a separate architectural vision document.

Rationale: An inconsistent approach is far less likely to produce the most suitable architectural vision. The resulting selection is also likely to have little credibility with the stakeholders.

■ **Ensure complete evaluation criteria.** Ensure that the architectural vision evaluation criteria include architectural concerns such as relevant quality requirements grouped by quality characteristic or quality attribute, development and life-cycle costs, schedule, and risk.

Rationale: Having too few evaluation criteria means that important concerns are left out and the most suitable architectural vision is much less likely to be selected.

■ **Avoid unwarranted assumptions.** Avoid making unwarranted assumptions regard decision models that do not reflect measurements or the situation at hand. Do not assume the:
  - Additivity of the decision criteria (i.e., do not assume an additive scale when only a nominal scale is appropriate)
  - Linearity (i.e., linear addition) of the measurements
  Rationale: The decision results will not be trustworthy and the most suitable architectural vision may not be selected.

■ **Use common sense when using decision methods to select the most suitable candidate architectural vision.** When reviewing the single resulting architectural vision for rationality, the following common sense checks may be appropriate:
  - If too many (or too few) candidate competing architectural visions meet the thresholds on evaluation criteria measurement scales, then:
    • Adjust the thresholds to narrow the field so that a selection can be made (if the thresholds are inappropriate)
    • Identify additional competing architectural visions (if the thresholds were appropriate but none of the existing architectural visions were adequate)
  - Check to determine if the results of filtering of candidate architectural visions based on the relative importance of the decision criteria is highly sensitive to the order of these criteria.
    Rationale: The order of the decision criteria should not be relevant, and if it is, then the decision results are not trustworthy.
  - Check to determine if the weighted-sum model yields only small variations among the competing candidate architectural visions.
    Rationale: There may be a scaling problem or linearity assumptions may be suspect.

■ **Take reuse into account.** When any candidate competing architectural visions incorporate reused components, ensure that the architectural vision evaluation criteria also include:
  - The impact of the components' reuse on the business process
  - The impact of licensing agreements, contracts, or similar acquisition vehicles
  - Other credible known risks associated with the reuse of architectural components
  Rationale: These evaluation criteria have proven important to the suitability of the reusable architectural components.

■ **Test reusable architectural component suitability.** Obtain evaluation copies of the reusable architectural components. If not already a part of the overall development test facilities, then create a reuse test facility in which to test these reusable architectural components. Produce an executable architectural prototype of the architectural vision that contains these reusable components. Test the reusable architectural components for their suitability within the rest of the architectural vision.

Rationale: Because "reusable" architectural components may be appropriate in certain contexts and not in other contexts, it is important to test their suitability within the context of the architectural vision in which they are incorporated. Testing provides much more credible information than vendor marketing literature and other such sources of information.

■ **Maintain the architectural vision.** Evolve the single selected or created architectural vision as the overall architecture evolves.

Rationale: The architectural vision provides a good summary of the system or subsystem architecture.

■ **Formally manage architectural risks.** As part of the risk management activity, identify architectural risks during the performance of this task. Add any architectural risks identified to the official endeavor risk management database.

Rationale: Architectural risks that are not identified or that are not added to the project risk management database when identified are much less likely to be effectively managed.

12.9 Pitfalls

Avoid the following common pitfalls associated with performing the *Select or Create the Most Suitable Architectural Vision* task and mitigate their negative consequences when they occur:

■ Architects use an inappropriate decision method.
■ Management provides inadequate decision resources.
■ Selection is viewed as purely a technical decision.
■ Stakeholders do not understand risks.
■ The decision makers are weak.

The following express these pitfalls in more detail, describing their negative consequences and the steps the architects can use to mitigate them:

■ **Architects use an inappropriate decision method.**
  – **Pitfall.** The architects use an inappropriate decision method (see Appendix D) to select the most suitable architectural vision.

    For example, suppose subjective factors (such as familiarity of the architectural approach to the organization and staff) are significant to the organization and the decision at hand. If these factors are not explicitly captured in the criteria for the decision procedure, they may only be ultimately recognized by a management-level *veto* at some level of management review.
  – **Negative consequences.** This pitfall often causes the following negative consequences:
    • The decision method may not result in the most suitable architectural vision.
    • The resulting architecture may have unnecessary defects, weaknesses, and risks.
    • The inappropriate decision method may require an unnecessarily large amount of resources (e.g., effort by the architects, time, and cost) that may in turn result in the violation of budget and schedule constraints.
    • The budget and schedule for subsequent architecture engineering may be inadequate due to the late discovery of these unnecessary architectural defects, weaknesses, and risks.
    • The resulting complete architecture may not meet the architecturally significant requirements.
    • The resulting system may not be acceptable to its stakeholders.
  – **Mitigations.** To avoid or mitigate the negative consequences of this pitfall:

- Include the *Determine the Selection Criticality, Determine the Required Selection Resources,* and *Determine the Evaluation Approach* steps of this task in the system architecture engineering method.
- Verify that these steps are properly performed (such as via quality assurance assessments).
- Obtain input from an expert in decision making and the alternative decision-making approaches.

- ■ **Management provides inadequate decision resources.**
  - – **Pitfall.** The architects are not given sufficient resources to properly evaluate and select the most suitable architectural vision on which to base the completed architecture.

    Selecting the most suitable architectural vision or determining that a better candidate architectural vision can be created are critical architectural decisions that justify a significant amount of time and effort to make.
  - – **Negative consequences.** This pitfall often causes the following negative consequences:
    - The selection process is incomplete and inadequate, potentially leading to hidden risks in the selected vision.
    - The most suitable architectural vision is not selected.
    - The resulting architecture may have unnecessary defects, weaknesses, and risks.
    - The budget and schedule for subsequent architecture engineering is inadequate due to the late discovery of these unnecessary architectural defects, weaknesses, and risks.
    - The resulting complete architecture may not meet the architecturally significant requirements.
    - The resulting system may not be acceptable to its stakeholders.
  - – **Mitigations.** To avoid or mitigate the negative consequences of this pitfall:
    - Incorporate the architectural vision selection method in the system architecture engineering method.
    - Properly document the architectural vision selection task in the project architecture plan and related conventions (such as the architectural vision selection report content and format standard, procedure, guidelines, and document template).
    - Include architectural vision selection in the project budget, master schedule, and work breakdown structure.

- ■ **Selection is viewed as purely a technical decision.**
  - – **Pitfall.** The architects (and possibly other stakeholders) believe that selecting the most suitable architectural vision is merely a technical decision.

    In addition to the ability of the vision to form the foundation of an architecture that meets the architecturally significant system requirements, it must also meet process requirements such as constraints on cost and schedule, as well as be acceptable to management and other major stakeholders. For example, the incorporation of a COTS architectural component assumes that there exists such a component with acceptable licensing from an acceptable source (such as a vendor with adequate financial stability, market share, and estimated business longevity).
  - – **Negative consequences.** This pitfall often causes the following negative consequences:
    - Nontechnical selection criteria are excluded from the selection process.
    - The architectural vision selection report is misleading.
    - The most suitable architectural vision is not selected.
    - The resulting architecture may have unnecessary defects, weaknesses, and risks.

- The budget and schedule for subsequent architecture engineering are inadequate due to the late discovery of these unnecessary architectural defects, weaknesses, and risks.
- The resulting complete architecture may not meet the architecturally significant requirements.
- The resulting system may not be acceptable to its stakeholders.
  - **Mitigations.** To avoid or mitigate the negative consequences of this pitfall:
    - Train the architects to include both technical and nontechnical selection criteria in the architectural vision selection method.
    - Have stakeholders (including managers) verify the completeness of the architectural vision selection method and the architectural vision selection report.

■ **Stakeholders do not understand risks.**
  - **Pitfall.** Architects and other stakeholders in authority do not understand and appreciate the architectural risks associated with the various candidate architectural visions.

    Note that some project managers and/or acquisition representatives are so risk averse that they create strong political and social pressures to not permit risks to be identified as risks. For example in severe cases, even use of the word *risk* becomes taboo and one instead hears words such as *issue* or *concern* used instead so that the real risk is not added to the risk management database of risks and is neither tracked nor managed.

    On the other hand, sometimes risk aversion may result in only the lowest-risk visions being considered, even if a much better vision having only moderate risk would be a better choice. Depending on the endeavor, relative risk level can trump technical superiority. Thus, this is not necessarily a "bad" outcome. For example, a technically inferior solution that has a low risk of missing its time-to-market window might well be preferred over a better solution with even a slightly higher risk missing that all-important marketing target. Being first to market is often more important than being best, especially if plans exist to improve quality and functionality over time.
  - **Negative consequences.** This pitfall often causes the following negative consequences:
    - An unnecessarily high-risk architectural vision is selected.
    - The associated risks are not managed, increasing the likelihood that they will occur.
  - **Mitigations.** To avoid or mitigate the negative consequences of this pitfall:
    - Include stakeholder risk-aversion in the selection criteria.
    - Add architectural risk management to the architecture engineering method, and provide training in architectural risk management to the architecture stakeholders.
    - Work closely with management to ensure that architectural risks are identified as such and managed like any other risk by the project risk management program.
    - Ensure that architectural risks are properly addressed during the vision selection process, being neither weighed too heavily or not heavily enough.
    - If appropriate, renegotiate some of the architecturally significant requirements to be able to modify the architectural vision so that it has less risk. In certain circumstances, it is better to renegotiate the requirements than it is to take what major stakeholders feel is an unacceptable architectural or business risk.

■ **The decision makers are weak.**
  - **Pitfall.** The people responsible for approving the most suitable architectural vision are not committed to the selection process, or are unresponsive, not architecturally knowledgeable, or not authorized to make the decision.
  - **Negative consequences.** This pitfall often causes the following negative consequences:

- The decision takes longer than necessary, thereby causing project deadlines to be missed and project cost to unnecessarily increase.
- The decision makers may not select the most suitable architectural vision.
- The decision makers may not accept the completed architecture derived from the selected architectural vision if they did not adequately understand the ramifications of their approval of the vision.
- **Mitigations.** To avoid or mitigate the negative consequences of this pitfall:
  - Clearly define in the architecture plan the roles, responsibilities, and authorities of the different architecture stakeholders. Ensure that this includes the role(s) that are responsible for approving the selection of the architectural vision. Ensure that this role not only has the responsibility, but also the authority to select the architectural vision.
  - Provide architecture training to the people responsible for selecting the architectural vision.
  - Ensure that the architectural vision selection report provides all the information needed on which to base the selection. This includes specifying the report contents in the project architecture engineering method and verifying the completeness of the report before its publication.

12.10 Summary

During the *Select or Create the Most Suitable Architectural Vision* task, the architecture teams obtain a single architectural vision for the system or subsystem architecture from the competing candidate visions.

12.10.1 Steps

When instantiating the *Select or Create the Most Suitable Architectural Vision* task, the following steps may be included, modified, or tailored out:

1. Determine the selection criticality.
2. Determine the required selection resources.
3. Determine the evaluation approach.
4. Evaluate the competing candidate architectural visions.
5. Select the most suitable architectural vision.
6. Optionally create the new most suitable architectural vision.
7. Approve the architectural vision.
8. Identify any new architectural risks and opportunities.

12.10.2 Work Products

When instantiating the *Select or Create the Most Suitable Architectural Vision* task, the following work products may be included, modified, or tailored out:

- Architectural concern versus candidate architectural vision matrix
- Architectural vision selection report

- Architectural vision document
- New and updated architectural risks and opportunities

12.10.3 Guidelines

Use the following guidelines (and associated rationales) when performing the *Select or Create the Most Suitable Architectural Vision* task:

- Ensure a commensurate approach.
- Ensure a consistent evaluation approach.
- Ensure complete evaluation criteria.
- Avoid unwarranted assumptions.
- Use common sense when using decision methods to select the most suitable candidate architectural vision.
- Take reuse into account.
- Test reusable architectural component suitability.
- Maintain the architectural vision.
- Formally manage architectural risks.

12.10.4 Pitfalls

Avoid the following common pitfalls associated with performing the *Select or Create the Most Suitable Architectural Vision* task and mitigate their negative consequences when they occur:

- Architects use an inappropriate decision method.
- Management provides inadequate decision resources.
- Selection is viewed as purely a technical decision.
- Stakeholders do not understand risks.
- The decision makers are weak.

Chapter 13

Task 8: Complete the Architecture and Its Representations

13.1 Introduction

During this MFESA task, the system or subsystem architecture team completes the architecture and its representations based upon the architectural vision selected or created (Figure 13.1).

Starting with the selected architectural vision as a foundation, the architecture team completes and documents the architectural models of the system or subsystem during this task. Because the architecture is based on the selected architectural vision of the system or subsystem, the vision must exist before the architecture team can begin this task. The architecture team begins by completing the logical and physical as well as static and dynamic architectural views of the system or subsystem. These views will largely consist of a consistent set of architectural models that were identified in the architectural vision. The architecture team then builds on these incomplete models by creating specialty-engineering focus areas, which typically address the architecture's support for individual architectural concerns (such as interoperability, performance, reliability, safety, and security). Next, the architecture team engineers the different types of interface architectural models. During the performance of these steps, the architecture team documents the architecture in architectural diagrams, models, architectural quality cases [Firesmith et al., 2006], and various architectural documents.

This task is a reusable MFESA method component and as such is intended to be tailorable to meet the specific needs of the endeavor. Therefore, it is to be expected that at least some of the task's objectives, preconditions, inputs, steps, postconditions, work products, guidelines, and pitfalls will be tailored during the creation of the endeavor-specific system architecture engineering method.

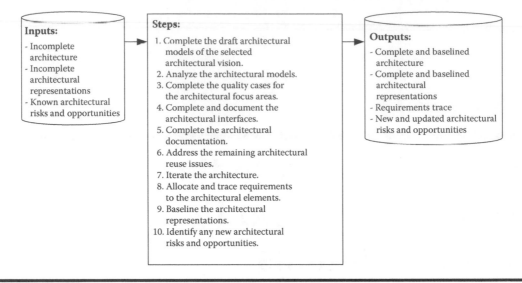

Figure 13.1 Summary of Task 8 inputs, steps, and outputs.

Tailoring of a method component includes adding missing content, modifying existing content, or removing existing content that is inappropriate, unnecessary, or not cost-effective to perform.

13.2 Goals and Objectives

The *Complete the Architecture and Its Representations* task has the following overall goals:

- Complete the system or subsystem architecture based on the selected or created architectural vision.
- Complete the representations of the architecture.

The typical subordinate objectives of this task are to:

- Complete the architectural models and focus areas.
- Complete the interface aspects of the architecture.
- Complete the reuse aspects of the architecture.
- Provide architectural representations that can be evaluated and accepted by its authoritative stakeholders.

13.3 Preconditions

The *Complete the Architecture and Its Representations* task can typically begin when the following precondition has been met:

- **The architectural vision is selected or created.** Task 7 has been completed so that a single architectural vision has been either selected or created.

■ **The first versions of the most important architectural models exist.** The partial draft logical and physical static and dynamic models for the selected architectural vision have been created.

13.4 Inputs

The *Complete the Architecture and Its Representations* task typically uses the following inputs:

■ **Incomplete architecture:** the existing incomplete architecture of the system or the architectural component being architected.
■ **Incomplete architectural representations:** the existing incomplete architectural representations for the system or the architectural component being architected, especially the:
  - Single selected or created architectural vision as documented in the architecture vision description document
  - Architectural views including associated models
  - Architectural focus areas
  - Architecture document
■ **Known architectural risk and opportunities:** a list of associated architectural risks and opportunities.

13.5 Steps

During the *Complete the Architecture and Its Representations* task, members of the relevant architecture team collaborate to perform the following steps in an incremental, iterative, concurrent, and time-boxed manner:

1. **Complete the draft architectural models of the selected architectural vision.** Complete the existing partial views and associated models of the architectural structures of the selected architectural vision by adding missing architectural elements and relationships between them. Add missing views and models of the architectural structures.
2. **Analyze the architectural models.** Analyze the completed architectural models to ensure that the architecture adequately supports the achievement of the architecturally significant requirements (e.g., interoperability, performance, reliability, safety, and security).
3. **Complete the quality cases for the architectural focus areas.** Work with the specialty engineering teams to complete the engineering of the architectural focus areas (such as availability, capacity, interoperability, performance, portability, reliability, safety, security, and usability) of the system or subsystem. Generate or complete the associated architectural quality cases documenting the architects' claims, arguments, and evidence.

   Note that MFESA architectural quality cases are taken from the Quality Assessment of System Architecture and their Requirements (QUASAR) method for assessing system requirements and architecture [Firesmith et al., 2006].
4. **Complete and document the architectural interfaces.** Complete the engineering of the architecture of the system or subsystem external and internal interfaces to ensure interoperability and intra-operability.

- **Ensure physical interoperability and intra-operability.** Select standard (or engineer new) *physical* interfaces including, for example:
  - Electrical connections (e.g., electrical plug type, power rating, number and configurations of prongs, and male versus female connection)
  - Electronic connections (e.g., number and configuration of pins as well as male versus female connection)
  - Physical connections and connectors (e.g., size and shape of surfaces as well as the number, type, location, orientation, and size of physical connectors such as bolts and screws)
  - Clearance for full range of motion of interconnected and/or potentially interfering mechanical subsystems
  - Power connections (e.g., physical linkage type such as belt cable, chain, and hose as well as specifics such as composition and dimensions)
  - Coordination to protect vulnerable subsystems from effects of normal operation of other systems (e.g., temporarily shrouding optical apertures from muzzle flash and particulates from rocket motor operation or provide shielding to protect subsystem from electromagnetic interference)
- **Ensure power interoperability and intra-operability.** Select standard (or engineer new) *power* interfaces including, for example:
  - Hydraulic interfaces (e.g., fluid composition as well as maximum, average, and minimum pressure, flow rate, and temperature)
  - Mechanical linkages (e.g., maximum, average, and minimum force and speed)
  - Pneumatic interfaces (e.g., gas composition as well as maximum, average, and minimum pressure, flow rate, and temperature)
  - Wired communication (e.g., proper logic voltages, frequency, and amperage)
  - Wired power (e.g., amperage ranges, voltage ranges, alternating versus direct current)
  - Wireless communications (e.g., radio, microwave, or laser, including frequency and power)
- **Ensure interoperability and intra-operability of operational environments.** Select standard (or engineer new) *mitigations* of the:
  - Thermal interactions among co-located subsystems
  - Electromagnetic interference (EMI) among co-located subsystems
  - Potential increase in platform observability (loss of stealth) due to observability of mechanical interface points
  - Mechanical interactions of coupled subsystems (vibration, jitter, and the like)
  - External thermal loads, electromagnetic radiation, and ionizing radiation effects
- **Ensure protocol interoperability and intra-operability.** Select standard (or engineer new) information exchange *protocols* including, for example:
  - Physical layer (layer 1) protocols such as Integrated Services Digital Network (ISDN) and RS-232
  - Data-link layer (layer 2) protocols, including Ethernet, Fiber Distributed Data Interface (FDDI), and Point-to-Point Protocol (PPP)
  - Network layer (layer 3) protocols such as Internet Control Message Protocol (ICMP), Internet Protocol (IP), Internetwork Packet Exchange (IPX), NetBEUI, and NWLink
  - Transport layer (layer 4) protocols such as NetBEUI, NWLink, Sequenced Packet Exchange (SPX), Transmission Control Protocol (TCP), and User Datagram Protocol (UDP)

- Session layer (layer 5) protocols such as Network File System (NFS)
- Presentation layer (layer 6) protocols such as American Standard Code for Information Interchange (ASCII), Moving Picture Experts Group (MPEG), and Secure Socket Layer (SSL)
- Application layer (layer 7) protocols such as Domain Name Service (DNS), File Transfer Protocol (FTP), Hypertext Transfer Protocol (HTTP), Hypertext Transfer Protocol Secure (HTTPS), Network News Transfer Protocol (NNTP), Simple Mail Transfer Protocol (SMTP), Simple Network Management Protocol (SNMP), Server Message Block (SMB), Telnet, X.400, and X.500

   – **Ensure syntax interoperability and intra-operability.** Select standard (or engineer new) information exchange *syntax* including, for example:
      - Communication mechanisms (e.g., synchronous, asynchronous)
      - Operation types (e.g., message, remote procedure call)
      - Operation signature (e.g., data types and data flow direction)
      - Information exchanges
      - Error handling
      - Associated operation quality characteristics
   – **Ensure semantics interoperability and intra-operability.** Select standard (or engineer new) information processing and exchange interface *semantics,* including units of measure, relevant state, meaning of operation names, data types, and exceptions.

5. **Complete the architectural documentation.** Complete the documentation of the architectural decisions regarding the system or subsystem — for example, using such documents, models, and diagrams as system or subsystem interface documents, interoperability white papers, vendor-supplied technical documents, UML or SysML models, context diagrams, hardware schematics, network diagrams, wiring diagrams, allocation diagrams, and system level configuration/class diagrams, layer diagrams, sequence diagrams, activity diagrams, and collaboration diagrams.

6. **Address the remaining architectural reuse issues.** Make the remaining architectural decisions related to the reuse of architectural elements including:
   – **Select the actual reusable components and their sources.** Based on the analysis of trade studies, select the actual reusable architectural components and their sources, such as product vendors, the government (GOTS), the military (MOTS), and open-source organizations.
   – **Acquire the actual reusable components from their sources.** Working with project management, acquire the selected reusable architectural components from their sources.
   – **Baseline the actual reusable components.** Working with the configuration management organization, identify the acquired reusable architectural components and any associated documentation as configuration items and place them under configuration control.
   – **Architect the wrapping of reused components.** Architect any wrappers or proxies for the reused components so that they fit within the existing architectural structures.

7. **Iterate the architecture.** Where appropriate, make changes to the architecture (including the selected architectural vision and the architectural interfaces) to fix defects and improve the architecture.

8. **Allocate and trace the requirements to the architectural elements.** Collaborate closely with the requirements team to help them derive appropriate new requirements for the current architectural component and its subordinate architectural elements. Allocate the requirements of the architectural component being architected to its subordinate architectural

elements (such as from system to subsystems, from current subsystem to newly identified sub-subsystems, or from current subsystem to architectural structure). Document this allocation in a forward and reverse requirements trace (e.g., as metadata in the requirements repository or tool, as metadata in an architecture repository or tool, or in a requirements traceability matrix).

9. **Baseline the architectural representations.** "Freeze" the architectural representations by placing them under configuration control so that the impact of future architecturally significant changes must be analyzed before making the changes is authorized by the configuration control board. This includes freezing architectural descriptions such as models, views, documents, and quality cases, as well as executable representations such as architectural prototypes and simulations.

10. **Identify any new architectural risks and opportunities.** Concurrently with the performance of the previous steps, identify and formally document any new architectural risks and opportunities that were discovered.

13.6 Postconditions

The *Complete the Architecture and Its Representations* task is typically complete for the current subsystem when the following postconditions hold:

■ **The architecture is complete.** The complete architecture of the system or architectural component being architected exists.

Note that there is an important difference between an architecture being complete and being done. The architecture is complete when all planned architectural decisions, inventions, trade-offs, assumptions, and associated rationales have been made. However, a complete architecture is likely never to be done until the system is retired (if even then) because changes to the complete architecture will continue to be made as the system and its requirements evolve.

■ **Architecture representations are complete.** The architecture representations of the system or architectural component being architected is complete and under configuration control.

■ **Architectural risks and opportunities are managed.** Any newly identified credible architectural risks have been analyzed and documented in the endeavor risk management repository. Any newly identified architectural opportunities have also been analyzed and documented.

13.7 Work Products

The *Complete the Architecture and Its Representations* task typically produces the following work products:

■ **Complete and baselined architecture:** the architecture of the system or architectural component is complete and baselined. All relevant architectural representations are properly completed and include all major architectural decisions, inventions, trade-offs, assumptions, and rationales. This includes the architectural models, the architectural focus areas (e.g., in architectural quality cases), the completion and documentation of the architectural interfaces, and the allocation and tracing of requirements to major architectural elements, especially the architectural components.

As illustrated in Figure 4.6, this task is performed concurrently with Task 9, *Evaluate and Accept the Architecture*. Because the architecture and its representations must pass their evaluations and be accepted before they can be baselined, the architecture and its representations cannot be completed until properly evaluated and accepted. Note also that the scope of completion is an individual release, and any changes to architecturally significant requirements will return the architecture and its representations to an incomplete state.

- **Complete and baselined architectural representations:** the complete and baselined architecture representations of the system or architectural component being architected, especially the updated:
  - Architectural diagrams and models
  - Architectural quality cases, including:
    - *Architectural claims* (e.g., address concerns and fulfill requirements)
    - Clear and compelling *architectural arguments* (e.g., architectural decisions, inventions, trade-offs, assumptions, and rationales)
    - Supporting official *architectural evidence*
  - Interface documentation
- **Requirements trace:** a tracing of the requirements from the current architectural component to its subordinate architectural elements as well as from these subordinate elements back through the current component to the requirements.
- **New and updated architectural risks and opportunities:** any new or modified architectural risks documented in the endeavor risk repository as a part of endeavor risk management activity. This also includes an updated list of architectural opportunities that the architecture teams may wish to take advantage of in the future.

13.8 Guidelines

Use the following guidelines (and associated rationales) when performing the *Complete the Architecture and Its Representations* task:

- Develop quality cases as a natural part of the architecture engineering process.
- Architect all relevant types of interfaces.
- Work with the requirements team to provide requirements traceability.
- Formally manage architectural risks.

The following descriptions express these guidelines in more detail and provide their associated rationales:

- **Develop quality cases as a natural part of the architecture engineering process.** Incrementally and iteratively develop architectural quality cases as a natural part of the architecture engineering process. Do not wait to develop the quality cases until immediately prior to an architectural evaluation that will verify that the architecture adequately supports the achievement of the associated quality concerns.

  Rationale: Producing architectural quality cases is an excellent way to ensure that the architecture adequately supports the system meeting its quality requirements and to document the architectural focus areas associated with quality characteristics and quality attributes. It is significantly more efficient and effective to develop these quality cases as the associated architectural decisions, inventions, trade-offs, assumptions, and rationales are made than it is to wait until just prior to an associated evaluation. There will typically be insufficient time and resources available to properly produce the quality cases in the days immediately prior to architectural quality evaluations.

■ **Architect all relevant types of interfaces.** Architect all of the relevant types of system and subsystem interfaces. Do not stop with message protocols and data syntax. Document these interfaces in the appropriate architectural representations. Ensure that the interfaces are documented to the level of detail needed to design and implement them.

Rationale: Many architectural defects reside in the interfaces between architectural components. Individual integrated product teams are often responsible for individual subsystems, but architecting the interfaces between subsystems requires the collaboration of multiple IPTs (and their associated architecture teams).

■ **Work with the requirements team to provide requirements traceability.** Work closely with the requirements team to trace the architecturally significant requirements to the architectural components that collaborate to implement them.

Rationale: Without proper traceability of the requirements to the architectural components, it is difficult to know which architectural components (and associated IPTs) are affected by changes to these requirements. Because the architecture teams (and not the requirements teams) are responsible for the allocation* of requirements to architectural components, it is more appropriate for the architecture teams to do the tracing.

■ **Formally manage architectural risks.** As part of the risk management activity, identify architectural risks during the performance of this task. Add any architectural risks identified to the official endeavor risk management database. Identify the fear of risk management as a risk, and a symptom of this dysfunction if the architects are being explicitly or implicitly forced to speak of architectural issues or concerns when the architects obviously mean risks.

Rationale: Architectural risks that are not identified or that are not added to the project risk management database when identified are much less likely to be effectively managed.

13.9 Pitfalls

Avoid the following common pitfalls associated with performing the *Complete the Architecture and Its Representations* task and mitigate their negative consequences when they occur:

■ Architecture engineering is finished.
■ Management provides inadequate resources.
■ The architectural representations lack configuration control.

---

* Note that the architecture teams should also collaborate with the requirements teams to help the requirements engineers derive new requirements that are appropriate at the level of the architectural component for which they are derived. During this process, the requirements teams and architecture teams may be virtual teams within integrated product teams, and the same person (if properly qualified) may well play the roles of both requirements engineer and architect.

The following express these pitfalls in more detail, describing their negative consequences and the steps that architects can use to mitigate them:

- **Architecture engineering is done.**
  - **Pitfall.** Some of the architecture stakeholders, especially management, believe that at some point in the development cycle, architecture engineering is done and no longer needs any resources (e.g., funding, schedule, or staffing). This may occur because the architecture stakeholders may:
    - Confuse the selected architecture vision with the completed system architecture
    - Assume that once the architecture is "complete," the architecture engineering work is also complete and the architectural resources (especially the architects) can be reallocated to other endeavors

      Note that although different levels of resources are needed at different times during the life cycle, some level of architecture engineering (and associated resources) need to be maintained as long as the system exists. In fact, system architects even have a role (if minor) to play during the retirement of the system.
    - Not understand the need to maintain the architecture and its representations as requirements evolve
    - Wish to allocate experienced architects to new endeavors and thereby support those projects while decreasing the cost of the current project
  - **Negative consequences.** This pitfall often causes the following negative consequences:
    - Without adequate dedicated resources, the architects become less productive and cannot effectively or efficiently perform their tasks.
    - Often, the junior architects are left to perform architectural maintenance because the experienced architects are reassigned to new system development projects. This is what the experienced architects want because they feel that developing new architectures is fun while maintaining that old architectures is boring.
    - Morale decreases among the architects as fewer architects must often carry a larger workload.
    - The architecture and its representations may not be completed or adequately maintained.
    - Architectural integrity is lost.
    - The probability increases that the endeavor budget is exceeded and deadlines are not met because the lack of architectural maintenance increases the likelihood of future expensive architectural problems.
  - **Mitigations.** To avoid or mitigate the negative consequences of this pitfall:
    - Ensure that all of the later architectural engineering tasks are included in the project method, budget, master schedule, and work breakdown structure.
    - Explicitly plan for one or more experienced architects to be involved in later stages as guides and mentors for the junior architects. This is also a good time to build junior architects into more experienced professionals.
    - Ensure that the architecture stakeholders, especially management, understand that the architecture is never finished, that the architecture and its representations need to be maintained, and that its integrity must also be maintained.
- **Management provides inadequate resources.**
  - **Pitfall.** Management does not provide adequate resources in terms of schedule, funding, and staffing to complete and maintain the system architecture and its representations.

- **Negative consequences.** This pitfall often causes the following negative consequences:
  - Without adequate dedicated resources, the architects become less productive and cannot effectively or efficiently perform their tasks.
  - Often, the junior architects are left to perform architectural maintenance because the experienced architects are reassigned to new system development projects.
  - Morale decreases among the architects as fewer architects must often carry a larger workload.
  - The architecture and its representations may not be completed or adequately maintained.
  - Architectural integrity is lost.
  - The probability increases that the endeavor budget is exceeded and deadlines are not met.
- **Mitigations.** To avoid or mitigate the negative consequences of this pitfall:
  - Ensure that all of the later architectural engineering tasks are included in the project method, budget, master schedule, and work breakdown structure.
  - Ensure that the architecture stakeholders, especially management, understand that the architecture is never finished, that the architecture and its representations need to be maintained, and that its integrity must also be maintained.
■ **The architectural representations lack configuration control.**
  - **Pitfall.** The architectural representations are not placed under configuration control.
    Note that the architecture and its associated architectural representations are very valuable resources that require proper configuration management. Although changes to the architecture and its representations can occur during the completion of the architecture over the course of a recursively incremental and iterative development cycle, this is especially a problem during architecture maintenance.
  - **Negative consequences.** This pitfall often causes the following negative consequences:
    - Architectural integrity is lost as changes are independently made to the architectural representations, thereby making the representations inconsistent.
    - Architects and other stakeholders lose trust in the architectural representations, and these representations become shelfware.
    - The architectural representations lose their value despite the large investment that was made in their production.
    - The architecture and its representations become more difficult to maintain.
  - **Mitigations.** To avoid or mitigate the negative consequences of this pitfall:
    - Baseline the verified and approved architectural representations by placing them under configuration control.
    - Analyze the impact of any proposed changes to the architecture and its representations before authorizing the changes to be made.

13.10 Summary

During the *Complete the Architecture and Its Representations* task, the architecture teams complete the system or subsystem architecture based on the selected or created architectural vision. They also maintain the system or subsystem architecture as the architecturally significant requirements change.

13.10.1 Steps

When instantiating the *Complete the Architecture and Its Representations* task, the following steps may be included, modified, or tailored out:

1. Complete the draft architectural models of the selected architectural vision.
2. Analyze the architectural models.
3. Complete the quality cases for the architectural focus areas.
4. Complete and document the architectural interfaces.
5. Complete the architectural documentation.
6. Address the remaining architectural reuse issues.
7. Iterate the architecture.
8. Allocate and trace the requirements to the architectural elements.
9. Baseline the architectural representations.
10. Identify any new architectural risks and opportunities.

13.10.2 Work Products

When instantiating the *Complete the Architecture and Its Representations* task, the following work products may be included, modified, or tailored out:

- Complete architecture
- Complete and baselined architectural representations
- Requirements trace
- New and updated architectural risks and opportunities

13.10.3 Guidelines

Use the following guidelines (and associated rationales) when performing the *Complete the Architecture and Its Representations* task:

- Develop quality cases as a natural part of the architecture engineering process.
- Architect all relevant types of interfaces.
- Work with the requirements team to provide requirements traceability.
- Formally manage architectural risks.

13.10.4 Pitfalls

Avoid the following common pitfalls associated with performing the *Complete the Architecture and Its Representations* task and mitigate their negative consequences when they occur:

- Architecture engineering is finished.
- Management provides inadequate resources.
- The architectural representations lack configuration control.

Chapter 14

Task 9: Evaluate and Accept the Architecture

14.1 Introduction

During this MFESA task, the architecture teams as well as other stakeholders evaluate the quality of the system or subsystem architecture so that architectural risks can be managed, compliance with architecturally significant requirements can be determined, and the architecture can be accepted by its authoritative stakeholders (Figure 14.1).

Note that this task and Task 2, *Identify the Architectural Drivers,* may be performed iteratively and concurrently because architectural risks and opportunities identified during this task may result in modifying or renegotiating architecturally significant requirements as part of Task 2. This overlapping of Tasks 2 and 9 can be seen in Figure 4.6: MFESA tasks by life-cycle phase.

This task is a reusable MFESA method component and as such is intended to be tailorable to meet the specific needs of the endeavor. Therefore, it is to be expected that at least some of the task's objectives, preconditions, inputs, steps, postconditions, work products, guidelines, and pitfalls will be tailored during the creation of the endeavor-specific system architecture engineering method. Tailoring of a method component includes adding missing content, modifying existing content, or removing existing content that is inappropriate, unnecessary, or not cost-effective to perform.

14.2 Goals and Objectives

The *Evaluate and Accept the Architecture* task has the following overall goals:

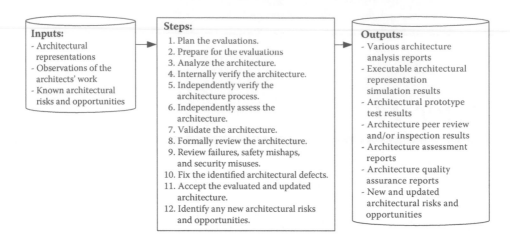

Inputs:
- Architectural representations
- Observations of the architects' work
- Known architectural risks and opportunities

Steps:
1. Plan the evaluations.
2. Prepare for the evaluations
3. Analyze the architecture.
4. Internally verify the architecture.
5. Independently verify the architecture process.
6. Independently assess the architecture.
7. Validate the architecture.
8. Formally review the architecture.
9. Review failures, safety mishaps, and security misuses.
10. Fix the identified architectural defects.
11. Accept the evaluated and updated architecture.
12. Identify any new architectural risks and opportunities.

Outputs:
- Various architecture analysis reports
- Executable architectural representation simulation results
- Architectural prototype test results
- Architecture peer review and/or inspection results
- Architecture assessment reports
- Architecture quality assurance reports
- New and updated architectural risks and opportunities

Figure 14.1   Summary of Task 9 inputs, steps, and outputs.

- Monitor and determine the quality, maturity, and completeness of the system or subsystem architecture and associated representations.*
- Monitor and determine the quality, maturity, and completeness of the process used to engineer the system or subsystem architecture.
- Provide information that can be used to determine the passage or failure of architectural milestones.
- Enable architectural defects, weaknesses, and risks to be fixed and managed before they negatively impact system quality and the success of the system development/enhancement project.
- Accept the system or subsystem architecture based on the results of the evaluations.

The typical subordinate objectives of this task are to:

- *Internally verify* the system or subsystem architecture so that:
  - Architectural defects are identified and corrected
  - Architectural risks are identified and managed
- *Independently assess* the system or subsystem architecture to determine compliance with architecturally significant product requirements.
- *Validate* that the system or subsystem architecture meets the needs of its critical stakeholders.
- *Formally review* the system or subsystem architecture by stakeholder representatives at one or more major project reviews.
- *Independently evaluate* the "as-performed" architecture engineering process to determine compliance with the documented architecture engineering method (e.g., as documented in the architecture plan, standards, procedures, and guidance).

---

* Note that where practical, the monitoring and determining of the quality of the system (product quality) and the quality of the process (process quality) should be done in a quasi-continuous manner or on a regular basis. Note also that because of the overlapping of the architecture engineering tasks, these goals can be achieved very early in the architecture engineering method. This applies to both the goals as well as the subordinate objectives of this task.

14.3 Preconditions

The *Evaluate and Accept the Architecture* task can typically begin when the following preconditions have been met:*

- **Architectural models exist.** Architectural models of relevant system structures exist to be evaluated.
- **Architectural vision exists.** The most suitable architectural vision has been selected or created.
- **Other architectural representations exist.** Some of the other architectural representations (e.g., architecture documents and architectural quality cases) exist.

14.4 Inputs

The *Evaluate and Accept the Architecture* task typically uses the following inputs:

- **Architectural representations:** architectural representations including:
  - Architectural descriptions:
    - Architecture models
    - Architecture documents
  - Executable architectural prototypes and simulations
- **Observations of the architects' work:** observations by evaluators of how the architects perform their work units (to help determine process quality).
- **Known architectural risks and opportunities.** a list of associated architectural risks and opportunities.

14.5 Steps

During the *Evaluate and Accept the Architecture* task, members of the relevant architecture team collaborate to perform the following steps in an incremental, iterative, concurrent, and time-boxed manner:

1. **Plan the evaluations.** The architecture team makes plans in preparation for performing the evaluations:
   - **Determine the evaluation needs.** The architecture team determines the needs driving the evaluation effort.†
   - **Determine the evaluation types.** Based on the needs for evaluation, the architecture team determines the types of evaluations to conduct.

---

* Note that because architectural evaluation is typically performed in a recursively incremental and iterative manner, the entire architecture and its representations need not be complete before evaluation begins The degree of necessary completeness and maturity varies by evaluation type.
† Planning the evaluations may also be considered by some to be a part of Task 1, *Plan and Resource the Architecture Engineering Effort.*

- **Determine the scope of the evaluations.** Based on the needs for evaluation and the types of evaluations selected, the architecture team determines the scope of the evaluations in terms of architectural concerns and components to be evaluated.
- **Determine the evaluation success criteria.** For each type of evaluation planned, the architecture team determines success criteria (e.g., quality and completeness metrics) for passing the evaluation.
- **Schedule the evaluations.** The architecture team schedules the planned evaluations.

2. **Prepare for the evaluations.** The architecture team prepares for the evaluations by developing and obtaining evaluation work products and resources:
   - **Develop evaluation work products.** For each planned evaluation, the architecture team develops needed evaluation work products such as evaluation models, simulations, prototypes, benchmarks, scenarios, simulators and stimulators, and instrumentation.
   - **Obtain evaluation resources.** For each planned evaluation, the architecture team obtains the resources needed to perform the evaluation, including the availability of evaluators, architects of the architectures being evaluated, evaluation tools (such as data analysis and visualization tools), funding, and schedule.

3. **Analyze the architecture.** On an ongoing basis, the architecture team analyzes their architectural graphical models, textual models, and executable architectural prototypes:
   - **Perform automated analyses of the models.** The architecture team performs static analyses of the architectural models. For example, this analysis can estimate system performance and reliability. Graphical debuggers can also be used to check for model consistency (and thereby produce the equivalent of "clean compile" of the models).
   - **Perform automated analyses of the executable models.** The architecture team performs dynamic analyses of the any executable architecture models with sufficient formality and existing tool support. For example, automated analyses can be performed on models implemented in SAE Architecture Analysis and Design Language (AADL) using its associated tools [Feiler et al., 2006].
   - **Perform manual analyses of architectural models.** The architecture team manually analyzes their architectural models for:
     - The degree to which they support their allocated and derived quality requirements
     - Their quality characteristics (such as completeness, conformance to standards, consistency, correctness, lack of ambiguity, readability, traceability to requirements, and understandability)
   - **Measure the architecture quality metrics.** The architecture team analyzes the architecture models to measure the associated level of architecture quality metrics.
   - **Analyze the executable architecture prototypes.** The architecture team analyzes their executable architecture prototypes via simulation and/or testing to verify that the architecture supports its architecturally significant requirements (e.g., quality requirements).

4. **Internally verify the architecture.** At appropriate times during architecture engineering, the architecture team hosts development organization internal evaluations to verify the quality, maturity, and completeness of their system or subsystem architectures. For example, the following techniques can be used to identify architectural defects and risks:
   - **Peer reviews.** The architecture team hosts informal peer reviews of their architecture by other architects within the project or development organization.
   - **Walk-throughs.** The architecture team hosts informal internal walk-throughs of their architecture by internal and representative external stakeholders such as other architects

within the project or development organization, requirements engineers, testers, technical leaders and managers, marketing, manufacturing, quality assurance engineers, and representatives from specialty engineering domains such as safety and security.

- **Formal inspections.** The architecture team hosts formal internal inspections (e.g., Fagan inspections) of their architecture by internal and representative external stakeholders such as other architects within the project and/or development organization, requirements engineers, testers, technical leaders and managers, marketing, manufacturing, quality assurance engineers, and representatives from specialty engineering domains such as safety and security.

- **Architecture evaluations.** The architecture team hosts evaluations of their architecture. Although the following architectural evaluation methods were designed for evaluating software architectures rather than system architectures, these methods can also be useful for evaluating the architecture of software subsystems and may even be applied to other subsystems* (i.e., subsystems containing hardware, data, personnel, manual procedures, facilities, etc.) if properly tailored or extended:

  • Active Design Review (ADR) [Parnas and Weiss, 1985]
  • Active Reviews for Intermediate Design (ARID) [Clements, 2000]
  • Architecture Evaluation Framework (AEF) [Lehto and Marttiin, 2005]
  • Architecture Level Prediction of Software Maintenance (ALPSM) [Bengtsson and Bosch, 1999]
  • Architecture Trade-off Analysis Methods (ATAM) [Kazman et al., 2000]
  • Cost Benefit Analysis Method (CBAM) [Nord et al., 2003]
  • Scenario-Based Architecture Reengineering (SBAR) [Bengtsson and Bosch, 1998]
  • Software Architecture Review and Assessment (SARA) [SARA WG, 2002]

5. **Independently verify the architecture process.** During architecture engineering, quality assurance independently determines compliance of the as-performed architecture process with the documented architecture engineering method (such as documented in the architecture plan, standards, procedures, and guidance).

6. **Independently assess the architecture.** Near the end† of system architecture engineering to support acquisition visibility and oversight, the architecture team hosts one or more development organization external assessments to determine compliance of the architecture with its architecturally significant requirements. For example, the following techniques can be used to perform all or part of these assessments:

- Modular Open System Approach (MOSA) using the Program Assessment and Review Tool (PART) [OSJTF, 2004]
- Network Centric Analysis Tool (NCAT™) [NCOIC, 2007]
- Open Architecture (OA) with the Open Architecture Assessment Tool (OAAT) [DAU, 2007]
- Quality Assessment of System Architectures and their Requirements (QUASAR) [Firesmith et al., 2006; Firesmith, 2007]‡

---

* In other words, this refers to subsystems containing hardware, data, personnel, manual procedures, equipment, materials, facilities, and the like.

† Because the system architecture is typically iteratively and incrementally engineered, these assessments are typically iteratively and incrementally held near the end of each build/block/release during which a significant amount of architectural engineering occurs.

‡ Note that although it was designed for independent assessment of system architectures (and their requirements), QUASAR can also be used for internal architecture evaluations (Step 5).

7. **Validate the architecture.** Near or at the end of system architecture engineering, the architecture team collaborates with representative stakeholders to validate that the architecture meets the stakeholders' needs.

8. **Formally review the architecture.** Near or at the end of system architecture engineering, the architecture team hosts one or more development organization external incremental milestone reviews to determine if the system or subsystem architecture is sufficient to support transition to the next phase of development:
   - System/subsystem architecture review (SAR)
   - System/subsystem design review (SDR)
   - System/subsystem preliminary design review (PDR)

9. **Review failures, safety mishaps, and security misuses.** As members of failure review boards, accident and safety incident investigation teams, and attack and security incident investigation teams, architects help to determine if the architecture contains a vulnerability that is a (partial) cause of a system failure, accident or safety incident, or attack or security incident.

10. **Fix the identified architectural defects.** The relevant architecture team fixes any defects in the architecture and the associated architectural representations that have been identified as a result of the evaluations.

11. **Accept the evaluated and updated architecture.** Upon correction of any architectural defects found during the architecture evaluations, the authoritative stakeholders accept (or reject) the architecture, whereby acceptance implies authorization to progress beyond the associated milestone. For example, authoritative stakeholders may include one or more of the following:
   - Acquisition managers (e.g., Program Office)
   - System development managers
   - Marketing managers
   - Organizational chief architects

12. **Identify any new architectural risks and opportunities.** Concurrently with the performance of the previous steps, identify and formally document any new architectural risks and opportunities that were discovered.

14.6 Postconditions

The *Evaluate and Accept the Architecture* task is complete for the current subsystem when the following postconditions hold:

- **Product quality is determined.** The quality of the *architecture* and its *representations* is determined and reported to its stakeholders. This includes the degree to which the architecture meets its architecturally significant:
   - Product requirements such as quality requirements and architecture constraints
   - Process requirements such as affordability (cost), schedule, and manufacturability requirements
- **Process quality is determined.** The quality of the *architecture engineering process* is determined and reported to its stakeholders.
- **The architecture and its representations are accepted.** The architecture and its representations have been accepted by their authoritative stakeholders.

■ **Architectural risks and opportunities are managed.** Any newly identified credible architectural risks have been analyzed and documented in the endeavor risk management repository. Any newly identified architectural opportunities have also been analyzed and documented.

14.7 Work Products

The *Evaluate and Accept the Architecture* task produces the following work products:

■ **Various architecture analysis reports:** reports that document the results of various types of analyses of the architecture and its representations.

■ **Executable architectural representation simulation results:** the documented results of running simulations using the executable architectural representations.

■ **Architectural prototype test results:** the documented test results from testing any architecture prototypes (e.g., to determine their support for any architecturally significant requirements).

■ **Architecture peer review and/or inspection results:** the documented results from any peer reviews or inspections of the architecture and its representations.

■ **Architecture assessment reports:** reports documenting the results of independent assessments of the architecture and its representations (such as QUASAR quality assessments of the system architecture).

■ **Architecture quality assurance reports:** quality engineering reports by the quality engineering team that document the results of evaluations of the architecture and its representation (quality control) or the process used to produce them (quality assurance).

■ **Known architectural risks and opportunities:** any new or modified architectural risks documented in the endeavor risk repository as a part of endeavor risk management activity. This also includes an updated list of architectural opportunities that the architecture teams may wish to take advantage of in the future.

14.8 Guidelines

Use the following guidelines (and associated rationales) when performing the *Evaluate and Accept the Architecture* task:

■ Use evaluations to support architectural milestones.
■ Evaluate continuously.
■ Internally evaluate models.
■ Perform architecture evaluation substeps.
■ Collaborate with the stakeholders.
■ Tailor software evaluation methods.
■ Perform independent architecture assessments.
■ Formally review the architecture.
■ Verify architectural consistency.
■ Perform cross-component consistency checking.
■ Perform *both* static and dynamic checking.
■ Set the evaluation scope based on risk and available resources.
■ Formally manage architectural risks.

The following descriptions express these guidelines in more detail and provide their associated rationales:

■ **Use evaluations to support architectural milestones.** Perform some of the evaluations prior to and in preparation for the architectural milestones. For example, perform at least one architectural evaluation on the architectural quality cases (e.g., architect-provided arguments and evidence) to independently verify that if the system is built in accordance with its architecture, then it will:
  − Adequately support its operational concept
  − Meet its architecturally significant requirements
  − Satisfy its critical stakeholders
  − Not have unacceptable architectural risks
  Rationale: When properly performed, the appropriate evaluations will provide sufficient information to determine whether the architecture passes or fails the architectural milestones.

■ **Evaluate continuously.** Perform the different kinds of evaluations on a continuous or regular basis. Do not wait until the architecture is complete to evaluate it.
  Rationale: Although the *Evaluate and Accept the Architecture* task is almost the "last" task in MFESA, the intent of this ordering is *not* for this step to be performed only near the "end" of the architecture engineering effort. Instead, architecture evaluation should typically be an ongoing effort, a primary goal of which is to identify architectural defects, weaknesses, and risks as early as is practical so that they can be fixed and managed before they negatively affect system quality and the success of the system development/enhancement project. The continuous and concurrent performance of this task is illustrated in Figure 4.6: MFESA tasks by life-cycle phase.

■ **Internally evaluate models.** The architecture team should perform internal architectural evaluations on an ongoing basis as they develop their architectural models.
  Rationale: Such informal internal evaluations are a cost-effective way to identify defects in the architectural models early during development before they cause major harm to the resulting system design and implementation.

■ **Perform architecture evaluation substeps.** Where appropriate and cost-effective, perform the following substeps for each evaluation:
  − Determine the goals and objectives of the evaluation.
  − Develop a plan for performing the evaluation, including the evaluation schedule and necessary resources.
  − Determine the appropriate evaluation method(s) and technique(s) to use.
  − Develop or update the executable models to incorporate new technologies and/or improve the accuracy or efficiency of the simulation or test.
  − Collect system external and internal data as input to the executable evaluation model.
  − Set model parameters to pre-simulation or pre-test values.
  − Execute the evaluation models to produce the simulation/test results.
  − Analyze the simulation/test results or compare the actual test results with the expected test results.
  − Prepare, publish, and distribute that evaluation report.
  Rationale: The preceding steps help produce a complete and effective evaluation.

■ **Collaborate with the stakeholders.** When appropriate, the architecture team should collaborate with internal and external stakeholders to incrementally verify their architecture and identify architectural risks. This collaboration can occur either at minor internal milestones (also known as "inch pebbles") during architecture engineering or during major system milestone reviews.

Rationale: The stakeholders can often contribute by identifying architectural defects and risks as well as by validating the architecture team's understanding of architecturally significant requirements.

■ **Tailor software evaluation methods.** If a *software* architecture evaluation method is used for more than the evaluation of a system's software components, then ensure that the method is properly tailored and extended to cover the different types the system architectural components on which it is used.

Rationale: More well-documented software architecture evaluation methods exist than system architecture evaluation methods. However, these methods have been designed to evaluate *software* architectures and there are some significant differences between the architectures of systems and the architectures of software applications or software architectural components such as types of architectural components, types of interfaces, and the increased size and complexity of both systems and their architectures. Thus, it is important to tailor the software architecture evaluation methods for use on systems if they are going to be used as system architecture evaluation methods.

■ **Perform independent architecture assessments.** The architecture should be independently assessed for conformance with the architecturally significant requirements and to ensure that all architectural risks have been properly managed prior to major architectural program milestones.

Rationale: The system architecture is sufficiently critical, especially its support for the architecturally significant quality requirements, to typically justify the resources needed to perform independent (quality) assessments.

■ **Formally review the architecture.** The architecture should be formally reviewed at every major project milestone following a build/block/phase during which the architecture undergoes significant engineering.

Rationale: Formal project reviews at major project milestones are good venues at which to provide stakeholders with an overview of the system/subsystem architecture, a review of its status, and an overview of any major architectural risks. Because project reviews are *not* an effective venue for performing serious technical evaluations of the system architecture, the architecture should be technically evaluated *prior* to the review and only the results of the evaluation presented at the review.

■ **Verify architectural consistency.** Check the consistency and integrity of the architecture of the architectural element being evaluated, including consistency:
  – Between components within the architectural element
  – With other "sister" architectural elements at the same level within the architectural hierarchy within the system (e.g., intra-operability)
  – With the architecture of the higher-level parent architectural element within the system
  – With the architectures of any lower-level architectural elements within the system
  – With the architecture of other systems (e.g., interoperability)
  – With reference architectures
  – With enterprise architectures

  Rationale: A system's architecture tends to be very large and complex with many structures, models and views of these structures, concerns and associated focus areas, and architectural decisions, inventions, trade-offs, assumptions, and rationales. It is extremely easy for the different parts of the architecture to become inconsistent with one another, and any inconsistencies can eventually lead to more serious defects during the design, implementation, and integration of the associated architectural components. It is therefore important to verify architectural consistency by checking for inconsistencies because the defects will be easier and faster to fix earlier rather than later.

- **Perform cross-component consistency checking.** Check the consistency of the architecture engineering methods being used to engineer the architectures of different architectural components within the system.

  Rationale: Large and complex systems are often decomposed into a hierarchy of subsystems and lower-level architectural components. During this decomposition, there is a danger that some of the architectural decisions made during the architectural engineering of lower-level architectural components will be inconsistent with the previous architectural decisions made at higher levels of the system architectural hierarchy. There is also a danger that these architectural decisions may conflict with decisions made for sibling components in the decomposition hierarchy. It is important that the evaluation is not entirely local to individual architectural components, but also assesses architectural consistency across architectural components (e.g., because multiple components must typically collaborate to meet the quality requirements).

- **Perform *both* static and dynamic checking.** Perform both static "syntax" checking of architectural models as well as dynamic "semantics" checking of the executable parts of the architectural models. This is like determining whether the architecture "compiles" as well as whether the architecture "executes" properly.

  Rationale: Most architecture evaluation methods concentrate on the evaluation and analysis of the static architectural structures, models, and views. They tend to under-emphasize the dynamic architectural structures, models, and views, perhaps because these are more difficult to understand and model properly. However, many major architectural defects are dynamic in nature and cannot be easily discovered without addressing the behavior of the relevant architectural elements.

- **Set the evaluation scope based on risk and available resources.** As illustrated in Figure 14.2, set the scope of the evaluations based on the associated architectural risks and the resources available to perform the evaluations. Do not set the scope of independent assessments purely on the tier of the architectural component within the system aggregation hierarchy.

  Rationale: There are never sufficient resources to perform all of the different types of evaluations on all architectural components. Thus, the evaluation effort must concentrate on where it will do the most good. Note also that size, complexity, and criticality of lower-tier architectural components may vary greatly, thereby influencing the size, formality, and contents of the associated system architecture engineering method.

- **Formally manage architectural risks.** As part of the risk management activity, identify architectural risks during the performance of this task. Identify the fear of risk management as a risk, and a symptom of this dysfunction is if the architects are being explicitly or implicitly forced to speak of architectural issues or concerns when the architects obviously mean risks. Add any architectural risks identified to the official endeavor risk management database.

  Rationale: Architectural risks that are not identified or that are not added to the project risk management database when identified are much less likely to be effectively managed.

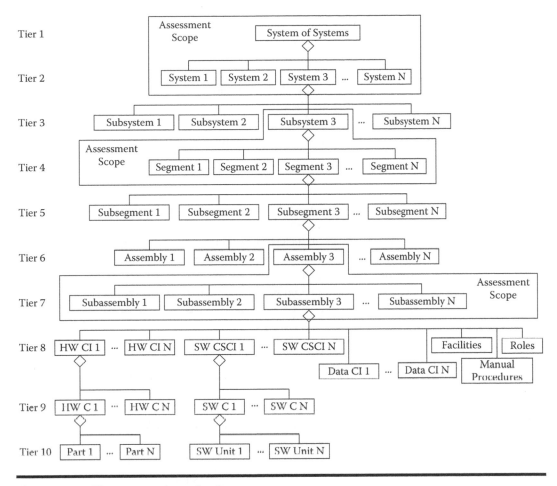

Figure 14.2   Three example evaluation scopes.

14.9 Pitfalls

Avoid the following common pitfalls associated with performing the *Evaluate and Accept the Architecture* task and mitigate their negative consequences when they occur:

- Disagreement exists over the need to perform evaluations.
- Consensus does not exist on the evaluation's scope.
- It is difficult to schedule the evaluations.
- Management provides insufficient evaluation resources.
- There are too few evaluations.
- There are too many evaluations.
- How good is good enough?
- Evaluations are not sufficiently independent.
- The evaluators are inadequate.
- Evaluations only verify the easy concerns.
- The architectural quality cases are poor.
- Stakeholders disagree on the evaluation results.

- ■ The evaluations lack proper acceptance criteria.
- ■ The evaluation results are ignored during acceptance.
- ■ The acceptance package is incomplete.

The following express these pitfalls in more detail, describing their negative consequences and the steps that architects can use to mitigate them:

- ■ **Disagreement exists over the need to perform evaluations.**
  - – **Pitfall.** The acquisition and development organizations disagree over the necessity to evaluate the architecture.
  - – **Negative consequences.** This pitfall often causes the following negative consequences:
    - • The architecture evaluations are either delayed or not held.
    - • The development organization does not provide sufficient resources to perform the evaluations (such as access to the architects, availability of the architects, access to the architectural representations, and access to the architecturally significant requirements).
    - • The architects do not provide the proper input for the evaluations being held, but rather supply only the documentation related to architecture engineering that happens to exist (such as architecture plans and architecture team organization charts), whether or not it is relevant.
    - • Less costly, but less powerful, evaluations are performed, which fail to identify significant architectural defects, risks, and opportunities.
  - – **Mitigations.** To avoid or mitigate the negative consequences of this pitfall:
    - • Perform the *Determine the Evaluation Needs* substep of the *Plan the Evaluations* step of the *Evaluate and Accept the Architecture* task.
    - • The acquisition organization includes architectural evaluations in the development contract.
    - • The acquisition organization mandates architectural evaluations as the method of verifying architecturally significant (e.g., quality) requirements.
    - • The acquisition organization verifies the existence of sufficient resources (such as availability of architects, project budget, and project schedule) to perform the evaluations.
    - • The acquisition organization uses incentives (such as contract award fees) to obtain development organization support to hold proper architectural evaluations.
- ■ **Consensus does not exist on the evaluation's scope.**
  - – **Pitfall.** Architects, their management, and the evaluators are not able to develop a consensus regarding the scope of the evaluations. Note that this might include disagreeing on the:
    - • Number of evaluations to perform
    - • Types of evaluations to perform, including:
      - ■ Informal evaluations, formal evaluations, or both
      - ■ Internal evaluations, independent evaluations, or both
      - ■ Peer reviews, walk-throughs, formal inspections, formal reviews, or quality case assessments
      - ■ Automated or manual analyses
      - ■ Analyses of executable architectural prototypes or simulations
    - • Architectural components to be evaluated
    - • Architectural concerns against which to evaluate the architectural components

- **Negative consequences.** This pitfall often causes the following negative consequences:
  - Insufficient evaluations are performed.
  - Some types of evaluations are not performed even though they should be.
  - The most effective and efficient types of evaluations are not performed.
  - Some architectural components that should be evaluated are not evaluated.
  - The architecture is not evaluated against some important architectural concerns.
  - Some architecture evaluations are delayed until consensus is eventually achieved.
  - The development organization does not provide sufficient resources to perform the evaluations (such as access to the architects, availability of the architects, access to the architectural representations, and access to the architecturally significant requirements).
  - The architects do not provide the proper input for the evaluations being held, but rather supply only the documentation related to architecture engineering that happens to exist (such as architecture plans and architecture team organization charts), whether or not it is relevant.
  - Architectural weaknesses and risks are not identified.
- **Mitigations.** To avoid or mitigate the negative consequences of this pitfall:
  - Perform the *Determine the Scope of the Evaluations* substep of the *Plan the Evaluations* step of the *Evaluate and Accept the Architecture* task.
  - The acquisition organization specifies the desired scope of the architectural evaluations in the development contract.
  - The acquisition organization mandates architectural evaluations as the method of verifying architecturally significant quality requirements.
  - The acquisition organization verifies the existence of sufficient resources (such as availability of architects, project budget, and project schedule) to perform the evaluations.
  - The acquisition organization uses contract award fees as an incentive to hold architectural evaluations.

■ **It is difficult to schedule the evaluations.**
- **Pitfall.** The architects, their management, and the evaluators are not able to properly schedule the evaluations with regard to master project, system, and subsystem schedules. This includes difficulty in deciding on schedules so that they fit the availability of the architects, evaluators, and potentially other stakeholders, as well as fitting an iterative, recursively incremental, concurrent, and time-boxed development cycle and its associated milestones.
- **Negative consequences.** This pitfall often causes the following negative consequences:
  - Evaluations are scheduled for the wrong reasons. For example, the evaluations may be scheduled based on the availability of architects as opposed to the proper maturity of architectural components being evaluated or the schedule of the major milestone reviews that the architecture evaluation results should feed into. For example, an architectural component, the architecture of which is not yet ready to evaluate, may be scheduled to be evaluated because its architects are not as busy as the architects of a component that should be evaluated.
  - Project delays caused by not being able to schedule evaluations when they should be conducted.
- **Mitigations.** To avoid or mitigate the negative consequences of this pitfall:
  - Work to gain a consensus between the architects, development management, and acquisition representatives on the positive effect that the performance of the

architecture evaluations has on the resulting quality of the architecture and its representations, thereby raising the priority of evaluations among all concerned.
- Include the *Schedule the Evaluations* substep of the *Plan the Evaluations* step of the *Evaluate and Accept the Architecture* task in the system architecture engineering method.
- Perform the *Schedule the Evaluations* substep.
- Verify the proper performance of the *Schedule the Evaluations* substep.

■ **Management provides insufficient evaluation resources.**
  - **Pitfall.** Management does not allocate sufficient resources (such as time, budget, and appropriate staffing) to properly perform the evaluations. Note that this is especially a problem when the evaluations are not included in the contract so that they end up not being properly funded and scheduled. Resources may be reduced or eliminated for the later evaluations to save costs and time to meet budget and schedule constraints.
  - **Negative consequences.** This pitfall often causes the following negative consequences:
    • Architects are not available to support the architecture evaluations.
    • Evaluators are not available to support the architecture evaluations.
    • Architecture evaluations unintentionally compete for the same resources as the architecture development tasks.
    • Training in how to perform the architecture evaluations is either not provided to evaluation stakeholders or is not provided at an appropriate time.
    • Architectural representations needed for the evaluations are either not produced, produced poorly, or are not produced in time for the evaluations.
  - **Mitigations.** To avoid or mitigate the negative consequences of this pitfall:
    • Work to gain a consensus between the architects, development management, and acquisition representatives on the importance that providing proper resources for the evaluations has on the resulting quality of the architecture and its representations.
    • Include the *Obtain Evaluation Resources* substep of the *Plan the Evaluations* step of the *Evaluate and Accept the Architecture* task in the system architecture engineering method.
    • Include evaluations in the careful planning of the architecture engineering activity at the beginning of the project. If one waits until the project is underway, then evaluations will be competing for either funds that have already been earmarked or for the management reserve. In either case, proper evaluations will not always be held even if everyone agrees that the evaluations are important.
    • Perform the *Obtain Evaluation Resources* substep.
    • Verify the proper performance of the *Obtain Evaluation Resources* substep.
    • Include architecture evaluations in the contract.

■ **There are too few evaluations.**
  - **Pitfall.** Either too few evaluations are performed or inadequate evaluations are performed. Note that this is often due to the cost of the evaluations and the limited availability of the architects and evaluators.
  - **Negative consequences.** This pitfall often causes the following negative consequences:
    • Some architectural defects, opportunities, and risks are not identified during the evaluations.
  - **Mitigations.** To avoid or mitigate the negative consequences of this pitfall:
    • Include the *Plan the Evaluations* step of the *Evaluate and Accept the Architecture* task in the system architecture engineering method.
    • Perform the *Plan the Evaluations* step.

- Verify the proper performance and adequacy of the *Plan the Evaluations* step.
- Include architecture evaluations in the contract.

■ **There are too many evaluations.**
  - **Pitfall.** The architects and the evaluators get caught in an "evaluation paralysis," whereby evaluations never stop. Note that this can occur if the quality of the architecture is really low, the architecturally significant requirements are constantly changing, or the system is so safety or business critical that more evaluations than are practical are scheduled by a strongly risk-adverse management.
  - **Negative consequences.** This pitfall often causes the following negative consequences:
    - The evaluations require too many resources.
    - The architects have less time and budget to spend on other architectural tasks.
    - The project budget is exceeded.
    - The project deadlines are missed.
  - **Mitigations.** To avoid or mitigate the negative consequences of this pitfall:
    - Schedule and time-box the performance of the architecture evaluations by setting evaluation deadlines.
    - Verify the sufficiency (completion) criteria of the architecture evaluations.
    - Ensure that the architecturally significant requirements are reasonably stable and baselined.
    - Ensure that the architecture and its architectural representations are ready to be evaluated before scheduling evaluations.
    - Report the progress of architectural evaluations during major milestone reviews.

■ **How good is good enough?**
  - **Pitfall.** There are inadequate architecturally significant requirements to drive the evaluations. Quality requirements may be totally missing, incomplete, and ambiguous. For example, quality requirements may lack clear thresholds. Note that one should never get to this point because the maturity and quality of the requirements should already have been evaluated (e.g., with a QUASAR requirements assessment). Architecturally significant requirements should drive the architecture; if these requirements are inadequate, then the architecture based on them cannot be ready for an evaluation.
  - **Negative consequences.** This pitfall often causes the following negative consequences:
    - The architects (and evaluators, including testers) are unable to determine if the architecture adequately supports its allocated and derived architecturally significant requirements.
    - The architects have no objective way of being sure when their architecture is good enough.
    - The architects have no objective way of performing architectural trade-offs.
    - The architecture is too good. If the specified requirements mandated more than was actually needed by the stakeholders, then excesses in the architecture unnecessarily increase cost and schedule.
    - The architecture is not good enough, being unable to meet all of its allocated and derived architecturally significant requirements.
    - The system or architectural component is considered infeasible because ambiguous requirements are interpreted to be more constraining than necessary.
  - **Mitigations.** To avoid or mitigate the negative consequences of this pitfall:
    - Train requirements engineers in how to properly engineer architecturally significant requirements, especially quantitative quality requirements.

- Verify these requirements to ensure that they have the appropriate characteristics.
- Validate these requirements with stakeholder representatives who have the authority to set the requirements especially quantitative quality requirements.
- Train the architects in how to properly model the architecture, select architectural visions, and perform architectural trade-offs.
- Verify the resulting architecture and architectural representatives.

■ **Evaluations are not sufficiently independent.**
  - **Pitfall.** All the evaluators report to the same management as the architects.
  - **Negative consequences.** This pitfall often causes the following negative consequences:
    - Management improperly pressures the evaluators to ignore or under-report architectural defects and risks.
    - Architectural defects and risks are improperly minimized for political or other non-technical reasons.
    - Architectural evaluations are ineffective and inefficient.
    - Architectural defects are not fixed in a timely manner.
    - Architectural risks are not managed.
    - The overall budget and schedule suffer when the architectural defects and risks are eventually identified.
  - **Mitigations.** To avoid or mitigate the negative consequences of this pitfall:
    - Ensure that at least some evaluators report to different management than that of the architects.
    - Ensure that at least some evaluations are performed "independently" of the developers and their management.
    - Incorporate this independence in the architecture engineering method, specifically in the architecture evaluation procedures and the evaluation team charters.

■ **The evaluators are inadequate.**
  - **Pitfall.** Members of the architecture evaluation teams lack the training and experience that is necessary to properly perform the evaluations.
  - **Negative consequences.** This pitfall often causes the following negative consequences:
    - The evaluations are ineffective because they do not identify some of the architectural defects and risks.
    - The evaluations are inefficient because the evaluators are less productive and the evaluations take longer.
    - The architectural defects and risks not found during the evaluations eventually increase project cost and cause project deadlines to be missed.
    - The system does not meet its architecturally significant requirements.
    - The system is not acceptable to its stakeholders.
  - **Mitigations.** To avoid or mitigate the negative consequences of this pitfall:
    - Develop and publish architecture evaluation standards, procedures, and guidelines.
    - Publish these conventions to the evaluation stakeholders, including the evaluators.
    - Develop training materials in how to properly evaluate the architecture and its representations.
    - Train the members of the architecture evaluation teams and other stakeholders involved in the evaluations.
    - Ensure that each architecture evaluation team includes at least one experienced architect and one experienced evaluator.

■ **Evaluations only verify the easy concerns.**
  – **Pitfall.** The evaluators only perform the easy evaluations (such as by evaluating the architecture against the architectural concerns that are easy to analyze and measure).
  – **Negative consequences.** This pitfall often causes the following negative consequences:
    • The architecture and its representations are not evaluated against quality requirements that are difficult or expensive (in terms of time and budget, for example) to evaluate.
    • Architectural defects and risks associated with these more difficult and expensive quality concerns are not identified and managed.
    • The probability increases that the architecture does not sufficiently support these concerns and that the defects will only be discovered late in the development process when the architecture (and associated design and implementation) are more difficult and expensive to modify.
    • The probability increases that the system will not meet all of its quality requirements.
    • The probability increases that the system will not be acceptable to its stakeholders.
  – **Mitigations.** To avoid or mitigate the negative consequences of this pitfall:
    • Include the *Obtain Evaluation Resources* substep of the *Plan the Evaluations* step of the *Evaluate and Accept the Architecture* task in the system architecture engineering method.
    • Ensure that adequate evaluation resources are obtained to perform the evaluations against all important quality concerns.
    • Ensure that the evaluation stakeholders understand that different systems and architectural components will have different quality concerns (i.e., that the importance of different quality characteristics and attributes varies from component to component).
    • Verify that the evaluations are performed against all important quality concerns.
■ **The architectural quality cases are poor.**
  – **Pitfall.** The architects do not make adequate cases that their architecture has proper quality and adequately supports the system achieving sufficient quality [Firesmith et al., 2006; Firesmith, 2007]. The architects' arguments that their architecture has sufficient quality may not be clear or sufficiently compelling to justify their claims that their architecture sufficiently supports its allocated and derived architecturally significant requirements. The evidence the architects provide to support their arguments may be insufficient or irrelevant.
  – **Negative consequences.** This pitfall often causes the following negative consequences:
    • The architecture may not have sufficient quality.
    • The architecture may not sufficiently support the system meeting its quality requirements.
    • Existing architectural weaknesses and risks may not be identified until late in the project.
    • The evaluation results cannot be used to meet architectural milestone success criteria. The architectural quality cases may have to be redone and the evaluation repeated to achieve closure.
  – **Mitigations.** To avoid or mitigate the negative consequences of this pitfall:
    • The evaluators or trainers train the architects in their evaluation responsibilities, including the proper components of the architectural quality cases they must create and present.
    • The evaluators or process engineers provide the architects with the relevant standards, procedures, guidelines, and templates for creating the architectural quality cases.

- The evaluators or trainers ensure that the architects understand the evaluation success criteria.
- A trainer or process engineer collaborates with the architects to ensure that the components of the architectural quality cases are naturally and properly produced as a natural part of the architecture engineering method.

■ **Stakeholders disagree on the evaluation results.**
- **Pitfall.** The evaluators fail to arrive at a consensus as to the evaluation results. The architects disagree with the evaluators on the evaluation results.
- **Negative consequences.** This pitfall often causes the following negative consequences:
  - It is difficult for the architects to act on the evaluation results.
  - It is difficult for management to act on the evaluation results.
  - It is difficult for representatives of the acquisition organization to act on the evaluation results.
  - The evaluation results cannot be used to obtain:
    - ■ Certification that the system is ready for operational use
    - ■ Authorization from the authorization authority to place the system into operational use
- **Mitigations.** To avoid or mitigate the negative consequences of this pitfall:
  - Ensure that the evaluation criteria are unambiguous and properly documented.
  - Repeat the evaluation, possibly with new evaluators.
  - Provide an escalation path for resolving disagreements over the evaluation results.

■ **The evaluations lack proper acceptance criteria.**
- **Pitfall.** There is no consensus on the acceptance criteria for the major architectural work products. Existing acceptance criteria are ambiguous or difficult to understand.
- **Negative consequences.** This pitfall often causes the following negative consequences:
  - It becomes impossible to objectively determine whether an architectural component passes an evaluation.
  - Time and effort are wasted in disagreements between the architects, the evaluators, and their management as to whether the architecture component has passed the evaluation.
  - There will be less motivation to fix architectural defects and manage architectural risks identified during the evaluation.
  - The evaluation results are less valuable as evidence supporting system certification and accreditation.
- **Mitigations.** To avoid or mitigate the negative consequences of this pitfall:
  - Include the development of architecture evaluation criteria in the system architecture engineering method.
  - Document the evaluation criteria in appropriate architecture work products such as architecture plans or architecture evaluation conventions (such as standards, procedures, and guidelines).
  - Verify the quality and appropriateness of the evaluation criteria.
  - Obtain early consensus on the proposed evaluation criteria.

■ **The evaluation results are ignored during acceptance.**
- **Pitfall.** The acceptance authority improperly ignores the evaluation results when determining whether to accept the system or its architecture. Note that the evaluation results should be a major driver of the acceptance of the system architecture.

- **Negative consequences.** This pitfall often causes the following negative consequences:
  - The probability increases that significant architectural defects and risks are ignored.
  - The probability increases that the *architecture* does not sufficiently support its allocated and derived architecturally significant requirements.
  - The probability increases that the *system* does not meet its architecturally significant requirements.
  - The probability increases that the system will not be acceptable to its stakeholders.
- **Mitigations.** To avoid or mitigate the negative consequences of this pitfall:
  - Ensure that the acceptance authority understands the implications of the evaluation results on the acceptability of the system and the success of the endeavor (in terms of cost and schedule constraints, for example).
  - Where necessary, elevate the issue to other stakeholders having sufficient authority to evaluate and mitigate this pitfall.

■ **The acceptance package is incomplete.**
- **Pitfall.** The architectural acceptance package is incomplete (e.g., missing important evaluation results). Note that important decisions should be based on all relevant data.
- **Negative consequences.** This pitfall often causes the following negative consequences:
  - Management has difficulty determining whether to accept the architecture or system.
  - The probability increases that architectures with significant defects and risks are accepted.
  - The probability increases that architectural defects are not fixed and architectural risks are not managed.
- **Mitigations.** To avoid or mitigate the negative consequences of this pitfall:
  - Define the architectural acceptance criteria and the contents of the architectural acceptance package as part of the architecture engineering method.
  - Verify the completeness of the architecture acceptance package prior to submitting it for acceptance.
  - Explain to management the importance of having a complete acceptance package before accepting the architecture.

14.10 Summary

During the *Evaluate and Accept the Architecture* task, the architecture teams monitor and determine the quality of the system or subsystem architecture and associated representations.

They monitor and determine the quality of the process used to engineer the system or subsystem architecture. They provide information that can be used to determine the passage or failure of architectural milestones. They enable architectural defects, weaknesses, and risks to be fixed and managed before they negatively impact system quality and the success of the system development/enhancement project. The teams also accept the system or subsystem architecture based on the results of the evaluations.

14.10.1 Steps

When instantiating the *Evaluate and Accept the Architecture* task, the following steps may be included, modified, or tailored out:

1. Plan the evaluations.
2. Prepare for the evaluations.
3. Analyze the architecture.
4. Internally verify the architecture.
5. Independently verify the architecture process.
6. Independently assess the architecture.
7. Validate the architecture.
8. Formally review the architecture.
9. Review failures, safety mishaps, and security misuses.
10. Fix the identified architectural defects.
11. Accept the evaluated and updated architecture.
12. Identify any new architectural risks and opportunities.

14.10.2 Work Products

When instantiating the *Evaluate and Accept the Architecture* task, the following work products may be included, modified, or tailored out:

- Various architecture analysis reports
- Executable architectural representation simulation results
- Architecture prototype test results
- Architecture peer review and/or inspection results
- Architecture assessment reports
- Architecture quality assurance reports
- New and updated architectural risks and opportunities

14.10.3 Guidelines

Use the following guidelines (and associated rationales) when performing the *Evaluate and Accept the Architecture* task:

- Use evaluations to support architectural milestones.
- Evaluate continuously.
- Internally evaluate models.
- Perform architecture evaluation substeps.
- Collaborate with the stakeholders.
- Tailor software evaluation methods.
- Perform independent architecture assessments.
- Formally review the architecture.
- Verify architectural consistency.
- Perform cross-component consistency checking.
- Perform *both* static and dynamic checking.
- Set the evaluation scope based on risk and available resources.
- Formally manage architectural risks.

14.10.4 Pitfalls

Avoid the following common pitfalls associated with performing the *Evaluate and Accept the Architecture* task and mitigate their negative consequences when they occur:

- Disagreement exists over the need to perform evaluations.
- Consensus does not exist on the evaluation's scope.
- It is difficult to schedule the evaluations.
- Management provides insufficient evaluation resources.
- There are too few evaluations.
- There are too many evaluations.
- How good is good enough?
- Evaluations are not sufficiently independent.
- The evaluators are inadequate.
- Evaluations only verify the easy concerns.
- The architectural quality cases are poor.
- Stakeholders disagree on the evaluation results.
- The evaluations lack proper acceptance criteria.
- The evaluation results are ignored during acceptance.
- The acceptance package is incomplete.

Chapter 15

Task 10: Maintain the Architecture and Its Representations

15.1 Introduction

During this MFESA task, the architecture teams maintain the architecture and its representations, while ensuring that the integrity of the system architecture and its representation does not degrade over time (Figure 15.1).

There is an unfortunate tendency for some stakeholders to believe that system architecture engineering is a one-time effort that, when completed, never has to be revisited. They feel that once the check is made in the architecture box, the architecture documentation can be placed on the shelf and forgotten. But although a good system architecture should be relatively stable and resistant to change, some change is unavoidable and the architecture is therefore never finished. The system architecture must change over time as the system evolves to meet new requirements and changes in its environment (such as changes to the systems with which it must interoperate). Similarly, as design and implementation work products are incrementally and iteratively developed, they may become incompatible with the architectural work products and these incompatibilities should be eliminated. Unless it is actively managed, the integrity of a system's architecture can degrade during both system development and maintenance. Ensuring the integrity of the system architecture is thus one of the key tasks performed during the engineering of the system architecture.

This task is a reusable MFESA method component and as such is intended to be tailorable to meet the specific needs of the endeavor. Therefore, it is expected that at least some of the task's objectives, preconditions, inputs, steps, invariants, work products, guidelines, and pitfalls will be tailored during the creation of the endeavor-specific system architecture engineering method.

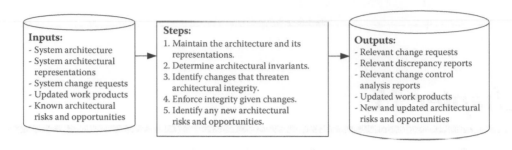

Figure 15.1 Summary of Task 10 inputs, steps, and outputs.

Tailoring of a method component includes adding missing content, modifying existing content, or removing existing content that is inappropriate, unnecessary, or not cost-effective to perform.

15.2 Goals and Objectives

The *Maintain the Architecture and Its Representations* task has the following overall goals:

- Maintain the architecture and its representations.
- Ensure the continued integrity and quality of the system architecture as the system evolves.

The typical subordinate objectives of this task are to:

- Ensure the elimination of inconsistencies within the system architecture and its representations.
- Ensure the elimination of inconsistencies between the system architecture, its representations, and the:
  - Architecturally significant requirements
  - Enterprise architecture(s)
  - Reference architecture(s)
  - Design of architectural components
  - Implementation of architectural components
- Ensure that the quality and integrity of the architectural representations does not degrade over time.
- Thereby, ensure that the system architecture and its representations do not degrade over time.

15.3 Preconditions

The *Maintain the Architecture and Its Representations* task can typically begin when the following preconditions have been met:

- **System architectural drivers and concerns are managed, baselined, and available.** The system architectural drivers and concerns are properly managed (requirements management), baselined (configuration management — configuration control), and available to the architecture teams.

- **System architectural representations are baselined.** The system architectural representations are under configuration control.
- **Enterprise and reference architectures are baselined and available.** The architectural representations for any relevant organizational enterprise architectures and reference architectures (such as product lines) are baselined and available to the architecture teams.
- **External system interfaces are baselined and available.** Any relevant interface documentation of external systems is baselined and available.
- **Range of potential additional interactions with external systems is assessed and bounded.** Complex systems, net-centricity, and emerging requirements inevitably lead to some uncertainty regarding the totality of external interactions. In such systems, the certainty system architects have been able to exploit by having an approved context diagram documenting external interfaces is lost. As a stopgap, it is worth attempting to characterize the range of interactions the system under synthesis might be called upon to support. If this characterization provides a fairly generous envelope to bound such interactions, synthesis may proceed without the need for a fixed set of external interfaces. Taking such a course represents an architecturally significant decision that will have a measurable effect on the synthesis activity.
- **Range of potential changes in content or interpretation of external interfaces is assessed and bounded.** Another side effect of the increase in complexity and the emergence of enterprise-level systems is that architects are expected to envision and capture the interactions with external systems that are similarly under synthesis or development. Once one acknowledges this state of flux, it is still incumbent upon the architect to make progress in the face of this kind of uncertainty. Again, the best one can do is to anticipate the range of interactions and how they might affect external interfaces in question. (Some advocates of the "agile development" approach recommend against anticipating future changes, but it is not clear how one can build a durable architecture without taking cognizance of uncertainty.)
- **Range of potential changes in human–machine task division is assessed and bounded.** The complexity of interactions among humans and human-created computational systems is dramatically increasing as levels of automation increase. For example, decision aids employed by tactical aircraft pilots must allow for human intervention into the decision process. This allows the human to insert actions and constraints into the decision loop that are derived from the situational awareness of the pilot, who is ultimately accountable for what happens onboard the aircraft. For any given system, a thorough understanding of the kinds of human–machine interactions, including the tempo and timelines of such interactions, may lead to the realization that the roles played by humans should be treated as architecturally significant considerations during development.

15.4 Inputs

The *Maintain the Architecture and Its Representations* task typically uses the following inputs:

- System architecture
- System architectural representations
- System change requests having architectural ramifications
- Updated work products incorporating changes with architectural ramifications
- Known architectural risks and responsibilities

15.5 Steps

During the *Maintain the Architecture and Its Representations* task, members of the relevant architecture team collaborate to perform the following steps in an incremental, iterative, concurrent, and time-boxed manner:

1. **Maintain the architecture and its representations.** Make changes to the architecture and its representations between major upgrades and releases of the system or its architectural components, including:
   - **Adaptive maintenance.** Perform adaptive maintenance on the architecture and its representations to make minor changes to meet new and modified architecturally significant requirements.
   - **Corrective maintenance.** Perform corrective maintenance on the architecture and its representations to correct defects between major upgrades.
   - **Perfective maintenance.** Perform perfective maintenance on the architecture and its representations to improve the quality (i.e., improve the level of one or more quality characteristics or quality attributes) of the system, its architectural components, or their architectural representations.
   - **Preventive maintenance.** Perform preventive maintenance on the architecture and its representations to prevent known defects in the system or its architectural components from causing failures.

2. **Determine architectural invariants.** Carefully identify and maintain a list of those parts of the system architecture that should not be changed because such architectural changes could risk failure to achieve architecturally significant requirements. Also, maintain a list of the technologies involving these architectural invariants.

3. **Identify changes that threaten architectural integrity.** Monitor both the proposed or actual changes to the system that can cause a loss in the integrity of the system architecture. Collaborate with the project configuration control board (CCB) to evaluate proposed changes to relevant baselined work products for their architectural ramifications. Monitor and identify relevant changes to the:
   - **Requirements.** Changes to the architecturally significant requirements and their associated architectural concerns may result in inconsistencies within the system architecture and its representations (such as inconsistencies between structures, models, views, and other architectural documents), as well as with any relevant enterprise and reference architectures.

     Note that if the system is to remain competitive and useful to its stakeholders over its lifetime, then there *will* be many requirements changes and some of these changes will be architecturally significant. This is a major reason why architects try to make their systems and architectures extensible and maintainable. Nevertheless, there will be useful changes that will require changes to the architecture and its representations. Maintaining architectural integrity does not mean that the architects should fight to keep their architectures frozen and thereby unviolated. Rather, the architects should work to ensure that:
     - When the architecturally significant changes are made, then they are worth making.
     - Before the changes are made, the authoritative stakeholders understand their architectural ramifications.
     - When changes are made, the architecture and its representations are updated and left in a consistent state.

- **System architecture.** Changes in one part of the system architecture (due, for example, to iterative, incremental, and parallel architectural engineering) may result in inconsistencies with other parts of the system architecture as well as with relevant enterprise and reference architectures.
- **Interfaces.** Changes to the interfaces to external systems with which the system must be or remain interoperable may result in inconsistencies between these interfaces and the system architecture.
- **Technologies.** Changes to the technologies incorporated into the system architecture may result in inconsistencies within the system architecture.
- **Enterprise architectures.** Changes in the enterprise architectures and their representations may result in inconsistencies between the system architecture and the changed enterprise architecture.
- **Reference architectures.** Changes in any relevant reference architectures and their representations may result in inconsistencies between the system architecture and the changed reference architecture.
- **Design.** Changes in or additions to the design of an architectural component might possibly violate architectural decisions, inventions, trade-offs, and assumptions.
- **Implementation.** Changes in or additions to the implementation of an architectural component might possibly violate architectural decisions, inventions, trade-offs, and assumptions.

4. **Enforce integrity given changes.** If an actual or proposed change is identified as causing an inconsistency, either within the system architecture or between the system architecture and something else, then initiate steps to either remove the inconsistency or prevent the inconsistency from occurring.
   - If a decision is made to accept a change that violates an architectural invariant, determine the effect on the entire architecture and/or the architectural component(s) in question. Consider whether there should be a revision to the list of invariants or simply a deviation or waiver for the architectural component in question. Assess the consequences of the choice made and document the result.

5. **Identify any new architectural risks and opportunities.** Concurrently with the performance of the previous steps, identify and formally document any new architectural risks and opportunities that were discovered.

15.6 Invariants

Because the *Maintain the Architecture and Its Representations* task is never completed until the system is retired, the task has no postconditions as such. However, the following *invariants** should hold at the end of each block or release:

■ **System architecture and architectural representations consistency.** The system architecture and its representations should remain internally consistent as well as consistent with one another as the system and its architecture evolve over time. Any changes to one should

---

* Note that the term invariant here refers to an invariant of the architecture engineering method, rather than an architectural invariant (i.e., an invariant of the system architecture).

trigger an assessment of their overall integrity, and any inconsistencies identified should be resolved by updating the no-longer-correct work products.

- **System architecture and requirements consistency.** The system architecture and its representations should remain consistent with the architecturally significant requirements. Any changes to one should trigger an assessment of their overall integrity, and any inconsistencies identified should be resolved by updating the appropriate work products (i.e., requirements or architectural representations).

- **System architecture and enterprise architecture consistency.** If a relevant organizational enterprise architecture exists, then the system architecture and the enterprise architecture and their representations should remain consistent. Any changes to one should trigger an assessment of their overall integrity, and any inconsistencies identified should be resolved by updating the work products that are no longer correct.

  Sometimes a system architecture update must be inconsistent with the enterprise architecture, either for a prolonged period (because enterprise architectures tend to change more slowly than system architectures) or permanently (because the system has to meet some critical need that was not foreseen at the enterprise level). The latter would be a case where work products on both sides are correct but inconsistent with each other. There should be some notion of a waiver or deviation to accommodate such circumstances.

- **System architecture and reference architecture consistency.** If a relevant reference architecture exists (e.g., for a product line containing the system), then the system architecture and the reference architecture and their representations should remain consistent. Any changes to one should trigger an assessment of their overall integrity, and any inconsistencies identified should be resolved by updating the work products that are no longer correct.

- **System architecture and external interfaces consistency (interoperability).** To maintain interoperability, the system architecture and its representations should remain consistent with any interfaces between the system and external systems. Any changes to one should trigger an assessment of their overall integrity, and any inconsistencies identified should be resolved by updating the relevant work products. Note that this may require negotiation with the architects of the external systems.

- **System architecture and design consistency.** The system architecture and its representations should remain consistent with the system design, including the design of all architectural components. Any potentially relevant changes to one should trigger an assessment of their overall integrity, and any inconsistencies identified should be resolved by updating the relevant work products.

- **System architecture and implementation consistency.** The system architecture and its representations should remain consistent with the system implementation, including the implementation (e.g., coding, fabrication, or manufacture) of all architectural components. Any potentially relevant changes to one should trigger an assessment of their overall integrity, and any inconsistencies identified should be resolved by updating the relevant work products.

15.7 Work Products

The *Maintain the Architecture and Its Representations* task produces the following work products:

- **Relevant change requests:** forms that document requested changes to the system. Change requests are produced during this task when the proposed changes elsewhere have architectural ramifications.

- **Relevant discrepancy reports:** reports that document discrepancies or defects within the system, whereby the discrepancies or defects have architectural ramifications. This naturally includes change requests identifying defects within architectural representations. It also includes other defects, the fixing of which will require changes to the architecture or its representations (such as defects within requirements or the documentation of reusable architectural components).

- **Relevant change control analysis reports:** any reports documenting the analysis of the impact of proposed changes to the system. They are produced during this task when the proposed changes have architectural ramifications.

- **Updated work products:** proposed changes to the architecture, its representations, and other work products may result in inconsistencies between them. Resolving these inconsistencies may necessitate updating the other work products, resulting in updated versions of the:
  - **Architecturally significant requirements.** Often, overly restrictive requirements may unnecessarily prohibit the reuse of appropriate architectural components. This may, for example, result in the elimination or weakening of some functional requirements, quality requirements, and architectural constraints.
  - **Updated system architecture and its representations.** These are changes to the system architecture or its representations. For example, changed requirements often necessitate the changing of the system architecture and its representations to meet these changes.
  - **Enterprise architecture and its representations.** The overall enterprise architecture and its representations, updated to be consistent with the architecture of the system and its representations.

    Note that usually the enterprise architecture outweighs and drives system architectures. Sometimes, however, changes to the architectures of important systems may drive changes to the enterprise architecture.

    Note also that making changes to the enterprise architecture and its representations will require negotiation with the architects of the enterprise architecture. Making changes to the enterprise architecture may have far-reaching consequences on many parts of the enterprise, including its other systems.
  - **Reference architecture and its representations.** The reference architecture of the system's product line and its architectural representations.

    Note that making changes to the reference architecture and its representations will require negotiation with the architects responsible for the reference architecture. Making changes to the reference architecture may have far-reaching consequences on the other systems within the product line.
  - **Reusable architectural component architectures and their representations.** When the architecture of a reusable architectural component is changed so that the component may be reused in the current system, then the component's architectural representations must be updated to remain consistent with the architecture.

    Note that making changes to the architecture of a reusable component will require negotiation with the architects responsible for maintaining the reusable component. Making changes to the component's architecture may have far-reaching consequences on the other systems reusing the same component. For example, it may no longer meet the requirements of the other systems and it may cause their architectural integrity to be lost. Inappropriately changing reusable architectures may also lead to safety mishaps and security misuses.

■ **Known architectural risks and opportunities:** any new or modified architectural risks documented in the endeavor risk repository as a part of endeavor risk management activity. This also includes an updated list of architectural opportunities that the architecture teams may wish to take advantage of in the future.

15.8 Guidelines

Use the following guidelines (and associated rationales) when performing the *Maintain the Architecture and Its Representations* task:

■ Maintain the architectural representations to maintain architectural integrity.
■ Consider the entire scope of the *Maintain the Architecture and Its Representations* task.
■ Consider the sources of architectural change.
■ Protect the architectural invariants.
■ Determine the scope of architectural integrity.
■ Train the architects and designers.
■ Formally manage architectural risks.

The following descriptions express these guidelines in more detail and provide their associated rationales:

■ **Maintain the architectural representations to maintain architectural integrity.** Ensure that the as-architected, as-implemented, and as-is architectures remain consistent so that the architecture does not degrade over time. Ensure that the architectural representations are maintained to:
  – Remain consistent with one another
  – Remain consistent with the architectural concerns as well as with the design and implementation
  – Support the design, implementation, test, and production activities
  Rationale: The architectural representations are a very valuable asset supporting maintenance of the system and future reuse of its architectural components. Such inconsistencies greatly lower the value of the architectural representations, and the considerable resources that were invested in the production of the architectural representations will be lost if they are not maintained as the system and its architecture evolve over time. Inconsistencies make it very difficult to communicate and maintain the system and its architecture. Inconsistencies also cause the architecture to degrade over time. More importantly, the representations cease to be reliable descriptions of the system, potentially opening the door down the road to system defects, up to and including complete failure.

■ **Consider the entire scope of the *Maintain the Architecture and Its Representations* task.** Consider the entire scope of architecture engineering when ensuring the architecture's integrity.
  Rationale: The *Maintain the Architecture and Its Representations* task can be performed during the:

- Initial engineering of the architecture
- Maintenance of the system
- Engineering of a major upgrade to the system

■ **Consider the sources of architectural change.** When assessing the potential impact of changes on system architectural integrity, consider the following sources of change that potentially indicate costly changes:

- Change in the concept of operations (ConOps) as the needs of organizations and their missions evolve
- Evolution in the depth of understanding and the interpretation of natural language specifications addressing requirements, interfaces, and the architecture
- New and altered requirements that are imposed over time
- Changes in the division of tasks between humans and automated subsystems
- Numerous accumulated changes, associated with end-user interaction and the appearance of the user interface
- Changes in the over-arching structure of the system, causing changes in its subsystems

Rationale: The above sources of architectural degradation may lead to architectural inconsistencies and thus violations of architectural integrity.

■ **Protect the architectural invariants.** Explicitly represent and maintain the invariants over the life cycle of the system. Keep the list of architectural invariants as short and as simple as practical (but no shorter). Although difficult to realize in practice, keep the architectural invariants technology neutral. Keep each pair of architectural invariants as logically independent of one another as is practical:

- Architectural invariants should not be contradictory.
- One architectural invariant should not be a consequence of another.

Rationale: Architectural integrity is lost when architectural invariants are violated. It is difficult to remember a large number of invariants. Because technology becomes obsolete, any architectural invariant based on technologies is likely to become violated when the technology becomes obsolete and is replaced. When invariants are logically independent, it is easier to change one without inadvertently violating another.

■ **Determine the scope of architectural integrity.** Ensure that the relevant stakeholders understand the full scope of effects regarding architectural integrity, which may include:

- Any reference architectures imposed by stakeholders
- Any enterprise architectures imposed by the business unit
- The architectures of external systems with which the system must interoperate
- The architectures of internal subsystems with which the subsystem under consideration must intra-operate

Rationale: Architectural integrity is sometimes lost because the architects only consider the architectural ramification of changes to their own architectural component or own system without considering the impact on external architectures such as enterprise architectures and reference architectures.

■ **Train the architects and designers.** Train the architects and designers to remain aware of the architectural ramifications of their proposed changes to the system.

Rationale: Architectural integrity is often lost because the architectural ramifications of changes is not analyzed prior to making those changes.

■ **Formally manage architectural risks.** As part of the risk management activity, identify architectural risks during the performance of this task. Add any architectural risks identified to the official endeavor risk management database.

Rationale: Architectural risks that are not identified or that are not added to the project risk management database when identified are much less likely to be effectively managed.

15.9 Pitfalls

Avoid the following common pitfalls associated with performing the *Maintain the Architecture and Its Representations* task and mitigate their negative consequences when they occur:

- The architectural representations become shelfware.
- Architecture engineering is finished.
- The architecture is not under configuration management.
- The architecture is not maintained.
- A "beautiful" architecture is frozen solid.
- There is inadequate tool support for architecture maintenance.

The following express these pitfalls in more detail, describing their negative consequences and the steps the architects can use to mitigate them:

- **The architectural representations become shelfware.**
  - **Pitfall.** The architectural representations are set aside as shelfware once they are completed and have been used to drive the initial design and integration testing.
  - **Negative consequences.** This pitfall often causes the following negative consequences:
    - The architectural representations are not maintained.
    - As the architecture and design continue to evolve, the architectural representations become obsolete and inconsistent with them.
    - The architectural representations are no longer useful or used.
    - The architectural representations lose their value.
    - The return on the very large investment made in the architecture is lost.
    - The architecture loses integrity, and the system consequently becomes difficult to maintain.
  - **Mitigations.** To avoid or mitigate the negative consequences of this pitfall:
    - Establish a track record of the architecture and its ramifications providing value to management and the other development teams. Ensure that this occurs early, before the architecture is completed, so that convincing others of the value of architecture engineering is not a cold sell when the budget is reduced.
    - Ensure that both managers and architects understand the:
      - Value and usefulness of the architecture and its architectural representations
      - Need to maintain them
      - Need to supply the necessary resources needed to maintain them
- **Architecture engineering is finished.**
  - **Pitfall.** Once the initial architecture is completed, managers believe that architecture engineering is finished and no longer needs any resources.
  - **Negative consequences.** This pitfall often causes the following negative consequences:
    - Architects are transferred to other projects.
    - Any remaining funding for architecture work is transferred to other project activities.

- The project schedule and work breakdown structure (WBS) do not contain time for any further architectural work.
- As the architecture and design continue to evolve, the architectural representations become obsolete and inconsistent with them.
- The architectural representations lose their value.
- The return on the very large investment made in the architecture is lost.
- The architecture loses integrity.
- The architectural representations are no longer useful and become shelfware.
- **Mitigations.** To avoid or mitigate the negative consequences of this pitfall:
  - Ensure that architecture maintenance is part of the up-front plans and budget.
  - Ensure that both managers and architects understand that the architecture and its representations will continue to change over time, even if these changes become smaller and less frequent as the architecture matures and stabilizes.
  - Ensure that the architectural engineering method contains tasks for maintaining the architecture and ensuring its integrity.
  - Ensure that management supplies the resources needed to maintain the architectural representations.

■ **The architecture is not under configuration management.**
- **Pitfall.** The architecture representations are not baselined and placed under configuration control, even after they are verified and approved.
- **Negative consequences.** This pitfall often causes the following negative consequences:
  - As the architecture and design continue to evolve, the architectural representations become obsolete and inconsistent with them.
  - The architectural representations lose their value.
  - The return on the very large investment made in the architecture is lost.
  - The architecture loses integrity.
  - The architectural representations are no longer useful and become shelfware.
- **Mitigations.** To avoid or mitigate the negative consequences of this pitfall:
  - Ensure that management understands that the architecture and its associated architectural representations are very valuable resources that deserve proper configuration management.
  - Baseline the architectural representations once they are verified and approved.
  - Place the baselined architectural representations under configuration control.
  - Ensure that the change control method has architects analyze and report the impact on the architecture of each change request that potentially has architectural ramifications before the change control board authorizes the change to be made.

■ **The architecture is not maintained.**
- **Pitfall.** The architectural representations are set aside once completed, forgotten, and never maintained (e.g., updated when the underlying architecture changes).
- **Negative consequences.** This pitfall often causes the following negative consequences:
  - Architectural integrity is lost as the architectural representations become inconsistent with each other as well as with the requirements, the architecture, and the designs and implementations that were driven by the architecture.
  - Architects and other stakeholders lose trust in the architectural representations and these representations become shelfware.
  - The architectural representations lose their value despite the large investment that was made in their production.

- The architecture and its representations become even more difficult to maintain.
- Architectural defects and risks increase.
- The probability increases that the system will not meet its architecturally significant requirements.
- The probability increases that the system will not be acceptable to its stakeholders.
  - **Mitigations.** To avoid or mitigate the negative consequences of this pitfall:
    - Include architecture maintenance in the system architecture engineering method.
    - Ensure that adequate resources are made available to maintain the architecture and its representations.
    - Verify that the architectural representations are maintained so as to remain consistent with each other as well as with the requirements, the architecture, and the designs and implementations that were driven by the architecture.

■ **A "beautiful" architecture is frozen solid.**
  - **Pitfall.** An architect believes some aspect of the architecture it too beautiful to change and therefore attempts to keep it frozen despite clear and compelling reasons why it should be changed. Similarly, the acquisition organization may unnecessarily mandate certain critical architectural decisions that are incompatible with the system meeting certain architectural requirements or concerns.

    For example, an architect might defend an architectural decision to use a service-oriented architecture (SOA) when clear evidence exists that such an architecture will not meet performance requirements.
  - **Negative consequences.** This pitfall often causes the following negative consequences:
    - Certain aspects of the architecture become inconsistent with the rest of the system.
    - The architecture contains weaknesses and its representations contain defects.
    - The architecture does not enable the system to meet its architecturally significant requirements.
  - **Mitigations.** To avoid or mitigate the negative consequences of this pitfall:
    - Rely on a team of architects rather than an individual architect to develop the architecture.
    - Independently verify the consistency of the architecture and its representations against the requirements, design, and implementation.
    - Focus making the initial models easily maintainable to reduce their cost, and thus any reluctance to modify them as the system matures. Where appropriate, reduce the amount of unnecessary details in architectural representations until later in development if they are not needed for early decision making and yet costly to maintain.

■ **There is inadequate tool support for architecture maintenance.**
  - **Pitfall.** Management does not supply adequate tool support for architectural maintenance.

    Note that it is very difficult to keep the various architectural models and their associated views and focus areas consistent as individual changes during maintenance occur. Unfortunately, it is also difficult or impossible to obtain adequate computer-aided system engineering (CASE) tools that track and ensure consistency across multiple models and associated documentation.

    Note also that lack of adequate tool support for architecture was a pitfall for architectural development as well. If adequate architectural tools were available for development, then adequate tools should be available for maintenance as long as the tool licenses are maintained.

- **Negative consequences.** This pitfall often causes the following negative consequences:
  - The architects must spend extra time and effort to manually maintain the architecture and its representations.
  - The architects must spend extra time to manually maintain architectural integrity.
  - Architectural integrity is lost as the architectural representations become inconsistent with each other as well as with the requirements, the architecture, and the designs and implementations that were driven by the architecture.
  - Architects and other stakeholders lose trust in the architectural representations, and the architectural representations become shelfware.
  - The architectural representations lose their value despite the large investment that was made in their production.
  - Architectural defects and risks increase over time.
  - The probability increases that the system will not meet its architecturally significant requirements.
  - The probability increases that the system will not be acceptable to its stakeholders.
- **Mitigations.** To avoid or mitigate the negative consequences of this pitfall:
  - Include architecture maintenance in the system architecture engineering method.
  - Ensure that adequate resources are made available to maintain the architecture and its representations.
  - Incorporate sufficient time into the schedule and funds into the budget to manually maintain the architecture and its representations.
  - Where practical, develop local software tools to support architectural maintenance.

15.10 Summary

During the *Maintain the Architecture and Its Representations* task, the architecture teams ensure the continued integrity and quality of the system architecture as the system evolves.

15.10.1 Steps

When instantiating the *Maintain the Architecture and Its Representations* task, the following steps may be included, modified, or tailored out:

1. Maintain the architecture and its representations.
2. Determine architectural invariants.
3. Identify changes that threaten architectural integrity.
4. Enforce integrity given changes.
5. Identify any new architectural risks and opportunities.

15.10.2 Work Products

When instantiating the *Maintain the Architecture and Its Representations* task, the following work products may be included, modified, or tailored out:

- Relevant change requests
- Relevant discrepancy reports

- Relevant change control analysis reports
- Updated work products:
  - Architecturally significant requirements
  - Updated system architecture and its representations
  - Enterprise architecture and its representations
  - Reference architecture and its representations
  - Reusable architectural component architectures and thier representations
- New and updated architectural risks and opportunities

15.10.3 Guidelines

Use the following guidelines (and associated rationales) when performing the *Maintain the Architecture and Its Representations* task:

- Maintain the architectural representations to maintain architectural integrity.
- Consider the entire scope of the *Maintain the Architecture and Its Representations* task.
- Consider the sources of architectural change.
- Protect the architectural invariants.
- Determine the scope of architectural integrity.
- Train the architects and designers.
- Formally manage architectural risks.

15.10.4 Pitfalls

Avoid the following common pitfalls associated with performing the *Maintain the Architecture and Its Representations* task and mitigate their negative consequences when they occur:

- The architectural representations become shelfware.
- Architecture engineering is finished.
- The architecture is not under configuration management.
- The architecture is not maintained.
- A "beautiful" architecture is frozen solid.
- There is inadequate tool support for architecture maintenance

Chapter 16

MFESA Method Components: Architectural Workers

16.1 Introduction

As illustrated in Figure 16.1, an MFESA system architecture engineering method consists of an integrated set of reusable method components of the following three metatypes: (1) *architectural workers*, (2) the *architectural work units* they perform, and (3) the *architectural work products* they produce. That is, an MFESA repository stores not only the different types of architectural work units (such as architectural tasks and techniques) and architectural work products (such as architectural models and documents), but also the different types* of architecture workers.

As illustrated in Figure 16.2, MFESA includes three types of architectural workers: (1) architecture teams, (2) architects, and (3) architecture engineering tools. All three types of architectural workers perform architectural work units to produce architectural work products. These architectural workers are defined as follows:

> **System architecture worker:** the role played by any person or thing when performing system architecture engineering work units to produce system architectural work products.
>
> **System architect:** the highly specialized role played by a system engineer when performing architecture engineering work units to produce system architectural work products.
>
> Note that an architect is a role played by a person and may (or may not) be a job title. The same person may play different roles, such as technical leader, requirements engineer, architect, and tester.

---

* Note that the MFESA repository contains *types* of architectural workers, not actual architectural workers. MFESA methods contain as-planned method components (e.g., types of architectural workers), whereas MFESA processes contain as-performed process components (e.g., actual workers, which are instances of types of architectural workers).

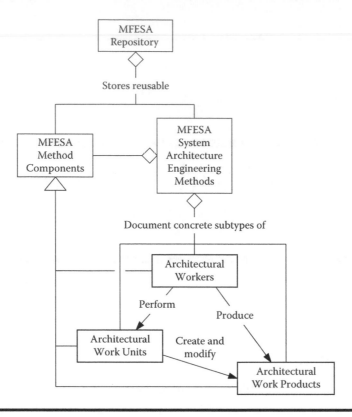

Figure 16.1   The three types of MFESA method components.

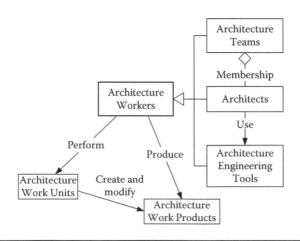

Figure 16.2   Three types of system architecture workers.

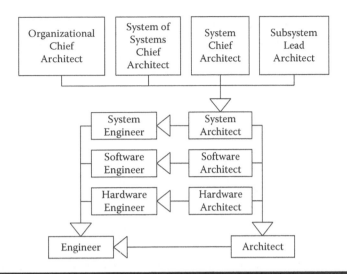

Figure 16.3 Types of architects.

> **Architecture team:** any team responsible for performing architecture engineering work units to produce architecture work products.
>
> Note that an architecture team may be an official team or merely a logical or virtual grouping that does not appear on any organizational chart. For example, a subsystem may be developed by an integrated product team (IPT) that unofficially and on an as-needed basis contains an ad hoc requirements team, an architecture team, a development team, and a test team.
>
> **Architecture engineering tool*:** any tool used to perform system architecture engineering work units and produce all or part of a system architectural work product.

This chapter continues with a discussion of system architects in section 16.2. It then progresses to a discussion of system architecture teams in section 16.3, followed by a discussion of system architecture engineering tools in section 16.4. A brief summary of the chapter is given in section 16.5.

16.2 System Architects

As illustrated in Figure 16.3, there are several types of system architects. These are roles played by individuals who have specific responsibilities and authorities to perform architecture engineering tasks. Although a specialized subtype of a system engineer, a system architect typically operates in a different fashion than a general engineer. Understanding this difference and how it complements and supports effective systems engineering and systems management is essential to effectively employing architects to engineer a quality system. In addition to defining system architects and

---

* Because they have a difficult time considering a tool a type of worker, some people may have difficulty with the use of the term "architectural worker" to mean the abstract superclass of architecture team, architect, and architecture engineering tools. However, by automating some of the work traditionally done manually by architects, architecture tools do perform some steps of architectural tasks to create and modify some architecture work products.

their responsibility, this section describes the profile of system architects as well as their interfaces as members of an architecture team and members of the overall project. Finally, this section addresses some guidelines and common pitfalls with regard to system architects.

16.2.1 Definitions

Figure 16.3 illustrates the various types of engineering architects and how they relate to more general engineers.

A major source of confusion and disagreement is the relationship between system engineers and system architects. Some believe that these are two quite different professions with quite separate areas of expertise and responsibility [Maier and Rechtin, 2002]. Some system engineers even go so far as to believe that many system architects are only able to draw pretty pictures and are often unable to create feasible architectures that do not need significant correction and improvement by chief system engineers.

Unlike those who consider system architecting primarily an art, we strongly believe it is actually a form of system engineering (hence this book's title, *Method Framework for Engineering System Architectures*). Thus, we believe that it is very difficult to be a good system architect producing successful system architectures if one does not first have a basic foundation in system engineering, just like a software architect needs to be a type of software engineer specializing in software architectures. We therefore believe that a system architect is a type of system engineer with further specialized training, experience, and expertise in system architecture engineering in the same manner that medical specialists (e.g., surgeons, gastroenterologists, and neurologists) must first obtain their education as a medical doctor (i.e., have their M.D.) and then have further study and apprenticeship (i.e., residency) in their area of specialization.

> **System architect:** the highly specialized role played by a system engineer when performing system architecture engineering tasks to produce system architecture engineering work products.

Note that while an architect is a role played by a system engineer, there are many skills and responsibilities associated with the position that are not typical of a system engineer. To that extent, descriptions of the system architecture engineering methods and profession from a strictly system engineering perspective tend to only emphasize those that are directly related to system engineering methods and professions. While a good system architect does need to have experience with system engineering, the reverse may not apply. A typical system engineer may not have the profile (i.e., training, expertise, experience, and aptitude) necessary to perform the architecture engineering tasks (such as big-picture mindset as well as training in architecture engineering tasks and techniques). Hence, we do not believe that just any "engineer" or "system engineer" without the profile noted below will be qualified to act as the system architect.

16.2.2 Types of System Architect

MFESA includes the following types of system architect:

> **Organizational chief architect:** any system architect who is responsible for the architecture engineering of all systems within an organization.

System-of-systems chief architect: any system architect who leads a system-of-systems architecture team and is responsible for the architecture engineering of a system-of-systems, family of systems, network of systems, or a product line of systems.

System chief architect: any system architect who leads the overall system architecture team and is responsible for the overall architecture engineering of a single system.

Subsystem lead architect: any system architect who leads a subordinate architecture team that is responsible for engineering the architecture of one or more architectural components (e.g., subsystem, software configuration item, hardware configuration item).

Software architect: any software engineer who is responsible for engineering all or part of the software architecture (i.e., the software aspects of a system architecture).

Hardware architect: any hardware engineer who is responsible for engineerirng all or part of the hardware architecture (i.e., the hardware aspects of a system architecture).

Note that the preceding are engineering roles and not management roles, even though the architect may be responsible for the architecture engineering budget and schedule, and although the same person may play both the architect and technical leader roles on a project.

16.2.3 Responsibilities

A system architect typically has the following responsibilities:

- **Primary responsibilities:**
  - **Determine and assess the impact of the architectural drivers and concerns:**
    - Using existing product and process requirements as inputs, system architects determine which requirements are architecturally significant (the architectural drivers).
    - System architects collaborate with requirements engineers to:
      - Derive architecturally significant requirements.
      - Ensure that all the architecturally significant requirements are properly engineered.
      - Ensure the feasibility of each architecturally significant requirement.
    - System architects help the requirements engineers prioritize the architecturally significant requirements for scheduling implementation.
    - System architects group the architecturally significant requirements into the associated architectural concerns.*
    - System architects also assess the impact of these architectural concerns on the system architecture.
  - **Develop the architecture and architectural representations.** Using the architectural drivers and concerns (i.e., the architecturally significant requirements) as inputs, system architects develop an appropriate architecture implementing these drivers and concerns. System architects also develop associated architectural representations modeling and documenting the system architecture.
  - **Analyze the architecture using the architectural representations.** Using the architectural representations, system architects analyze the system architecture to determine

---

* Note that these may be architectural drivers and concerns for the entire system or architectural drivers and concerns derived for and allocated to individual subsystems or other architectural components.

certain quantitative or qualitative properties about the system. Further, the architects relate the results of these analyses to support project management decisions, build engineering baselines, and for other activities.

- **Evaluate the architecture and the architectural representations.** Using the architectural representations, goals, and quality models as inputs, system architects (as well as others) evaluate the system architecture to determine the degree to which it addresses the architectural concerns and meets the architecturally significant requirements. This evaluation includes the generation of appropriate evidence to support the result of the evaluation to neutral stakeholders (such as in support of audits, reviews, and when certifying the evaluation to the chief architect).

- **Maintain the architecture and the architectural representations.** System architects are responsible for ensuring that the current architectural work products are maintained during their period of use. However, this responsibility does not mean that products that are no longer useful to the current and future product life-cycle phases need to be maintained — rather, only to those products that are or will be used will be maintained. Architects may also identify products that are no longer needed so that they can be removed from the repository of architectural work products.

- **Ensure architectural integrity.** System architects are responsible for ensuring that the architecture remains consistent, both internally consistent within the architecture and its representations (such as within and among different architecture models and documents) as well as externally consistent with other work products (such as requirements, design, implementation, testing, manufacturing, and deployment). This also includes maintaining certain architectural invariants while making architecturally significant changes.

- **Be an information collector.** System architects must actively identify, elicit, describe, and catalog architecturally significant information. Many types of such information will need to be captured and tracked during performance of the architecture engineering tasks. This includes information such as architectural drivers, reuse candidates, standards, and existing architectural representations.

- ■ **Organizational responsibilities:**
  - **Lead the architectural activities.** System architects are leaders of architecture activities. The chief and lead architects must provide leadership skills appropriate to the situation. Different situations can call for directive and transactional activity, guidance and facilitation, or simple production honcho responsibilities. This leadership is of great importance when senior architects mentor and work with junior architects on common products. Also, architects "in the field" may end up leading stakeholders by example when working in an integrated product team that is considering architectural issues.
  - **Manage the performance of the architecture engineering tasks.** System architects manage the performance of the system architecture engineering tasks, including planning, scheduling, and task reporting. The architect must play a manager's role with respect to architectural activities. Thus, architects must interact with management and other functional areas as needed to manage the performance of the architecture engineering method, acquire resources, select modeling methods, select tools, generate plans, and prioritize and schedule activities.
  - **Be an architecture advocate.** System architects should be strong advocates for the importance of the system architecture and the need for it to be properly engineered, including development and maintenance.

- **Be a stakeholder advocate.** System architects should be advocates for the needs of the stakeholders. This role is specifically cross-cutting in that the stakeholders need advocates when performing organizational tasks related to the architecture (such as choosing a quality model that will support users, testers, and client goals). Architects should also be advocates for stakeholder needs when making architectural decisions, inventions, and trade-offs (such as advocating for the needs of security certifiers as well as for the needs of users when considering alternate architectural visions).
- **Instantiate and tailor the architecture engineering method.** System architects should collaborate with process engineers and management to create, mandate, and use an appropriate system-specific or organizational-specific architecture engineering method. With MFESA, that includes selecting the appropriate architecture engineering method components, tailoring these components (if necessary), and integrating them into a consistent architecture engineering method.
- **Select and acquire the architecture engineering tools.** System architects are responsible for ensuring that they have the tools to accomplish the tasks that are appropriate to the life cycle and type of system being developed. The tools for a simple project may be normal documents, presentations, and spreadsheets, while more complex systems may require specialized computer-aided system/software engineering (CASE) tools to aid the architects in managing complexity.
- **Train the architecture stakeholders.** System architects act as trainers (or advocates of system architecting processes) for non-architects within the organization. Training need not be a formal activity; the architect may explain a specific model and analysis to a stakeholder to support a specific decision.
- **Evaluate the architecture method and process.** System architects are responsible for ensuring that architectural quality assessment tasks are performed as assigned. This responsibility includes helping to determine the contents of the project quality model as well as determining and implementing the assessment model.
- **Interface and collaborate with the architecture stakeholders.** System architects must actively identify, elicit, describe, and catalog information that is architecturally significant to the stakeholders of the system of interest. Many types of information will be identified in the several tasks: architectural drivers, reuse candidates, standards, and candidate system and subsystem architectures, among others.

In addition to the preceding basic system architecture engineering responsibilities, a *chief* system architect typically has the following responsibilities:

- **Take responsibility for the architectural work products.** The chief system architect takes responsibility for the architecture products. This taking of responsibility means that non-professionals can trust the work products to meet the appropriate level of rigor, to protect the public welfare, and to stand up to the scrutiny of peers. In some disciplines, this taking of responsibility may be performed by "stamping" or "affixing a seal" to the work product (as may be required for safety or regulatory needs). Even in an unlicensed fashion, the chief architect is responsible to ensure that the work products were developed appropriately and reflect the required or requested quality.
- **Supervise the subordinate architects.** The chief system architect typically leads the top-level system architecture team, which is responsible for the architecture of the entire system. In this case, the chief system architect oversees and must provide appropriate supervision to

both his or her subordinate architects who are members of the top-level architecture team, as well as the architects who are members of lower-level architecture teams that are responsible for engineering the architectures of individual subsystems and possibly other architectural components. Often, the membership of the top-level system architecture team is made up of the lead subsystem architects who lead the lower-level subsystem architecture teams. In this situation, the chief system architect directly supervises the lead subsystem architects and indirectly supervises the architects who are only members of lower-level architecture teams and who are directly supervised by the lead system architect who directly leads these lower-level teams.

On the other hand, the project's specific organizational charter or work breakdown structure (WBS) may not have an overall top-level system architecture team and each subsystem may be developed by an integrated product team (IPT) that is led by a person who also plays the role of lead subsystem architect. In this case, there may be no chief system architects, and the lead architects involved only operate as peers and collaborate to make system-level architectural decisions, inventions, and trade-offs. This latter situation is not recommended because the individual subsystem architects tend to do things differently, thereby causing inconsistencies between subsystem methods and processes as well as unproductive arguments between subsystem architects.

■ **Communicate the architecture to its stakeholders.** The chief system architect is responsible for communicating the system architecture, its representations, and the status of system architecture work units to their stakeholders, both within and without the development organization. Although much of this communication is often delegated to lower-level architects, in cases of architectural controversies (such as inconsistencies between subsystem architectures or inconsistencies between the architecture and the requirements, design, and implementation), the chief architect has final responsibility and authority to communicate the official architectural models as well as other architectural decisions, inventions, trade-offs, and assumptions. Furthermore, when communicating this information to the stakeholders, chief architects are also responsible for verification and validation of the architecture work products with the stakeholders.

16.2.4 Authority

Accountability must involve both responsibilities and the authority to carry out these responsibilities. Thus, a system architect typically has (or should have) the authority, including funding authority and the ability to initiate and stop work, to:

■ Develop architecture engineering methods.
■ Determine the architectural work products to produce, including models, documents, and architectural prototypes.
■ Select and acquire architecture engineering tools.
■ Develop the architecture and its representations.
■ Initiate evaluation of off-the-shelf architectural components.

16.2.5 Tasks

A system architect typically performs both architecture engineering tasks, as well as architecture-related tasks that are part of other disciplines:

- **Architecture engineering tasks:**
    1. Plan and resource the architecture engineering effort.
    2. Identify the architectural drivers.
    3. Create the first versions of the most important architectural models.
    4. Identify opportunities for the reuse of the architectural elements.
    5. Create candidate architectural visions.
    6. Analyze reusable components and their sources.
    7. Select or create the most suitable architectural vision.
    8. Complete the architecture and its representations.
    9. Evaluate and accept the architecture.
    10. Maintain the architecture and its representations.
- **Other tasks (and associated disciplines):**
    - Evaluate requirements (quality assurance).
    - Baseline architectural representations (configuration management).
    - Manage architectural risks (risk management).

16.2.6 Profile

System architects should typically have the following personal characteristics, expertise, training, and experience. However, the actual characteristics may well vary from project to project and from architect to architect based on the type of architect and the specific responsibilities allocated to the architect.

16.2.6.1 Personal Characteristics

System architects should typically have the following personal characteristics:

- An abstract thinker (i.e., a "big-picture" person) who:
    - Is comfortable working strategically at high levels of abstraction
    - Can see the big picture without being caught up in diversionary details
    - Can recognize the essential underlying issues in complex situations
    - Can see beyond the obvious issues to patterns or connections of issues that are not obviously related
- Able to perform engineering trade-offs between:
    - Conflicting architecturally significant requirements
    - Competing architectural visions, decisions, inventions, and trade-offs
- Able to perform trade-offs between nonquantitative propositions (i.e., qualitative, moral, ethical, or societal value) that may have significant:
    - Technical impact
    - Stakeholder interest
- Able to use "common sense" to ensure the rationality of architecture products:
    - Identify if the product of a specific method or technique is appropriate, given what is known about the system.
    - Determine alternate methods and products to support decisions if initial results are not appropriate.
- Able to identify architecturally significant requirements and their ramifications on the various architectures

- Able to clearly grasp the context of the system and its architecture
- Able to comfortably multitask (e.g., perform multiple architecting tasks concurrently)
- Able to make important decisions and trade-offs, given incomplete and conflicting knowledge
- Highly self-directed, being able to both manage and (re)prioritize the multiple concurrent and competing challenges, issues, ambiguities, and contradictions that necessarily occur during the production of the system architecture
- Strong analytical problem-solving skills
- Is influential with the confidence and force of personality to stand up to non-architects who improperly attempt to dictate the architecture
- Excellent verbal and written communication skills, and thus able to explain and document the system architecture for its diverse stakeholders
- Solid team-building skills
- Objectivity to assess a set of competing architectural decisions, inventions, and trade-offs without undue loyalty to or pride of authorship in any one architectural vision

16.2.6.2 Expertise

System architects should typically have the following expertise:

- Expert practical knowledge of:
  - System architecture tasks, techniques, and tools
  - Component-based development (CBD) concepts such as components, component models, component infrastructures, and interfaces
  - Hardware, information, database, and software architecting
  - The theory, practice, and tools of systems engineering
  - Architecture modeling languages (e.g., SysML, UML, AADL), including the ability to create, read, and explain the meanings of the associated models and diagrams
  - Architecture modeling tools
- Solid practical knowledge of:
  - Networks and network connection devices
  - The major infrastructure and middleware technologies used to implement system architectures, as well as the associated vendor organizations and their commercially available products
  - How to develop architectures of software components that will be distributed across multiple, heterogeneous platforms (such as hardware, operating system, server software, browsers, and the like)
  - Object-oriented concepts such as abstraction, encapsulation, inheritance, and polymorphism
  - Object, use case, and process/functional modeling
  - The application domain
  - Requirements engineering concepts and techniques
  - The major reusable system mechanisms and patterns
  - How to produce a common architecture for a family of related systems (e.g., a product line)
- Basic practical knowledge of the:
  - Business enterprise of the customer's organization
  - Integration and system-level testing theory, practice, and tools

- Configuration identification, configuration control, and associated configuration management tools

16.2.6.3 Training

System architects should typically have the following training:

- A bachelor's degree or better in systems engineering, software engineering or computer science, electrical engineering, domain-specific engineering (such as aerospace engineering, communications), or the equivalent
- Practical classroom as well as hands-on on-the-job training in:
  - The project architecture engineering method (especially in terms of relevant tasks, techniques, and work products) and the relationship between architecture engineering and other project disciplines and activities
  - Architecture modeling techniques, languages, and tools
  - Architectural component technologies, infrastructures, and interfaces
  - Technical writing
  - Group facilitation and stakeholder interviewing techniques
  - The application domain
  - Any relevant reusable architectural work products, including enterprise architectures and product line reference architectures

16.2.6.4 Experience

System architects should typically have the following experience:

- A minimum of three- to five-years of experience successfully producing software and network architectures during similar endeavors
- A minimum of three- to five-years of experience successfully designing and implementing data, hardware, or software components on similar endeavors

16.2.6.5 Interfaces

Many disciplines are required to successfully deliver a complex system, among them project management, systems engineering, process engineering, project control, and quality assurance. Further, organizational assets such as a systems engineering process group (SEPG) and independent verification and validation (IV&V) (including external operational test) bring slightly different assets to aid in project delivery.

This section outlines the collaboration boundaries between the different disciplines and groups as typically experienced on a project. On a specific project, these collaboration points would need to be established based on the actual project organization. Although these collaboration points could be called interfaces, that designation may be too formal for some projects, such as on a smaller project where the architect also performs the chief engineer role. Ultimately, the ownership or responsibilities are project specific and set up by the various project or team charters, legal and environmental dictates, and inherent structures of the organization or company developing the system.

System architects typically must interface and collaborate with the following roles, teams, and organizations:

■ **Project management:**
- Collaborates with system architects to provide resources (schedule, budget, and personnel) to the system architecture teams
- Is the ultimate arbiter of risks not externally assigned to another organization or team

■ **Systems engineering:**
- Collaborates with system architects to provide resources for engineering processes
- Collaborates with system architects to define and select the system engineering life cycle

■ **Hardware engineering and software engineering:**
- Identify technology maturity, technology risk, and technology-specific aspects of the architecture
- Build the design and implementation based on the system architecture
- Are ultimately responsible for the correctness of the design relative to the architecture

■ **Project control:**
- Monitors compliance of the architecture engineering work with regard to financial and schedule goals
- Ensures that architecture tasks are performed within their respective allocated resources

■ **Quality engineering:**
- Ensures that architectural work products are created in accordance with their specifications or standards
- Ensures that processes are performed in accordance with the associated method
- Ensures that missed process steps that control quality are re-examined/re-accomplished

■ **Systems engineering process group (SEPG):**
- Provides the project with archetypal engineering best practices
- Collaborates with the architecture team to create an appropriate tailored method using MFESA
- Provides the architects with information concerning architectural styles and patterns

■ **Independent verification and validation (IV&V):**
- Assures the stakeholders that tailorings made are consistent with goals
- Assures the stakeholders that quality assurance is functioning
- Assures the stakeholders that products are created in accordance with their given methods
- Assures the stakeholders that architectural products are created in accordance with their specifications
- Assures the stakeholders that products are appropriate for their purpose

■ **Subsystem, specialty, hardware, and software architecting:**
- Collaborate with system architects to engineer architectures and their representations
- Relay architectural, design, or implementation issues that impact architecturally significant requirements to higher architects
- Translate and elaborate requirements and architectural styles from the system architecture to their sub-architecture

■ **Integration engineering:**
- Feeds issues and "as-built" changes back to the system architects so that they can update the architecture

■ **Test engineers:**
- Collaborate with system architects to identify integration tests as well as test scenarios using behavioral threads defined in the architecture

16.2.7 Guidelines

Use the following guidelines (and associated rationales) regarding system architects:

- Provide proper authorization.
- Empower a chief architect.
- Develop listening and elicitation skills.
- Encourage professional participation.
- Wear multiple hats.

The following descriptions express these guidelines in more detail and provide their associated rationales:

- **Provide proper authorization.** The architecture tasks must be properly authorized within the organization. Typically, this is achieved through charters, documented agreements, and documented understandings (such as memoranda of agreement).
  Rationale: Without proper authorization, cooperation and involvement of appropriate stakeholders, as well as the ability of architects to advocate, may be compromised.
- **Empower a chief architect.** The chief architect should be empowered for the organization to execute the roles and responsibilities of the chief architect, as well as to be involved in project wide decision bodies (either an observer or voting member).
  Rationale: Successive best-practice reports indicate that projects that have problem architectures also tend to lack a dedicated chief architect.
- **Develop listening and elicitation skills.** Architects must actively listen to stakeholders and eliciting architecturally relevant information.
  Rationale: Successful architects gather information from many sources when creating and verifying their architecture and their associated representations.
- **Encourage professional participation.** Architects need to keep current on issues relating to architecture; this is done through participation in professional societies, reading professional journals and papers, attendance at conferences, and presentation of experiences to the profession to encourage continued evolution of the practice.
  Rationale: Technology, application domains, and architecture engineering techniques are rapidly evolving, and without adequate professional participation, the architect is unlikely to be able to effectively take advantage of this evolution.
- **Wear multiple hats.** People playing the architect role must often wear multiple hats (i.e., play multiple roles such as subsystem architect, subsystem technical leader, and subsystem requirements engineer on the same project). Thus, architect need not be a job title.
  Rationale: There is typically insufficient staff to assign single roles to individual people. However, it is critical that the people playing the architect role be qualified to do so.

16.2.8 Pitfalls

Avoid the following common pitfalls associated with system architects and mitigate their negative consequences when they occur:

- The architect knows best.
- The architect is weak willed.

Note that these pitfalls are not unique to architects, and good management is needed for technical roles as well as other roles. The following express these pitfalls in more detail, describing their negative consequences and the steps that architects can use to mitigate them:

- **The architect knows best.**
  - **Pitfall.** An individual architect believes that he or she always knows best, even with regard to disciplines other than architecture engineering.

    The architect insists on having the authority to direct non-architects who are experts in areas in which the architect may (or may not) have significant experience. This typically includes requirements engineers, designers, integrators, testers, safety and security engineers, and even technical and nontechnical managers. Examples of this would be an architect who engineers architecturally significant requirements, does manufacturing design, or improperly directs engineers to redo engineering analyses until they support the architect's architectural decisions, inventions, trade-offs, or assumptions. Architects must not unduly bias the independent analysis of other experts in their own domains.
  - **Negative consequences.** This pitfall often causes the following negative consequences:
    - Project morale decreases as others come to resent the intrusions by the architectural *prima donna*.
    - Turf battles break out between the architecture team and other teams such as the requirements team, design team, and specialty engineering teams.
    - The architects create architecturally significant requirements (such as quality requirements) that are of poor quality (i.e., they do not exhibit the important characteristics of good requirements such as verifiability and lack of ambiguity).
    - The results of engineering analyses cannot be trusted.
  - **Mitigations.** To avoid or mitigate the negative consequences of this pitfall:*
    - The architects should closely collaborate with the members of other relevant teams.
    - Engineering requirements is the responsibility of the requirements teams, not the architecture teams. Thus, the architects should not unilaterally establish architecturally significant requirements (such as quality requirements) for the system. Although the architect should operate as an advocate and collector of such requirements, the architect should not unilaterally impose them on the project.
    - In cases of conflict, the architect should advocate for the introduction of relevant requirements and provide assessments to the appropriate decision-making bodies or individuals to encourage positive change in the project.
- **The architect is weak willed.**
  - **Pitfall.** The architects allow non-architects to improperly dictate architectural decisions, inventions, trade-offs, and assumptions.

    This is the polar opposite of the "architect knows best" pitfall. In this case, the architect is not maintaining positive control over the architecture and its representations and is allowing these work products to be changed by others. Whereas changing product and process requirements may legitimately drive architectural changes, it is the architects who are responsible for making these changes. And while management may legitimately choose to overrule the architect for business and other reasons, it is the architect's responsibility to make management aware of the ramifications of these changes.

---

* Note that when an architect is the root cause of a pitfall, it is typically unproductive to rely on that architect to mitigate the pitfall. In this case, higher-level architects or management must step in to address the problem.

- **Negative consequences.** This pitfall often causes the following negative consequences:
  - The architecture loses its integrity and internal consistency.
  - The architecture contains poor decisions, inventions, trade-offs, and assumptions.
  - The morale within the architecture teams is decreased.
  - Trust in the architect is lost.
- **Mitigations.** To avoid or mitigate the negative consequences of this pitfall:
  - When hiring or assigning architects, ensure that the prospective architect has sufficient experience, confidence, and influential force of personality to stand up to non-architects who improperly attempt to dictate the architecture.
  - Provide a means for the architect to escalate the problem (e.g., to a chief architect or technical leader).

16.3 System Architecture Teams

For the large majority of systems, the size and complexity of the system architecture are too large for it to be effectively and efficiently engineered by a single architect. System architecture engineering is thus a group activity, performed by one or more system architecture teams. These teams may be officially created and formally included in the project organization chart, or they may be informal teams that come into existence on an ad-hoc basis and only exist when members of other teams come together to perform system architecture engineering tasks.

Even if a single person could engineer an entire system architecture, it is unlikely that the single person could effectively and efficiently perform all the architecture engineering tasks. Thus, although systems should have a single chief system engineer to provide leadership and ensure the creation of a single consistent architectural vision, system architectures should be engineered by a cross-functional team of architects and other related subject matter experts.

16.3.1 Types of Architecture Teams

The size and complexity of systems vary greatly. System architectures may also be influenced by reference architectures at the enterprise, system-of-system, and product-line levels. Finally, specialty engineering areas such as reliability, safety, security, and software also greatly influence a system's architecture. As illustrated in Figure 16.4, these factors lead MFESA to include numerous types of system architecture teams.

The following provide definitions of these different types of architecture teams:

System architecture team: any team responsible for developing and maintaining all or part of a system's architecture.

System-of-systems architecture team: a system architecture team responsible for developing and maintaining the overall architecture of a system of systems, network of systems, or family of systems.

Reference architecture team: a system architecture team responsible for developing and maintaining the common part of the architectures of a product line of systems.

Top-level system architecture team: the single system architecture team responsible for developing and maintaining the entire overall top-level architecture of a single system.

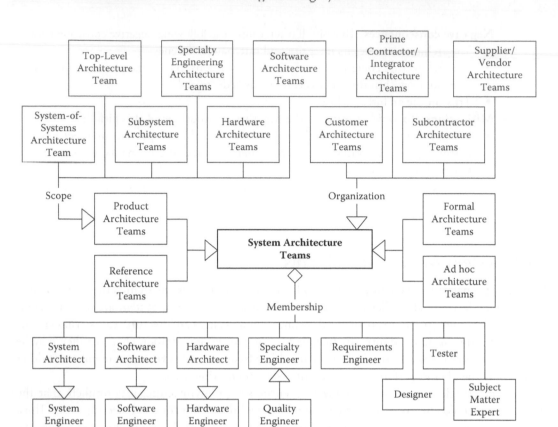

Figure 16.4 Types and memberships of architecture teams.

Note that the top-level system architecture team should lead and must collaborate closely with the system's subsystem architecture teams and specialty engineering architecture teams.

> **Subsystem architecture team:** a system architecture team responsible for developing and maintaining the architecture of a single subsystem.
>
> **Specialty engineering architecture team:** a system architecture team responsible for developing and maintaining those aspects of the system architecture related to the area of specialty engineering (i.e., focus areas).

General system and subsystem architecture teams may not have sufficient expertise in specialty engineering areas. If experts in these areas are not members of the general architecture teams, then specialty engineering architecture teams may be formed. If such specialty engineering architecture teams exist, they must collaborate very closely with the system and subsystem architecture teams to ensure that the architectural decisions, inventions, trade-offs, and assumptions are consistent across teams.

Note that technically we are referring to teams responsible for individual *focus areas* rather than separate *architectures*. Thus, reliability, safety, security, and software can be considered focus areas. Within MFESA there is no separate *reliability architecture*, *safety architecture*, or *security architecture*.

Hardware architecture team: a system architecture team responsible for developing and maintaining the hardware architecture of a single system or subsystem (i.e., the hardware subset of the system architecture).

Software architecture team: a system architecture team responsible for developing and maintaining the software architecture of a single system or subsystem (i.e., the software subset of the system architecture)

Another way to classify architecture teams relates to the top-level organizations for which they work:

Customer architecture team: a system architecture team that is a part of the organization that acquires the system.

Prime contractor or integrator architecture team: a system architecture team that is a part of the organization responsible for developing the system and integrating its major architectural components.

Subcontractor architecture team: a system architecture team that is a part of the organization that develops a major architectural component for the prime contractor or integrator of the system.

The question arises as to whether (1) the subcontractors should be allowed to develop the architectures of their own architectural components in a relatively independent manner using their own architecture engineering methods, or (2) they should collaborate closely with the prime contractor to develop a common system architecture using a common architecture engineering method. Although the first option is more common, is easier to manage contractually, and works reasonably well for modeling the views of static aggregation structures, system behavior tends to cut across the prime contractor/subcontractor interfaces in unanticipated ways, resulting in intra-operability defects that are typically not identified until integration testing or later.

Supplier or vendor architecture team: a system architecture team that is a part of the organization that produces off-the-shelf architectural components and supplies them to the prime contractor or integrator for incorporation into the system.

16.3.2 Responsibilities

The responsibilities of an architecture team are essentially a roll-up of the architectural responsibilities of its individual members. To summarize, an architecture team typically has the following responsibilities:

- **Understand the architecturally significant requirements.** Each system architecture team must identify and understand the relevant requirements that will drive the part(s) of the system architecture for which they are responsible. To do this, they will need to evaluate all the system requirements that have been allocated to their parts of the system, including both higher-level requirements as well as new requirements derived specifically for them. They will need to differentiate the requirements that are architecturally significant from those that only influence downstream disciplines such as design, implementation, and testing.
- **Engineer the architecture and its representations.** This responsibility is the most commonly viewed aspect by outsiders to the discipline. In this area of responsibility, the team

must be able to generate architectural visions and architectures, evaluate differences between candidate architectures, and maintain the architecture and its representations throughout the life cycle. Ultimately, this is the bread-and-butter work of the system architects. While this is not the whole of the work, this practice is the most visible to stakeholders and generates the direct value to the system. Competency here must be demonstrated quickly to keep sufficient attention and effort for other supporting architecture development activities.

- **Have sufficient training, experience, and expertise in architecture engineering.** This responsibility is the most commonly relied upon aspect by outsiders to the discipline. In this way, the architecture team needs to have sufficient personnel who are capable to work with the various architectural models, representations, and views that will support effective system development.

   Note that this responsibility is largely allocated to management, but the individual teams also need to work with management to ensure that they are adequately staffed.

- **Understand the domain and its architectural patterns.** This responsibility covers many areas of the endeavor that the architecture team must be ready to address. First, the team must be familiar with the common architectural styles of the system they are architecting, which includes common decompositions, nomenclature, typical requirements allocations, and the like. This can also include the ability to decompose a system in such a way as to take advantage of architectural reuse, reuse of architectural components, or to modify a common architectural style to support reuse in an innovative fashion.

- **Promote the need for sufficient system architectural development resources.** This responsibility is important in many ways. Project management requires sufficient insight to allocate sufficient resources to the architecture teams. Further, each team must tailor its method and processes to be appropriate to the scale and type of system to be architected. Even with the appropriate tailoring and resources, the team must ensure that it is capable of meeting the goals of the architecture by obtaining sufficient training, acquiring adequate architecture tools, and continuing the momentum with architecture efforts throughout the life cycle.

16.3.3 Membership

The membership of an architecture team typically includes people playing one or more of most of the following roles:

- **Chief or lead system architect.** The chief system architect is responsible for the overall architecture of the system and typically leads the top-level system architecture team, whereas a lead system architect is responsible for leading a lower level subsystem architecture teams.

- **Supporting system architects.** Each architecture team should typically include two or more supporting system architects so that sufficient experience is available to engineer multiple candidate architectural visions and sufficient peer review is available to identify any associated architectural risks and defects.

- **Software architects.** Each architecture team should typically include one or more software architects if the component being architected involves either a significant number of software architectural subcomponents or if software implements a significant amount of the functionality allocated to the component.

- **Hardware architects.** Each architecture team should typically include one or more hardware architects if the component being architected involves a significant number of hardware architectural subcomponents.

■ **Specialty engineering area subject matter experts (SMEs).** Each architecture team should typically include SMEs, especially architects, who are experts in any *relevant* specialty engineering areas such as human factors, performance, reliability, safety, and security. Although it is optimum for the architecture to have these SMEs be full members of the architecture team, another approach would be to have these SMEs be either part-time members or merely made available when the architecture team needs their input.

■ **Application domain SMEs.** Each architecture team should typically include SMEs, especially architects, who are experts in any *relevant* application domain areas such as avionics, biology, chemistry, engines, sensors, software, telecommunications, and weapons. Although it is optimum for the architecture to have these SMEs be full members of the architecture team, another approach would be to have these SMEs be either part-time members or merely made available when the architecture team needs their input. It is desirable, but not always possible, to find SMEs who are also capable architects. If skills and experience in the subject domain and engineering discipline are highly segregated, then team leadership must ensure that both sets of skills and experience are fully staffed and work well together within the team.

■ **Liaison in and to the requirements engineering team.** Each requirements team should typically include an architect for either the current or parent architectural component to ensure that the architecturally significant requirements are engineered, technically feasible, and sufficiently understandable to the other members of the architecture team. In addition to identifying architectural risks, the architect in the requirements team may also identify architectural opportunities to use existing components or technologies that enable the system to better meet the requirements.

Each architecture team may also include a requirements engineer to ensure that the architecture sufficiently meets its architecturally significant requirements. This is especially important when some of these requirements are ambiguous and need to be improved (e.g., made to be feasible, validatable, and verifiable).

■ **Liaison to the system test team.** Each architecture team should typically include a member of the system test team for either the current or parent architectural component to ensure that the architectural component is sufficiently testable or at least verifiable. This member is also responsible for feedback to the team of any findings or changes made during test and integration that need to be reflected in the architecture and its representations. The architecture representations may diverge from the architecture being implemented if this role is neglected (e.g., due to funding cuts during the later development cycle stages). The value of the significant investment in architecture engineering will then be lessened, and the models will be less useful for subsequent analysis.

16.3.4 Collaborations

Architecture teams typically collaborate and interact with the following teams and organizations:

■ **Requirements teams.** Collaborate with the requirements teams to:
  - Ensure that architecturally significant requirements are engineered.
  - Ensure that architecturally significant requirements are identified as such (e.g., according to architectural concern).
  - Allocate the requirements to lower-level architectural components.
  - Evaluate the requirements for feasibility and architectural ramifications.

- **Design teams.*** Evaluate the component designs to ensure that they are consistent with the component architectures.
- **Implementation teams.** Evaluate the component implementations (such as hardware fabrications and software source code) to ensure that they are consistent with the component architectures.
- **Integration teams.** Collaborate with the integration teams to ensure that the integration process is consistent with the relevant architectural (such as aggregation) structures.
- **Testing teams.** Collaborate with the test teams to ensure that integration and specialty engineering testing is consistent with the relevant architectural (such as aggregation) structures, models, views, and focus areas.
- **Quality engineering teams.** Collaborate with the quality engineering teams to ensure that the:
  - Project quality model properly defines all architecturally significant quality characteristics and attributes (quality planning)
  - Quality of the architecture engineering method (as planned) and associated process (as performed) are verified (quality assurance)
  - Quality of the architecture work products is verified (quality control)
- **Process engineering team (also known as the system engineering process group or SEPG).** Collaborate with the process engineering team to:
  - Ensure that appropriate architecture engineering method(s) are developed
  - Capture lessons learned from the actual use of the architectural engineering methods for purposes of process improvement
- **Training team.** Collaborate with the training team to develop and provide training:
  - In the architectural engineering method for those involved in architecture engineering
  - On the system architecture for stakeholders (such as new architects as well as system acquirers, developers, maintainers, operators, and users) of the architecture
- **Project or program management team.** Collaborate with the management teams to:
  - Establish requirements for architect recruitment, competency, and training
  - Interview, hire, and promote the architects
  - Obtain via mandate the appropriate responsibility, authority, and reporting channels for the architects
  - Staff the architecture teams
  - Obtain funding for the:
    - Performance of architectural engineering tasks
    - Acquisition of architecture engineering tools
  - Schedule architectural tasks and milestones
- **Configuration management team.** Collaborate with the configuration management team to:
  - Identify architectural components (and other work products) as configuration items (configuration identification)
  - Identify baselines involving architectural components (configuration identification)
  - Manage multiple versions and variants of architectural components and other architectural work products (version control)

---

* When integrated product teams (IPTs) are used, there may not be any separate, officially designated design and implementation teams. In this case, the design team and implementation team refer to the IPT when performing design and implementation tasks.

- Place architectural work products under configuration control (configuration control)
- Control changes to such architectural work products (configuration control)
- Report the status of architectural baselines and configuration items (configuration status reporting)
- Audit the identity, version, and completeness of architectural baselines (configuration auditing)

■ **Risk management team.** Collaborate with the risk management team to:
  - Identify architectural risks
  - Analyze the probability and severity of architectural risks
  - Control architectural risks via mitigations and transfer of risks
  - Monitor the status of architectural risks and the effectiveness of their mitigations

■ **Measurement and metrics team.** Collaborate with the measurement and metrics team to:
  - Determine the appropriate architecture metrics to collect
  - Collect (or support the collection of) these architecture metrics
  - Analyze (or support the analysis of) these architecture metrics
  - Report (or support the reporting of) these architecture metrics

■ **Human factors team.** Collaborate with the human factors team to:
  - Ensure that the architecture takes into account human (such as user, operator, and maintainer) limitations and capabilities
  - Verify that the architecture does not exceed human limitations and capabilities

■ **Reliability team.** Collaborate with the reliability team to:
  - Ensure that the architecture sufficiently supports the system meeting its reliability requirements
  - Verify that the architecture sufficiently supports the system meeting its reliability requirements

■ **Safety team.** Collaborate with the safety team to:
  - Select and incorporate appropriate security patterns, mechanisms, and safeguards into the architecture
  - Evaluate the architecture and its representations for safety vulnerabilities and its support for meeting the safety requirements

■ **Security team.** Collaborate with the security team to:
  - Select and incorporate appropriate security patterns, mechanisms, and countermeasures into the architecture
  - Evaluate the architecture and its representations for security vulnerabilities and its support for meeting the security requirements

16.3.5 Guidelines

Use the following guidelines (and associated rationales) regarding system architecture teams:

■ Develop a single shared vision.
■ Earn architecture team respect.
■ Development architects should not be replacement architects.
■ Subsystem architects should be members of the system architecture team.
■ Differentiate architecture team members from collaborators.

The following descriptions express these guidelines in more detail and provide their associated rationales:

- **Develop a single shared vision.** All members of the architecture team need to share a single vision once that vision has been selected from the set of candidate visions.

  Rationale: The best architectures are the product of a single architectural vision. The best vision is rarely the first vision identified. Note that this is much easier said than accomplished.

- **Earn architecture team respect.** The members of the architecture team as well as their management should actively work to gain the respect of the stakeholders of the architecture. The architecture team should identify problems that they can solve and risks that they can mitigate so that they can demonstrate the value of their work products and analyses. Then they will begin to be invited to meetings and their opinions and recommendations will be listened to. They should avoid the use of architectural jargon when talking to nontechnical stakeholders.

  Rationale: An architecture team and its members must be respected by the stakeholders of their architecture if architectural integrity is to be maintained by lower-level architecture teams, designers, implementers, and technical leaders.

  Note that there are many factors that influence respect for the architecture team, and some of these factors are outside the control of the architecture team. Nevertheless, the team should do what it can to gain and maintain that respect. This is especially important because architecture teams are new to many organizations, which often do not know how to best utilize them. Others may resent the architecture team as a perceived encroachment on their "turf."

- **Development architects should not be replacement architects.** The architecture team that originally engineered the architecture of a system should probably not be the team tasked to architect the system's replacement.

  Rationale: The architecture team that engineered a presently successful architecture is often technically the optimal team to maintain that architecture and to insure its integrity.* However, they are rarely the best team for engineering the architecture of a revolutionary replacement to the system. The architecture team that built the prior version of the system may well be too wedded to their previous architecture and locked into undocumented assumptions that are no longer true.

- **Subsystem architects should be members of the system architecture team.** It is useful to have the lead architects of the *subsystem* architecture teams be supporting members of the *system* architecture team.

  Rationale: This facilitates collaboration and generates buy-in by the lower-level subsystem architecture teams.

- **Differentiate architecture team members from collaborators.** It is important to decide which people should be full-time (voting) members of an architecture team, which people should be as-needed part-time (nonvoting) members of the architecture team, and which people should be external (nonvoting) collaborators with the architecture team. It is also

---

* This is often difficult to achieve in practice because good architects are rare and are typically reassigned to develop new systems. Good architects also prefer new challenges and rarely enjoy performing architecture maintenance duties.

useful to codify such relationships explicitly, lest part-timers or non-members come to think of themselves as voting members.

Rationale: This prevents unnecessary disagreements and arguments between full-time voting members and others. It therefore also increases efficiency and the meeting of project deadlines.

16.3.6 Pitfalls

Avoid the following common pitfalls associated with system architecture teams and mitigate their negative consequences when they occur:

- There is only one architect.
- The architects are isolated.
- The architects have inadequate experience.

Note that these pitfalls are not unique to architecture teams. The following express these pitfalls in more detail, describing their negative consequences and the steps the architects can use to mitigate them:

- **There is only one architect.**
  - **Pitfall.** All of the architecture engineering is assigned to a single architect. Most systems are too large and complex to have their architectures engineered by a single architect. The architecture work is typically too extensive for a single architect to complete within schedule constraints. Whereas it is good to have a lead architect to provide a consistent and coherent architectural vision, it is also important to have an architectural team to ensure that multiple viewpoints are considered and that architectural defects are identified. Although no one should want to end up with an architecture that was obviously "designed by committee," the system architecture is too important to let it reside in a single architect's hands.
  - **Negative consequences.** This pitfall often causes the following negative consequences:
    - The engineering of the architecture and its representations do not meet schedule deadlines.
    - There is inadequate brainstorming.
    - There is insufficient iteration of the architecture and its representations.
    - There is difficulty producing adequate competing architectural visions.
  - **Mitigations.** To avoid or mitigate the negative consequences of this pitfall:
    - Use small, focused architecture teams.
    - Make the chief system architect the leader of the top-level system architecture team.
    - Ensure that the architects on the architecture team collaborate.
    - Avoid using chief system architects who suffer from the "architect knows best" syndrome.
    - If diverse views cannot be developed within the architecture team (such as on a small project with limited budget and staff), the architects should solicit input from external stakeholders and demonstrate professional respect for and consideration of their ideas. Not only does this tend to yield superior architectures and their representations, it also supports the previously mentioned objective of gaining respect from the stakeholders for the architects.

■ **The architects are isolated.**
  - **Pitfall.** The architects and architecture teams attempt to engineer the system architecture or the architecture of a major architectural component in isolation. The architects do not collaborate with other stakeholders in the architecture. For example, the architects do not collaborate with other teams such as the requirements, design, and test teams. This may happen either by the choice of the architects if they fail to appreciate their role on the broader project context or by people in other specialty engineering areas who implement decisions without including the architects in their review process. The latter problem can be addressed somewhat by conventions such as review checklists, but acceptance by the organization ultimately comes with an earned reputation for value-added contributions. Note that this is the group version of the "architect knows best" architect pitfall.
  - **Negative consequences.** This pitfall often causes the following negative consequences:
    • The architecture is inconsistent with the requirements, design, or implementation.
    • Architectural integrity is lost.
    • The architecture is not feasible to implement.
    • The architects lose their influence with the rest of the project teams.
    • The architecture loses its influence on the requirements and over the design, implementation, testing, manufacturing, and deployment of the system.
  - **Mitigations.** To avoid or mitigate the negative consequences of this pitfall:
    • Ensure that the architects closely collaborate with external stakeholders, subject matter experts, and the members of other related teams (such as the requirements team, management team, quality team, safety team, security team, and test team). Note that this may necessitate explicit organizational changes to foster improved communication.
    • Use evaluations to verify that the architecture and architectural representations are consistent with the work products produced by these external individuals and teams and are useful for communicating with them.

■ **Architects have inadequate experience.**
  - **Pitfall.** An architecture team is staffed with architects who do not have sufficient training, experience, or expertise to effectively and efficiently perform the architecture engineering tasks. This often occurs during the development of very large systems because the development organization does not have sufficient architects and cannot hire enough to make up the difference. Designers may be given "field promotions to the rank of architect" (i.e., be "Peter principle-d") when they lack adequate training, experience, and expertise, and they may not even have the right aptitude and attitude for the job. This is especially a problem if the experienced architects do not have sufficient time to provide the necessary mentoring and oversight for these apprentice architects.
  - **Negative consequences.** This pitfall often causes the following negative consequences:
    • The architecture and its representations contain defects, weaknesses, and risks.
    • The architecture is inconsistent with the requirements, design, or implementation.
    • Architectural integrity is lost.
    • The architecture is not feasible to implement.
    • Architecture team productivity is decreased.
    • The architecture tasks take longer than anticipated to complete, leading to budget overruns and missed schedule deadlines.
    • The architecture loses its influence on the requirements and over the design, implementation, testing, manufacturing, and deployment of the system.

- **Mitigations.** To avoid or mitigate the negative consequences of this pitfall:
  - Provide the architects with significant training in system architecture engineering.
  - To the extent practical, hire (or transfer from other projects) experienced system architects to become members of the architecture teams.
  - Have one or more trained expert system architects provide consulting and mentoring to the architecture teams.

16.4 Architectural Tools

The MFESA metamodel treats architectural tools as a type of architectural worker. MFESA groups tools with architects and architecture teams to the abstract metaclass architecture worker because all three can perform (parts of) architecture engineering work units to produce architectural work products. Note that many of the following tools can be and are used by non-architects for non-architectural purposes. Nevertheless, they are listed here to help document the many types of tools needed by the architects to successfully perform their tasks.

16.4.1 Example Tools

As illustrated earlier in Figure 16.2, architects use tools in their performance of work units that result in creating and updating architectural work products. Key system or subsystem architectural decisions are modeled or otherwise documented via these work products that are collectively referred to as architectural representations. Architectural representations may be implemented in a variety of formats, as noted previously in sections 5.7 and 5.9. Because the various architectural representations typically include different models or views of the same underlying architecture, they are interdependent and it is very important that these representations are mutually consistent. Maintaining a reasonable uniformity of approach and mutual consistency among the many types of architectural representations that are likely to exist in a large systems architecture engineering effort can be daunting. This is particularly so when coordinating and interchanging representations among multiple, geographically dispersed architecture engineering teams. Manual efforts to maintain uniformity and consistency of architectural representations may become unwieldy. This is where a judicious selection of architectural tools may assist.

It should be noted that the consideration and selection of architectural tools can be a significant effort. For large projects, it is likely that trade studies with substantial effort to find a "way forward" may need to be undertaken. In some cases where a tool chain is contemplated that encompasses both system and software architecture engineering activities, significant training and buy-in may be needed across several different organizations. Where buy-in is sought, elements of organizational behavior change are likely involved in addition to training efforts. Tool selection efforts can be even more time consuming when organizational behavior change is needed for tool buy-in. In these cases, particular attention should be paid to ensure that sufficient effort levels have been considered in Steps 3 and 4 of MFESA Task 1, *Plan and Resource the Architecture Engineering Effort.*

> **Architecture tool:** any tool that assists with the production, coordination, or maintenance of architectural work products.

This definition is meant to be very broad and is not limited to tools that are software products. The definition is also not meant to imply that only automated, integrated tools are included, nor that only single tool environments are included. The "coordination" aspect may imply an inclusion of some unspecified mechanism that provides a means for data interchange and/or representation interchange/sharing/repository. The definition's coordination aspect also implies some mechanism to establish consistency between the various work products within the architectural representation. The "maintenance" aspect within the definition implies that some unspecified mechanism exists for controlling versions of the work products as the architecture representations mature.

Each program will need to establish appropriate limits for its system architecting tools capabilities. A one-size-fits-all approach to tool selection is typically not workable, if for anything else other than budget reasons. These established capabilities can then serve as a filter to help identify an initial subset of available tools that will undergo further detailed evaluation. Capabilities appropriate for your program may range from being quite extensive (such as found in enterprise architecture tools that encompass business process considerations as drivers of technical architecture) to being very basic but effective (such as manual whiteboards and digital cameras). It is unlikely for a program to select and use just one architectural tool. Indeed, several different types of architectural tools are likely needed, some of which are described below.

For either end of the tool capability spectrum, there are various sources for good sets of questions that should be answered during evaluations of system architecting tools. These same questions can also be asked prior to tool evaluations when establishing the limits of one's program's tools capabilities. The following Web sites contain some representative architecture tool evaluation questions that may be useful for both of these activities. These sites are neither meant to be an endorsement nor exhaustive. However, these sites contain representative evaluation question lists with fairly wide scope that emphasize the more capable end of the tool spectrum. One can tailor these lists of questions to establish limits for architecture tool capabilities that are suitable to a specific program's needs (Web links were active October 2007):

- INCOSE's Systems Architecture Tools Survey page: http://www.paper-review.com/tools/sas/read.php
- Institute for Enterprise Architecture Development's Tool Overview page: http://www.enterprise-architecture.info/EA_Tools.htm
- The Open Group's chapter discussing Tools for Architecture Development: http://www.opengroup.org/architecture/togaf8-doc/arch/chap38.html

16.4.2 Types of Architecture Tools

The diversity of the architectural tasks typically implies that the architects and architecture teams require multiple architecture tools. A far too common mistake is to only provide a graphical modeling tool for the production of graphical architectural models. The following definitions of potential architecture tools with associated discussion is intended to provide the reader with a sense of the scope of architecture tools that need to be evaluated, selected, obtained, and configured to properly support the architecture engineering tasks.

> **Whiteboard:** a (typically) mechanical or electronic medium for capturing graphic or text representations of architectural concepts, decisions, inventions, trade-offs, and rationales.

Common use of whiteboards is made during the brainstorming efforts that take place during architectural engineering. Whiteboards are very effective at capturing and communicating rich concepts during the formative stages of the system architecture — but the rationale for arriving at those concepts also needs to be captured at this time. While electronic versions of whiteboards exist, the manual versions appear to be much more widely employed, in part due to their informality, ease of use, and low cost. Computer-based whiteboard environments are available but do not seem to be nearly as widespread.

The whiteboard's informal ease of use is perhaps simultaneously both its greatest strength and drawback — very creative ideas appear to flow from whiteboard sessions. However, capabilities such as version control, repository and data interchange, and coordinating consistency with other architectural work products need to be accomplished via highly manual means, prone to error, and perhaps being skipped altogether as programmatic "firefighting" episodes increase in frequency and duration during the life cycle. The availability of digital cameras and whiteboard hardcopy printing features may provide some mitigation of these capability shortfalls, but the whiteboard may still have insufficient tool capability, depending upon one's specific program's tool capability needs. Programs with classified architectural information present further challenges when whiteboards are used.

Nonetheless, the whiteboard has been, and will likely continue to be, a very key tool to stimulate the creative tasks within architectural engineering efforts, particularly in small group brainstorming sessions. A specific program may need to address how to capture the pertinent information and rationale from whiteboard sessions and put it in a form that may be version controlled, placed in a repository for collaboration, and coordinated for consistency with other work products.

> **Image capturing device:** a device used to capture and preserve an image representation of a model, a decision, invention, algorithm, rationale, or the like, that is generated by means other than the device itself during the creation of architectural work products.

Examples of these are digital cameras, document scanners, and whiteboards with printing capability. Note that the definition includes a "preserve" perspective but does not mandate that this preservation method be electronic storage. That is, hardcopy forms of image preservation are also included within the definition's scope. Of course, data sharing and ongoing image maintenance flexibility will have limited scope when the means of image preservation is, for example, a hardcopy printout of a whiteboard. Nevertheless, these tools are very effective at capturing and preserving very rich concepts and ideas that were created during architectural engineering work sessions. Very little, if any, specialized training is usually needed to use these devices as architectural tools.

Note that the definition of an image capturing device implies that the image being captured was generated by means other than the device itself. Thus, tools such as general-purpose drawing tools and computer-aided design tools, which capture images generated by the tools themselves, are not included in this more limited scope definition of an image capturing device. The whiteboard with printout capability could be considered an exception, but is also included in this more limited definition of an image capturing device if one views the printing capability as distinctly separate from the whiteboard's "creative surface."

Some image capturing devices may be restricted or not permitted in certain classified environments, rendering them ineffective for use as an architectural tool in these instances.

> **Word processor:** a software application primarily focused on the creation and maintenance of textual documents and is typically run on general-purpose computers or workstations. Graphical and tabular capabilities may also be included in these application programs but these capabilities are not the primary purpose for these tools.

Word processors may range from simple text editors to full-blown document management formatting systems. While these tools may have graphical and tabular capabilities, the feature sets and capabilities are far more geared toward creating and formatting text and associated layout than toward creating drawings or tabular spreadsheets. As an architectural tool, word processors are very capable for capturing the rationale used for making architectural decisions, reporting on the results of a trade study, or for creating white papers that describe architectural inventions. Many commercially available and open source word processors have data import and export capability that assists with sharing information collaboratively. Support may also exist to help enable consistency updates of word processing architectural work products with other architectural representations.

> **Spreadsheet:** a software application primarily focused on the creation and maintenance of tabular data with associated built-in tabular computational capabilities.

Spreadsheet applications are typically run on general-purpose computers or workstations. Graphical and text formatting capabilities may also be included in these application programs but these capabilities are not the primary purpose for these tools.

Spreadsheets contain capabilities that are highly optimized for creation and ongoing maintenance of computational as well as noncomputational tables. General text and graphical capabilities may be included in spreadsheet applications, but these are not the primary focus of these tools. As an architectural tool, some spreadsheet applications may have computational capability of sufficient power to model certain behavioral aspects of the system being architected. Many commercially available spreadsheets have data import and export capabilities that assist with sharing information collaboratively. Support may also exist to help enable consistency updates of architectural work products modeled by spreadsheet applications with other architectural representations.

> **General-purpose drawing tool:** a software application primarily focused on the creation and maintenance of domain-independent graphical representations.

General-purpose drawing tools are typically run on general-purpose computers or workstations. Text and tabular formatting capabilities may also be included in these application programs, together with some computational ability, but these capabilities are not the primary purpose for these tools.

General-purpose drawing tools are domain independent and as such are typically not primarily optimized for a particular specialized graphical application (such as mechanical drafting and dimensioning applications). These applications may export data to interchange graphical objects, but the underlying data that describes these objects may not contain descriptions of specialized object behavior due to the general-purpose nature of the application. General-purpose drawing tools can assist in the creation of architectural work products that illustrate architectural structures and components or that describe architectural behavior for the system being engineered.

> **Graphical modeling tool:** a software application primarily focused on the creation and maintenance of visual images that model certain structural and behavioral aspects of system architectures using a modeling language with standardized syntax and semantics.

Graphical modeling tools are typically run on general-purpose computers or workstations. Graphical and textual modeling semantics may be included but the tool emphasis is on graphical representation using a standardized notation and language.

Graphical modeling tools provide standardized languages (such as UML, SysML, and AADL), visual representations, and semantics that can be used to model aspects of system architectural behavior. They can play an important role in capturing and documenting a common understanding of how the system structures are modeled. Typically, both static architecture structures and dynamic system behavior can be modeled using these tools. Some of the commercially available tools include support for "executable architectures," which permits dynamic simulation of the architecture being modeled (see "Simulation Tools" below). Many commercially available tools support common data interchange formats that promote collaboration as well as coordination and consistency across various architecture work products.

Some of the common modeling languages such as UML (Unified Modeling Language) focus more toward software rather than systems modeling. SysML (System Modeling Language) is a generalization of UML and adds some modeling constructs to UML that better accommodate system architectural notions. AADL adds additional modeling capabilities of interest to real-time system software engineers. AADL tools may typically be able to interchange data with UML-based tool products, thus extending overall modeling capability to include real-time software behavior. Finally, DODAF (the United States Department of Defense Architecture Framework) graphical and textual models were designed to support interoperability between systems and are often mandated by defense contracts [DODAF, 2007a–c].

> **CAD/CAM (computer-aided design/computer-aided manufacturing):** a combination of software applications and possibly specialized hardware platforms primarily focused on the creation and maintenance of two- and three-dimensional models of various parts of a system's physical architectural structures. The physical models correspond to system structures that are to be directly manufactured by a manufacturing process.

CAD/CAM systems are typically highly specialized tools for use in specific domains that require physical parts to be manufactured to specific geometries, usually in high volumes. These systems can be used, for example, to design and manufacture a simple stamped clip, a molded plastic toy, an intricate fuel pump with exacting tolerances, a complex multilayer circuit board, an assembly of an entire jet engine, etc.

When a CAM system is used to create parts of a system, CNC (Computer Numerical Control) codes are used to transfer the design information that was captured in the CAD stage. These CNC codes use an industry-standardized format that permits the use of various types of commercially available CAD tools to drive the manufacturing platform. The CNC capability, together with full-featured two- and three-dimensional modeling capability, are among the main differentiators of CAD tools and general-purpose drawing tools.

> **Simulation tools:** any software applications primarily focused on imitating specific types of architectural structures by predicting how they will behave in the context of specific scenarios.

Simulation tools provide architectural simulations and thus are a kind of executable representation of parts of the system architecture. Referring to Figure 5.16 in Chapter 5, one sees that an architecture's executable representations may come in two flavors: (1) executable architectures and (2) architectural simulations. Note that the definition above for simulation tools excludes architectural prototypes, which, in the context of this publication, are considered as being separate from architectural simulations.

Simulation tools may model the behavior of key system components such as software, algorithms, human actors, hardware, networks, processes, and the like, or some combination of these. The range of tool capability can be quite broad. For example, simulation tools that predict the behavior of a system's network structure might include extensive capability for handling hundreds of network elements with thousands of users. In contrast, low-end simulation tools, for example, may be able to predict certain aspects of behavior for small, individual software components. A system may include a significant number of application-specific integrated circuits (ASICs) for which architectural simulation tools with extensive capabilities are needed. If business processes need to be simulated, tools with significant capabilities are likely required. Additionally, human user interfaces, complex mathematical algorithms, chemical reactions, or discrete event systems in general may need to be simulated. Certain tools may be available with an integrated corresponding modeling language, or separate tools that provide appropriate modeling languages may need to be chosen.

Due to the availability of such a wide range of program needs and corresponding tool capabilities, it is important for the program to establish early on which parts of the system architecture embody behaviors that are important enough to simulate with tools during architecture engineering. Then, an exercise to establish the limits of simulation tool capabilities should be done as part of the planning tasks, much like that suggested for the architectural tool definition in section 16.4.1. In particular, attention should be paid to ensure that sufficient effort levels have been considered for simulation tool identification, selection, and evaluation in Steps 3 and 4 of MFESA Task 1, *Plan and Resource the Architecture Engineering Effort*.

> **Configuration management tool:** any software application primarily used to establish control and maintenance of the integrity of architectural baseline and their constituent architectural work products.

For this definition, "baseline" means a set of correlated, consistent architectural work products that have been formally reviewed and accepted by the involved parties as representing the system architecture as of a specific point in time. A baseline should be changed only through a program's formal change management procedures. Configuration management tools are likely to have significant roles within these procedures.

Within the industry, configuration management, or CM, is quite an overloaded term, potentially conjuring up different meanings to different individuals. For some, CM might be synonymous with the concept of version control. To others, CM's meaning might be so broad as to encompass software development procedures, released software baselines, hardware configurations, and data management and problem management procedures. CM and change management may be synonymous yet to others. At a minimum, software configuration management (SCM) has a whole host of potential differences in meanings that are unique to software (e.g., is the ability to build specific versions of software from source code part of CM?).

Thus, it is very important for each program to define what it specifically means by CM before it can begin to identify the scope of configuration management tools that are appropriate for the program. For the purposes of this publication, we have limited the definition scope for configuration management tools to those software applications that deal with controlling the baselines specifically of system architectural work products. Note that even this relatively narrow scope is likely to demand intricate CM tool capability because architectural work products in general need to be consistent; thus, a single change in one might need to ripple through many others. This intricacy is above and beyond any that is needed for the CM tool to maintain baseline integrity across all work products with respect to a specific set of system requirements.

Each program will still need to define other CM tool capability limits that are appropriate for the specific circumstances. As an example, certain programs may need to decide whether they need a CM tool that is capable of managing, in some way, hand-made sketches of architectural structures drawn on whiteboards that are either printed or photographed with an image capturing device. As another example, a program that decides that CM is to include activities in the sustainment part of the program life cycle may need to limit candidate CM tools to those that provide capabilities that include integrated support for problem management.

An exercise to establish the limits of CM tool capabilities should be done as part of the planning tasks, similar to the exercises discussed for simulation tools and the architectural tool definition in section 16.4.1. In particular, attention should be paid to ensure that sufficient effort levels have been considered for CM tool identification, selection, and evaluation in Steps 3 and 4 of MFESA Task 1, *Plan and Resource the Architecture Engineering Effort*.

> **Requirements engineering tool:** any software application primarily used to engineer requirements (such as support for the identification, analysis, management, specification, and verification of requirements and other requirements work products).

From a system architecture engineering standpoint, we are interested in tools that support the engineering of architecturally significant requirements and that support the forward and backward tracing between requirements work products and architectural work products.*

Within the context of this book, the requirements engineering tool's relationship management capability must also include support for managing the life cycle of the requirements themselves (e.g., requirements addition or deletion throughout the life of the program).

The full life-cycle nature of our definition of requirements engineering tools exposes some potential difficulties in specific tracing tool capabilities that one will likely need to consider, particularly for software-intensive systems. Depending on the system domain addressed by one's project, the nature of the requirements themselves, or the nature of the process and environment one's team uses to subsequently derive them from system requirements, may not facilitate requirement definition within a single tracing tool. That is, defining requirements for a particular domain may be much better accommodated within one's development environment tool rather than in the system requirements tracing tool for a variety of reasons. This situation demands that additional attention be paid to requirements tracing methods to ensure maintenance and allocation of integrity of bi-directional tracing between derived system requirements and test and verification procedures — particularly when nonfunctional requirements are involved. This is not a simple issue to address, and doing so can be very time consuming, inefficient, and error prone if left to manual methods by default. It is thus important that these issues are considered early on to ensure that one plans for sufficient effort levels to identify, select, and evaluate requirements tracing tools in Steps 3 and 4 of MFESA Task 1, *Plan and Resource the Architecture Engineering Effort*.

> **Information architecting tool:** any software application primarily used to produce and maintain the high-level strategic structure and organization of the systems' data

---

* Note that the selection of requirements engineering tools is a responsibility of the requirements teams, not the architecture teams. The architects will need access to the requirements engineering.

components along with associated decisions and rationale used for establishing these data components.

Currently, the term "information architecture" may often be used in the context of structuring and organizing information for Web site designs. Within this book, information architecture is not meant to be limited to the context of Web site design, but rather is intended to encompass a broader meaning that includes the structure and organization of system and enterprisewide, top-level data components. Information architecture in this broader sense involves tools with capabilities in other subdomains of expertise, such as those that support information and data modeling, information engineering, database design and management, etc. Thus, information architecting tools, when referred to in this book, may include tools with capabilities that support Web site development as a subset of broader capabilities, but we do not intend to mean that the tool capabilities are limited solely to Web site design.

Software-intensive systems are even more likely to drive the need for information architecting tools with capabilities that span a fairly broad spectrum. Further challenges arise due to tools that target different subdomains of expertise within information architecture that may use different modeling notations. For example, information modeling and software data modeling may use UML notation, database modeling may use IDEF-1X or other IDEF-variant notation, and XML may be used to describe the information architecture for enterprisewide and Web-based applications.

Indeed, it is not likely that a single tool will provide the necessary broad scope of information architecture tool capabilities. It is observed that the breadth of incorporating these information architecture tool capabilities within a single tool or single tool suite begins to blur the distinction between information architecture tools and tools that are described as being enterprise architecture tools. A particular project may not need or want to select tools that provide such a wide breadth of capabilities. Therefore, an exercise to establish the limits of information architecture tool capabilities should be done as part of the planning tasks, similar to the exercises discussed for simulation tools and the architectural tool definition in section 16.4.1. In particular, attention should be paid to ensure that sufficient effort levels have been considered for information architecture tool identification, selection, and evaluation in Steps 3 and 4 of MFESA Task 1, *Plan and Resource the Architecture Engineering Effort*.

> **Business process modeling tool:** any software application primarily used to produce and maintain an elaboration of the activities and information an enterprise uses to achieve its business goals.

Business process modeling (BPM) tools are likely to use graphical representations of the activities performed in an enterprise. There are a variety of modeling notations that may be used, depending on the specific tool. For example, UML activity diagrams, sequence diagrams, use case diagrams, and use case descriptions may be used to represent portions of, and the entirety of, a business process. Alternatively, Business Process Modeling Notation (BPMN) may be used to describe whole processes or portions thereof [BPMN, 2004]. Additionally, BPM tools may support the use of the Web Services-Business Process Execution Language (WS-BPEL) for creating executable models of the business processes (such as tools that generate WS-BPEL executable business models from BPMN graphical models) [WS-BPEL, 2007].

While general-purpose drawing tools can be used to perform business process modeling, we do not include them as examples of BPM tools in our present discussion. Within this book, a BPM tool is expected to have some specific capabilities that focus on supporting BPM-specific management and concepts. Some example BPM tool differentiators include (but are not limited to) the ability to:

- Create models of business processes and their relationships
- Create a catalog of the defined business processes as searchable documentation
- Search through all defined business processes for purposes of identifying similar processes
- Collect and report on business process measurements from actual process usage

The technologies incorporated within BPM tools are still maturing as of the writing of this book, and many tools still provide extensions to relevant standards (such as UML) for addressing BPM capabilities. This results in the increased potential for tools to provide proprietary data formats leading to more difficult tool integration efforts.

Because the decision to use BPM within a project has a direct effect on how system architects will describe top-level system behaviors, it is important to make early decisions and to communicate decisions across the project regarding the use of BPM and the BPM tool path forward. The longer these decisions and communications are postponed, the more likely it is that behavioral artifacts will be created by architecture teams in formats that are likely to be inconsistent with each other and with the ultimate BPM tool decision. Potentially, significant amounts of rework with corresponding increases in project costs and schedule may negate the benefits that were perceived from allowing programmatic milestone pressures to drive an "early" start to architecture team efforts (i.e., having architecture teams start their work before the use of BPM and related modeling tool decisions are made). Due to the potential increases in cost and schedule in this situation, there is an accompanying increased likelihood that any inconsistent artifacts (which were produced before BPM tool decisions are made) will not be updated for consistency. This, of course, can lead to outdated artifacts that are more difficult to use and maintain. Therefore, an exercise to establish the use of BPM and requirements for the associated BPM tool capabilities should be done as part of the planning tasks similar to the exercises discussed for simulation tools and the architectural tool definition in section 16.4.1. In particular, attention should be paid to ensure that sufficient effort levels have been considered for BPM tool identification, selection, and evaluation in Steps 3 and 4 of MFESA Task 1, *Plan and Resource the Architecture Engineering Effort*.

Mass/size/geometry modeling tool: any software application used to model the mass, size, and geometry of the system and its subsystems.

Mass/size/geometry (MSG) modeling tools are typically specialized for targeted domains and industries. For example, there are likely to be very different *mass modeling* capabilities within a tool that is targeted for modeling buildings in the construction industry versus a tool that is targeted for modeling airplane or spacecraft parts in the aeronautical/aerospace industries, even though the objective of these capabilities for both tools is to provide mass modeling.

It is possible that some of the other types of tools selected for use within a particular project (i.e., general-purpose drawing tools or specialized CAD/CAM tools) can provide some elements of MSG modeling capabilities that one will find suitable for use. Again, this is quite dependent on the specific project's domain and industry.

Due to this fairly wide range of project needs for MSG modeling and the corresponding MSG modeling tool capabilities, it is important for the project to establish early on which parts of the system architecture need to be modeled with MSG modeling tools during architecture engineering. Then, an exercise to establish the limits of MSG modeling tool capabilities should be done as part of the planning tasks, much like was done for the architectural tool definition in section 16.4.1. In particular, attention should be paid to ensure that sufficient effort levels have been considered for MSG modeling tool identification, selection, and evaluation in Steps 3 and 4 of MFESA Task 1, *Plan and Resource the Architecture Engineering Effort*.

> **Software architecture tool:** any software application primarily used to produce and maintain the architectural work products that elaborate the software's high-level strategic structures, organization, externally visible properties, and that provides capabilities for managing their interrelationships along with associated decisions and rationale used for defining such entities.

It should be noted that the above definition means to exclude general-purpose drawing tools from being considered as software architecture tools. Although such tools can be used to visually represent aspects of the software architecture, these general-purpose tools usually lack specific capabilities for managing interrelationships between software architectural elements. On the other hand, some general-purpose graphical modeling tools may include such capabilities and therefore may be considered included within the above definition of a software architecture tool.

Software architecture tools generally support representations that are created via a software-focused common modeling notation/language such as UML (Unified Modeling Language), SysML (System Modeling Language), or AADL (Architectural Analysis and Design Language). Some of the more capable tools may also include simulation features and support for simulation of executable architectures.

It is common for software architecture tools to provide capabilities that go well beyond elaboration of a software system's high-level structures and organization as implied by the present definition. Indeed, many available tools will also support very detailed and low-level software design constructs that approach the code level. Some of these tools may even generate code from such designs.

The commercial market for software architecture tools is currently quite dynamic. Tool capabilities are rapidly expanding in commercially available products. It is not uncommon today to identify software architecture tools that provide an entire software engineering environment that includes capabilities for requirements development, tracing, and management. The prospective growth in use of service-oriented architecture (SOA) approaches within the engineering of software-intensive systems has driven the inclusion of capabilities such as business process modeling, support for creating Web services, simulation and behavioral modeling, and the like within tools that had previously been considered as focused only on producing software architectural representations and designs. Indeed, the boundaries are becoming increasingly blurred between various types of systems architecting tools such as enterprise architecture tools and software architecture tools.

The choice of the software architecture tool for use within a software-intensive project can have a fundamental effect on the detailed activities of the project's software development community. These tools tend to establish the means of sharing architecture and design information across the development community. Thus, they become an integral driver of the project's detailed development and test procedures. The tool's ability to function across the relevant geographical project environment and the willingness of the organization-wide development community to buy into use of the tool are as important to consider for tool selection as are the tool's raw technical capabilities and associated costs. Thus, attention should be paid to ensure that sufficient effort levels have been considered for software architecture tool identification, selection, and evaluation in Steps 3 and 4 of MFESA Task 1, *Plan and Resource the Architecture Engineering Effort*.

Enterprise architecture tool: any software application primarily focused on the use of an architecture framework to structure the production and maintenance of associated architectural work products that elaborate the current and/or future structure and behavior of an organization's processes, information systems, personnel and organizational subunits, aligned with the organization's core goals and strategic direction.

Enterprise architectures (EA) typically use reference architectures as frameworks to define specific sets of views that may be tailored for specific instantiations within a given project or program of related projects. The architecture frameworks supported by these tools may be any of several standards that are openly available within the community, or may be a custom or proprietary framework.

EA tools tend to be software products embodying very significant capabilities. These products indeed may consist of a suite of several tools that have been described in other definitions elsewhere in this section. The set of features included within EA tools are typically intended to provide a wide spectrum of information technology-centric capabilities. Some commercially available EA tools may make claims of including diverse architectural capabilities such as business process modeling, information architecting, database architecting, requirements tracing, simulation, systems and software architectures, support for service-oriented architectures (SOA), etc. Some EA tools also may include support for various modeling languages or notations such as BPMN (Business Process Modeling Notation), BPEL (Business Process Execution Languages), UML (Unified Modeling Language), SysML (System Modeling Language), and the like.

The selection, training, and use of a tool as significant as an EA tool has dramatic impacts on the activities of the systems engineering, system architecture engineering, and the development and test communities within a project. These considerations are similar to those described for the business process modeling tool above. At a minimum, one will need to evaluate the appropriateness of use of a specific architecture framework within a specific project. In some cases, use of a specific framework may be mandated, such as the use of DODAF within U.S. Department of Defense programs or use of the FEA framework within U.S. government agency IT programs. Because the decision to use an EA tool within a project has a direct effect on how system architects will describe top-level system behaviors, it is important to make early decisions and communications across the project regarding the use of an EA tool as well as the path forward for the tool's use within the program. The longer these decisions and communications are postponed, the more likely it is that architectural work products will be created by architecture teams in formats inconsistent with each other and/or inconsistent with the ultimate EA tool decision. Potentially, significant amounts of rework with corresponding increases in project costs and schedule may be realized that negates benefits that may have been realized from allowing programmatic milestone pressures to drive an "early" start to architecture team efforts (i.e., having architecture teams start their work before the selection of the EA tool is made). Due to the potential increases in cost and schedule in this situation, there is an accompanying increased likelihood that any inconsistent work products (which were produced before the EA tool decision is made) will not be updated for consistency. This, of course, can lead to outdated work products that are more difficult to maintain. Therefore, an exercise to establish the need for using an EA tool within the project and corresponding requirements for EA tool capabilities should be done as part of the planning tasks similar to the exercises discussed for simulation tools and the architectural tool definition in section 16.4.1. In particular, attention should be paid to ensure that sufficient effort levels have been considered for EA tool identification, selection, and evaluation in Steps 3 and 4 of MFESA Task 1, *Plan and Resource the Architecture Engineering Effort.*

> **Decision support tool:** any software application primarily focused on assisting complex decision-making activities.

Decision support (DS) tools can be used to assist with making complex decisions by allowing identification and weighting of selection criteria and providing capabilities to analyze the data collected for the criteria. These tools may be used, for example, to run several "what-if" iterations for complex decisions that involve multiple, weighted criteria, thus facilitating the decision process. These tools typically utilize formal decision-making techniques, an overview of which is given in Appendix D.

DS tools may be part of larger decision support systems, which can be considered a specific class of computerized information systems that support business and organizational decision-making activities. Thus, decision support systems may also include significant tool entities such as databases, etc.

DS tools are typically available and chosen for use within specific business or technical domains. For example, a DS tool or system may be focused for supporting decisions in the medical or health field, the manufacturing domain, the project or program management domain, and the like.

Due to the wide spectrum of potential application of decision support tools within a specific program, it is important to plan for sufficient time and resources to investigate the need for these tools and how they will be used within that specific program. There is also the need to plan for sufficient time and resources to evaluate and select specific tools. This planning should be done in Steps 3 and 4 of MFESA Task 1, *Plan and Resource the Architecture Engineering Effort.*

16.4.3 Relationships

Depending on the needs of a specific project, different architectural tools may be needed by various architectural teams. Some tools may be shared by multiple teams.

The architecture repositories store data structures created by a specific tool. For example (see Figure 16.5), Architecture Tool 3 might represent a requirements tracing tool used by both the Enterprise System Architecture Team 3 and the Software Architecture Team N. Tool 1 might represent an enterprise architecture (EA) tool used by the EA Team 1, Tool M might represent a software architecture tool used by the Software Architecture Team N. One should create a similar relationship diagram or spreadsheet for the team and tool relationships used within a specific program.

Not explicitly shown, but assumed, is that one of the architectural tools illustrated in Figure 16.5 is a configuration management (CM) tool. The CM tool is responsible for supporting the important tasks involved with tracking and managing the various work product baselines that are spread across the various architecture repositories.

16.4.4 Guidelines

Use the following guidelines (and associated rationales) regarding architecture tools:

- Let the modeling method drive tool selection.
- Use trade studies and pilot projects to evaluate new tools.
- Provide training before tool use.
- Obtain management commitment.

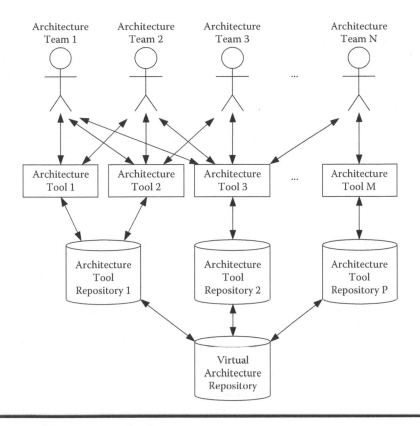

Figure 16.5 Architecture repositories.

The following descriptions express these guidelines in more detail and provide their associated rationales:

- **Let the modeling method drive tool selection.** A major tool evaluation criterion should be the degree to which the tool supports the performance of the system architecture engineering methods. Ensure that the project and architecture teams understand the architecture engineering method and are ready to acquire the tools before purchasing them.

  Rationale: A major part of the architects' work consists of engineering architectural models to provide useful views into the important architectural structures of the system. It is important to be able to effectively and efficiently develop, analyze, document, and verify the correctness and consistency of these models. Unfortunately, no single tool enables the architects to engineer all the potentially useful models. As has been widely known since at least the late 1980s in the CASE community, it is important to let the development method drive the selection of the tool(s) to support the method. Among other things, this involves determining the models to engineer, the associated modeling languages, and then determining the appropriate tools to support the engineering of these models. The tools should support the architects' work rather than drive the work the architects will perform.

- **Use trade studies and pilot projects to evaluate new tools.** Do not overly rely on the word of tool vendor salespeople or their marketing brochures when evaluating architecture tools. Perform proper trade studies comparing potential tools *and* their vendors. Take into account the market share and financial stability of the vendors, who may choose to discontinue tool support once a project has acquired a tool. Obtain evaluation copies of the most appropriate tools and try them out using typical architectural models. If practical, try the architecture tools out on small pilot projects before using them on major, business-critical projects.

Rationale: Many tools prove not to be as effective and usable when actually used in practice. The tool may not scale up to the number of users needed or the size of the architectural models. Tool sets are often not as well integrated as claimed, or tools may not be as easy to integrate as planned.

Buying a tool is often equivalent to buying a business process. If the tool does not fit the current (or planned) architecture engineering method, one may be forced to adapt one's method to match the method inherent in the tool.

■ **Provide training before tool use.** Provide tool users (and potentially other stakeholders) with adequate tool training, including both general vendor-supplied training as well as (if possible) hands-on training using data from the project (or similar) system architecture. Make the availability of tool training a tool evaluation criterion.

Rationale: All significant architecture tools require some level of user training before proficiency can be efficiently reached.

■ **Obtain management commitment.** Obtain management commitment to supply the resources needed to:
  - Evaluate the prospective tools.
  - Install, configure, and integrate the selected tools.
  - Train users (i.e., primarily the architects but also other stakeholders) in the proper use of the tools.
  - Plan for resources and schedule time to deal with tool pitfalls that are almost guaranteed to occur.

Rationale: As of this writing, there are no known tool suites that fully integrate the system and software architecture activities with the downstream design activities. It is almost a certainty that translation difficulties will be encountered when flowing architectures from, say, an enterprise architecture tool into a system/software architecture/design tool. Difficulties in tracing requirements between tools will exist. Reproducing work products within both tool environments may need to be done as a partial means of mitigation for this type of situation. If architectural work products were initiated within one tool but there is a transition subsequently planned into another tool, time and resources should be planned to deal with a longer than desired overlap period.

Identify, as early as possible, those architecture teams that will proceed with producing system architectural work products before tool decisions are finalized. This may be due to geographical spread of the various teams or due to a subcontractor relationship. It also may result due to unplanned events, such as discovering that a previously selected architectural tool is not adequate for the project's needs.

The amount of resources and schedule time needed to sufficiently deal with these types of known pitfalls are easy to underestimate or to completely ignore due to their distasteful character. However, most experienced chief system architects are well aware of the likely occurrence of these kinds of pitfalls and should aggressively address them by realistic schedule plans or by appropriately registering project risks with mitigation plans as per the project's risk management plan.

16.4.5 Pitfalls

Avoid the following common pitfalls associated with architecture tools and mitigate their negative consequences when they occur:

■ Beware of vendors selling silver bullets.
■ Architecture tools are poorly integrated and lack interoperability.
■ Architecture tools have poor support for multiple modeling languages.
■ Architecture tools have poor support for all models.
■ Low-level software design tools are used as system architecture metrics.
■ Architecture tools lack support for architectural models.
■ Management underestimates the effort and resources needed.
■ A fool with a tool is still a fool.

The following express these pitfalls in more detail, describing their negative consequences and the steps that architects can use to mitigate them:

■ **Beware of vendors selling silver bullets.**
  – **Pitfall.** It is not uncommon for tool vendors to sell their tools as if they were silver bullets able to kill the architecture engineering werewolf. Unfortunately, no single tool or toolset currently supports all the architecture engineering tasks, and it is unlikely that any will in the foreseeable future. Many of these tools were originally designed to support low-level software design (e.g., UML modeling tools), and not intended for high-level system architecting.
  – **Negative consequences.** This pitfall is almost certainly to be realized on most projects and often causes the following negative consequences:
    • A suboptimal tool is purchased.
    • The return on investment on the money spent acquiring the tool is lessened.
    • The productivity and effectiveness of the architects is lowered.
    • The architecture teams may need to replace the tool and transition the data from the original tool to the replacement tool, which typically costs far more and takes far longer than anticipated.
    • The architecture budget is exceeded and architecture milestones are missed.
  – **Mitigations.** To avoid or mitigate the negative consequences of this pitfall:
    • Take everything said by a tool vendor salesman with a large grain of salt. Architects, evaluators of architectural tools, and those with purchasing authority should not naively believe everything said by vendors of architectural tools.
    • Pay more attention to the tool's users.
    • Contact the users via tool users groups.
    • Ask the vendor to supply references.
    • Obtain a trial version of the tool for evaluation purposes.
    • Test the tool under realistic conditions.
■ **Architecture tools are poorly integrated and lack interoperability.**
  – **Pitfall.** Architecture tools suffer from a general lack of integration and interoperability. They tend to be poorly integrated with requirements tools and design tools. They are also not easily interoperable in that they often use incompatible interface protocols.

- **Negative consequences.** This pitfall is almost certainly to be realized on most projects and often causes the following negative consequences:
  - Poor integration and intra-operability between architectural tools makes it difficult to trace architectural decisions, inventions, trade-offs, and assumptions back to the derived and allocated architecturally significant requirements.
  - It is also difficult to automatically generate design from the architecture and to trace the architecture to the design.
  - Finally, it is difficult to ensure architectural integrity throughout design.
- **Mitigations.** To avoid or mitigate the negative consequences of this pitfall:
  - Buy tools that are integrated and interoperable to the degree practical, given other tool selection constraints.
  - Select tools that store architectural information in useful forms (such as XML or open relational databases) that support interoperability and reuse.
  - Select architectural tools that are extensible (e.g., via scripting languages).
  - Develop your own programs to integrate the tools.

■ **Architecture tools have poor support for multiple modeling languages.**
- **Pitfall.** The architecture tools do not adequately support the multiple modeling languages being used to produce all architecture models.

  Unless they are general-purpose (such as a word processor or whiteboard), the architecture tools tend to only support a single modeling language, even when multiple modeling languages are needed (such as SysML for static modeling, AADL for dynamic simulation, and DODAF for a combination of interoperability and contract compliance).
- **Negative consequences.** This pitfall often causes the following negative consequences:
  - This lack of support for multiple modeling languages tends to force the use of multiple, incompatible modeling tools.
  - The architectural models become inconsistent
  - The quality of the architectural representations suffers.
  - The architects' productivity suffers.
- **Mitigations.** To avoid or mitigate the negative consequences of this pitfall:
  - Develop simple applications to port data between the tools and to check the consistency of tools developed using different tools.
  - Increase verification of consistency between models written using different modeling languages.

■ **Architecture tools have poor support for all models.**
- **Pitfall.** The project uses graphical modeling tools that do not support the modeling of all relevant architectural structures. To support the production of all useful models, one may thus be forced to either restrict oneself to the use of general-purpose drawing tools or else use the non-integrated combination of a graphical modeling tool for the models it supports together with a general-purpose drawing tool for the remaining models. Note that use of a whiteboard for initial model development and brainstorming naturally supports all models known to the architecture team, although it hardly provides adequate documentation of the models, even if combined with an image capturing device.
- **Negative consequences.** This pitfall often causes the following negative consequences:
  - The architects must use multiple incompatible tools to create all models.
  - Inconsistencies between the different models are not automatically identified and corrected.

- • The quality of the architectural representations suffers.
- • The architects' productivity suffers.
- − **Mitigations.** To avoid or mitigate the negative consequences of this pitfall:
  - • To the extent practical, use architectural modeling tools such as CAD/CAM tools.
  - • Use whiteboards for initial informal modeling.
  - • Use general-purpose drawing tools when necessary to create models that are not supported by architectural tools.
  - • Use popular design tools to capture appropriate models.
  - • Use multiple architecture tools to support the different architecture models.
  - • Prefer tools that store data in readable formats such as XML.
  - • Develop simple applications to port data between the tools and to check the consistency of tools developed using different tools.

■ **Low-level software design tools are used as system architecture tools.**
  - − **Pitfall.** The project attempts to use low-level software design tools as system architecture tools.

    This often occurs and may be to some degree unavoidable because of the relative lack of system modeling tools and the existence of large numbers of software design tools.
  - − **Negative consequences.** This pitfall often causes the following negative consequence:
    - • This typically forces the architects to redefine parts of the associated modeling language (e.g., UML) to represent architectural elements rather than software design elements. For example, the architects may be forced to redefine UML classes to represent subsystems rather than object-oriented design or programming language constructs. The use of such nonstandard definitions can cause significant confusion to stakeholders who understand the correct meaning of the redefined concepts.
  - − **Mitigations.** To avoid or mitigate the negative consequences of this pitfall:
    - • To the extent practical, keep the new meanings of the symbols analogous to their original meanings.
    - • If practical, switch to a newer version of the modeling language if it better supports system architecture engineering.
    - • Properly document and publish any changes to the meanings of the modeling languages.

■ **Architecture tools lack support for architectural metrics.**
  - − **Pitfall.** There is little or no tool support for architectural metrics. Note that this is only indirectly a tool problem. It is primarily a metrics problem because no well-known validated metrics exist to measure architecture quality and completeness as well as architect productivity. Although some projects attempt to use lower-level design metrics (such as metrics designed for object-oriented software), few projects have developed and used a significant amount of architectural metrics. Tool vendors have therefore not incorporated architectural metrics into their graphical modeling architecture tools. The sole counterexample of this pitfall is simulation tools that provide simulation performance metrics.
  - − **Negative consequences.** This pitfall often causes the following negative consequences:
    - • There are no metrics providing objective data on the:
      - ■ Quality and completeness of the architecture and its representations
      - ■ Productivity of the architects
    - • It is difficult for stakeholders to know with high confidence the:
      - ■ Quality and completeness of the architecture and its representations
      - ■ Productivity of the architects
      - ■ Efficiency and effectiveness of the architecture engineering method

- **Mitigations.** To avoid or mitigate the negative consequences of this pitfall:
  - Obtain input from experienced architects and project metrics subject matter experts.
  - Start simple by collecting only a small set of the most obvious and easily collectable quality, completeness, and productivity architectural metrics.
  - Automate metrics collection to the degree practical.
  - Rely on expert opinion and evaluations where metrics are unavailable and validate the results of metrics analysis.

■ **Management underestimates the effort and resources needed.**
  - **Pitfall.** The effort and resources needed to introduce new architecture tools into the project is underestimated. This includes both the selection and use of new tools as well as the replacement of one tool by another, which is typically a much larger effort than planned even if the replacement tool comes with the ability to automatically translate architectural model information from the original tool's format to the new tool's format. This also includes (or may include) the resources needed to:
    - Evaluate the prospective tools.
    - Install, configure, and integrate the selected tools.
    - Train users (i.e., primarily the architects but also other stakeholders) in the proper use of the tools.
    - Transition data from obsolete and unsupported tools to new replacement tools.
  - **Negative consequences.** This pitfall often causes the following negative consequences:
    - The budget and schedule allocated to architecture tools are inadequate (and consequently the overall budget and schedule for architecture engineering are inadequate).
    - The architecture tool evaluation task is inadequately performed.
    - The architects select inadequate or inappropriate architecture tools.
    - The selected tools take longer than expected to install, configure, and integrate.
    - Tool integration remains incomplete.
    - There are insufficient funds to develop training materials for architecture tool use.
    - The architects take much longer than expected to transition data from obsolete and unsupported tools to new replacement tools.
    - The architecture engineering work is over budget and does not meet schedule deadlines.
  - **Mitigations.** To avoid or mitigate the negative consequences of this pitfall:
    - Include the *Evaluate and Select the Architecture Engineering Tools* step of the *Plan and Resource the Architecture Engineering Effort* task in the project architecture engineering method.
    - Base the budget and schedule estimation on real data and input from others who have performed the same work.
    - Take care to not underestimate the required effort and resources because of management pressure to cut costs and shorten schedules.

■ **A fool with a tool is still a fool.**
  - **Pitfall.** As noted by INCOSE Fellow David Oliver, "A fool with a tool is an empowered fool." Management tries to make up for a lack of adequately trained and experienced architects by acquiring some architecture modeling tool touted by its vendor as an architectural silver bullet. This common, if insulting, aphorism warns us that no architecture tool can replace an experienced and well-trained architect. Also, there is much more to system architecture engineering than the production of architecture models.

- **Negative consequences.** This pitfall often causes the following negative consequences:
  - Although the graphical architectural models may look good, they may be incomplete and of poor quality.
  - The architectural models may be syntactically correct but semantically incorrect or meaningless.
  - The architecture and its representations may contain defects, miss opportunities for improvement, and have unnecessary risks.
- **Mitigations.** To avoid or mitigate the negative consequences of this pitfall:
  - Identify and manage the potential misuse and inadequacy of the tools as a project risk.
  - First, determine the most suitable architecture engineering method. Second, determine the appropriate architecture views and models. Third, determine the appropriate modeling language(s). Fourth, evaluate competing architectural tools. Fifth and finally, select the appropriate architecture engineering tools.
  - Provide the architects with proper training in system architecture engineering, including architecture view, models, and modeling languages.

16.5 Architecture Worker Summary

A system architecture worker is a person or thing playing any role when performing architecture engineering work units to produce architecture work products. The MFESA metamodel includes the following three abstract types of architecture workers: (1) system architects, (2) architecture teams, and (3) architecture tools.

16.5.1 System Architects

System architects are system engineers with the specialized training, experience, and expertise to perform system architecture engineering tasks using system architecture engineering techniques. As illustrated in Figure 16.3, there are chief enterprise architects, chief system architects, and lead architects in addition to the general system architects.

The primary responsibilities of system architects are to determine and assess the architectural drivers and concerns, develop the architecture and its representations, analyze and evaluate architectures using its representations, maintain the architecture and its representations, and ensure the continued integrity of the architecture and its representations. System architects also lead the architectural activities, manage the performance of the architectural tasks, advocate the importance of architectural engineering, advocate on behalf of the architecture's stakeholders, develop the architecture engineering methods, select the architecture engineering tools, and train the architecture's stakeholders. System architects both interface and collaborate with the architecture's stakeholders, including project management, systems engineering, hardware and software engineering, project control, quality assurance, process engineering, independent verification and validation (including testing), as well as with other architects. In addition to these architectural responsibilities, the architects also need sufficient authority to carry out their architectural duties.

To be successful, system architects need to conform to a specific profile of personal characteristics, expertise, training, and experience.

16.5.2 System Architecture Teams

Most systems are so large and complex that they require more than a single system architect. Instead, such systems are best developed using one or more system architecture teams. As illustrated in Figure 16.4, there are many types of system architecture teams, the membership of which may include people playing many different roles. The scope of an architecture team may be a single system, a reference architecture for a product line of systems, or an enterprise architecture for an enterprise's network of systems. The teams may work for the acquisition organization, for the system prime contractor or integrator, or for subcontractors or vendors supplying some of the system's architectural components.

System architecture teams must understand the nature of the system's architecturally significant requirements and generate, evaluate, and maintain the system architecture and its representations based on these requirements. To do this, the teams must contain sufficient members with adequate training, expertise, and experience in the discipline of system architecture engineering, including techniques and tools for creating the representations of the architecture. Finally, the teams must communicate the need for sufficient system architectural development resources for them to effectively and efficiently fulfill their architecting responsibilities.

System architecture teams should typically be cross-functional, having members with expertise in system architecture engineering, hardware and software architecture engineering, specialty engineering areas, and the application domain of the system or architectural component being developed or updated. Architecture teams also benefit from having members from disciplines such as requirements engineering and testing to act as liaisons with their associated teams.

16.5.3 Architecture Tools

Crucial to a project's success is the choice and effective use of a set of architectural tools to support the generation and maintenance of the architecture work products. The effort required just to identify and evaluate tool candidates and then select, acquire, and install the tools to be used can be daunting and damaging to the program schedule and budget if not adequately investigated and planned. It is critical that project and team plans, budgets, and schedules realistically account for these activities.

It is important to identify the limits of tools capabilities needed within the context of the project. If a tool is chosen that has many more capabilities than are needed, there is a risk of overspending for unnecessary capabilities as well as a schedule risk due to the potentially heavier training and performance requirements associated with a more complex tool.

Identify, as early as possible, the resources and schedule duration and milestones that represent architectural tool investigations and evaluations. Pay particular attention to the time needed to produce the system architectural work products before selecting the corresponding tools. Highlight these schedule dependency areas and develop plans on how to "normalize" the architectural products once final tool decisions are made.

Allow for worst-case resource assignments and schedule durations when planning to integrate architectural work products that are created within one tool and that will need to be used, accessed, or otherwise imported across various other architectural tools. At the time of this publication, it is somewhat typical for *system* architecture teams that create higher-level architectural representations to use *different* tools than *software* architecture teams will use to produce software architectural representations. Architecture teams that are geographically dispersed or that are part of different organizations may also use different architectural tools. Whenever project life-cycle activities are identified where different architectural tools will be used to create, iterate, or maintain the architectural baselines, plan for sufficient resources and time to deal with issues such as requirements tracing and for ensuring the consistency of the architectural work product baselines.

There are many known potential pitfalls to avoid in the selection and use of architectural tools. Some of these are almost certain to materialize on all projects. Develop risk mitigation strategies for these "almost certain" pitfalls to determine early on how to deal with them when they occur.

Chapter 17

MFESA: The Metamethod for Creating Endeavor-Specific Methods

17.1 Introduction

The previous chapters described the individual, reusable MFESA method components from which effective and efficient project-specific methods for engineering system architectures can be created. These method components included descriptions of the different types of MFESA work products, work units, and workers. However, because the method components are interrelated, they were not completely described individually and independently. For example, all the MFESA work products were not collected into a single chapter, but instead were documented along with the tasks that produce them. In fact, to improve the understandability and ensure the consistency of the individual method components, the MFESA method components have instead been described as if they have all been selected and integrated to create the largest and most complete MFESA system architecture method.

But in practice, all of the MFESA method components will almost never be needed or cost-effective to use within a single method. The selected method components will also typically need to be tailored to meet the specific needs of the project. Thus, the system architects collaborating with the project technical leader and process engineers must choose which of the method components to use and how they should be tailored to create an effective and efficient project-specific method for engineering the architecture of the system to be developed or updated. This method for creating a method is the topic of this chapter.

As the name implies, MFESA is a method framework for engineering system architectures. As illustrated in Figure 17.1, MFESA consists of the following four highly related parts:

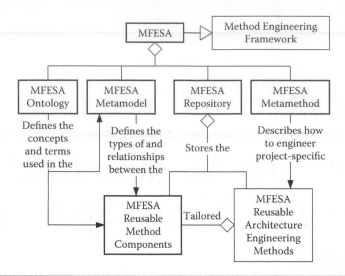

Figure 17.1   The four primary MFESA components.

1. **MFESA ontology of concepts and terminology:** an information model defining a consistent set of interrelated concepts and terms underlying system architecture engineering.
2. **MFESA metamodel:** a model of the fundamental abstract types of method components (see Figure 17.2) in the repository of reusable method components (i.e., architectural work products, architectural work units, and architectural workers).
3. **MFESA repository:** a repository storing both a consistent class library of reusable method components for creating project-specific methods for engineering system architectures as well as any reusable methods constructed from those components.
4. **MFESA metamethod:** a method for constructing effective and efficient project-specific methods for engineering system architectures (i.e., the method for creating methods).

Note that thus far the book has addressed the first three MFESA components. This chapter is devoted to the fourth MFESA component — the MFESA metamethod.

17.2 Metamethod Overview

A metamethod is a method for constructing a method. The fourth part of MFESA, the MFESA metamethod, is thus the method for constructing an MFESA method, that is, a method composed of MFESA method components. As a method itself, the MFESA metamethod can be viewed as being composed of its own component tasks and work products. During the second step of the first MFESA task, the relevant system architects collaborated closely with process engineers, technical leaders, experts in architectural engineering, and experts in MFESA to engineer one or more efficient and effective team-, project-, or organization-specific methods for engineering system architectures.

As illustrated in Figure 17.3, the MFESA metamethod consists of 12 distinct tasks typically performed in an iterative, incremental, parallel, and time-boxed manner. Because systems, endeavors, and organizations can vary so much, an assessment is initially made to determine the endeavor's specific needs for system architecture engineering methods. Because a single common method

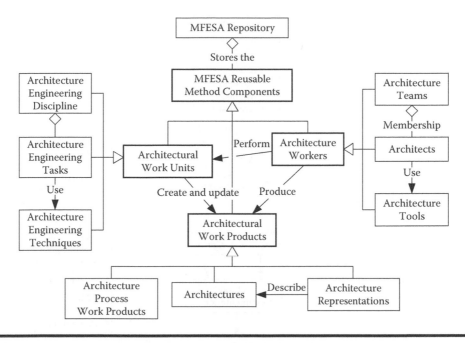

Figure 17.2 The MFESA metamodel of reusable abstract method component types.

may not be sufficiently flexible to meet all needs, the number of system architecture engineering methods to construct is determined next. Then for each method needed, a determination is made of the appropriate type of reuse (i.e., method component reuse or method reuse). If one or more potentially appropriate reusable methods already exist, then the most suitable method is selected and tailored to meet the specific needs of the endeavor. On the other hand, if no existing method is appropriate, then a new method is constructed by selecting appropriate components, tailoring these components, and integrating these components to form a new method suitable for the specific needs of the endeavor. Finally, for each reused existing method and each newly constructed method, the method is appropriately documented and then verified to ensure that it has the needed characteristics (e.g., internally consistent, adequate, not excessive, and properly documented).

17.3 Method Needs Assessment

When creating one or more system architecture engineering methods, the starting point should be to assess the needs driving their creation. The team of architects, process engineers, technical leaders, and MFESA experts must first determine the characteristics that will influence the size, contents, formality, and number of appropriate architecture engineering methods to instantiate. This will include determining and analyzing the following:

■ **System characteristics.** The system architecture engineering method(s) to be created will depend on several fundamental characteristics of the system being engineered:
  – **System number.** Will more than one system be developed or updated? Roughly how many systems will be developed or updated? Will these be highly related systems (e.g., systems within a product line), or will the systems be unrelated (e.g., systems produced

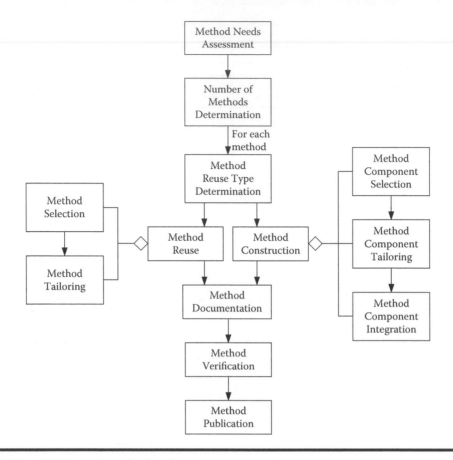

Figure 17.3   MFESA metamethod tasks.

by a system development organization for multiple, unrelated customer organizations). As the number of relevant systems being engineered increases and as these systems decrease in similarity, it becomes more likely that a single system architecture engineering method will *not* be sufficiently flexible to be appropriate for all of them. Thus, the engineering of multiple systems may (but need not) justify the creation of multiple system architecture engineering methods to architect them.

 — **Requirements size.** Roughly how many architecturally significant requirements will drive the architecture engineering effort? Roughly how large will the required feature set be? The optimum size, complexity, and rigor of a system architecture method should typically increase with the number of architecturally significant system requirements and with the number of system features. A system with an extremely large number of requirements may (but need not) also justify the creation of multiple system architecture engineering methods to architect different parts of the system that vary in size, complexity, or criticality due to the numbers and types of requirements allocated to them.

 — **System size.** Roughly how large will the system be, and consequently, roughly how large will the system's architecture be in terms of number of tiers or number of architectural components? The answer to these questions can be based on the size of an existing architecture (if the new system is an upgrade to an existing system), a reference architecture (if the system to be architected is part of a product line and must conform to the associated reference architecture), or the proposed architecture (if the system is totally new but an initial proposed architecture exists).* The optimum size, complexity, and rigor of the system architecture method should typically increase with the size of the system and its architecture. A very large system may (but need not) also justify the creation of multiple system architecture engineering methods to architect different parts of the system that vary in size, complexity, or criticality.

- **System complexity.** Roughly how complex will the system be in terms of the interrelationships between its proposed architectural components? Roughly how complex will the system be in terms of the number of different technologies proposed for incorporation into the architecture? The optimum size, complexity, and rigor of the system architecture method should typically increase with the complexity of the system. A very complex system may (but need not) also justify the creation of multiple system architecture engineering methods to architect different parts of the system that vary in their complexity.

- **System criticality.** Roughly how critical will the system be to its stakeholders? Roughly how business critical will the system be? Roughly how safety or security critical will the system be? The optimum size, complexity, and especially rigor of the system architecture method should typically increase with the criticality of the system.

■ **Endeavor characteristics.** The system architecture engineering method(s) to be created will depend on several fundamental characteristics of the associated endeavor (e.g., project, program of related projects, and enterprise):

- **Endeavor types.** To ensure adequate consistency across multiple architectures and their associated architectural representations, the rigor and formality of the optimal system architecture engineering method(s) tend to increase as the scope of the methods increases from an individual project to a program of related projects to an entire enterprise. Similarly, engineering the architecture of a new system tends to require a more formal and rigorous system architecture engineering method than does an endeavor that merely performs a minor upgrade to an existing system and its architecture. As the scope of the methods increases from an individual project to a program of related projects to an entire enterprise, more systems are being engineered and the generation of multiple systems may (but need not) justify the creation of multiple system architecture engineering methods.

- **Contract types.** There are many different types of contracts, such as cost plus fixed fee, fixed price, contract between acquisition and development organizations, and contract between prime contractor versus subcontractor. These different contract types and their provisions may have a major impact on the architecture engineering method.

  Does the system development project involve a contract between the acquisition and development organizations, or between a prime contractor/system integrator and subcontractors? If so, what type of contract is it? For example, is the contract fixed price or cost plus fixed fee? The optimum size, complexity, and especially rigor of the system architecture engineering method should typically increase with the number and formality of the contractual relationships.

---

* Some method was probably used to develop this proposed architecture. However, that is not reason to believe that the same method should be used to develop the actual delivered architecture. For example, the proposed architecture may have been developed by a totally separate organization, possibly even the acquisition organization. The quality and completeness of the actual architecture and its representations may also need to be much better than that of the proposed architecture.

The impact of fixed price versus cost plus fixed fee contracts on the system architecture engineering method is not obvious. For example, all things being equal, a fixed price contract might engender the temptation to have a more modest method than a cost plus fixed fee contract due to the tighter funding constraints. On the other hand, a fixed price contract may necessitate using a more robust method to help avoid costly rework late in the program.

- **Schedule constraints.** What is the severity of the schedule constraints on the development project? When developing small systems that are not business- or safety-critical, severe schedule constraints may be seen as grounds for using a lightweight, low-rigor, agile architecture engineering method. However, when developing large, complex, business- and safety-critical systems, strong schedule constraints may be seen as grounds for a heavyweight, high-rigor architecture engineering method because time and effort invested in architecture engineering often pays for itself in terms of schedule during implementation, integration, and testing.

- **Budget constraints.** How severe are the budget constraints on the development project? When developing small systems that are not business- or safety-critical, severe budget constraints may be seen as grounds for using a lightweight, low-rigor, agile architecture engineering method. However, when developing large, complex, business- and safety-critical systems, strong budget constraints may be seen as grounds for a heavyweight, high-rigor architecture engineering method because time and effort invested in architecture engineering often pays for itself several times over during implementation, integration, and testing.

■ **Organizational characteristics.** The system architecture engineering method(s) to be created will depend on several fundamental characteristics of the associated organizations (e.g., development, acquisition, and accreditation):

- **Organizational culture.** What is the culture of the organization performing system architecture engineering? Are the acquisition and development organizations comfortable with being bleeding-edge, early users of new technology and processes, or are they more risk averse and comfortable with traditional technologies and processes? The optimum size, complexity, and especially rigor of the system architecture method may need to increase to support organizations that are unfamiliar with the method and thus need extra guidance. Enterprises having multiple organizations with different cultures may (but need not) justify the creation of multiple system architecture engineering methods for the different culture types.

- **Experience and expertise.** To what degree do the acquisition and development organizations have experience and expertise with system architecture engineering, method engineering, and MFESA? Higher expertise in system architecture engineering may justify smaller, simpler, less rigorous system architecture engineering methods because many of the processes have been mastered and inculcated into the organizational body of knowledge. However, the more familiar the organizations are with the concepts of method engineering, the more likely that they will be able to successfully create multiple, project-specific architecture engineering methods.

- **Organizational separation.** Large and complex systems are often developed by a number of development organizations. There may be a prime contractor with multiple subcontractors responsible for developing different subsystems for which they have expertise (e.g., hardware, software, or application domain). Instead of also engineering subsystems, the prime contractor may play the role of system integrator of the architectural components supplied by subcontractors and vendors. When multiple organizations are involved, it often becomes impossible to mandate that they all use the same development methods, including system architecture engineering methods. To obtain the benefits of a subcontractor's or vendor's expertise and experience in the application domain (such as radar, telecommunications, and sensor fusion), the prime contractor or system integrator must often either permit the subcontractor or vendor to continue using their traditional methods or else pay heavily for the production and use of standardized architectural representations. On the other hand, mandating a single common standardized method may also add costs due to the need to provide training in the new method, lower architect productivity due to the lack of experience with the standard method, and higher defect rates also due to lack of experience with the standard method. Thus, organizational separation, especially when reinforced by contractual barriers, may (but need not) justify the use of multiple system architecture engineering methods by different organizations.

- **Geographic separation.** As different parts of the system development organization become widely separated geographically, it becomes more difficult to communicate verbally and more important to obtain the benefits of standardization, both of architectural work products (especially architectural representations) and the architectural work units that produce them. Geographic separation of developers, including architects, implies the greater need for consistency of the system architecture engineering method(s) so that developers can benefit from the use of standard architecture engineering work products and work units.

■ **Method constraints.** The system architecture engineering method(s) to be created may be constrained to have certain characteristics to conform to the other methods to be used on the endeavor:

- **Mandated compliance.** Must the method conform to specific international, government, military, or *de facto* standards? For example, this could include compliance with the use of:
  - Architectural Analysis and Design Language (AADL) [Feiler et al., 2006]
  - Department of Defense Architecture Framework (DODAF) [DODAF, 2007a–c]
  - Integration Definition for Information Modeling (IDEF1X) [FIPS, 1993]
  - Modular Open Systems Approach (MOSA) [OSJTF, 2004]
  - OMG Systems Modeling Language (OMG SysML) [OMG, 2007]
  - Recommended Practice for Architectural Description of Software-Intensive Systems (IEEE 1471) [IEEE, 2000]
  - Software Engineering: Metamodel for Development Methodologies (ISO/IEC 24722) [ISO/IEC, 2007]
  - Standard for a Software Quality Metrics Methodology (IEEE 1061) [IEEE, 1998]
  - Software Engineering – Product Quality — Part 1: Quality Model (ISO/IEC 9126) [ISO/IEC, 2001]
  - Systems Engineering — System Life Cycle Processes (ISO/IEC 15288) [ISO/IEC, 2002]
  - Unified Modeling Language Specification (ISO/IEC 19501) [ISO/IEC, 2005]
- **Development/life cycle.** The system architecture method will need to conform to the overall development/life cycle of the endeavor. Will this cycle be waterfall or an iterative, incremental, parallel, and time-boxed cycle?
- **Agile versus document driven.** The system architecture method will need to conform to the overall level of agility or formality of the rest of the methods used on the endeavor.

The preceding considerations all deal with the size, complexity, formality, and number of the system architecture engineering methods to construct or reuse. They do not, however, provide any specific guidance as to the specific method components that would be appropriate to include in these methods. This will occur during the following *Method Reuse* and *Method Construction* tasks.

17.4 Number of Methods Determination

During the *Method Reuse Type Determination* task, members of the architecture team collaborate with members of the process team and management team to determine the number of individual system architecture engineering methods that are appropriate for the endeavor. The number will be based on the results of the needs assessment.

17.5 Method Reuse Type Determination

During the *Method Reuse Type Determination* task, members of the architecture team collaborate with members of the process team and management team to determine whether to reuse existing methods stored in the MFESA repository or to construct new methods from the method components stored in the MFESA repository.

This and subsequent steps are performed for each method that has been determined to be needed.

17.6 Method Reuse

During the *Method Reuse* task, performed to reuse an appropriate, existing, reusable system architecture engineering method stored in the MFESA repository, members of the architecture team collaborate with members of the process team and management team to perform the following subtasks in an incremental, iterative, concurrent, and time-boxed manner:

1. **Method selection.** Based on the results of the method needs assessment, select the appropriate reusable system architecture engineering method stored in the MFESA repository.

   Note that except for the maximally sized and complex method containing all of the method components, the official MFESA repository currently contains only method components, and it may be some time until more reusable methods become available for selection and tailoring.

2. **Method tailoring.** Based on the results of the method needs assessment, tailor the selected method to meet the identified needs of the project. This may include tailoring out unnecessary method components, modifying the remaining method components, and adding additional useful method components.

17.7 Method Construction

During the *Method Construction* task, performed to construct an appropriate system architecture engineering method out of the reusable method components stored in the MFESA repository, members of the architecture team collaborate with members of the process team and management team to perform the following subtasks in an incremental, iterative, concurrent, and time-boxed manner:

1. **Method component selection.** Based on the results of the method needs assessment, select the appropriate method components from the MFESA method component repository. Two main approaches are most often used to select the appropriate method components to be reused, although in reality, the constructors of the method often iterate rapidly between these two approaches:
   - **Work product first.** The relevant architectural work products to engineer are identified first. This work often tends to concentrate on deciding on the appropriate architectural models to engineer. Then, based on these architectural work products to produce, the associated work units (especially tasks and associated techniques) to be performed are identified. Finally, appropriate workers (i.e., architects, architectural teams, and architectural tools) are identified to perform the work units to engineer the desired work products.
   - **Work unit first.** Based on the method needs assessment, the major work units (especially tasks) to perform are identified first. Then, the associated work products are selected, especially the appropriate architectural models to engineer. Finally, as before, appropriate workers (i.e., architects, architectural teams, and architectural tools) are identified to perform the work units to engineer the desired work products.
2. **Method component construction.** Based on the results of the method needs assessment, construct any method components that do not already exist within the MFESA repository. Because of the relative completeness of the MFESA repository with regard to method components, it is assumed that this substep will rarely be needed.
3. **Method components tailoring.** Based on the results of the method needs assessment, tailor the selected method components to meet the identified needs of the project. This includes tailoring out unnecessary parts of the selected method components that are not cost-effective to perform, modifying parts of the selected method components, and adding additional useful information to existing selected method components. Because of the relative completeness of the method components within the MFESA repository, It Is assumed that most of the tailoring will be in the form of tailoring out inappropriate parts of existing method components as opposed to tailoring in missing parts of components.
4. **Method component integration.** Integrate these method components into the system architecture engineering method(s). This includes ensuring the referential integrity of references between method components (i.e., no *dangling* references to missing method components or *orphan* method components).

17.8 Method Documentation

During the *Method Documentation* task, members of the architecture team collaborate with members of the process team and management team to perform the following subtasks in an incremental, iterative, concurrent, and time-boxed manner for each system architecture engineering method reused or constructed:

1. **Documentation type determination.** Determine the appropriate type of documentation needed to communicate the system architecture engineering method(s) to their stakeholders:
   - **Plans.** For example, this could include documenting the architecture engineering methods in the system engineering management plan or system architecture engineering plan.

- **Conventions.** For example, this could include documenting the method(s) in standards, procedures, guidelines, templates, examples, and tool manuals.
- **Training materials.** This would include training materials for teaching stakeholders the method(s) for engineering system architectures.

2. **Documentation media determination.** Determine the appropriate medium for communicating the system architecture engineering method(s) to their stakeholders. For example, one or more of the following types of media can be used to store the method(s):

- **Paper or electronic documents.** For example, the system architecture engineering method could be documented in the traditional form of one or more static documents.
- **Web site.** For example, the system architecture engineering method could be documented in the form of Web pages, which would better support browsing the method components by following hyperlinks between them.
- **Process engineering tool repository.** For example, the system architecture engineering method could be stored in the repository of a process engineering tool and accessible via the tool's user interface.

3. **Method documentation.** Document the system architecture engineering method(s) in the appropriate documentation type and using the determined media.

17.9 Method Verification

During the *Method Verification* task, members of the architecture team collaborate with members of the stakeholders to verify that the system architecture engineering methods have the appropriate characteristics, including completeness, correctness, referential integrity, and readability. For example:

1. **Internal peer review.** Informally perform an internal peer review of the documented method(s). Iterate them as necessary to correct any defects found.
2. **External evaluation.** Have the process team perform an external evaluation of the documented method(s) to ensure correctness and consistency with endeavor process conventions. Iterate them as necessary to correct any defects found.

17.10 Method Publication

During the *Method Publication* task, members of the architecture team publish the verified and approved system architecture engineering method(s) to their stakeholders. Depending on the media and location of the stakeholders, the documented method(s) can be published either directly (either as part of other documents or as stand-alone documents) or by making them available (e.g., by placing them within a public directory or announcing their availability via a Web site or tool).

17.11 Guidelines

Use the following guidelines (and associated rationales) when performing the MFESA metamethod to produce system architecture engineering methods:

- Ensure consistency.
- Construct system architecture engineering methods early.
- Iterate as lessons are learned.
- Ensure the right level of formality.

The following descriptions express these guidelines in more detail and provide their associated rationales:

- **Ensure consistency.** To the extent practical, ensure that the system architecture engineering methods reused or constructed are consistent with the rest of the methods used on the endeavor (i.e., project, on the program of related projects, or enterprise). Work with the management team and process engineering team to ensure that:
  - The work products produced include those needed by other disciplines.
  - Other disciplines and activities can use the work products produced by system architecture engineering.

  Rationale: Much unnecessary effort and argument can be saved if the different architecture teams on an endeavor use the same architecture tasks and techniques to produce the same architecture work products in the same way. System architecture engineering is highly interrelated with other disciplines. Note that this is considerably easier to achieve when the different architecture teams are collocated within the same organization than when they are part of different organizations (such as different cost centers, prime contractor, subcontractors, and product vendors) or are geographically dispersed. It is also more difficult to achieve when the top-most system architecture team does not have its methods and practices ready at the beginning of the endeavor and the lower-level architecture teams must start work and cannot wait for guidance from the highest-level team.

- **Construct system architecture engineering methods early.** To the extent practical, the higher-level architecture teams should collaborate with the process engineering team and endeavor technical leader to engineer the endeavor's system architecture engineering method(s) as early as is practical during the system development cycle.

  Rationale: An initial version of the architecture and its representations must often be submitted as part of a proposal to win the development contract. The top-most architecture team needs to supply the endeavor-wide architecture engineering method to the lower-level architecture teams so that the work products they produce are compatible and do not need to be modified for the sake of consistency later on in the middle of the project. The top-level architecture engineering models and vision need to be able to influence the requirements early to support early requirements negotiation (e.g., to enable the reuse of architecture components).

- **Iterate as lessons are learned.** To the extent practical, do not let the documentation of the system architecture engineering method(s) become shelfware. Instead, update the method as appropriate, especially early on in the endeavor, to ensure that they are truly suitable for the endeavor and provide an effective and efficient way to engineer its system architectures.

  Rationale: Although it is important to create a system architecture engineering method very early during development, it is very difficult to produce an optimal method in terms of effectiveness and architect productivity before one has had the opportunity to try it out and iterate it. Thus, it is important to integrate lessons learned during the performance of the method.

- **Ensure the right level of formality.** Use the characteristics identified during the *Method Needs Assessment* task to ensure that the system architecture engineering method(s) have a level of rigor and completeness that is appropriate for the situation in which they will be used.

   Rationale: A leaner (less complete), more agile (less formal) system architecture engineering method might be more appropriate for architecting certain system components, whereas a more comprehensive, rigorous method may be more appropriate for architecting other system components.

17.12 Pitfalls

Avoid the following common pitfalls associated with method engineering and mitigate their negative consequences when they occur:

- There is a lack of expertise in method engineering.
- It is difficult to keep the constructed method *internally* consistent.
- It is difficult to keep the constructed method *externally* consistent.
- There is a lack of adequate tool support.

The following express these pitfalls in more detail, describing their negative consequences and the steps the architects can use to mitigate them:

- **There is a lack of expertise in method engineering.**
  - **Pitfall.** Most projects do not have sufficient process engineers or system architects with significant expertise in the method engineering of endeavor-specific methods.
  - **Negative consequences.** This pitfall often causes the following negative consequences:
    - Too much time is lost trying to produce an appropriate project-specific system architecture engineering method.
    - The resulting project-specific architecture engineering method is not appropriate for the project because it includes some inappropriate method components, does not include other appropriate components, and the components it does include are not properly tailored to meet the needs of the project.
    - The resulting architecture engineering method is not properly documented.
  - **Mitigations.** To avoid or mitigate the negative consequences of this pitfall:
    - Read technical articles and conference papers on method engineering.
    - Obtain training in method engineering.
    - Acquire consulting services from an organizational process team, from external process engineering experts, or from process tool vendors.
- **It is difficult to keep the constructed method *internally* consistent.**
  - **Pitfall.** It is very difficult to keep newly constructed system architecture engineering methods internally consistent. This can be due to several factors, such as the large number of potentially reusable method components from which to select, the difficulty in properly tailoring these method components, and the fact that method components often reference other method components. When selected method components refer to other method components that were not selected, the selected method component can be said to have *dangling* references.

- **Negative consequences.** This pitfall often causes the following negative consequences:
  - The users of the inconsistent method waste time dealing with the inconsistency.
  - Time and effort are wasted because architects or stakeholders perform unnecessary system architecture engineering tasks or unnecessary steps within selected tasks.
  - Time and effort are wasted because architects or stakeholders create unnecessary system architecture engineering work products or unnecessary parts of selected work products.
- **Mitigations.** To avoid or mitigate the negative consequences of this pitfall:
  - Use method engineering tools that guarantee method referential integrity when generating the project-specific architectural engineering method.
  - During the *Method Verification* task of the MFESA metamethod, perform either manual or automated checks of method referential integrity, which ensures the equivalent of a method "clean compile."

■ **It is difficult to keep the constructed method *externally* consistent.**
- **Pitfall.** The constructed system architecture engineering method(s) become inconsistent with the rest of the development methods on the endeavor. This includes becoming inconsistent with the overall development plan as well as with various conventions describing the proper performance of other disciplines such as requirements engineering and system testing.
- **Negative consequences.** This pitfall often causes the following negative consequences:
  - The outputs of the requirements engineering method are incomplete (e.g., missing architecturally significant requirements) because the requirements engineering method does not adequately address these requirements or does not have system architects as liaisons with the requirements teams or reviewers of the requirements work products.
  - The outputs of the system architecture engineering method are inappropriate for their users because the method does not produce the necessary inputs to the system design, implementation, integration, or test activities.
  - The architecture does not meet its true (but possibly unspecified) architecturally significant requirements.
  - The system does not meet its architecturally significant requirements.
- **Mitigations.** To avoid or mitigate the negative consequences of this pitfall, ensure that:
  - Coordination and consensus occurs between the architects, process engineers, and the developers of the endeavor's overall system engineering method and activity-specific methods during the performance of the *Method Component Selection* and *Method Component Tailoring* tasks of the MFESA metamethod.
  - The system architecture engineering methods are evaluated for consistency with the endeavor's overall system engineering method and activity-specific methods during the *Internal Peer Review* and *External Evaluation* steps of the *Method Verification* task of the MFESA metamethod.

■ **There is a lack of adequate tool support.**
- **Pitfall.** Method engineering is performed manually because of the lack of production-quality tool support for method creation, tailoring, storage, verification, and maintenance beyond word processors or possibly Web page editors.
- **Negative consequences.** This pitfall often causes the following negative consequences:
  - The amount of time and effort required to create, tailor, store, verify, and maintain is often significantly underestimated.

- The quality of the engineered method is relatively low in terms of method consistency and appropriateness for the project.
- The size and usability of the engineered method decreases as the amount of process engineering work increases. For example, method components are less completely described and fewer links between them are built in and maintained.
  - **Mitigations.** To avoid or mitigate the negative consequences of this pitfall:
    - Use word processors for storage and updating.
    - Use Web page editors for publication with XML for storage.
    - Use a public domain tool. One example of such a tool is the Eclipse Process Framework (EPF).*

17.13 Summary

As the name implies, MFESA is a method framework for engineering system architectures that consists of the following four parts:

1. **MFESA ontology:** a structured glossary of system architecture engineering concepts and terminology that is built around the MFESA metamodel.
2. **MFESA metamodel:** defines the fundamental abstract types of reusable method components, including the *architectural workers* (roles, teams, and tools) that perform *architectural work units* (tasks and techniques) to produce *architectural work products* (e.g., architecture, architectural representations, and other documents).
3. **MFESA repository:** a logical repository of reusable method components that eventually may also store reusable methods.
4. **MFESA metamethod:** a method defining the tasks for constructing appropriate project-specific architecture engineering methods.

When using the MFESA metamethod, the architecture teams collaborate with the process and technical management teams to perform the following tasks:

- **Method needs assessment:** the method engineering task during which these teams collaborate to assess the stakeholders' needs for engineering one or more system architecture engineering methods.
- **Number of methods determination:** the method engineering task during which these teams collaborate to determine the number of system architecture engineering methods to create or reuse.
- **Method reuse type determination:** the method engineering task during which these teams collaborate to determine whether to reuse an appropriate existing system architecture engineering method (if any exists) or to create a new project-specific system architecture engineering method.

---

* Note that this is *not* an endorsement, but merely one such tool that is fairly well known and can be used as an example of what such tools are like.

- **Method reuse:** the method engineering task during which these teams collaborate to select and tailor an appropriate existing system architecture engineering method to make it appropriate and project specific.
  - **Method selection:** the method engineering task during which these teams collaborate to select an appropriate system architecture engineering method for reuse, possibly with tailoring.
  - **Method tailoring:** the method engineering task during which these teams collaborate to tailor the selected system architecture engineering method to make it more appropriate for the project.
- **Method construction:** the method engineering task during which these teams collaborate to construct one or more new appropriate project-specific system architecture engineering methods.
  - **Method component selection:** the method engineering task during which these teams collaborate to select existing reusable method components from the MFESA repository for reuse within a new system architecture engineering method(s) to be constructed.
  - **Method component construction:** the method engineering task during which these teams collaborate to construct one or more reusable method components because no such appropriate components currently exists in the MFESA repository.
  - **Method component tailoring:** the method engineering task during which these teams collaborate to tailor either the selected or newly constructed method components for reuse within a new system architecture engineering method(s) to be constructed.
  - **Method component integration:** the method engineering task during which these teams collaborate to integrate the selected or constructed tailored method components to construct the new, appropriate project-specific system architecture engineering method(s).
- **Method documentation:** the method engineering task during which these teams collaborate to properly document the reused or newly constructed, appropriate project-specific system architecture engineering method(s).
- **Method verification:** the method engineering task during which these teams collaborate to verify the quality and appropriateness of the reused or newly constructed project-specific system architecture engineering method(s).
- **Method publication:** the method engineering task during which these teams publish the documentation describing the verified system architecture engineering method(s) to the architects and other stakeholders.

Chapter 18

Architecture and Quality

18.1 Introduction

In Chapter 5, "MFESA: The Ontology of Concepts and Terminology," we very briefly addressed how quality characteristics and quality attributes can be used to engineer quality requirements that can be grouped together to form quality concerns, the architectural ramifications of which can be documented in the form of quality focus areas. In that chapter, we also touched on how the system architecture had a major impact on the degree to which the system could meet its quality requirements and suggested the use of architectural quality cases as a way for the architects to demonstrate that their architecture sufficiently supports those important architecturally significant requirements. These are some of the reasons why we stated in Chapter 3, "Principles of System Architecture Engineering," that system architecture engineering must concentrate on ensuring system quality. These are also the reasons why MFESA has incorporated these quality concepts and work products into its repository of reusable method components.

To provide architects and other stakeholders with a firmer foundation in quality and its relationships to system architecture, this chapter discusses the following in significantly more detail:

- **Quality models.** What is system quality, and how do we get the system stakeholders to agree on the meaning of system quality? What are the important quality characteristics (types of system quality) and quality attributes (parts of quality characteristics)? What are quality measurement scales and the methods used to measure quality?
- **Quality requirements.** What are quality requirements? What do quality requirements look like, and what are their component parts? How can the quality requirements be made objective and verifiable? How do the quality requirements drive the architecture?
- **Quality-related architectural patterns.** What architectural patterns support what quality characteristics and attributes? What are the architectural decisions and associated trade-offs, assumptions, and rationales that are commonly used to achieve levels of quality characteristics and attributes sufficient to meet quality requirements?

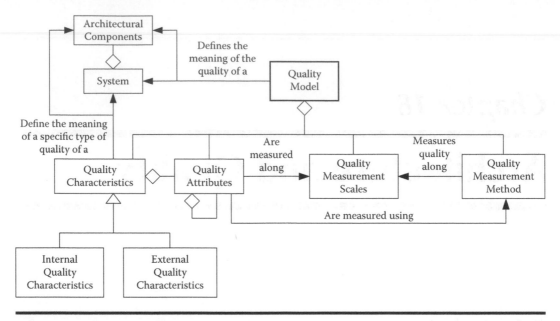

Figure 18.1   The components of a quality model.

■ **Architectural quality cases.** What are architectural quality cases? How do the system architects make their case that their architecture has sufficient quality? What are the component parts of architectural quality cases? How can stakeholders effectively and efficiently assess whether the architecture has sufficient quality?

18.2 Quality Model Components and Their Relationships

The most important architectural concerns are based on the quality characteristics defined in the project quality model [Barbacci et al., 1995; ISO/IEC, 2001; Firesmith et al., 2006]. The architecture should sufficiently support the achievement of the required levels of system quality. Thus, a primary responsibility of the architects is to ensure that the system and subsystem architectures sufficiently support the meeting of the allocated and derived quality requirements that mandate minimum levels of these quality characteristics and their related quality attributes.* As illustrated in Figure 18.1, a quality model consists of a consistent set of relevant quality characteristics, their component quality attributes, and the associated quality measurement scales and quality measurement methods used to measure the degree to which these quality attributes exist within the system and its architectural components.

The most important terms found in Figure 18.1 are defined as follows:

> **Quality:** the degree to which a system or architectural component has useful and desirable characteristics as defined by the quality characteristics and quality attributes of a quality model.

---

* Architects must also perform engineering trade-offs between these quality requirements and other requirements, and these trade-offs are an important part of the architecture.

Quality model: a hierarchical model for defining, specifying, and measuring the different types of quality of a system or architectural component in terms of the model's component quality characteristics, quality attributes,* and associated quality measurement scales and methods.

Quality characteristic: a high-level characteristic or property of a system or architectural component that characterizes an aspect of its quality. Quality characteristics are also known as quality factors, quality attributes, and "ilities." Note that unlike quality attributes, quality characteristics are typically at such a high level of abstraction that they are not measurable along quality measurement scales using quality measurement methods.

Quality attribute: a major component (aggregation) of a quality characteristic or of another quality attribute that is directly measurable.

Quality measurement scale (a.k.a. quality measure): a measurement scale that defines numerical values used to measure the degree to which a system or architectural component exhibits a specific quality attribute.

Quality measurement method (a.k.a. quality metric): a method, function, or tool for making a measurement along a quality measurement scale.

The example illustrated in Figure 18.2 should help clarify the distinction between quality characteristics and attributes, there are two orthogonal subtypes of performance attributes: performance problem types and performance solution types. Jitter, latency, response time, schedulability, and throughput are the performance problem type quality attributes, whereas mandated threshold, failure detection, failure reaction, and failure adaptation are the performance solution type quality attributes. Because of the multiple inheritance involved (i.e., two orthogonal subtypes of quality attributes), individual performance requirements tend to specify a combination of both subtypes. For example, such requirements could specify a maximum threshold for response time or what must be done to detect or react to latency failures.

Similarly and as illustrated in Figure 18.3, the quality characteristics safety and security as subtypes of defensibility can also be modeled as having two similar subclasses of quality attributes. Both safety and security deal with unauthorized harm to valuable assets. Safety deals with accidental (unintentional) harm, whereas security deals with malicious (intentional) harm such as loss of confidentiality, loss of integrity, denial of service, repudiation of transactions, and unauthorized access to services. Note that, like performance requirements, safety and security quality requirements can be formed by taking the cross-product of the two subclass hierarchies of quality attributes. For example, architects may have to address prevention of the occurrence of unauthorized harm, detection of the occurrence of abuse, reaction to the existence of a danger, etc.

Historically, computer and information security have been defined in terms of integrity (the malicious loss of which is one type of harm), confidentiality (the malicious loss of which is another type of harm), and denial of service (both a type of attack and an associated malicious type of harm). Sometimes, security is defined additionally in terms of non-repudiation (the malicious loss of which is yet another type of harm) or accountability (the malicious loss of which is also a type of harm). Note that the definitions given here for safety and security are more expansive in that they

---

* Note that this book follows [ISO/IEC, 2001] in its differentiation between quality characteristics and quality attributes, although it does provide a much richer and more complete quality model. Note also that not every book makes this distinction and some use the term "quality attribute" for both characteristics and attributes [Clements et al., 2003].

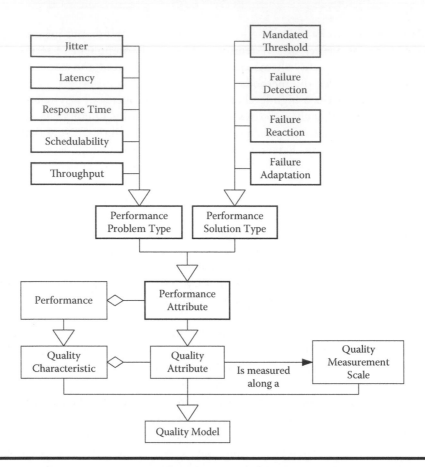

Figure 18.2 Performance as an example quality characteristic with associated attributes.

go beyond harm to include the occurrence of abuse (accident, attack, and incident), the existence of external abusers, the existence of system-internal vulnerabilities, the existence of danger (hazard and threat), and risk. Similarly, these definitions go beyond prevention to include detection, reaction, and even adaptation.

As illustrated in Figure 18.1, quality characteristics can be subtyped into internal and external quality characteristics:

> **Internal quality characteristic (a.k.a. white box quality characteristic):** a quality characteristic that characterizes an internally visible (i.e., white box) quality of a system or architectural component when it is in the process of being developed, modified, or retired. Internal quality characteristics are visible to and are primarily of interest to the system's developers and maintainers.
>
> **External quality characteristic (a.k.a. black box quality characteristic):** a quality characteristic that characterizes an externally visible (i.e., black box) quality of a system or architectural component when it is deployed and in service in its operational environment. External quality characteristics are visible to and are primarily of interest to the users, orperators, administrators, and sustainers of an existing system or architectural component.

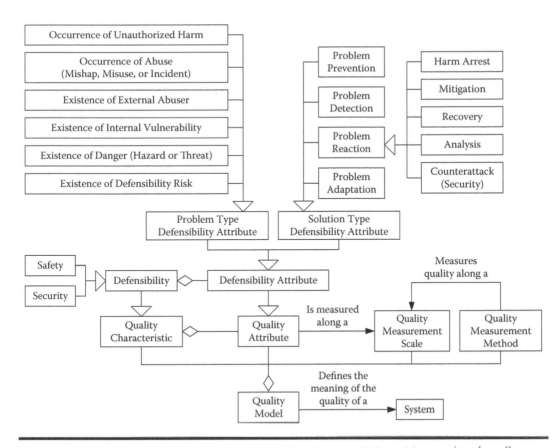

Figure 18.3   Safety and security as example quality characteristics with associated attributes.

A project may reuse an international- or industry-standard quality model, employ an organizational quality model, or develop and use a project-specific quality model. Ordinarily, during the quality planning task of quality engineering, the quality team in collaboration with the system requirements and architecture teams as well as other system stakeholders (such as acquirers, managers, maintainers, operators, and users) should select and tailor (or develop) the project's quality model. If this does not occur in time to support the engineering of the quality requirements, then the system requirements team in collaboration with the quality and architecture teams as well as other stakeholders should select and tailor (or develop) the quality model during the requirements planning task of requirements engineering. Finally, if the quality model does not exist in time to support the engineering of the system architecture, then the architecture team in collaboration with the quality and requirements teams as well as other stakeholders should select and tailor (or develop) the project's quality model during the MFESA Task 1, "Plan and Resource the Architecture Engineering Effort."

Although MFESA does not require the use of any specific quality model, the following classification hierarchy of quality characteristics is provided as a starting point for the development of a consensus project or organizational quality model. It can be extended with additional quality characteristics as appropriate. The following internal and external quality characteristics have been found to be important when architecting software-intensive systems. Their definitions, hierarchical classification relationships, component quality attributes, and associated quality measures are provided in order to lessen the time needed to develop a new quality model.

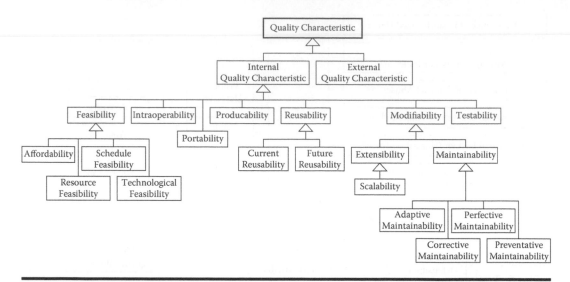

Figure 18.4   An example partial hierarchy of important internal quality characteristics.

The following classification hierarchy of quality characteristics is intentionally quite large. While no list can be truly exhaustive, we provide a fairly complete list. However, these characteristics will never be equally important on all projects, and no system or individual architectural component is ever likely to need quality requirements for all of them. The reason why the following hierarchy is so complete is that it should be easier for the requirements engineers and architects to identify the relevant ones from the given hierarchy than it will be for them to determine needed quality characteristics from scratch. The definitions have also been provided to help clarify the many meanings of quality and promote communication between the architects and the stakeholders.

18.3 Internal Quality Characteristics

As illustrated in Figure 18.4, a relatively complete classification hierarchy can contain quite a large number of internal quality characteristics. Although several other quality models also exist [ISO/IEC, 2001; Barbacci et al., 1995], this book uses the classification hierarchy of the quality model found in Firesmith and the OPEN Process Framework Repository because of its completeness and depth in terms of quality characteristics [Firesmith et al., 2006; OPF, 2008]. Note, however, that no attempt has been made to be exhaustive with regard to quality attributes, measurement scales, and measurement methods.

The following quality characteristics are listed in the order presented in Figure 18.4: from top to bottom and from left to right, with indentation representing the level in the hierarchy. This listing is not intended to be in order of importance as the criticality of the quality characteristics will vary from one system and architectural component to another.

■ **Feasibility:** the internal quality characteristic that is the degree to which it is possible to develop and sustain the system or architectural component within the practical constraints of existing budget, schedule, and technology.

The subtype *quality characteristics* (classification) of feasibility include affordability, resource feasibility, schedule feasibility, and technological feasibility:

- **Affordability:** the subtype of feasibility that is the degree to which the cost of the system or architectural component lies within budgetary constraints. The *quality attributes* (aggregation) of affordability include:
  - **Acquisition cost:** the part of affordability that is the cost of acquiring the system or architectural component.
  - **Development cost:** the part of affordability that is the cost of developing the system or architectural component to the point where it is ready for full-scale production.
  - **Manufacturing cost:** the part of affordability that is the cost of manufacturing instances of the system or architectural component.
  - **Support cost (a.k.a. sustainment cost):** the part of affordability that is the cost of supporting the system or architectural component once manufactured and deployed in its operational environment.
  - **Retirement cost (a.k.a. the disposal cost):** the part of affordability that is the cost of retiring the system or architectural component once it is no longer needed.
- **Resource feasibility:** the subtype of feasibility that is the degree to which the system or architectural component can be developed using existing resources (such as available staffing as well as development, manufacturing, and maintenance facilities).
- **Schedule feasibility:** the subtype of feasibility that is the degree to which the system or architectural component can be developed (initial development, initial small-scale production, or full-scale production) within schedule constraints.
- **Technological feasibility:** the subtype of feasibility that is the degree to which the system or architectural component can be developed using existing technologies.

■ **Intra-operability:** the internal quality characteristic that is the degree to which an architectural component operates effectively with other architectural components within the same system by successfully providing data, information, material, and services to those components and using data, information, material, and services provided by those components. See the external quality characteristic "interoperability" for more details because interoperability and intra-operability are analogous subtypes.

■ **Portability:** the internal quality characteristic that is the degree of ease* with which the software system or architectural component can be transferred to a specified platform (i.e., hardware and software environment). Three subtype *quality characteristics* (classification) of portability are:

- **Hardware portability:** the subtype of portability that is the degree of ease with which the software system or architectural component can be ported to new specified underlying hardware.
- **Operating system portability:** the subtype of portability that is the degree of ease with which the software system or architectural component can be ported to a new specified operating system.

---

* The phrase "the degree of ease" refers to the amount of effort required to do something, whereas the phrase "the degree to which" refers to the extent to which something occurs. The difference between these two phrases determines the type of scale of measure used to set the threshold on the associated quality requirements. Large changes typically require a new development cycle (a.k.a., block, iteration, or increment) rather than maintenance during an existing cycle.

- **Middleware portability:** the subtype of portability that is the degree of ease with which the software system or architectural component can be ported to new specified middleware (whereby middleware is software such as database systems, transaction monitors, content management systems, Web servers that sit "in the middle" between the operating system and application software).

Note that portability refers to changing software's allocation and is not the same thing as *future reusability* or *transportability*, which refers to the degree of ease with which a physical system or architectural component can be moved from one physical location to another.

■ **Producibility (a.k.a. manufacturability):** the internal quality characteristic that is the degree of ease with which the system or architectural component can be produced or manufactured in quantity (i.e., during full-scale production). Producibility is typically measured in terms of unit manufacturing cost.

■ **Reusability:** the internal quality characteristic that is the degree to which a system or architectural component supports reuse in more than one system. The subtype *quality characteristics* (classification) of reusability include:

- **Current reusability (a.k.a. internal reusability):** the subtype of reusability that is the degree of ease with which externally produced components can be incorporated with little or no modification into the current system or architectural component.

- **Future reusability (a.k.a. external reusability):** the subtype of reusability that is the degree of ease with which the architectural component currently being produced will be able to be incorporated with little or no modification in the future into the architecture of other specified systems.

■ **Modifiability (a.k.a. changeability):** the internal quality characteristic that is the degree of ease of successfully changing the system or architectural component once it has been initially developed and deployed. Modifiability is typically measured in terms of financial cost, staff effort, or calendar time needed to make a specific modification using available resources. The subtype *quality characteristics* (classification) of modifiability include extensibility and maintainability:

- **Extensibility (a.k.a. extendability and expandability):** the subtype of modifiability that is the degree of ease with which the system or architectural component can be significantly enhanced to meet specified future goals or changing requirements (such as new functionality or additional capacity).

  • **Scalability:** the subtype of extensibility that is the degree of ease with which a system or architectural component can be extended to significantly increase its existing capacities (such as increased workload).

- **Maintainability:** the subtype of modifiability that is the degree of ease with which the system or architectural component can be minorly modified between major releases. The subtype *quality characteristics* (classification) of maintainability include:

  • **Adaptive maintainability:** the subtype of maintainability that is the degree of ease with which the system or architectural component can be adapted to meet minor changes to the (allocated and derived) requirements.

  • **Corrective maintainability:** the subtype of maintainability that is the degree of ease with which minor defects within the system or architectural component can be localized, corrected, and the corrections verified.*

---

* Corrective maintenance corrects defects or weaknesses that have already caused failures, whereas preventive maintenance corrects defects on weaknesses that have not yet caused failures.

- **Perfective maintainability:** the subtype of maintainability that is the degree of ease with which minor improvements to the system or architectural component can be made (e.g., by refactoring the architecture or using a more appropriate architectural pattern).
- **Preventative maintainability:** the subtype of maintainability that is the degree of ease with which changes to the system or architectural component can be made to prevent future failures.

The three *quality attributes* (aggregation) of modifiability are:

- **Analyzability:** the part of modifiability that is the degree of ease with which defects, deficiencies, and causes of failures can be diagnosed and localized to the components to be modified.
- **Fixability:** The part of modifiabililty that is the degree of ease with which identified and analyzed defects, deficiencies, and causes of failure can be fixed.
- **Verifiability:** the part of modifiability that is the degree of ease with which changes to a system or architectural component can be verified as having been correctly made and to be without unexpected and undesirable side effects.

■ **Testability:** the internal quality characteristic that is the degree of ease with which the system or architectural component facilitates the creation and execution of successful tests (i.e., tests that cause failures that can help identify underlying defects). The *quality attributes* (aggregation) of testability include:

- **Controllability:** the part of testability that is the degree of ease with which the system or architectural component can be:
  - Placed into the proper pretest state.
  - Stimulated with the test message, data, or exception.
- **Observability:** the part of testability that is the degree of ease with which the system or architectural component can be observed to:
  - Be in the proper pretest state.
  - Provide the proper output to its clients, peers, and servers (e.g., returned values, output messages, output requests for data, and exceptions).
  - Be in the proper post-test state.

18.4 External Quality Characteristics

As illustrated in Figure 18.5, a relatively complete classification hierarchy of quality characteristics can also contain quite a large number of external quality characteristics. Although several other quality models also exist [Barbacci et al., 1995; ISO, 2001], this book uses the classification hierarchy of the quality model found in Firesmith and the OPEN Process Framework Repository because of its completeness and depth in terms of quality characteristics [Firesmith et al., 2006; OPF, 207]. Although the following hierarchy is relatively complete in terms of quality characteristics, no attempt has been made to be exhaustive with regard to quality attributes and associated measurement scales and measurement methods.

The following quality characteristics are listed in the order presented in Figure 18.5: from top to bottom and from left to right with indentation representing the level in the classification hierarchy. This listing is *not* intended to be in order of importance, as the criticality of the quality characteristics will naturally vary from one system and architectural component to another.

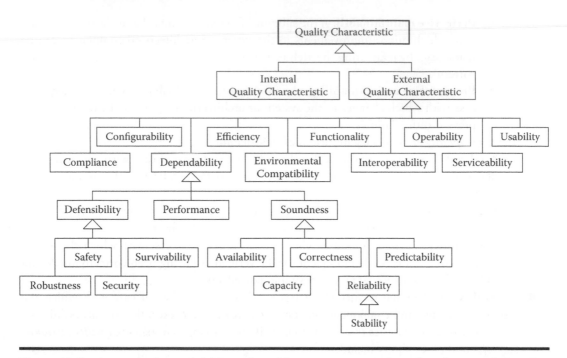

Figure 18.5 An example partial hierarchy of important external quality characteristics. (For simplicity sake, this figure omits certain quality characteristics described in the text.)

The commonly important external quality characteristics illustrated in Figure 18.5 are defined as follows:

- **Compliance:** the external quality characteristic that is the degree to which the system or architectural component adheres to (complies with) laws, regulations, international and industrial standards, and similar prescriptions. The subtype *quality characteristics* (classification) of compliance include legal compliance (international and national laws and regulations), standards compliance (international, national, and industrial standards), and organizational policy compliance.
- **Configurability:** the external quality characteristic that is the degree to which the system or architectural component can be configured in multiple ways without redesign and implementation. The subtype *quality characteristics* (classification) of configurability include:
  - **Internationalization (a.k.a. globalization and localization):** the subtype of configurability that is the degree to which the system or architectural component can be (or is) configured to function appropriately in a global environment. Internationalization may include configuration in terms of:
    - Native languages such as words, spelling, grammar, punctuation, language idioms, definitions, and character sets
    - Culture such as the use of appropriate colors, symbols, and product names
    - Currencies including real-time currency conversion
    - Legal issues such as import/export laws, tariff and sales tax calculations, customs documentation, trademarks, and privacy laws
    - Formats of contact information such as name, address, and phone number

- Country-specific security countermeasures (such as encryption and digital signatures)
- **Personalization:** the subtype of configurability that is the degree to which the system or architectural component can be (or is) configured so that individual users can be presented with a unique user-specific experience.
- **Subsetability:** the subtype of configurability that is the degree to which the system or architectural component can be released in multiple variants, each of which implements either a usable partial initial subset of operational capabilities (e.g., as part of an incremental development cycle) or a different subset of a common set of requirements (i.e., each variant is a subset of a primary complete variant).
- **Supersetability:** the subtype of configurability that is the degree to which the system or architectural component exists in multiple variants, each of which implements a different superset of the common set of requirements (i.e., some requirements are unique to each variant).

■ **Dependability:** the external quality characteristic that is the degree to which legitimate stakeholders can reasonably rely on the system or architectural component to behave as they expect. The subtype *quality characteristics* (classification) of dependability include defensibility, performance, and soundness:
- **Defensibility:** the subtype of dependability that is the degree to which the system or architectural component defends valuable assets* from unauthorized harm. Defensibility can be subtyped into robustness, safety, security, and survivability:
  - **Robustness (a.k.a. resiliency):** the subtype of defensibility that is the degree to which the system or an architectural component tolerates potentially harm-causing events or conditions and recovers from them. The *quality attributes* (aggregation) of robustness include:
    - **Tolerance:** the part of robustness that is the degree to which the system or architectural component continues to provide essential mission-critical services despite potentially harm-causing events or conditions. The *quality attributes* (classification) of tolerance include:
      - **Environmental tolerance:** the subtype of tolerance that is the degree to which the system or architectural component tolerates potentially harm-causing *environmental conditions* (e.g., salt spray causing corrosion or radiation randomly degrading information processing or memory).
      - **Error tolerance:** the subtype of tolerance that is the degree to which the system or architectural component tolerates the presence of erroneous input (such as incorrect, stale, or out-of-order data). Note that erroneous input is typically due to human error, although it may also be due to sensor failure, timing delays, and the like.
      - **Fault tolerance:** the subtype of tolerance that is the degree to which the system or architectural component tolerates the presence or execution of defects, whereby a defect (a.k.a. a fault and bug) is an underlying flaw in a work product (i.e., a work product that is

---

* Note that valuable assets can include the system and its parts (e.g., architectural components) as well as people, property, and the natural environment.

inconsistent with its requirements, policies, goals, or the reasonable expectations of its customers or users).

- **Failure tolerance:** the subtype of tolerance that is the degree to which the system or architectural component tolerates the occurrence of failures, whereby a failure is the execution of a defect that causes an inconsistency between an executable work product's actual (i.e., observed) and expected (e.g., specified) behavior. Note that failures are externally visible (black box), whereas defects are only internally visible (white box). A defect may or may not cause a failure, depending on whether or not the defect is executed and whether or not exception handling prevents the failure from occurring.

■ **Recoverability:** the part of robustness that is the degree to which the system or architectural component automatically recovers from a fault or failure

- **Safety:** the subtype of defensibility that is the degree to which the system or architectural component prevents or reduces the probability or severity of, detects, and properly reacts to:

  ■ Unauthorized *unintentional* (i.e., *accidental*) harm to valuable assets
  ■ Mishaps (such as accidents and safety incidents)*
  ■ Hazards (i.e., conditions that can cause mishaps). Note that hazards typically include the existence of valuable assets that can be accidentally harmed, vulnerabilities that enable the system or architectural component to cause or allow the accidental harm to occur, and abusers (such as environmental conditions or users) that can accidentally trigger a mishap.
  ■ Safety risks (i.e., probability of accidental harm multiplied by maximum or average credible harm severity)

- **Security:** the subtype of defensibility that is the degree to which the system or architectural component prevents or reduces the probability or severity of, detects, and reacts to *civilian*:†

  ■ Unauthorized *intentional* (i.e., *malicious*) harm to valuable assets
  ■ Misuses (e.g., attacks and security incidents)
  ■ Threats (i.e., conditions that can cause misuses). Note that threats typically include the existence of valuable assets that can be maliciously harmed, vulnerabilities that enable the system or architectural component to cause or allow the malicious harm to occur, and misusers (such as malware or civilian attackers) with the means, motives, and opportunities to attack the system that can maliciously trigger a misuse.
  ■ Security risks (i.e., probability of malicious harm multiplied by maximum or average credible harm severity)

The subtype *quality characteristics* (classification) of security include communications security (COMSEC), computer security (COMPSEC), emissions security (EMSEC or Tempest), information security (INFOSEC), network security (NETSEC), operations security (OPSEC), personal security (PERSEC), and physical security (PHYSEC).

---

* As illustrated in Figure 18.3, an abuse (defensibility) can be either a misshap (safety) or a misuse (security). Similarly, a danger (defensibiity) can be either a hazard (safety) or a threat (security).
† The military counterpart to security is survivability.

- **Survivability:** the subtype of defensibility that is the degree to which the system or architectural component prevents or reduces the probability or severity of, detects, and reacts to *military*:
  - Unauthorized intentional (i.e., malicious) harm to valuable assets
  - Misuses (e.g., attacks and survivability incidents)
  - Threats (i.e., existence of vulnerable assets and military attackers with means, motives, and opportunities to militarily attack the system)
  - Survivability risks (i.e., probability of malicious harm multiplied by maximum credible harm severity)

  The reader should note that the term "survivability" is used here in a manner different from that which is sometimes used in the information security community. Although defined similarly to security, survivability is used in the military sense of being able to sustain significant military damage and still function. Two examples of very survivable systems are the B-29 aircraft during World War II and the A-10 Warthog, both of which would sometimes return to base despite having been perforated numerous times by enemy bullets and shrapnel.
- **Performance:** the subtype of dependability that is the degree to which the system or architectural component operates within its designated temporal constraints. The *quality attributes* (aggregation) of performance include:
  - **Jitter:** the part of performance that is the degree to which the variability of the time intervals between periodic actions controlled by the system or architectural component remains within its designated constraints.
  - **Latency:** the part of performance that is the degree to which the time that the system or architectural component takes to execute specific tasks (such as system operations and use case paths) from end to end is within acceptable time limits.
  - **Response time:** the part of performance that is the degree to which the time it takes for the system or architectural component to initially respond to a client request for a service is within acceptable time limits.*
  - **Schedulability:** the part of performance that is the degree to which the system or architectural component events and behaviors can be accurately scheduled.
  - **Throughput:** the part of performance that is the degree to which the system or architectural component is able to complete an operation and provide a service within acceptable time limits.
- **Soundness:** the subtype of dependability that is the degree to which the system or architectural component behaves properly and predictably when and for as long as needed.
  - **Availability (a.k.a. readiness):** the subtype of soundness that is the degree to which the system or architectural component is ready to function without failure in one or more specified ways at any time during a specified period of time under normal conditions or circumstances given its operational profile.

    Availability is typically measured as the percentage of the time that the system or architectural component can be used in a specific way or the probability that the system or architectural component is functioning properly at an arbitrary point in time. When multiple instances of a system exist (such as multiple aircraft of a given

---

* Note the subtle difference between latency and response time. Because systems seldom provide intermediate responses, beware of confusions where the term response time is used when latency is meant, especially in requirements and when the latency is relatively large.

type), availability can also be measured in the percentage of these systems that are available for operational use at any arbitrary point in time.*

- **Capacity:** the subtype of soundness that is the degree to which the system or architectural component:
    - ■ Can successfully handle a large workload or number of things at a single point in time or during a specific interval of time.
    - ■ Gracefully degrades its capabilities as the fundamental limits are approached.

    Capacity is typically measured in terms of the number of things handled (e.g., number of users) or the number of events (e.g., transactions) per unit of time.

    The subtype quality characteristics (classification) of capacity include bandwidth, material flow, processing, and storage capacity.

- **Correctness:** the subtype of soundness that is the degree to which the system or architectural component and its outputs are free from defects. The *quality attributes* (aggregation) of correctness include:
    - ■ **Accuracy:** the part of correctness that is the magnitude of defects (i.e., the deviation of the actual or average value from their true value) in the stored and output data of the system or architectural component. Note that accuracy can refer to different types of data such as quantitative data (e.g., incorrect values of numbers), textual or character data (e.g., number of incorrect characters in a string), audio data (e.g., the acoustic quality of an audio file), or video data.
    - ■ **Precision:** the part of correctness that is the degree of dispersion of the stored and output quantitative data around their average values of the system or architectural component.
    - ■ **Timeliness:** the part of correctness that is the degree to which data of the system or architectural component remains current (i.e., up to date).

- **Reliability (a.k.a. continuity):** the subtype of soundness that is the degree to which the system or architectural component continues to function without failure in one or more specified ways during a specified period of time under normal conditions or circumstances given its operational profile.

    Reliability is frequently measured in terms of the conditional probability that an architectural element remains functioning for a given duration or the mean time to first failure (MTFF) or mean time between failures (MTBF).

    - ■ **Stability:** the subtype of reliability that is the degree to which the system or architectural component continues to deliver *mission-critical* services during a given time period† under a given operational profile regardless of any failures whereby the:
        - – Failures may prevent the system from delivering less critical services.
        - – Failures limiting the delivery of mission-critical services occur at unpredictable times.

---

* Note that when estimating availability (and reliability), the amount and quality of system sustainment, including maintenance and support, must be taken into account.

† Stability is a type of reliability. Whereas reliability is typically measured in terms of mean time between failures (MTBF), stability is typically measured in terms of mean time between mission-critical failures (MTBMCF). Stability is closely related to robustness, which is a type of defensibility. Whereas robustness refers to how well a system defends itself in terms of its tolerance of negative events and conditions and in terms of its recoverability once failure occurs, stability refers to how often mission-critical failures occur.

- Root causes of such failures are difficult to identify efficiently.
  - **Predictability:** the subtype of soundness that is the degree to which system or architectural component events and behaviors are deterministic and can therefore be predicted from a given set of inputs when in a given state or environment.
- **Efficiency:** the external quality characteristic that is the degree to which the system or architectural component effectively uses (i.e., minimizes its consumption of) the resources upon which it depends. The subtype *quality characteristics* (classification) of efficiency include:
  - **Computing efficiency:** the subtype of efficiency that is the degree to which the system or architectural component effectively uses (i.e., minimizes its consumption of) computing resources. For example, computing efficiency can be further subtyped into communications bandwidth efficiency, memory efficiency, and processor efficiency.
  - **Consumables efficiency:** the subtype of efficiency that is the degree to which the system or architectural component effectively uses (i.e., minimizes its consumption of) consumable materials. For example, consumables efficiency can be subtyped into fuel efficiency, oxidizer efficiency, and raw materials efficiency.
  - **Energy efficiency:** the subtype of efficiency that is the degree to which the system or architectural component effectively uses (i.e., minimizes its consumption of) energy. For example, energy efficiency can be subtyped into cooling efficiency, electrical efficiency, heating efficiency, and mechanical efficiency.
  - **Size efficiency:** the subtype of efficiency that is the degree to which the system or architectural component effectively uses (i.e., minimizes its consumption of) its size. For example, size efficiency can be subtyped into mass/weight efficiency, density efficiency, dimension efficiency, area efficiency, and volume efficiency. Note that an architectural component's maximum dimensions, volume, and weight are often critical considerations when dealing with aircraft and satellite systems.
- **Environmental compatibility:** the external quality characteristic that is the degree to which the system or architectural component can be used and functions correctly under specified conditions of the physical environment (e.g., air pressure, dust, radiation, vacuum, and water pressure) in which it is intended to operate.
- **Functionality*:** the external quality characteristic that is the degree to which the system or architectural component implements all the functions (or functional requirements) allocated to it.
- **Interoperability:** the external quality characteristic that is the degree to which the system or architectural component operates (i.e., interfaces and collaborates) effectively with specified (types of) external systems by successfully providing data, information, material, and services to those systems and by successfully using data, information, material, and services provided by those systems.† The subtype *quality characteristics* (classification) of interoperability include:
  - **Physical interoperability:** the subtype of interoperability that is the degree to which the system or architectural component physically connects with specified (types of)

---

* Functionality is a controversial quality characteristic that is included in some quality models and not in others. In many ways, it is different from the others and is often implicitly specified in functional requirements rather than explicitly specified in functionality quality requirements.

† A system may not know the specific *client* systems to which it provides services and data. The system may depend on *server* systems, which may be replaced by other systems providing the same or equivalent services and data. The system architecture should minimize unnecessary coupling between the system and other systems with which it interoperates.

interfaces with specified (types of) external systems. Some example subtype *quality characteristics* (classification) of physical interoperability include:

- **Electrical connection interoperability** in terms of electrical plug type, power rating, number and configurations of prongs, male versus female connection, etc.
- **Electronic connections interoperability** in terms of number and configuration of pins, male versus female connection, etc.
- **Physical connections interoperability** in terms of size and shape of surfaces as well as the number, type, orientation, and size of physical connectors such as bolts and screws.
- **Power connections interoperability** in terms of cable, chain, and hose.

- **Energy interoperability:** the subtype of interoperability that is the degree to which the system or architectural component correctly exchanges and uses energy of the appropriate types and levels with specified (types of) external systems. Some example subtype *quality characteristics* (classification) of energy interoperability include:
  - **Hydraulic power interoperability** in terms of fluid type, maximum and minimum pressures, etc.
  - **Mechanical linkage energy interoperability** in terms of linkage type (e.g., belt, cable, or chain) and average and maximum force.
  - **Wired communication energy interoperability** in terms of proper logic voltages, frequency, and amperage of electricity such as logic low level (0–1 volts), logic high level (3.5–5.0 volts), 40 megahertz (MHz), and 100 milliamps (mA).
  - **Wired power interoperability** in terms of proper voltages including ranges, AC (alternating current) frequency including ranges, and maximum amperage of electricity such as 120 VAC or 270 VDC (volts of direct current) (250–280 VDC), 60 Hz, and 100 milliamps or 150 kilowatts maximum.
  - **Wireless communication energy interoperability** in terms of proper electromagnetic frequency (e.g., proper spectrum of radio waves, microwaves, or visible light) such as 1850–1990-MHz microwaves for cellular telephone transmission or C-band for satellite transmission, broadcast or laser, minimum/maximum signal strength.

- **Material interoperability:** the subtype of interoperability that is the degree to which the system or architectural component correctly exchanges and uses materials of the appropriate types and amounts with specified (types of) external systems. Some subtype *quality characteristics* (classification) of material interoperability include:
  - **Chemical interoperability** in terms of the exchange of chemicals (e.g., petrochemicals and pesticides and their intermediate ingredients)
  - **Fuel interoperability** in terms of the exchange of fuels
  - **Water interoperability** in terms of the exchange of water of the correct purity, pH, or salinity

- **Protocol interoperability:** the subtype of interoperability that is the degree to which the system or architectural component correctly uses the same interface (and communications) protocols as the specified (types of) external systems. Note that protocol interoperability increases when architects select open interface standards (i.e., interfaces, the protocols of which are specified by international, national, or industry public standards). Some subtype *quality characteristics* (classification) of protocol interoperability include compatibility of protocols at the following layers:
  - **Physical layer protocols:** Layer 1 protocols such as Integrated Services Digital Network (ISDN) and RS-232.

- **Data-link layer protocols:** Layer 2 protocols such as Ethernet, Fiber Distributed Data Interface (FDDI), and Point-to-Point Protocol (PPP).
- **Network layer protocols:** Layer 3 protocols such as Internet Control Message Protocol (ICMP) and Internet Protocol (IP).
- **Transport layer protocols:** Layer 4 protocols such as Transmission Control Protocol (TCP) and User Datagram Protocol (UDP).
- **Session layer protocols:** Layer 5 protocols such as Network File System (NFS).
- **Presentation layer protocols:** Layer 6 protocols such as American Standard Code for Information Interchange (ASCII), Moving Picture Experts Group (MPEG), and Secure Socket Layer (SSL).
- **Application layer protocols:** Layer 7 protocols such as Domain Name Service (DNS), File Transfer Protocol (FTP), Hypertext Transfer Protocol (HTTP), and Hypertext Transfer Protocol Secure (HTTPS).
  - **Syntax interoperability:** the subtype of interoperability that is the degree to which the system or architectural component correctly communicates data having the correct syntax (like data types such as text, integer, date, and money, including associated attributes, ranges, and default values) with specified (types of) external systems.
  - **Semantic interoperability:** the subtype of interoperability that is the degree to which the system or architectural component communicates requests and data via specified (types of) interfaces of the specified (types of) external systems in a manner in which both systems interpret the syntax in a single standard way to gain the same meanings. Semantic interoperability includes such things as units of measure (e.g., differentiating between English and metric units or differentiating between Canadian and United States dollars).
- ■ **Operability:** the external quality characteristic that is the degree of ease of operating (as opposed to using) the system or architectural component. Operability is typically measured in terms of financial cost or staff effort per unit time to operate the system or component under specified conditions.
- ■ **Serviceability:** the external quality characteristic that is the degree of ease of servicing the system or architectural component (such as adding or replacing consumable materials such as fueling an aircraft). Serviceability is typically measured in terms of financial cost, staff effort, or staff effort per unit time to service the system or component in specified ways under specified conditions.
- ■ **Usability:** the external quality characteristic that is the degree to which the human user interface of the system or architectural component enables a specified group of users achieve specified goals in a specified context of use. The quality attributes (aggregation) of usability include:
  - **Accessibility:** the part of usability that is the degree to which the system is useful to users with disabilities (e.g., color blindness, deafness, loss of mobility, etc.).
  - **Attractiveness (a.k.a. engageability, preference, and stickiness*):** the part of usability that is the degree to which users find the system or architectural component to be attractive or appealing, engage their attention, provide a positive user experience, be preferable to its alternatives, make users want to continue to use it, and make them return to use it in the future.
  - **Credibility (a.k.a. trustworthiness):** the part of usability that is the degree to which users are confident with and have trust in the system, including that its output and

---

* The term "stickiness" is typically used with reference to Web pages and refers to how long users remain at (i.e., remain "stuck" to) given Web pages.

behavior are correct, content is authoritative, owner's motives are trustworthy, and developers were competent.*

- **Differentiation:** the part of usability that is the degree to which the system or architectural component differentiates itself from competing products.
- **Ease of entry:** the part of usability that is the degree of ease with which users can start using the system or architectural component (e.g., can log on and begin using their desired functionality without waiting an excessive amount of time to be identified, be authenticated, and navigate to the point where they can start performing their tasks).
- **Ease of remembering:** the part of usability that is either the degree to which occasional users can remember how to use the system or architectural component to perform common tasks or the degree to which regular users can remember how to use the system to perform infrequent tasks.
- **Effectiveness:** the part of usability that is the quality attribute characterizing the degree of ease with which the system or architectural component enables users to successfully achieve their goals.
- **Effort minimization:** the part of usability that is the degree to which the system or architectural component minimizes the amount of effort users (and operators) must expend to achieve their goals (in relation to the correctness and completeness with which these goals are achieved).
- **Error minimization:** the part of usability that is the degree to which the system or architectural component minimizes the number of errors that its users make.
- **Learnability:** the part of usability that is the degree to which representative users can learn to use the system or architectural component to achieve their goals (such as to find desired content and to perform their tasks).
- **Navigability:** the part of usability that is the degree of ease with which the system or architectural component enables users to easily move through the user interface or documentation to achieve their goals, such as finding desired content or services (such as finding Web applications such as Web sites using search engines).
- **Retrievability:** the part of usability that is the ease with which the system or architectural component enables users to obtain information in a form that is useful to them (such as print out a paper report or make a copy of a multimedia file in the correct format and media).
- **Suitability (a.k.a. appropriateness):** the part of usability that is the degree to which users find that the system or architectural component is suitable for supporting the performance of their tasks.
- **Transportability:** the part of usability that is the degree of ease with which a physical system or architectural component can be transported from one physical location to another using available means of transportation.
- **Understandability:** the part of usability that is the degree to which users find the human interfaces and output of a system or architectural component to be clear, legible, unambiguous, and comprehensible (especially during unusual situations).
- **User satisfaction:** the part of usability that is the degree to which users are satisfied with the system or architectural component and consider it to be beneficial to them.

---

* Note the subtle difference between dependability and credibility. Dependability is more objective , while credibility is more subjective. Although they tend to go together, a system or architectural component can be dependable without being credible, and vice versa.

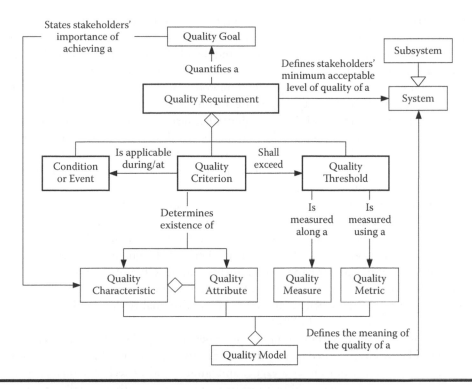

Figure 18.6   Quality requirements are based on a quality model.

18.5 Quality Requirements

As illustrated in Figure 18.6, a quality requirement is specified in terms of one or more conditions, a quality criterion, and quality thresholds on some quality measurement scales [Firesmith, 2005]. Having quality requirements that themselves exhibit high quality is a critical prerequisite for the production of a high-quality system architecture that sufficiently supports the systems' meeting of these quality requirements. Building on Figure 18.6, this section defines quality requirements and shows how they relate to the components of the quality model.

> **Condition/triggering event.** An optional *condition* that states under what conditions the quality requirement must be met or an *event* that triggers when the requirement must be met. For example, quality requirements specifying high performance and reliability may only hold during normal conditions, but not under degraded mode operations.
>
> **Quality criterion.** A *quality criterion* is a system-specific description that provides evidence either for or against the existence of a given quality characteristic or attribute.* Quality criteria significantly contribute toward making the high-level quality

---

* Under certain circumstances, a quality criterion may be related to more than one quality attribute. For example, the quality criteria of defensibility requirements (e.g., safety, security, and survivability requirements) typically address both a defensibility problem subfactor (e.g., unauthorized harm to valuable assets, danger, defensibility event, or risk) and a defensibility solution subfactor (e.g., prevention, detection, and reaction). Thus, a safety requirement may specify that a system must prevent accidental harm, detect an accident, or react a specific way to the detection of an accident.

characteristics and attributes detailed enough to be unambiguous and verifiable. When quality criteria are adequately specific, they lack only the addition of quality metrics to make them sufficiently complete and detailed to form the basis for detailed quality requirements. There are many more quality criteria than quality characteristics and attributes because there are typically numerous possible criteria per quality characteristic and attribute. Quality criteria are also more domain-specific and less reusable than quality characteristics and attributes because they are specific descriptions of specific systems and architectural components. To deal with the large number of criteria and to make them reusable, quality criteria can often be parameterized in the quality models, and specific instances of the parameterized classes of criteria can then be used to produce quality criteria.

Quality threshold. A *quality threshold* specifies a minimum level of quality along a quality measurement scale. Thus, the threshold is measured in units of measure based on the quality measure of the quality model for the quality characteristic or quality attributes associated with the quality criteria of the quality requirement. For example, throughput performance requirements may specify a minimum acceptable quality threshold of a certain number of transactions per second. Similarly, a reliability requirement may specify a minimum acceptable quality threshold of a certain mean time between failures (MTBF).

Perhaps the most important thing to remember about quality requirements is that they should have clearly stated quality thresholds. Without quality thresholds, it is not a quality requirement but rather a vague, ambiguous, and unverifiable quality goal. From a system architecture quality assessment standpoint, without quality requirements with associated quality thresholds, it is impossible for system:

■ Architects to *properly* make engineering trade-offs between competing quality requirements

■ Architects and system architecture assessors to *know* if the system architecture sufficiently supports its derived and allocated quality requirements and is therefore good enough

■ Testers to determine test completion criteria for verifying quality requirements

18.5.1 Example Quality Requirements

The following are some example quality requirements written using the previously described recommended format:

■ **Availability requirement.** Upon occurrence of event W when in [state | mode] X, the system shall [provide service | perform function] Y at least Z% of the time.
　　Condition = "Upon occurrence of event W when in [state | mode] X"
　　Quality criterion = "the system shall [provide service | perform function] Y"
　　Threshold = "at least Z% of the time" (i.e., be available at least Z% of the time)

■ **Performance requirement (throughput attribute).** While in [state | mode] X, the system shall [provide service | perform function] Y at least Z times per [hour | minute | second | millisecond].
　　Condition = "While in [state | mode] X"
　　Quality criterion = "the system shall [provide service | perform function] Y"
　　Threshold = "at least Z times per [hour | minute | second | millisecond]"

- **Reliability requirement.** While in [state | mode] X, the system shall [provide service | perform function] Y with a mean time between failure of at least Z [years | days | hours | minutes | seconds | milliseconds] of continuous operation.

  Condition = "While in [state | mode] X"

  Quality criterion = "the system shall [provide service | perform function] Y"

  Threshold = "with a mean time between failure of at least Z [years | days | hours | minutes | seconds | milliseconds] of continuous operation"

- **Safety requirement (accident prevention attributes).** While in [state | mode] X, the system shall prevent the occurrence of accidents of type Y with a mean time between such accidents of at least Z [years | days | hours].*

  Condition = "While in [state | mode] X"

  Quality criterion = "the system shall prevent the occurrence of accidents of type Y"

  Threshold = "with a mean time between such accidents of at least Z [years | days | hours]"

- **Security requirement (attack detection).** While in [state | mode] X, the system shall detect the occurrence of attacks of type Y with a probability of at least Z%.

  Condition = "While in [state | mode] X"

  Quality criterion = "the system shall detect the occurrence of attacks of type Y"

  Threshold = "with a probability of at least Z%"

- **Stability requirement.** While in [state | mode] X, the system shall ensure that the mean time between the failure of *mission critical functionality* Y† is at least 5000 hours of continuous operation.

18.6 Architectural Quality Cases

Quality cases (also known as assurance cases) are a generalization of safety cases, which are commonly used in the safety community [Bishop and Bloomfield, 1999]. A quality case makes the developer's case that their system exhibits the necessary level of a specific type of quality (i.e., quality characteristic). Quality cases are useful for assessing the quality of work products and supporting certification and accreditation.

> **Quality Case:** A cohesive collection of *claims*, *arguments*, and *evidence* that makes the developers' case that their work product(s) have a *sufficient* level of some *quality characteristic* or *attribute*.

Because quality can be supported and lost during all phases of the development and manufacturing cycle, a convincing quality case must be complete, containing compelling information from all activities of the development cycle, including requirements, architecture, design, implementation, integration, testing, manufacture, and deployment.

---

* Note that requirements are often *not* written in the order of condition, criterion, and threshold. For example, this requirements might be written as, "While in [state | mode] X, the system shall ensure that the mean time between accidents of type Y is at least Z [years | days | hours]," or as, "the system shall ensure that the mean time between accidents of type Y shall be at least Z [years | days | hours] when it is in [state | mode] X." An advantage to using a standard format is that it increases the likelihood that all parts of the requirement are included.

† Note that Y needs to be properly defined in the project glossary.

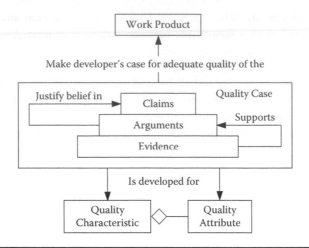

Figure 18.7   The three components of a general quality case.

18.6.1 Quality Case Components

As illustrated in Figure 18.7, a quality case consists of claims, arguments, and evidence.

> **Claims:** the developers' assertions that their work products sufficiently support the system or architectural component achieving a sufficient level of some quality characteristic. System or architectural component quality is defined in terms of the quality characteristics and quality attributes defined in the official project quality model. An additional implied claim is that these work products are themselves sufficiently complete and of sufficient quality.
>
> **Arguments:** clear, compelling, and relevant developer arguments justifying the assessors' belief in the developers' claims. Arguments consist of the developers' decisions, inventions, trade-offs, assumptions, and associated rationales.
>
> **Evidence:** adequate credible substantiation supporting the developers' arguments. Examples of such evidence include official project diagrams, models, requirements specifications and architecture documents; requirements repositories; analysis and simulation reports; test results; and demonstrations witnessed by the assessors.

18.6.2 Architectural Quality Case Components

Architectural quality cases are a specialized type of quality case, restricted to making the architects' case that their architecture enables the system to exhibit the required level of a specific quality characteristic or quality attribute. As illustrated in Figure 18.8, an architectural quality case is therefore restricted to architectural information and consists of the following three types of components: (1) the architects' claims that their architecture sufficiently supports the associated quality characteristic or attribute,* (2) their clear and compelling arguments justifying stakehold-

---

* This typically means that the architecture sufficiently supports the system or architectural component meeting their associated quality requirements. Note that a sufficient architecture does *not* guarantee that the systems or architectural component meets these requirements because defects may exist in the design, implementation, integration, and deployment.

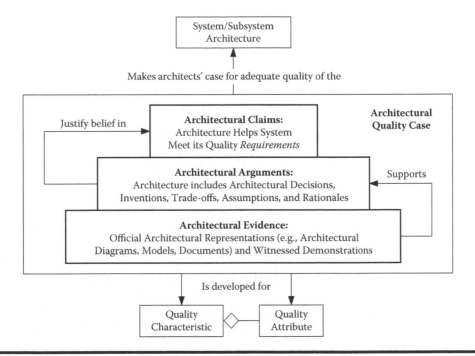

Figure 18.8 The three components of architectural quality cases.

ers' belief in these claims, and (3) their official evidence adequately supporting these arguments. The key difference between a complete quality case and an architectural quality case is that the architectural quality case is restricted to *architectural* claims, arguments, and evidence.

In summary, the architectural claims are that the architecture sufficiently supports the system meeting its quality requirements. The architectural arguments consist of the architectural decisions, inventions, and trade-offs that the architects have made, as well as any associated assumptions and rationales. Finally, the architectural evidence consists of official architectural representations and relevant demonstrations witnessed by the assessors of the architectural quality cases.

As illustrated in Figure 18.9, architectural quality case diagrams typically consist of the following three layers:

1. **Architectural claims (quality characteristic or attribute):** the claim made by the architects that their architecture sufficiently supports the system meeting its quality requirements for a single quality characteristic or relevant set of quality attributes.
2. **Architectural arguments:** the arguments made by the architects that their architecture justifies belief in their claims and consisting of the relevant parts of their architecture (i.e., the relevant architectural decisions, inventions, trade-offs, assumptions, and rationales).
3. **Architectural evidence:** the official architectural representations and witnessed demonstrations that support the architects' arguments.

There exist several challenges with dealing with real-world quality cases, including architectural quality cases. They are typically quite big, containing a large amount of information in the form of text, diagrams, and references to existing documentation. It is therefore easy

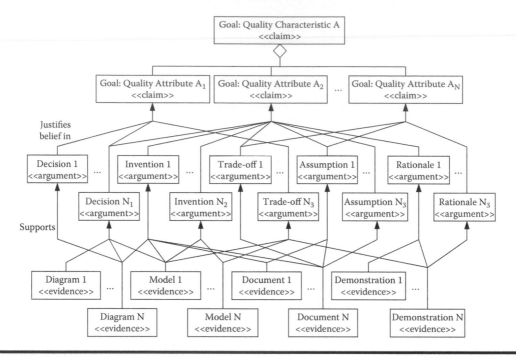

Figure 18.9   Architectural quality case diagram notation.

for stakeholders to get lost when reading and assessing architectural quality cases. What is needed is a practical way to communicate, summarize, and index the essentials of each quality case. Quality case diagrams fill this need for guides to and indexes of individual quality cases [Firesmith et al., 2006]. A quality case diagram is a layered UML class diagram that labels and summarizes the main parts of a single quality case. As illustrated by Figure 18.9, quality case diagrams consist of:

- **Rectangular nodes.** The nodes are rectangular class icons labeled with the names of (and stereotypes for) claims, arguments, and evidence. The nodes are arranged in a layered format with claims on the top, arguments in the middle, and evidence on the bottom.
- **Arrows.** The arcs are arrows documenting the relationships between the nodes. This includes aggregation relationships between claims, "justifies belief in" associations from arguments to the claims they justify, and "supports" associations from evidence to the arguments supported by the evidence.

18.6.3 Example Architectural Quality Case

The following is a general representative example architecture quality case for the quality attribute protocol interoperability and for a specific example subsystem X. As with any architecture quality case, it consists of the architects' claims, arguments, and evidence:

- **Architectural claims.** The subsystem X architects make the following claims concerning the sufficiency of their architecture to meet its derived and allocated reliability-related goals and requirements:

- **Protocol interoperability goals.** Subsystem X architecture sufficiently supports the achievement of the following protocol interoperability goals:
  - Subsystem X correctly uses the interface protocols of all relevant external systems.
  - Subsystem X will use open interface standards (i.e., industry standard protocols) when communicating with all external systems.
- **Protocol interoperability requirements.** Subsystem X architecture sufficiently supports the achievement of the following protocol interoperability requirements:
  - The subsystem shall use open interface standards (i.e., industry standard protocols) when communicating with external systems across all key interfaces identified in document X.
  - The subsystem shall use the Ethernet over RS-232 for communication across interface X with external system Y.
  - The subsystem shall use HTTPS for communicating securely when performing function X across interface Y with external system Z.

■ **Architectural arguments.** The subsystem X architects present their following arguments to the assessors:
- **Layered architecture.** The subsystem uses a layered architecture.
  Rationale: Interface layer supports interoperability.
- **Modular architecture.** The subsystem architecture is highly modular.
  Rationale: Architecture includes modules (proxies) for interoperability.
- **Wrappers and proxies.** The subsystem architecture includes proxies that wrap the interfaces to external subsystems.
  Rationale: Proxies localize and wrap external interfaces.
- **Service-oriented architecture (SOA).** The subsystem service-oriented architecture uses XML, SOAP, and UDDI to publish and provide Web services over the Internet to external client systems.
  Rationale: SOA improves interoperability. Standard languages and protocols support interoperability between heterogeneous systems.

■ **Evidence.** The subsystem X requirements engineers make the following evidence available to the assessment team and are able to present it at the assessment meeting if called on to do so:
- **Context diagram.** A context diagram shows the external interfaces requiring interoperable protocols.
- **Architectural class diagram.** One or more architectural class diagrams show modularity and the location of proxies and Web services that can be used to increase the interoperability.
- **Allocation diagram.** One or more allocation diagrams show the allocation of software modules to hardware as well as the modularity of the architecture.
- **Layer diagram.** A layer diagram shows the architectural layers of the software.
- **Activity/collaboration diagrams.** UML activity and collaboration diagrams show the existence, location, and usage of proxies and wrappers as well as the source and use of services.
- **Interoperability white paper.** The interoperability white paper describes how the system or architectural component supports interoperability, including protocol interoperability.
- **Vendor-supplied technical documentation.** Technical documentation supplied by the vendor of COTS products shows the products' support for SOA and associated protocols.

In practice, there are often a very large number of claims, arguments, and sources of evidence, each of which may involve significant amounts of textual and graphical information. Because it

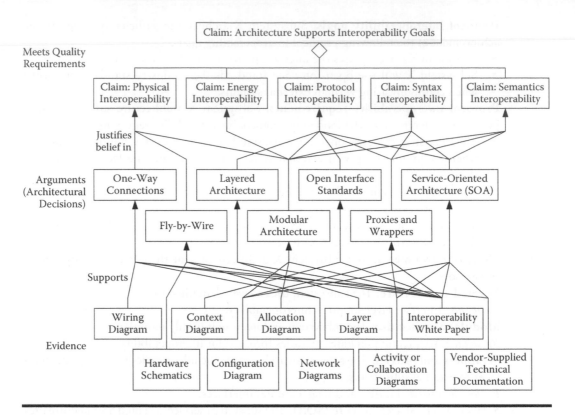

Figure 18.10 Example architectural quality case diagram.

can become quite difficult to read and analyze all of this information without getting lost, we have developed quality case diagrams to summarize and act as an index to the actual quality case. Figure 18.10 is an example of such a quality case diagram.

18.7 Architectural Quality Case Evaluation Using QUASAR

The QUality Assessment of System Architectures and their Requirements (QUASAR) method is, as the name implies, a method developed to assess the quality of system architectures and their associated architecturally significant quality requirements [Firesmith et al., 2006]. QUASAR can be used during MFESA Task 9, *Evaluate and Accept the Architecture,* Step 5 *Independently Assess the Architecture.* QUASAR is based on the:

■ Production of requirements and architecture quality cases as a natural part of system requirements engineering and system architecture engineering
■ Independent assessment of these quality cases by assessment teams staffed by people who are
trained and experienced in system requirements engineering, system architecture engineering, quality cases, the application domain, and quality assessments

As illustrated in Figure 18.11, the QUASAR method for assessing the quality of system architectures and their requirements consists of the following three phases:

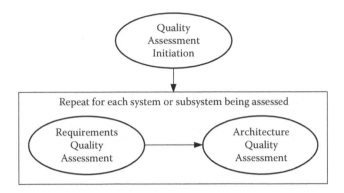

Figure 18.11 The three phases of the QUASAR method.

1. **Quality Assessment Initiation (QAI).** The objectives of the QUASAR QAI phase are to:
   - Prepare the requirements and architecture teams from the development organization and the assessment team from the acquisition organization so that they can effectively and efficiently perform the requirements and architecture quality assessments.
   - Develop a consensus on the scope, schedule, and location of the quality assessments.
   - Tailor and update the QUASAR assessment method and its associated training materials.
2. **Requirements Quality Assessment (RQA).** The objectives of the QUASAR RQA phase are to use requirements quality cases to:
   - Independently assess the quality and maturity of the architecturally significant requirements that drive the architecture and form the foundation for the architecture quality assessment.
   - Help the requirements engineers identify requirements defects and weaknesses so that they can be corrected, the system and its architecture can be improved, and the probability of project success can be increased.
   - Identify requirements risks so that they can be managed.
   - Ensure that the architecture teams will be prepared to support the coming architecture quality assessment.
   During this phase, the requirements team presents and defends their requirements quality cases, and the assessment team actively assesses the sufficiency of the quality-related requirements to meet the quality goals of the stakeholders. The architecture team also presents a representative sample of the kinds of information that they intend to include in their architecture quality cases. The assessment team provides an initial outbrief of the assessment results, followed by a more formal assessment report.
3. **Architecture Quality Assessment (AQA).** The objectives of the QUASAR AQA phase are to use architectural quality cases to:
   - Independently assess the quality and maturity of the architecture in terms of its support for its derived and allocated architecturally significant requirements (especially its quality requirements).
   - Help the architects identify architectural defects and weaknesses so that they can be corrected, the system and its architecture can be improved, and the probability of project success can be increased.
   - Identify architectural risks so that they can be managed.

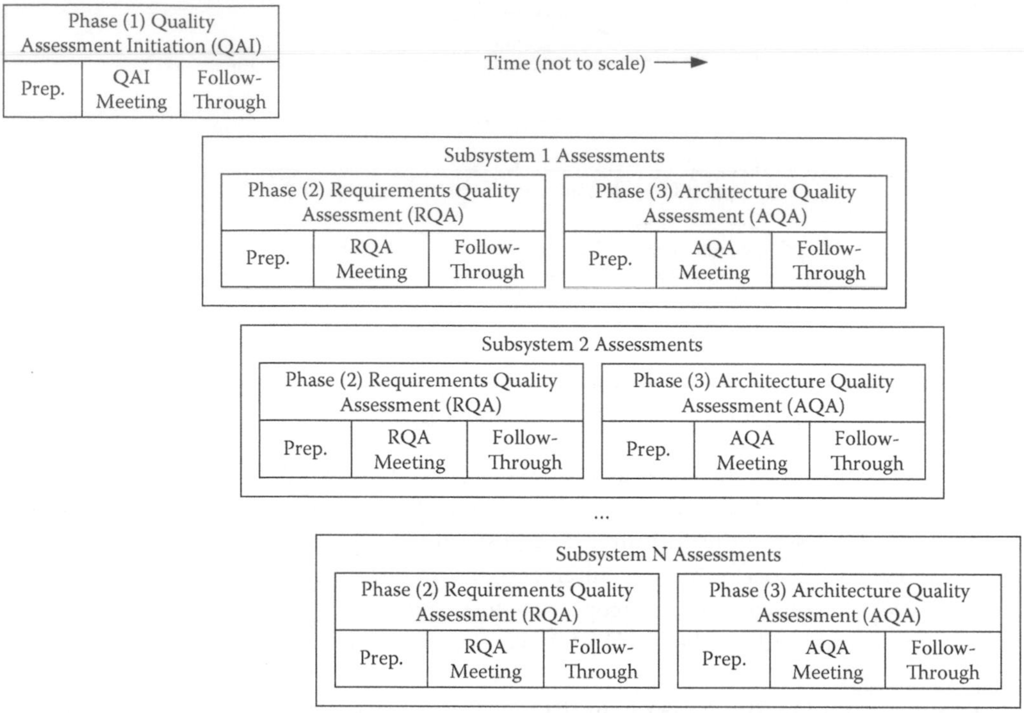

Figure 18.12   QUASAR tasks.

During this phase, the architecture team presents and defends their architecture quality cases, and the assessment team actively assesses the sufficiency of the architecture to meet its derived and allocated quality-related requirements. As with the requirements assessment, the assessment team provides an initial outbrief of the assessment results, followed by a more formal assessment report.

As illustrated in Figure 18.12, each of the three QUASAR phases consists of three similar tasks:

1. **Preparation.** The teams provide and review information to prepare for the coming meetings.
2. **Meeting.** Meetings where plans are collectively made (QAI) or quality assessments are performed (RQA and AQA).
3. **Follow-through.** Outbriefs are given and either meeting minutes or reports documenting the results of the assessment meetings are produced and published.

Figure 18.13 illustrates the primary responsibilities of the assessment, requirements, and architecture teams involved in the assessments. The symmetry between the left and right sides of the diagram shows the similarity between the responsibilities of the requirements teams and the responsibilities of the architecture teams.

Architects and assessors have the following major responsibilities:

■ **Architects** should *make their case* to assessors:
  – The architects *should* know the architecturally significant requirements driving their architecture.

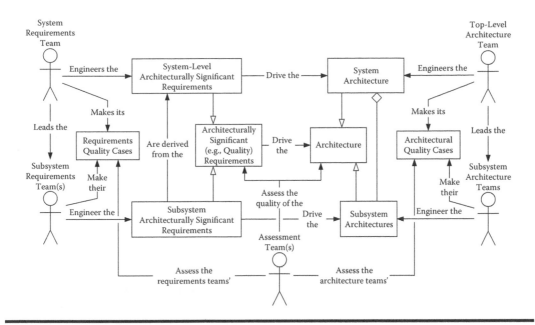

Figure 18.13   QUASAR team responsibilities.

- The architects *should* know the architectural decisions, inventions, trade-offs, and assumptions they made and their associated rationales.
- The architects *should* know where they documented their architectural decisions, inventions, trade-offs, assumptions, and rationales.
- **Assessors' responsibilities.** The QUASAR assessors should *actively* probe the requirements and architectural quality cases to determine the following:
  - **Are the claims sufficient?** Do the claims include all architecturally significant requirements and thus all architectural concerns? Do the claims address all the relevant quality characteristics and quality attributes?
  - **Are the arguments sufficient?** Are the arguments sufficient to justify belief in the claims? Are the arguments correct and complete? Are the arguments clear and compelling? Do the arguments address all relevant quality concerns, including all significant architecturally significant requirements? Do the arguments address all the relevant quality characteristics and quality attributes? Do the arguments include all the relevant architectural inventions, decisions, trade-offs, assumptions, and rationales?
  - **Is the evidence sufficient?** Is the evidence relevant? Is the evidence provided sufficient to support the associated arguments? Are the evidentary documents official, maintained, identified, and controlled, or are they quick-and-dirty unofficial docments produced just for the assessment and not intended to be controlled?
  - **Are the quality cases appropriate for the current point in the schedule?** Are the quality case claims, arguments, and evidence appropriate for the current point in the schedule?* Are any parts of the quality cases inappropriately missing or incomplete?

---

* This is especially important if a recursively incremental, iterative, parallel, and time-boxed development cycle in which the architecture evolves over time.

Table 18.1 QUASAR Assessment Results Matrix

	SYS		AC 1		AC 2		AC 3		AC 4		AC 5		AC 6			AC n	
	R	A	R	A	R	A	R	A	R	A	R	A	R	A		R	A
QC 1	G	Y	G	G	G	G	G		G		X	X					
QC 2	G	G	G	Y	G	G	G		G								
QC 3	Y	Y	X	X	G	Y	Y		X	X							
QC 4	G	G	G	G	G	G	G		G								
QC 5	G	G	G	Y	G	G	G		R		X	X	X	X			
QC 6	G	G	Y	Y	G	Y	G		X	X						X	X
QC 7	R	R	X	X	R	R	R		R								
QC 8	G	G	G	G	G	G	G		G								
QC 9	G	G	G	G	X	X	X	X	G		X	X				X	X
QC m	G	G	X	X	G	G	G		X	X							

Systems are typically so large and complex that it is impossible to effectively and efficiently assess their quality in a single assessment. That is why pairs of requirements quality assessments and architecture quality assessments are typically performed in an incremental (and sometimes iterative) manner, an architectural component at a time. A way is therefore needed to summarize the results of these multiple assessments.

As illustrated in Table 18.1, a QUASAR assessment results matrix is a way to document a summary of the results of these multiple assessments. Rows contain the assessment results for individual quality characteristics (or possibly quality attributes). Columns contain the assessment results for the system or individual architectural components (e.g., subsystems), whereby the columns labeled with an "R" contain the results of requirements quality assessments and columns labeled with an "A" contain the results of architecture quality assessments. To produce a "stoplight chart" that provides a clear overview while preserving the results of the individual assessments, the cells in the matrix are filled in with:

■ **Green (or boldface G if black-and-white printing is intended).** The system or architectural component passes the assessment because of:
  – **Requirements.** All of the quality-related requirements have sufficient quality and are sufficiently complete and mature to drive the architecture.
  – **Architecture.** The architecture has sufficient quality, sufficiently supports the quality-related requirements, and is sufficiently complete and mature to drive the design, implementation, integration, and testing.
■ **Yellow (or boldface Y if black and white printing is intended).** The system or architectural component conditionally passes the assessment because of:
  – **Requirements.** Almost all the quality-related requirements have sufficient quality and are sufficiently complete and mature to drive the architecture. The requirements team can easily and quickly correct the requirements defects and deficiencies.

- **Architecture.** Almost all the architecture has sufficient quality, sufficiently supports the quality-related requirements, and is sufficiently complete and mature to drive the design, implementation, integration, and testing. The architecture team can easily and quickly correct the architectural defects and deficiencies.
■ **Red (or boldface R if black and white printing is intended**). The system or architectural component clearly fails the assessment because of:
    - **Requirements.** Several of the quality-related requirements do *not* have sufficient quality or are missing or *not* sufficiently mature to drive the architecture. The requirements team *cannot* easily and quickly correct the requirements defects and deficiencies.
    - **Architecture.** The architecture clearly does *not* have sufficient quality, will *prevent* the system from supporting at least some of quality-related requirements, and is *not* sufficiently complete and mature to drive the design, implementation, integration, and testing. The architecture team *cannot* easily and quickly correct the architectural defects and deficiencies.
■ **White (or empty).** The assessment is planned but has not yet taken place.
■ **X (or crossed out).** This quality characteristic or attribute is either not relevant or the relevancy is insufficient to justify an assessment. Therefore, no assessment is planned.

The example QUASAR Assessment Results Matrix in Table 18.1 can be read as follows:

■ **Requirements quality assessments completed.** Requirements quality assessments have been held at the system level as well as for architectural components 1, 2, 3, and 4.
■ **Architecture quality assessments completed.** Architecture quality assessments have been held at the system level as well as for architectural components 1 and 2. The architecture quality assessment has not yet taken place for architectural components 3 and 4.
■ **Quality assessments not relevant.** The following quality assessments are not relevant and will not take place:
    - Architectural component 1 will not be assessed for its support for quality characteristics 3, 7, and m.
    - Architectural components 2 and 3 will not be assessed for their support for quality characteristic 9.
    - Architectural component 4 will not be assessed for its support for quality characteristics 3, 6, and 7.
    - Architectural component 5 will not be assessed for its support for quality characteristics 1, 5, and 9.
    - Architectural component 6 will not be assessed for its support for quality characteristic 5.
    - Architectural component n will not be assessed for its support for quality characteristics 6 and 9.
■ **Failed quality assessments.** The following architectural components have failed the following assessments (i.e., support for certain quality characteristics has been evaluated as red):
    - The system has not passed its requirements quality assessment for quality characteristic 7, thereby causing architectural components 2, 3, and 4 to fail their requirements quality assessments for the same quality characteristic. Without good system-level quality requirements for quality characteristic 7 to allocate to architectural components 2, 3, and 4, it was practically impossible for their requirements teams to derive appropriate lower-level quality requirements for quality characteristic 7.

- Similarly, the system and architectural component 2 have failed their architectural quality assessments because of the lack of adequate requirements against which to assess their architectures.
- Architectural component 4 has also not passed its requirements quality assessment for quality characteristic 5.

■ **Conditionally passed quality assessments.** The following architectural components have conditionally passed the following assessments (i.e., support for certain quality characteristics has been evaluated as yellow):

- The system has only conditionally passed its architecture quality assessment for support for quality characteristic 1 because the assessment team had questions that could only be answered during the assessments of architecture components 3 and 4, which have not yet taken place. The system has also conditionally passed the architecture quality assessment for quality characteristic 3.
- Architecture component 1 has conditionally passed its requirements quality assessment for quality characteristic 6 and has conditionally passed its architecture quality assessment for architectural components 2, 5, and 6. The questionable nature of the architecture component's requirements for quality characteristic 6 contributed to the questionable nature of its architectural support for the same quality characteristic.
- Architecture component 2 has conditionally passed its architecture quality assessment for quality characteristics 3 and 6.
- Architecture component 3 has conditionally passed its requirements quality assessment for quality characteristic 3.

18.7.1 Work Products

The following work products are typically produced during the Quality Assessment Initiation (QAI) phase of QUASAR:

■ **QAI preparation task.** The following work products are typically developed and provided to meeting attendees prior to the QAI meeting:

- **QAI agenda.** The top-level assessment, requirements, and architecture teams collaborate to develop and publish the QAI meeting agenda, which includes the meeting dates, times, location(s), and schedule of meeting topics.
- **QUASAR materials.** The assessment team delivers the following to the top-level requirements and architecture teams:
  - QUASAR training materials
  - QUASAR standards and procedures
  - QUASAR templates for the RQA and AQA reports

■ **QAI meeting task.** The following work products are typically developed during the QAI meeting:

- **Attendee notes.** The individual members of the top-level assessment, requirements, and architecture teams take detailed notes during the meeting to be used as input to the QAI outbrief and minutes.
- **Action item list.** The assessment scribe makes a list of action items identified during the QAI meeting.

■ **QAI follow-through task.** The following work products are typically developed following the QAI meeting:

- **QAI outbrief.** The assessment team develops and presents a short outbrief describing the top-level results of the QAI meeting to the requirements team, architecture team, and other invited stakeholders.
- **QAI minutes.** The assessment team develops and publishes meeting minutes documenting the results of the QAI, including the initial scope and schedule of the assessments as well as any agreed upon tailorings of the QUASAR method.
- **Tailored QUASAR materials.** The assessment team tailors and redistributes the tailored:
  - QUASAR training materials
  - QUASAR standards and procedures
  - QUASAR templates for the RQA and AQA reports

The following work products are typically produced during the Architecture Quality Assessment (AQA)* phase of QUASAR:

- **Architecture Quality Assessment (AQA) preparation task.** The following work products are typically developed and provided to meeting attendees prior to the AQA meeting:
  - **AQA agenda.** The assessment team and architecture team collaborate to develop and publish the AQA meeting agenda, which includes the meeting dates, times, location(s), and schedule of meeting topics (e.g., quality characteristics and architectural quality cases to be covered).
  - **AQA presentation materials.** The architecture team develops and presents to the assessment team the materials that the architecture team intends to present at the AQA meeting, including slides providing:
    - A summary of the architecturally significant requirements
    - A high-level overview of the architecture
    - Architectural quality cases, including quality case diagram, claims, arguments, and evidence
- **AQA meeting task.** The following work products are typically developed during the AQA meeting:
  - **Assessor notes.** The individual members of the assessment team take detailed notes during the meeting to be used as input to the AQA outbrief and report.
  - **Action item list.** The assessment scribe makes a list of action items identified during the AQA meeting.
- **AQA follow-through task.** The following work products are typically developed following the AQA meeting:
  - **AQA outbrief.** The assessment team develops and presents a short outbrief describing the top-level results of the AQA to the architecture team and other invited stakeholders.
  - **AQA report.** The assessment team develops and publishes a detailed report documenting the results of the AQA.
  - **Updated QUASAR assessment results matrix.** The assessment team updates the QUASAR assessment results matrix with the results of the AQA.

---

* Similar work products are produced during the preceding Requirements Quality Assessment (RQA) phase, but are outside the scope of this book.

18.8 Guidelines

Use the following guidelines (and associated rationales) when dealing with quality models, quality cases, and quality assessments:

- Use a single common quality model.
- Ensure proper quality requirements.
- Develop architectural quality cases.
- Obtain adequate resources.

The following descriptions express these guidelines in more detail and provide their associated rationales:

- **Use a single common quality model.** Obtain a consensus across the project, including among the acquisition organization and development organizations (such as prime contractor or system integrator and subcontractors) concerning the single common quality model to use. Ensure that everyone uses the same quality characteristics and quality attributes with the same names and meanings. Obtain a consensus between the architects and the assessors regarding the relative importance of these quality characteristics and quality attributes on an architectural component-by-component basis.

    Rationale: Much time and effort can be wasted by stakeholders arguing about the meanings of the individual quality characteristics, their attributes, and their measurement scales. These meanings must be unambiguous for the associated quality requirements to be unambiguous so that the architects and assessors can agree as to the results of the quality assessments. Although several of the major quality characteristics may well be important across all architectural components, the relative criticality of different quality characteristics and attributes varies from architectural component to architectural component.

- **Ensure proper quality requirements.** Technical management, the requirements teams, and the architecture teams must collaborate to ensure that the quality requirements have the proper characteristics of good requirements. This includes, but is not limited to, ensuring that the quality requirements are clear, complete, feasible, verifiable, and unambiguous. Ensure that the quality requirements have well-defined and unambiguous thresholds along some appropriate measurement scale.

    Rationale: The architects need proper quality requirements to drive the development of the architecture, to perform engineering trade-offs between competing quality characteristics and attributes, and to know when the architecture is sufficient to meet the architecturally significant requirements. Testers also need proper quality requirements to properly test the architecture during integration testing, system testing, and specialty engineering testing (such as reliability, safety, and security testing).

- **Develop architectural quality cases.** Managers, technical leaders, quality engineers, and quality assessors (such as acquisition representatives) need to ensure that the architecture teams develop architectural quality cases for all relevant quality characteristics and associated quality attributes as a natural part of the architecture engineering process. This includes all quality characteristics and attributes for all architectural components (e.g., subsystems) having architecturally significant requirements.

    Rationale: The architecture teams need architectural quality cases to demonstrate (i.e., make their case) that their architecture sufficiently supports the achievement of its associated quality requirements. Quality engineers and quality assessors need architectural quality cases to properly determine whether the architecture sufficiently supports the quality requirements. Certifiers need quality cases to provide them with some of the arguments and evidence they need to certify that the system meets its quality requirements (e.g., is sufficiently safe and secure to use).

■ **Obtain adequate resources.** Obtain sufficient resources (e.g., staffing, schedule, and access to architects and evidence) for developing the architectural quality cases and for performing the associated architecture quality assessments. Make the architectural quality cases and assessments a part of the contract and document their development and performance in plans and schedules.

Rationale: Without adequate resources, it becomes difficult for requirements engineers to engineer quality requirements, for the architects to develop architectural quality cases, and for assessors to properly assess architectural support for the quality requirements.

18.9 Pitfalls

Avoid the following common pitfalls associated with *Architecture and Quality* and mitigate their negative consequences when they occur:

■ The endeavor lacks a quality model.
■ The endeavor uses a poorly defined and poorly documented quality model.
■ The definitions of the quality characteristics and attributes are inadequate.
■ There is disagreement over the relative importance of the quality characteristics and attributes.
■ The quality requirements lack thresholds.
■ The quality cases contain inappropriate evidence.
■ Disagreement exists over the need to perform quality assessments.
■ Disagreement exists over the scope of the quality assessment.
■ There is no quality case index and summary.

The following express these pitfalls in more detail, describing their negative consequences and the steps the architects can use to mitigate them:

■ **The endeavor lacks a quality model.**
  – **Pitfall.** The endeavor does not use even a single quality model to name and define quality characteristics and their associated attributes and measurement scales.
  – **Negative consequences.** This pitfall often causes the following negative consequences:
    • The stakeholders disagree over which quality characteristics and attributes to use, the meanings of these quality characteristics and attributes, the associated quality measurement scales to use, and the quality measurement methods to use.
    • The quality requirements become ambiguous.
    • Architects do not know how to properly perform architectural trade-offs between quality characteristics and attributes.
    • Testers do not know how to test for sufficient architectural (and system) support for quality requirements (e.g., how many test cases are sufficient).

- The project schedule slips and therefore funding is wasted as stakeholders argue over which quality characteristics and attributes to use, the meanings of these quality characteristics and attributes, the associated quality measurement scales to use, and the quality measurement methods to use.
  - **Mitigations.** To avoid or mitigate the negative consequences of this pitfall:
    - Use a single *well-documented* quality model such as the one in this book or another standard quality model.
    - Representatives from the acquisition and development organizations collaborate to produce and develop a consensus regarding the project quality model.
    - Incorporate the quality model into the endeavor development method and associated documentation (e.g., quality engineering documentation). Where appropriate, also incorporate the quality model into architectural plans (e.g., if not documented elsewhere).
    - Develop training materials that teach the project quality model and how to use it to engineer quality requirements, how to select architectural patterns for improving architectural support for the quality characteristics and attributes, and how to assess the architecture against its derived and allocated quality requirements.

■ **The endeavor uses a poorly defined and poorly documented quality model.**
  - **Pitfall.** The endeavor quality model is poorly defined and documented:
    - The model is incomplete, missing useful characteristics, attributes, and measurement scales.
    - The quality characteristics and quality attributes are poorly defined or documented (e.g., the definitions are ambiguous, too general and inclusive, or too specific and not inclusive enough).
    - The relationships between the quality characteristics and quality attributes are poorly or incorrectly defined or documented.
    - The quality attributes do not have associated measurement scales or the measurement scales are poorly defined or documented.
    - The quality measurement methods for the measurement scales are poorly defined or documented.
    - The quality characteristics, attributes, measurement scales, and measurement methods are ambiguous.
  - **Negative consequences.** This pitfall often causes the following negative consequences:
    - The stakeholders disagree over which quality characteristics and attributes to use, the meanings of these quality characteristics and attributes, the associated quality measurement scales to use, and the quality measurement methods to use.
    - Requirements engineers cannot engineer unambiguous quality requirements.
    - The ambiguous quality requirements make it difficult for the architects to know when their architecture is good enough and to properly perform architectural trade-offs between the quality characteristics and attributes.
    - Testers do not know how to test for sufficient architectural (and system) support for quality requirements.
    - The project schedule slips and therefore funding is wasted as stakeholders argue over which quality characteristics and attributes to use, the meanings of these quality characteristics and attributes, the associated quality measurement scales to use, and the quality measurement methods to use.
  - **Mitigations.** To avoid or mitigate the negative consequences of this pitfall:

- Use a *well-documented* quality model such as the one in this book (which is an updated version derived from that in [Firesmith et al., 2006; OPF, 2008]) or possibly another standard quality model such as [IEEE, 1998] or [ISO/IEC, 2001].*
- Representatives from the acquisition and development organizations collaborate to produce and develop a consensus regarding the project quality model.
- Incorporate the quality model into the endeavor development method and associated documentation (e.g., quality engineering documentation). Where appropriate, also incorporate the quality model into architectural plans (e.g., if not documented elsewhere).
- Verify the quality and completeness of the endeavor quality model.
- Develop training materials that teach the project quality model and how to use it to engineer quality requirements, how to select architectural patterns for improving architectural support for the quality characteristics and attributes, and how to assess the architecture against its derived and allocated quality requirements.

■ **The definitions of the quality characteristics and attributes are inadequate.**
- **Pitfall.** The definitions of the quality characteristics and quality attributes are incomplete or ambiguous.
- **Negative consequences.** This pitfall often causes the following negative consequences:
  - Although the stakeholders use the same quality characteristics and quality attributes, they disagree over the meanings of these quality characteristics and attributes. The stakeholders who disagree may belong to requirements and architecture teams or the acquisition and development organizations.
  - The quality requirements become ambiguous, and the stakeholders disagree on their meanings and whether the architecture adequately supports these requirements.
  - The project schedule slips and therefore funding is wasted as representatives from the acquisition and development organizations argue over the meanings of the terms and the implications of these meanings on appropriate verification approaches and the degree to which the architecture passes quality assessments.
- **Mitigations.** To avoid or mitigate the negative consequences of this pitfall:
  - Use a well-documented quality model such as the one in this book or another standard quality model.
  - Representatives from the acquisition and development organizations shoud collaborate to produce and develop a consensus regarding the project quality model.
  - Brief definitions of the relevant quality characteristics and quality attributes should be enhanced with additional text further explaining their meanings as well as examples.
  - Training materials should be developed that teach the project quality model and how to use it to engineer quality requirements, architectural patterns for improving architectural support for the quality characteristics and attributes, and how to assess the architecture against its derived and allocated quality requirements.

■ **There is disagreement over the relative importance of the quality characteristics and attributes.**
- **Pitfall.** The acquisition and development organizations disagree over the relative importance of the quality characteristics and their associated quality attributes and quality requirements.

---

* This book uses the quality model from the OPEN Process Framework Repository Organization (OPFRO) because of its completeness, both in terms of the number of quality characteristics and attributes as well as the relationships between them. You may have valid reasons to use other quality models.

- **Negative consequences.** This pitfall often causes the following negative consequences:
  - The two organizations disagree over the scope of the quality assessments.
  - The assessments are incomplete in that architectural quality cases for certain appropriate quality characteristics are not produced and not assessed.
- **Mitigations.** To avoid or mitigate the negative consequences of this pitfall:
  - Base the scope of the quality cases on the relevant customer and derived quality requirements allocated to the architectural component being assessed. Note that this requires a corresponding assessment of the requirements to ensure that quality requirements are properly engineered.
  - Document in the assessment method the escalation process for dealing with disagreements. The acquisition organization should have final say (i.e., "the customer is always right").

■ **The quality requirements lack thresholds.**
  - **Pitfall.** The quality requirements do not have explicit thresholds on specific measurement scales.
  - **Negative consequences.** This pitfall often causes the following negative consequences:
    - The quality requirements become ambiguous.
    - The lack of explicit thresholds makes it difficult for the architects to know when their architecture is good enough and to properly perform architectural trade-offs between competing quality characteristics and attributes.
    - It is difficult or impossible to know whether the arguments sufficiently justify belief in the quality case claims and whether the evidence sufficiently supports the arguments.
  - **Mitigations.** To avoid or mitigate the negative consequences of this pitfall:
    - Train the requirements engineers in how to engineer unambiguous and verifiable quality requirements.
    - Verify the quality requirements to ensure that they contain explicit thresholds.
    - Include architects (and testers) in the requirements verification teams.
    - Validate the quality requirements with the stakeholders to ensure that their thresholds are correct.

■ **The quality cases contain inappropriate evidence.**
  - **Pitfall.** The architects provide evidence that is not relevant. Evidence that is part of a quality case should show the actual current state of the architecture, not the architects' good intentions concerning the future of the architecture. Thus, the following are *not* acceptable types of evidence: an architecture plan, architecture engineering method documentation (e.g., standards, procedures, and templates), architecture team organizational charts (despite Conway's law), the project work breakdown schedule (WBS), master schedule, and the local component schedule. To be credible, the documentation should be official project documentation and not temporary informal slides developed specifically for an architecture quality assessment.
  - **Negative consequences.** This pitfall often causes the following negative consequences:
    - The evidence does not support the arguments.
    - The architectural quality case does not justify belief that the architecture sufficiently supports the achievement of the quality requirements.
    - Stakeholders lose trust in the architects and their architecture.
  - **Mitigations.** To avoid or mitigate the negative consequences of this pitfall:
    - Explicitly define what is meant by proper evidence in the conventions documenting architectural quality cases.

- Provide examples of proper and improper evidence.
- Provide training in how to develop and provide proper evidence.
- Provide initial consulting to the architects to ensure that the intended evidence is proper.
- Verify the quality of the evidence before commencing the assessment. Either postpone the architecture assessment until such time as the quality of the evidence is deemed sufficient or fail the assessment, thereby providing a strong incentive to provide better evidence in the future.

■ **Disagreement exists over the need to perform quality assessments.**
  - **Pitfall.** The acquisition and development organizations disagree over the necessity to perform independent quality assessments. The development organization feels that their current informal evaluation process is adequate, even when it is not.
  - **Negative consequences.** This pitfall often causes the following negative consequences:
    - Independent quality assessments are either delayed or not held.
    - The development organization docs not provide sufficient resources to hold an effective assessment (e.g., access to the architects, the architectural representations, and the architecturally significant requirements).
    - The architects do not spend the time needed to provide proper architectural quality cases but rather supply only existing documentation, whether or not it is relevant.
  - **Mitigations.** To avoid or mitigate the negative consequences of this pitfall:
    - Ensure that independent architectural quality assessments are included in the development contract, project budget, and project schedule.
    - Mandate architectural quality assessments as the method of verifying architecturally significant quality requirements.
    - Use contract incentives (e.g., award fees) to convince the developers to support proper quality assessments of the architecture.

■ **Disagreement exists over the scope of the quality assessment.**
  - **Pitfall.** The acquisition and development organizations disagree over the scope of the assessments in terms of the:
    - Selection (i.e., number and identity) of the architectural components to assess
    - Number and identity of the quality characteristics and quality attributes against which to assess (i.e., which quality cases to assess)
    - Tailoring of the assessment method (i.e., which tasks to perform, which assessment work products to produce, and who is responsible for performing these tasks and producing these work products)
  - **Negative consequences.** This pitfall often causes the following negative consequences:
    - The project schedule slips and therefore funding is wasted as representatives from the acquisition and development organizations argue over the scope of the architecture quality assessments.
    - Some of the architectural components that should be assessed are not assessed.
    - The assessments are incomplete in that architectural quality cases for certain quality characteristics are neither produced nor assessed.
    - The architecture quality assessment method is not optimum to meet the needs of all stakeholders.
  - **Mitigations.** To avoid or mitigate the negative consequences of this pitfall:
    - Determine the number and identity of the architectural components to assess based on the priority of the allocated and derived requirements, the criticality of the

architectural component, project risk due to architectural failures, and the available resources with which to perform the assessments.

- Determine the number and identity of the quality characteristics and quality attributes against which to assess based on the priority of the allocated and derived quality requirements and associated architectural concerns.
- Obtain consensus and documented agreement concerning the scope of the assessments early in the development cycle before a significant amount of the architecture is engineered.

■ **There is no quality case index and summary.**
- **Pitfall.** Although their architectural quality cases are very large and complex, containing many claims, arguments, and evidence, the architects do not provide any index to or summary of their quality cases.
- **Negative consequences.** This pitfall often causes the following negative consequences:
  - Assessors, maintainers, and other stakeholders of the architectural quality cases cannot verify the sufficiency and completeness of the quality cases.
  - Assessors unnecessarily take excessive time to prepare for architectural quality assessments.
  - Assessors find fewer architectural defects, weaknesses, and risks.
- **Mitigations.** To avoid or mitigate the negative consequences of this pitfall:
  - Incorporate quality case diagrams into the architecture engineering method.
  - Architects should develop quality case diagrams for each architectural quality case.

18.10 Summary

Quality models are used to define the meanings of different types of quality, to engineer associated quality requirements, to enable architects to engineer architectures supporting these requirements; and, to enable architects to develop associated architectural quality cases. A quality model is composed in terms of a classification hierarchy of quality characteristics, their quality attributes (i.e., parts of characteristics), associated quality measurement scales, and quality measurement methods. It is critically important to have and use an appropriate quality model (e.g., the quality model includes all the quality characteristics and attributes that are important on the project).

Quality requirements specify required amounts of a given quality characteristic or quality attribute. These requirements can be specified in terms of a set of conditions that state when the requirement holds, a quality criterion that documents system or subsystem support for the quality characteristic or quality attribute, and a quality threshold that specifies a minimum amount of the quality characteristic or quality attribute. Cohesive sets of quality requirements are collected into architectural concerns that drive the engineering of the architecture.

Architects create architectural quality cases to document how their architecture supports its allocated and derived quality requirements. Architectural quality cases consist of (1) *claims* that the architecture meets its quality requirements; (2) clear, compelling, and relevant *arguments* justifying belief in these claims; and (3) supporting relevant and credible evidence. The architectural arguments consist of the architects' decisions, inventions, trade-offs, assumptions, and associated rationales. The architectural evidence consists largely of project official architectural representations (e.g., models and documents), although it may also include demonstrations witnessed by assessors.

Because of the critical importance of documenting architectural support for the quality requirements, it is important to assess the quality, completeness, and maturity of the architectural quality cases. The QUality Assessment of System Architectures and their Requirements (QUASAR) is one method that has been successfully used to perform these assessments.

Chapter 19

Conclusions

19.1 Introduction

MFESA is a method framework for producing project-specific system architecture engineering methods. It incorporates a consistent set of architectural engineering best practices based on sound engineering principles. When used properly, MFESA helps architects and architecture teams meet the many challenges facing them.

After summarizing the MFESA method framework, including its four main components, and emphasizing its ten reusable tasks, this chapter lists the book's most important points to remember. It concludes with a brief discussion of the authors' future plans for MFESA.

19.2 Summary of MFESA

MFESA is not a single, general-purpose system architecture engineering method. Instead, it is a system architecture engineering method framework for generating appropriate project-specific methods. The following subsections briefly discuss its four major components. Because of their central importance within the repository of reusable method components, this chapter also summarizes the ten reusable MFESA tasks.

19.2.1 MFESA Components

As illustrated in Figure 19.1, MFESA consists of the following four primary components:

1. **MFESA ontology of concepts and terminology:** an information model defining a consistent set of interrelated concepts and terms underlying system architecture engineering.
2. **MFESA metamodel:** a model of the foundational abstract types of method components (see Figure 4.3) in the repository of reusable method components (i.e., architectural work products, architectural work units, and architectural workers).

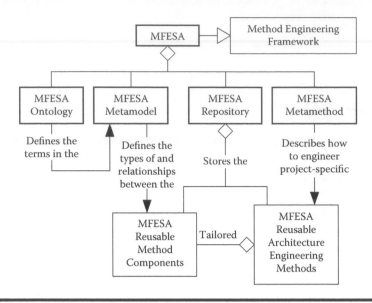

Figure 19.1   The four primary components of MFESA.

3. **MFESA repository of reusable method components:** a repository containing a consistent class library of reusable components for creating project-specific methods for engineering system architectures. The repository may also store reusable MFESA-compliant methods.

4. **MFESA metamethod:** a method for creating effective and efficient project-specific methods for engineering system architectures (i.e., a method for creating methods).

19.2.2 Overview of the MFESA Tasks

The ten MFESA tasks are central to the repository of reusable method components from which architects and process engineers construct their project-specific system architecture engineering methods. Whereas it is the architecture work products that are the critical outputs of any MFESA method, it is the tasks that tie together the work products and the work units that produce them. As illustrated in Figure 19.2, the MFESA repository includes the following tailorable tasks that are typically performed in a highly recursively incremental, iterative, parallel, and time-boxed manner:

Task 1: Plan and resource the architecture engineering effort. Plan the overall architecture engineering effort and obtain the necessary resources to engineer the system architecture.

Task 2: Identify the architectural drivers. Identify the architecturally significant product and process requirements that have been derived for and/or allocated to their system or architectural component and categorize them into a set of architectural concerns.

Task 3: Create the first versions of the most important architectural models. Create the first partial versions of the most important architectural models. This typically consists of a consistent set of partial draft logical and physical, as well as static and dynamic models of the system or architectural component based on the architectural drivers, the associated architectural concerns, and the opportunities for reuse.

Task 4: Identify opportunities for the reuse of architectural elements. Identify and initially analyze potential opportunities for the reuse of architectural decisions and inventions such as styles, patterns, and the relevant parts of existing architectural structures.

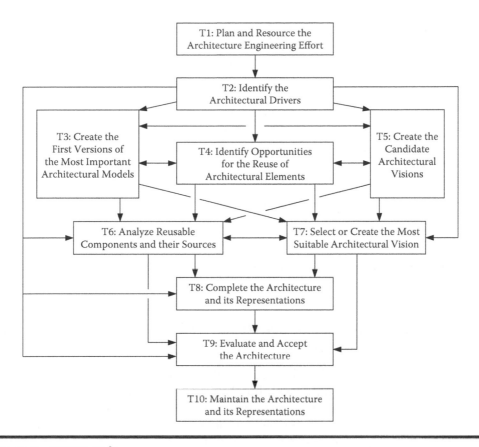

Figure 19.2 MFESA tasks.

Task 5: **Create the candidate architectural visions.** Use the initial architectural models and opportunities for reuse to create a set of competing candidate architectural visions for their system or architectural components that support meeting the derived and allocated architectural drivers and associated architectural concerns.

Task 6: **Analyze reusable components and their sources.** Identify and evaluate potentially reusable physical architectural components (and their sources) for reuse within the candidate architectural visions.

Task 7: **Select or create the most suitable architectural vision.** Select the most suitable architectural vision for the system or architectural components from the competing candidate architectural visions. If combining consistent parts of multiple competing candidate architectural visions would yield an even more suitable architectural vision, then do so to obtain the single new architectural vision.

Task 8: **Complete the architecture and its representations.** Complete the system architecture and its representations based on the architectural vision selected or created.

Task 9: **Evaluate and accept the architecture.** Evaluate the quality of the system or subsystem architecture so that architectural risks can be managed, compliance with architecturally significant requirements can be determined, and the architecture can be accepted by its authoritative stakeholders.

Task 10: **Maintain the architecture and its representations.** Maintain the architecture and its representations. Ensure that the integrity of the system architecture does not degrade over time.

19.3 Key Points to Remember

When architecting systems, the following key points made by this book should be remembered and followed:

- System architecture and system architecture engineering are critical to success.
- MFESA is *not* a system architecture engineering method, but rather a framework for constructing appropriate, project-specific system architecture engineering methods.
- Quality is key.
- Architectural quality cases make the architects' case that their architecture sufficiently supports the architecturally significant requirements.
- It is critical to capture the rationales for architectural decisions, inventions, and trade-offs.
- Architects should keep their work at the right level of abstraction.
- Reuse has a major impact on system architecture engineering.
- Architecture engineering is never finished.
- Beware of ultra-large systems of systems.

19.3.1 System Architecture and System Architecture Engineering Are Critical

The system architecture is critical to the success of the system and system development project because of its major impact on the cost and schedule of the project and the functionality and quality of the system. It is critical to the communication between the system stakeholders. It drives the downstream effort, including design, implementation, integration, and integration testing. Finally, it has a major impact on reuse within the system.

The use of system architecture engineering is also critical in engineering a good system architecture. As shown by a survey of 45 defense systems, a project's use of architectural engineering best practices has a major positive impact on the project's success [Elm et al., 2007]. The need for an appropriate system architecture engineering method is due to the extreme size and complexity of many of today's systems, the lack of sufficient experienced and trained architects, and the rapid change in technologies incorporated into systems, especially the rapid increase in the amount of software and the percentage of system functionality provided by software. Without adequate system architecture engineering methods, architectural defects tend to be discovered late in the development cycle, often during system integration when they are difficult and expensive to fix.

Because a project's use of architectural engineering best practices has a major positive impact on project success, system architecture engineering and the system architecture teams should not be raided for resources when the going gets rough (i.e., when projects exceed their budgets and schedules).

19.3.2 MFESA Is Not a System Architecture Engineering Method

Although it can be used to generate general-purpose reusable system architecture engineering methods, this is not what MFESA was designed for. Because of the major differences that exist between systems and the projects and organizations that engineer them, no single system architecture engineering method is optimal (or even appropriate) for all system development projects. This is why MFESA is *not* a system architecture engineering method. Rather, it can be viewed as a metamethod, which is a method for generating situation-specific methods for engineering system architecture engineering.

The power of MFESA comes largely from its extensive repository of reusable architectural engineering method components. Just like the developers of modern object-oriented programming languages engineer the well-defined classes in their class library to be consistent and useful as a foundation for programming software applications, the developers of the MFESA repository have created an equally consistent library of architecture engineering method components to act as a foundation for producing architecture engineering methods. Because typical object-oriented programming languages such as Java and C++ are general purpose, their associated class libraries only contain very generic low-level classes, forcing programmers to extend the preexisting class library by using inheritance to produce many new subclasses of the existing classes. However, because MFESA is restricted to system architecture engineering, its repository of method components is much more complete so that architects and process engineers typically do not need to extend the repository with many new method components.

Just as software engineers select appropriate programming language classes as the base classes on which to engineer software programs, system architects and process engineers select the appropriate method components from which to construct their architecture engineering methods. Programmers typically need to extend their base classes via inheritance, thus creating ever more complex classes as they move down the inheritance tree. On the other hand, architects and process engineers will typically achieve flexibility by tailoring out the unnecessary method components and parts of method components to engineer system architecture methods consisting of smaller and simpler method components. Finally, whereas programmers use linkers to integrate their classes into complete software applications, architects and process engineers link their tailored method components into complete system architecture engineering methods.

When used properly, MFESA achieves two opposing goals. MFESA obtains the benefits of *standardization* because the resulting system architecture engineering methods are constructed from a common repository of preexisting reusable method components that were designed to be consistent and work well together. MFESA also obtains the benefits of *flexibility* by using selection and tailoring to create methods that are appropriate and project specific. Thus, MFESA achieves the benefits of both standardization and flexibility.

19.3.3 Quality Is Key

The word "quality" itself is subject to qualification. While people generally presume that when one says "quality workmanship" one means "high-quality workmanship," the word "quality" actually demands a sound definition. What does "high quality" mean, after all? MFESA provides a mechanism by which system qualities are systematically characterized by decomposition into quality characteristics and measurable attributes. These characteristics and their attributes require stakeholders and architects to define their intentions operationally across the entire span of qualities affecting the ultimate system. From the conscious management of system qualities, the architects can determine which among many quality characteristics best suit the intended purpose of the system. Obviously, this perspective has a sweeping influence in the decision-making processes affecting aspects of the system, large and small.

Only by the addition of measures by which the extent of any of these qualities may be judged can it be determined to what extent any of them serves to drive the system's ultimate capabilities. Likewise, only with a method that is geared to surfacing the measurable aspects of these attributes as a matter of conscious consideration, negotiation, and management by stakeholders do they become anything other than mere words. MFESA provides tools by which qualities of the system are made explicit, observable, and manageable aspects of the system architecture.

19.3.4 Architectural Quality Cases Are Important

The use of architectural quality cases is central to MFESA. Architectural quality cases are based on the concept of safety cases from the safety community. They are generalizations of safety cases because they address all relevant quality characteristics and quality attributes. They are also specializations of safety cases because they include only architecturally relevant information. MFESA architectural quality cases are taken from the QUality Assessment of System Architectures and their Requirements (QUASAR) method for assessing system requirements and architecture [Firesmith et al., 2006].

Well-documented and complete architectural quality cases consist of:

- **Claims.** Architecture quality cases should contain claims that the architecture sufficiently supports meeting architectural concerns.
- **Arguments.** Architecture quality cases should contain clear and compelling arguments (including architectural decisions, inventions, trade-offs, assumptions, and rationales) that are sufficient to justify belief in these claims.
- **Evidence.** Architecture quality cases should also contain sufficient evidence (architectural representations) to support these arguments.

System architects on development projects are often extremely busy and have no time for work that is not useful and cost-effective. Architectural quality cases should *not* be viewed as an administrative add-on to their already burdened schedule. Rather, the development and maintenance of architectural quality cases should be a natural part of their architectural engineering method. By developing architectural quality cases, architects obtain highly valuable documentation that supports:

- **Architecture documentation.** An architectural quality case captures how and to what degree an architectural component meets a quality concern (i.e., a cohesive set of architecturally significant requirements). Architectural quality cases collect together critically important architectural information that is typically scattered across architectural models and views, thereby documenting focus areas specific to individual quality characteristics and attributes and making them understandable.
- **Architecture assessment.** Architectural quality cases provide the foundation for assessments of the architecture's support for the quality requirements derived and allocated to the architectural component being assessed.
- **Architecture risk management.** A commonly occurring architectural risk is that the architecture will not adequately support its derived and allocated quality requirements. By developing and assessing architectural quality cases, such architectural risks are identified and managed.
- **Certification and accreditation.** Architectural safety and security cases provide input to the overall safety and security cases needed for safety and security certification and accreditation. If certification and accreditation are required in the future for other quality characteristics, the associated architectural quality cases can supply needed information.

19.3.5 Capture the Rationales

Hopefully, it is obvious by now that documenting the rationale for architectural decisions, inventions, trade-offs, and assumptions is critical. Architectural rationales make decisions, inventions, trade-offs, and assumptions understandable. Rationales provide backward traceability, a "trail of bread crumbs" that enables new architects to see where the existing architecture came from. As requirements and assumptions change, rationales help the architects assess their impacts. The existence of rationales greatly improves maintainability and extensibility, while minimizing the accidental future incorporation of architectural changes that weaken the architecture or make it inconsistent and therefore defective. Keeping a structured information base of architectural decisions serves as a system and organizational "memory" and supports architectural quality cases, architectural evaluation, architectural maintenance, and ensuring architectural integrity.

19.3.6 Stay at the Right Level

Architects should restrict their work to the correct level in the architecture. That said, understanding the nature of *architecting* becomes an essential skill. Throughout MFESA (and most sources related to system architecture), *communication* is shown to be of paramount concern. Stakeholders, including the acquisition representatives, requirements engineers, system architects, and others, influence the cascade of decisions that flow generally from the top down. Clarity as to the level and the role of the decision maker is essential. The topic is too sweeping to address in any detail here, but from the MFESA perspective, we advocate that architects not allow themselves to indulge in the temptation to inappropriately constrain the design by focusing their attention on nonarchitectural concerns. Except to ensure architectural integrity, architects should concentrate on their own architectural scope (e.g., the system and its architectural components) and not try to do the work of other disciplines and teams.

19.3.7 Reuse Significantly Affects Architecture Engineering

Reuse has a major effect on the system architecture engineering effort. Although reuse has become a byword for efficiency and cost savings, the reaction of experienced technical people is typically skepticism. The MFESA perspective on reuse is that in no way can reuse be taken as an assumption of goodness. Depending on the manner in which it is addressed, reuse can have either positive or negative effects on the system and its architecture. On the positive side, legacy resources can, of course, have the potential to provide significant cost and time savings for the architecture team and the project as a whole. On the other hand, reusable assets improperly identified, analyzed, and reused can have a significant negative impact on cost, schedule, and the meeting of requirements, especially quality requirements specifying minimum acceptable levels of critical quality characteristics and attributes.* Reuse must always take place under consideration of "fitness to purpose."

During the development of MFESA, proper incorporation of reuse into system architecture engineering was one of the main drivers. MFESA primarily deals with reuse during the following tasks and steps:

Task 4: Identify opportunities for the reuse of architectural elements. The architects identify and initially analyze the potential opportunities for the reuse of architectural decisions

---

* The Ariane 5 disaster caused by the improper reuse of a software component from the Ariane 4 is a good example.

and inventions such as styles, patterns, and the relevant parts of existing architectural structures. As the task name implies, the architects address reuse during all the steps of this task.

Task 6: Analyze reusable components and their sources. The architects identify and evaluate potentially reusable physical architectural components and their sources for reuse within the candidate architectural visions. As the task name implies, the architects also address reuse during all the steps of this task.

Task 8: Complete and maintain the architecture. The architects complete and maintain the architectural models and other architectural representations of their system or subsystem architecture based upon the architectural vision selected. During this task, the architects perform the *Address the Remaining Architectural Reuse Issues* step, which includes the following:

- **Select actual reusable components and their sources.** Based on the analysis of trade studies, the architects select the actual reusable architectural components and their sources such as product vendors, the government (GOTS), the military (MOTS), and open-source organizations.

- **Acquire actual reusable components from their sources.** Working with project management, the architects acquire the selected reusable architectural components from their sources.

- **Baseline actual reusable components.** Working with the configuration management organization, the architects identify the acquired reusable architectural components and any associated documentation as configuration items and place them under configuration control.

- **Evaluate individual acquired reusable components.** The architects evaluate (e.g., via testing) the individual acquired reusable components to ensure that they have the required characteristics (such as functionality and levels of quality characteristics and quality attributes).

- **Architect wrapping of reused components.** The architects architect any wrappers or proxies for the reused components so that they fit within the existing architectural structures.

19.3.8 Architecture Is Never Finished

A system's architecture begins with its initial concept consisting of one or more candidate competing architectural visions. The architecture then grows through the selection of the winning vision and completion of the corresponding architecture. During the life cycle of the system, architecturally significant system requirements are added, modified, or removed, and the system architecture must evolve to meet these changes. All or part of the system architecture may be incorporated into reusable reference architectures and live on, even after the original system is retired.

Different architectural tasks having different objectives and requiring different levels of effort are performed at different times.

Thus, although architects strive to produce a stable architecture that provides a solid foundation supporting extensibility and modifiability, the architecture is never actually finished. Neither is the work of the system architects, even if the individual architects making up the system architecture team also change over time.

19.3.9 Beware of Ultra-large Systems of Systems

The challenges facing architects increase greatly in both number and magnitude as they move from individual systems to systems of systems and then on to ultra-large systems of systems. As pointed out in Figure 3.1, these three general categories of size and complexity have a major impact on the success of the system architecture engineering effort. Individual first-generation, general-purpose standards and methods were reasonably able to cope with the architecture of the smaller and simpler individual stand-alone systems. As size and complexity increases to the larger and more complex systems as well as systems of systems, one needs to use a second-generation method framework to produce a situation-specific method if one hopes to be reasonably successful in terms of the architecture's support for producing a system on time, within budget, and with the needed functionality and quality. In a word, one needs MFESA.

However, as we move on to architecting ultra-large systems of systems, *no one* yet knows the best ways to create system architecture engineering methods because no one has had any success in producing such ultra-large systems of systems on time, within budget, and with the needed functionality and quality. It may turn out that using a method framework such as MFESA is the best approach; after all, such systems of systems are even more likely to need a multiple set of situation-specific architecture engineering methods. On the other hand, an entirely new approach may be needed to architect such ultra-large systems of systems. Only considerably more time, research, and experience will tell. The only thing that is certain is that we will be engineering such systems of systems and they will have the largest and most complex architectures that we have ever attempted to engineer.

19.4 Future Directions

Predicting the future is a risky business. Like other authors who have tried to do this before us, we risk looking rather foolish a few years from now when the reader will be able to tell just how close we came either to or from the mark. First, we look at the direction we see system architecture engineering is tending toward. Then we do the same with MFESA, listing two directions we would like to see it take.

19.4.1 The Future Directions of System Architecture Engineering

It is always difficult to foresee the future, even based on clear trends in the past and present. Whereas our vision is often close to 20/20 when gazing into the past, vision rapidly blurs when looking into the future. Although authors of this book, we may well possess no better crystal ball than many of its readers. And yet, the temptation to peek behind the curtain of the future is irresistible. We briefly look in the following two directions:

1. Trends in systems and system engineering
2. Trends in system architecture engineering, architects, and tools

19.4.1.1 Trends in Systems and System Engineering

To predict the future directions of system architecture engineering, it is best to look for relevant trends in systems and system engineering. There are many such trends that are having major impacts on architecture engineering, and the impact of these trends will only increase as these trends continue and accelerate:

- **System size and complexity.** As discussed in Chapter 2 *System Architecture Engineering Challenges*, many systems are growing in size and complexity as they increase in functionality and must interoperate with growing numbers of external systems. The traditional, general, reusable system architecture engineering methods and standards currently in use are less effective and are becoming increasingly insufficient for engineering the architectures of such systems. Even more powerful system architecture engineering method frameworks such as MFESA may well not scale up to the ultra-large systems currently planned or under development.

- **Systems of systems (SOS).** The trend is from individual systems to systems of systems to ultra-large systems. These may be cohesive systems of systems engineered by a single development organization for a single acquisition organization such as a military aircraft, its ground support systems, and its training systems and flight/maintenance simulators. They may also be product lines of related systems or highly configurable systems engineered by a single development organization for different acquisition organizations and user communities such as different models or variants of automobile or aircraft. They may also be ultra-large systems of systems independently engineered by different development organizations for different acquisition and user organizations that evolve quasi-independently of each other such as major transportation systems of systems (e.g., aircraft, air traffic control systems, and airports). Note that in this later case, there is no single development project, no single set of system requirements to drive the overall SOS architecture, no single master schedule, certainly no single set of consistent system (and architecture) engineering methods, and no overall SOS chief architect or top-level architecture team.

- **One-size-fits-all to customized.** Initially, systems have tended to be either unique or one-size-fits-all if more than one instance of the system was needed. However, the strong trend for future systems is to be customized (e.g., personalized for the different users or internationalized for users from different countries). Systems are also tending to be produced in multiple variants as part of product lines. This greatly increases the importance of configurability (to multiple configurations or variants) and extensibility (as new needed configurations are identified).

- **Increasing incorporation of software.** As mentioned in Chapter 2 *System Architecture Engineering Challenges*, software is becoming a critically important part of almost all systems, so much so that the adjective software-intensive when used to describe systems is becoming redundant and unnecessary. The amount of software in systems rapidly increasing (see Figures 2.2, 2.3, and 2.4), and software is being used to implement an ever increasing percentage of the functionality of systems (see Figure 2.5). Software is also becoming the primary glue to integrate subsystems within a system and make separate systems interoperable.

- **Increasing duration of system operational lifecycle period.** Whereas some consumer systems (e.g., computers, cars, and televisions) are intended to be replaced every few years, many systems are now deployed in service for one or more decades. Over such long periods of time, system capabilities will need to evolve if the system is to remain usable and relevant. The system life cycle does not end with deployment, and the evolution of deployed systems to include a range of unforeseen capabilities is thus becoming increasingly important to systems engineering.

- **Localized assembly line to globalized supply chain.** Historically, many systems were developed by a single, geographically-localized development organization using what amounted to an assembly line approach where locally developed architectural components were integrated into the growing systems (i.e., in-house development). To concentrate on their core competencies and take advantage of the specialized expertise of other companies, many development organizations have transformed themselves into system prime contractors that rely on a set of subcontractors and vendors for the production of some of the system's subsystems (i.e., outsourcing). These prime contractors are now evolving further into system integrators depending even more on suppliers including both subcontractors and vendors to provide almost all of the system's components. To decrease the costs of both skilled and unskilled labor, system integrators are also turning to overseas suppliers (i.e., off-shoring) such as Indian suppliers for complex components (including software) produced by white-collar workers and Chinese suppliers for simple components produced by blue-collar workers. As we move to ultra-large systems of systems for which there are no single system integrators, we will see systems that are easier to integrate with each other. We may even begin to see systems that actively integrate themselves into the constantly evolving system of systems.

■ **Green-field to reuse.** Several decades ago, systems engineers often worked on green-field development projects where systems and their architectural components were developed (and architected) from scratch. Then during the 1980s, development organization internal reuse and the reuse of simple low-level components became a highly fashionable goal. Now, most systems are engineered by a system integrator using COTS, MOTS, GOTS, and open source reusable components. Most systems are also updated versions of existing systems or product lines reusing major components and existing (reference) architectures.

19.4.1.2 Trends in System Architecture Engineering, Architects, and Tools

The preceding trends in systems and system engineering are producing the following associated trends in system architecture engineering, system architects and their teams, and architecture tools.

■ **Individual model size and complexity.** As systems increase in size and complexity, their architectures will also naturally increase in size and complexity. Although clearly true about the number of individual architectural decisions, inventions, trade-offs, assumptions, and rationales, this is especially true of the architectural structures and their associated models, views, and focus areas.

Future architecture engineering methods must scale well to properly address this increasing size and complexity. The training of system architects will need to become more comprehensive to cover the breadth of the structures, models, and focus areas, while architectural tools will need to be enhanced to enable the architects to easily browse, navigate, and search them. Finally, there will need to be system of system and enterprise architects in addition to system, software, and hardware architects.

■ **Number of types of models, views, and focus areas.** Historically, the number of different types of architectural structures and associated models and views has been relatively limited, based on the limitations of modeling languages and the experience of the chief architects. However, the number of model types has been increasing as architects have discovered the need for more views, as software and data models and modeling languages have been added to the hardware models and modeling languages, and as single modeling languages and architecture documentation standards are being replaced by multiple languages and standards. Similarly, quality models have been increasing in size and complexity, thereby leading to increases in the number of quality characteristics and attributes recognized as the basis for quality requirements and their associated architectural concerns.

As with individual model size and complexity, architecture engineering methods must scale to properly handle this increase in the number of different types of architectural models, views, and focus areas. This has led to the need for more powerful tool support and tool integration for both the larger number (and completeness) of modeling languages as well as the need to ensure consistency and integrity of these numerous larger and more complex models and views.

■ **Emphasis from functionality to qualities**. Traditionally, system static architectural structures have been based on functional decomposition with individual architectural components being largely functionally cohesive. As systems have grown larger, more complex, and more software intensive, there has been a tendency to move away from such purely functional decomposition static architectures, at least as far as software is concerned. Due to limitations in power, weight, cooling, and cost, such systems now tend to incorporate more powerful embedded computers that execute more diverse software that performs numerous functions on numerous types of data. For example, modern aircraft can be thought of as many computers flying in close formation, just as modern automobiles can be considered to be a large number of computers driving in close formation. The allocation of software architectural components to hardware components is also made more complex as individual computers must often execute software that has multiple levels of safety and security criticality, where the failure of lower criticality software cannot be allowed to cause the failure of higher criticality software. All of this has made the allocation of software to hardware much more complex and systems significantly less functionally decomposed.

At the same time as functional decomposition has been growing less dominant, there has been an increasing awareness of the critical need for systems to meet their quality requirements and the fact that required levels of quality characteristics and attributes are typically some of the most important drivers of the architecture. In turn, this has led to the understanding of the importance of quality models and the need for quality focus areas and the use of architectural quality cases. Although many quality characteristics are typically important architectural concerns, the size of systems and the trend towards ultra-large systems of systems has led to a realization of the fundamental importance of interoperability and intraoperability.

Future architecture engineering methods must adequately and appropriately address quality via the incorporation of quality models, quality focus areas, and architectural quality cases. System architects will need to be properly trained in these topics, and architectural tools will need to be extended to properly incorporate support for them.

■ **Life cycle architecture engineering**. The long-term value of an architecture and its representations becomes more important as the system lifespan of many systems increases and the frequency at which systems must be modified to meet new requirements and evolve to meet new business challenges and opportunities also increases. During architecture engineering, this has led to an emphasis on system modifiability including both extensibility and maintainability. This has also led to the extension of system architecture engineering methods to include maintaining the system architecture and its representations as well as ensuring that the integrity of the architecture is not lost over time.

■ **Requirements driven to evolutionary -capabilities based**. Because of the importance of the architecturally-significant requirements, especially the quality requirements, the need to engineer these requirements before spending too much time engineering architectures has been long recognized. This has led some projects to attempt to complete the requirements before starting the architecture, that is, to use a Waterfall development cycle at least for the requirements and architecture phases. On the other hand, the trend towards reuse and the incorporation of reusable (e.g., COTS, MOTS, GOTS, and open source) architectural components within architectures has led to the concurrent engineering of the requirements and architecture. The modeling necessary to engineer safety-and security-related requirements as well as the associated architecture has also led to a concurrent development of the requirements and the architecturally-significant requirements.

The increasing size and complexity of many modern systems has led to the recursively incremental (e.g., tier-by-tier and subsystem-by-subsystem) engineering of both the requirements and architecture. Similarly, the increasing numbers of architectural defects introduced due to increasing size and complexity of system architectures has led to iterative development cycles. Increasing system evolution due to longer operational life spans combined with rapidly changing requirements has also led to the need for incremental and iterative development cycles. *Finally, the impossibility of knowing detailed technical requirements prior to architecture, design, and implementation has resulted in the replacement of such detailed user requirements with general operational capabilities.*

The preceding changes to requirements engineering have changed the type and scope of the architecturally-significant requirements driving the architecture as well as the development cycle by which the architecture is engineered. This, in turn, has brought corresponding changes to architectural engineering methods. Architects, their managers, and other architectural stakeholders need to understand these ramifications on architecture engineering methods. Finally, architectural tools need to support the collaboration of architects and architecture teams, both with each other as well and with other stakeholders such as requirements engineers and their teams, including the development and verification of numerous subsystem-specific subsets of the requirements.

■ **Decomposition versus composition.** Traditionally, individual systems were typically developed using a top-down functional decomposition approach. Over time, this has tended to be replaced by either reuse-driven bottom-up composition or system of system outside-in integration. Specifically, system and software architectural decisions, inventions, trade-offs, and assumptions have become increasingly interdependent and iterative. System quality characteristics and attributes are heavily influenced by the system and software architecture, and many of these qualities are cut across multiple architectural components and are thus not readily decomposed and allocated in a simple flow-down manner.

These trends in decomposition, composition, and integration must be addressed by future architecture engineering methods. More specifically, architecture engineering tasks need to be decoupled from system development "waterfall" phases so that they can be performed in a recursively incremental, iterative, and concurrent manner. Architects and other stakeholders in the architecture need to be made aware of the effects decomposition, composition, and integration have on development/life cycles, milestones, and architecture engineering.

■ **Emphasis from hardware to software.** Systems are providing ever-increasing amounts of functionality, and the capabilities being delivered are becoming increasingly complex with time. This increased functionality and other capabilities are being delivered more and more via software. The dramatic increase of software complexity will drive corresponding advances in software tools and environments. The growth in complexity of system capabilities also makes it harder to understand the details of those capabilities, particularly early-on in the system's development cycle. When coupled with increasing system lifecycle durations, system developers are depending more heavily on software as the primary means to provide the needed evolution of system capabilities as those capabilities become better-understood and as system-user wants and needs evolve over long periods of time. As systems grow in size and complexity, there is increased risk of developing architectures having inadequate qualities when engineering methods are used which relegate software architecture decisions to downstream activities in a manner that is largely decoupled from major system architecture decisions.

■ **Emphasis on model consistency.** As the size, complexity, and number of architectural models has increased, it has become harder and harder to keep them consistent. This is especially a problem because of the lack of architecture tools that scale to the necessary model sizes and that support the development, management, and verification of the different architectural models. The consistency problem is also made more pressing because of the recent recognition of the importance of quality models, quality focus areas, and architectural quality cases. This kind of consistency verification and preservation will need to be supported by architectural tools if the integrity of the models and focus areas is not to be lost.

■ **Formalism versus understandability.** There has been a general trend in system architectural work products to move from simple text and "cartoon" diagrams to the use of standard, primarily graphical modeling languages such as UML. There has also been a trend in increasing formality of these architectural modeling languages (e.g., from UML 1 to UML 2 and from UML to AADL). Although this trend has decreased ambiguity and increased the ability of using these models to support simulation and dynamic analysis, there has also been a recognition that many stakeholders of the architecture are not willing to spend a long time learning multiple modeling languages and model types.

The trend towards more formal modeling (e.g., to provide the ability to analyze models to verify their consistency and behavior) will likely continue until architects have the models and associated tools needed to help them engineer better architectures with fewer defects and risks. On the other hand, there will probably be a trend towards providing the right architectural work products and information at the right level of abstraction in the right format and media for the right stakeholders.

19.4.2 The Future Directions of MFESA

19.4.2.1 MFESA Organization

It is hoped that an organization can be founded to maintain MFESA, especially its repository of reusable method components as free and open-source process resources. We foresee the creation of additional method components, the improvement of existing components, and additional ancillary information (e.g., an informational Web site and example work products).

19.4.2.2 Informational Web Site

This book is currently the primary documentation of the MFESA method framework. While a professionally published book is a good way to document, teach, and advertise the existence of MFESA, a paper book has significant inherent limitations. It is difficult to update regularly and rapidly. It also provides little support for browsing beyond the table of contents, index, and the logical and standardized organization of its content. MFESA could benefit greatly from an electronic format.

For this reason, the authors intend to obtain the resources needed to turn the contents of this reference manual into an informational Web site or Wiki. The individual concepts and terms in

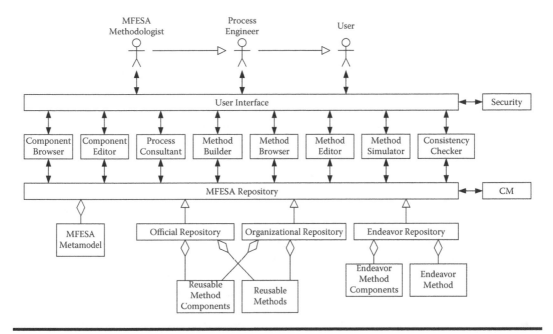

Figure 19.3   Future integrated MFESA toolset.

the MFESA ontology could each be documented in their own Web pages. Similarly, each individual architecture engineering method component could be easily documented in the form of a standardized Web page. These Web pages would contain hyperlinks to other related concepts and method components. Such a Web site would enable easy browsing as well as easy updating and maintenance of the content.

19.4.2.3 Method Engineering Tool Support

It is one thing to have good documentation of the individual reusable method components that can be used to create project-specific system architecture engineering methods. However, even with an informational Web site, there will still be a significant amount of manual labor involved in selecting the appropriate method components, tailoring them to meet the specific needs of the project, integrating them together to produce a project-specific method, and verifying the consistency of the resulting method. This cries out for simple tool support.

Figure 19.3 illustrates one possible integrated toolset that could be built to support MFESA and enable its use. It could consist of the following parts:

- **User interface.** The user interface would provide a common interface to the underlying MFESA tools as well as security (e.g., user identification, authentication, and authorization).
- **Component browser.** This tool would enable users to browse the MFESA method components.
- **Component editor.** This tool would enable authorized architects and process engineers to edit the contents of existing method components and extend the appropriate repository by creating new method components.
- **Process consultant.** After eliciting information about the system, the endeavor, and the development/architecture organizations, this tool would advise authorized architects and

process engineers on the appropriate method components to select to build endeavor-specific system architecture engineering methods.

- **Method builder.** The tool would enable authorized architects and process engineers to produce appropriate endeavor-specific system architecture engineering methods.
- **Method browser.** This tool would enable users to browse the existing reusable methods as well as endeavor-specific methods. Note that although anyone can browse the existing reusable methods and method components, only certain people may be authorized to use the method builder and method editor to create and modify actual project methods.
- **Method editor.** The tool would enable authorized architects and process engineers to edit existing system architecture engineering methods stored in the appropriate MFESA repositories.
- **Consistency checker.** This tool would check the consistency of new or modified methods. For example, it would ensure that there would be no "dangling references" between method components and no "orphan" method components.
- **MFESA repository.** This type of repository would store reusable method components, reusable methods, and the endeavor-specific method.
- **Official repository.** The is the MFESA organization's official repository containing the standard reusable method components and standard reusable system architecture engineering methods.
- **Organizational repository.** This is an organization's repository containing the organization's copies of the standard reusable method components as well as the organization's own proprietary method components and methods.
- **Endeavor repository.** An endeavor repository contains the endeavor-specific system architecture engineering method and its components.

19.5 Final Thoughts

The scale and scope of modern systems has expanded the scope of the term "architecture" beyond any meaning previously associated with it. Buildings, bridges, and other structures of the modern era have been relatively tractable in terms of their architectures. The scope of modern systems and ultra-large systems, however, has made architecture cognitively intractable, which means that the scope and complexity of a system cannot be understood by an individual person. There is no single picture that tells the entire story or provides sufficient information to reveal both the visible exterior of the system and the complexities below the surface. The building of the Greek Colossus of Rhodes provides an example of a success in building a giant statue that strained the very edges of the ancient Greeks' capacity for architecting. Prior to that Colossus, others had been built and failed for lack of understanding the full extent of the endeavor — the issues hidden beyond easy visibility. These failed efforts are examples of the human incapacity to fully understand the size and scope of a large-scale endeavor. Two mottoes for those who understand this principle are "Be very careful" and "Use the best means available." And that may well mean using MFESA to produce an appropriate system architecture engineering method for the job at hand.

In this light, it is reasonable to mention the effect of software in the world of systems. From within the systems development world, software has opened up great and seductive doorways to capabilities previously unknown and unimagined. This could easily be described as the positive side of the software story. These doorways, however, lead to paths unknown. Software is an intoxicating commodity that can, in the minds of some, bear all the weight of even the most poorly expressed requirements. Software's very *softness* creates a problem — the problem of "liquid logic" — often a kind of "mortar" slathered on bricks laid unsoundly. To some, the term "software engineering" is an oxymoron. The challenges confronting the systems development community, however, do not really warrant such easy and facetious commentary. Software engineering *should be* among the most serious concerns in the systems engineering communities. After all, software engineering is a kind of systems engineering where the system consists exclusively or almost exclusively of software. Software engineering should therefore inherit the best practices known within *any* engineering discipline, including all the principles and methods to control the implementation of complex systems and their complex structural elements. Because of our discovery of the power of software, the temptation to "just build it" often overwhelms efforts to sensibly rein in the "imagineering" that results in complex capabilities that are not truly feasible. Ironically, some of the "pain" associated with the revelation of the power of software is that the power has highlighted the importance of integral systems engineering and is fueling a resurgence of needed vitality within the systems community.

In light of the sea change associated with new and more powerful software-intensive systems, it becomes clear that modern engineering demands comprehensive methods to provide coherence to large-scale and ultra-large-scale systems — thus architecture, thus MFESA. And so the term "architecture" evolves with us, carrying with it the dynamic stability needed to remind us of guiding principles that will permit success instead of failure, promoting clarity over chaos.

Appendix A

Acronyms and Glossary

A.1 Acronyms

AADL: Architecture Analysis and Design Language
BPMN: Business Process Modeling Notation
CBD: Component-Based Development
CFD: Control Flow Diagram
COMPUSEC: Computer Security
COMSEC: Communications Security
COTS: Commercial-Off-The-Shelf
DFD: Data Flow Diagram
DODAF: Department of Defense Architecture Framework
EMSEC: Emissions Security (Tempest)
ERD: Entity Relationship Diagram
GOTS: Government-Off-The-Shelf
FEA: Federal Enterprise Architecture
INFOSEC: Information Security
MFESA: Method Framework for Engineering System Architectures
MODAF: U.K. Ministry of Defense Architecture Framework
MOTS: Military-Off-The-Shelf
NAF: NATO Architecture Framework
NAVSEA: U.S. Navy Naval Sea System Command
NETSEC: Network Security
OTS: Off-The-Shelf
SAE: System Architecture Engineering
SEMP: System Engineering Management Plan
SOA: Service-Oriented Architecture
SysML: System Modeling Language
T: Task

UML: Unified Modeling Language
WP: Work Product

A.2 Glossary

Abstraction: a model that captures essential behavior and characteristics while ignoring diversionary details.

Accessibility: the attribute of usability that is the degree to which the system is useful to users with disabilities (e.g., color blindness, deafness, etc.).

Accuracy: the attribute of correctness that is the magnitude of defects (i.e., the deviation of the actual or average value from their true value) in the stored and output data of the system or architectural component.

Adaptive maintainability: the subtype of maintainability that is the degree of ease with which the system or architectural component can be adapted to meet minor changes to the allocated and derived requirements.

Affordability: the subtype of feasibility that is the degree to which the cost of the system or architectural component lies within budgetary constraints.

Aggregation structure: a static logical or physical composition hierarchy of a system, its component parts, their component parts, etc.

Allocated requirement: a requirement of a system or system component that has been assigned either completely or partially to a system component at the next lower level in the system aggregation structure (e.g., allocated from a system to one of its direct subsystems).

Analyzability: the attribute of modifiability that is the degree of ease with which defects, deficiencies, and causes of failures can be diagnosed and localized to the components to be modified.

Architect: the role played by a person when performing the tasks of architecture engineering.

Architectural assumption: any assumption having significant architectural ramifications.

Architectural baseline: a uniquely identified set of formally reviewed and approved architectural work products that serve as a basis for further development or maintenance and that can be changed only through formal change control procedures.

Architectural component: an architectural element within a system or subsystem static physical configuration structure; a major, cohesive, structural element consisting of data, equipment, a facility, firmware, hardware, material, a manual procedure, a role played by a person, software, or a tool (or some combination thereof, such as a system, subsystem, assembly, subassembly, or configuration item).

Architectural concern: a cohesive collection of architectural drivers.

Architectural constraint: an architectural decision that is mandated as a requirement.

Architectural decision: any decision with any significant architectural ramifications.

Architectural description: a non-executable architectural representation.

Architectural driver: an architecturally significant requirement. An architectural driver may be either (1) a product requirement (e.g., associated with a quality characteristic or attribute, a major functional feature, a major interface set, or a major data type, or an architectural constraint), or (2) a programmatic process requirement (e.g., cost or schedule) allocated to an architectural element.

Architectural element: an atomic node (i.e., part) of an architectural structure.

Architectural engineering: the subdiscipline of systems engineering consisting of all tasks performed to develop and maintain a system architecture and associated consistent and cohesive set of architectural representations.

Architectural focus area: the cohesive set of all architectural decisions, inventions, and trade-offs related to a specific architectural concern, regardless of the architectural view, model, or structure where they are documented or found.

Architectural heuristic: a rule of thumb, insight, lesson learned, or guideline concerning architectural engineering. For example, when decomposing a system into subsystems, maximize subsystem cohesion and minimize coupling between the subsystems.

Architectural integrity: internal consistency within the architecture and its representations as well as external consistency with the architecturally significant requirements.

Architectural invention: a new solution to an architectural problem within the context of one or more architectural elements.

Architectural mechanism: a major architectural decision or invention, often an element of an architectural pattern.

Architectural model: an architectural representation that models a single system structure in terms of the structure's architectural elements and the relationships between them.

Architectural pattern: a well-documented reusable solution to a commonly occurring architectural problem within the context of a given set of existing architectural decisions, inventions, engineering trade-offs, and assumptions.

Architectural prototype: any architectural representation that (1) prototypes an aspect of the system with sufficient dynamic semantics to be executable, and (2) is used to verify or validate the quality or appropriateness of one or more architectural decisions, inventions, trade-offs, or assumptions.

Architectural quality case: a quality case making the architects' case that their architectural work products have a sufficient level of some quality characteristic.

Architectural rationale: the rationale behind one or more architectural decisions, inventions, or trade-offs.

Architectural representation: a cohesive set of information documenting a part of a system architecture.

Architectural risk: any risk primarily related with the architecture. As with any other type of risk, architectural risks are defined in terms of harm severity (the maximum credible amount of harm to valuable assets) multiplied by the probability that the harm will occur. In this case, the harm is related to architectural defects or vulnerabilities.

Architectural simulation: any executable architectural representation (e.g., dynamic model) used to simulate the behavior of the system.

Architectural structure: an abstraction of a system consisting of a cohesive set of architectural elements connected by associated relationships that captures a set of related architectural decisions, inventions, trade-offs, assumptions, and rationales.

Architectural style: a top-level architectural pattern that provides an overall context in which lower-level architectural patterns exist.

Architectural trade-off: an analysis used to decide between competing architectural visions or decisions based on their support for architecturally significant product or process requirements.

Architectural view: any architectural representation describing a single architectural structure that consists of one or more related models of that structure.

Architectural viewpoint: a specification of the conventions (e.g., standards, procedures, guidelines, and templates) for constructing, analyzing, and documenting an architectural view.

Architectural vision: a conceptual overview of the system or subsystem's architectures consisting of a cohesive and consistent set of architectural vision components (i.e., architectural decisions).

Architectural vision component: documentation of one of the more important actual or potential architectural decisions, inventions, or trade-offs addressing one or more architectural concerns.

Architecturally significant: describing a product or process requirement that has significant architectural ramifications (i.e., significantly influences architectural decisions).

Architecturally significant requirement: a requirement that has significant architectural ramifications and therefore drives architectural decisions.

Architecture: the most important, pervasive, top-level, strategic decisions, inventions, engineering trade-offs, and their associated rationales and assumptions about how a system and its elements will meet their derived and allocated requirements. An architecture is primarily concerned with the structures of the system in terms of its architectural elements, their associated characteristics and behavior, and how they collaborate together to achieve their allocated requirements. Design is contrasted with architecture in that design is the less important, nonpervasive, low-level, tactical decisions, inventions, engineering trade-offs, and their associated rationales and assumptions about how a system and its elements conform to the architecture while meeting their derived and allocated requirements. Therefore, MFESA does not combine the disjoint terms "architecture" and "design" into the confusing term "architectural design."

Architecture document: a document that describes the system architecture or the architecture of one of its architectural components, including views and models of its structures as well as other architectural decisions, inventions, trade-offs, assumptions, and their associated rationales.

Architecture team: a team of architects that is responsible for engineering one or more related architectures.

[quality case] Argument: a clear, compelling, and relevant developer argument justifying the assessors' belief in the developers' claims.

Attractiveness: the attribute of usability that is the degree to which users find the system or architectural element to be attractive or appealing, engage their attention, provide a positive user experience, be preferable to its alternatives, make users want to continue to use it, and make them return to use it in the future; a.k.a., engagability, preference, and stickiness.

Authoritative stakeholder: a legitimate stakeholder with the authority to provide requirements and accept architectural work products.

Availability: the subtype of soundness that is the degree to which a system or architectural component is ready to function without failure in one or more specified ways at any time during a specified period of time under normal conditions or circumstances given its operational profile. Availability can be measured as the proportion of time that the system or architectural component can be used or the probability that the system or architectural component is functioning at an arbitrary point in time.

Capacity: the subtype of soundness that is the degree with which the system or an architectural component can successfully handle a large number of things at a single point in time or during a specific interval of time.

[quality case] Claim: a developer claim that their work product(s) sufficiently support the system or architectural component having sufficient quality.

Class diagram: an object-oriented diagram documenting classes of system components and the aggregation, association, and classification relationships between them.

Class model: a model of a class structure, usually represented by one or more class diagrams and ancillary text.

Class structure: a system structure consisting of classes of system components and the aggregation, association, and classification relationships between them.

Cohesion: the degree to which the subcomponents of an architectural component collaborate to fulfill the requirements allocated to the architectural component (i.e., the degree to which the subcomponents belong together).

Collaboration: to cooperatively work together to accomplish some end.

Compliance: the external quality characteristic that is the degree to which the system or an architectural component adheres to laws, regulations, international and industrial standards, and similar prescriptions.

Configurability: the external quality characteristic that is the degree to which the system or an architectural component can be configured in multiple ways. Configurability can be classified into internationalization, personalization, subsetability, and variability.

Consistency: the quality characteristic characterizing the degree to which components of the system or an architectural component have the same architectural styles, implement the same architectural patterns, or use the same architectural mechanisms.

Constraint: an architectural, design, or implementation decision that is mandatory and therefore treated as a requirement restricting the architecture, design, or implementation.

Control flow: (1) a flow of control between system functions (logical data flow); (2) a flow of control between system components (physical data flow).

Control flow diagram: (1) a structured analysis diagram documenting all or part of a logical control flow model, (2) a diagram documenting all or part of a physical control flow model.

Control flow model: (1) a model of a logical control flow structure, usually represented by one or more logical control flow diagrams and ancillary text; (2) a model of a physical control flow structure, usually represented by one or more physical control flow diagrams and ancillary text.

Control flow structure: (1) a logical dynamic system structure consisting of system functions and the control flows between them; (2) a physical dynamic system structure consisting of system components and the control flows between them.

Controllability: the attribute of testability that is the degree of ease with which the system or architectural component can be placed into the proper pretest state stimulated with the test message or data.

Corrective maintainability: the subtype of maintainability that is the degree of ease with which defects within the system or architectural component can be corrected.

Correctness: the subtype of soundness that is the degree to which the system or an architectural component and its outputs are free from defects. *See* accuracy, precision, and timeliness.

Coupling: the degree to which an architectural component depends on (i.e., fan-out) and is depended on (i.e., fan-in) other architectural components.

Credibility: the attribute of usability that is the degree to which users are confident with and have trust in the system, including that its output and behavior are correct, content is authoritative, owners' motives are trustworthy, and developers were competent (a.k.a., trustworthiness).

Current reusability: the subtype of reusability that is the degree of ease with which externally produced components can be incorporated with little or no modification into the current system or architectural component.

Data flow: (1) a flow of data between system functions (logical data flow); (2) a flow of data between system components (physical data flow).

Data flow diagram: (1) a structured analysis diagram documenting all or part of a logical data flow model; (2) a diagram documenting all or part of a physical data flow model.

Data flow model: (1) a model of a logical data flow structure, usually represented by one or more logical data flow diagrams and ancillary text; (2) a model of a physical data flow structure, usually represented by one or more physical data flow diagrams and ancillary text.

Data flow structure: (1) a logical dynamic system structure consisting of system functions and the data flows between them; (2) a physical dynamic system structure consisting of system components and the data flows between them.

Defect: a defect in the architecture, design, implementation, integration, or installation (e.g., configuration) of a system or subsystem. Defects often cause faults and failures when executed.

Defensibility: the subtype of dependability that is the degree to which a system or architectural component defends valuable assets from unauthorized harm.

Dependability: the external quality characteristic that is the degree to which legitimate stakeholders can reasonably rely on a system or architectural component to behave as they expect.

Derived requirement: any requirement that is not levied directly on the development organization be external sources (e.g., the customer organization) but rather is engineered from one or more higher-level requirements.

Differentiation: the attribute of usability that is the degree to which the system or architectural component differentiates itself from competing products.

Domain modeling: the modeling of an application domain such as avionics or telecommunications, either as part of requirements engineering or architecture engineering.

Ease of entry: the attribute of usability that is the ease with which users can start using the system or architectural component (e.g., can log on and begin using their desired functionality without waiting an excessive amount of time to be identified, be authenticated, and navigate to the point where they can start performing their tasks).

Ease of location: the attribute of usability that is the ease with which users can find desired content or services of the system or architectural component (e.g., finding Web applications such as Web sites using search engines).

Ease of navigation: the attribute of usability that is either the degree to which occasional users can remember how to use the system or architectural component to perform common tasks or the degree to which regular users can remember how to use the system to perform infrequent tasks.

Effectiveness: the attribute of usability that is the quality attribute characterizing the degree to which the system or architectural component enables users to successfully achieve their goals; a.k.a. operability.

Efficiency: the external quality characteristic that is the degree to which the system or an architectural component effectively uses (i.e., minimizes its consumption of) the resources upon which it depends. Efficiency can be subtyped into bandwidth efficiency, consumables efficiency, CPU efficiency, power efficiency, and storage efficiency, whereby these quality characteristics have the obvious definitions implied by their names.

Effort minimization: the attribute of usability that is the degree to which the system or architectural component minimizes the amount of effort users (and operators) must expend to achieve their goals (in relation to the accuracy and completeness with which these goals are achieved).

Electrical interoperability: the subtype of interoperability that is the degree to which the system or architectural component correctly exchanges and uses energy of the appropriate types and levels with specified [types of] external systems.

Emergent behavior: a system behavior that requires the collaboration of multiple system components and therefore cannot directly be performed by any single system component.

Emergent characteristic: a system characteristic that results from the collaboration of multiple system components and therefore is not directly the characteristic of any single system component.

Endeavor: a venture undertaken by collaborating roles, teams, or organizations to perform work units during one or more development cycles or phases to produce one or more related work products (e.g., systems). Endeavors include projects, programs of related projects, and enterprises that engineer systems and may therefore need one or more system architecture engineering methods.

Energy interoperability: the subtype of interoperability that is the degree to which the system or architectural component correctly exchanges and uses energy of the appropriate types and levels with specified [types of] external systems.

Environmental compatibility: the external quality characteristic that is the degree to which a system or architectural component can be used and functions correctly under specified conditions of the physical environment in which it is intended to operate.

Environmental tolerance: the subtype of tolerance that is the degree to which essential mission-critical services continue to be provided despite potentially harm-causing *environmental conditions* (e.g., salt spray causing corrosion or radiation randomly changing the value of a bit within memory).

Error: (1) an improper input to the system or system component input, whether made by an external person, device, or system; (2) a mistake made by a human that causes the existence of a defect or fault.

Error minimization: the attribute of usability that is the degree to which the system or architectural component minimizes the number of errors that its users make.

Error tolerance: the subtype of tolerance that is the degree to which essential mission-critical services continue to be provided despite the presence of erroneous input (e.g., incorrect, stale, or out-of-order data). Note that erroneous input is typically due to human error, although it may also be due to sensor failure, timing delays, etc.

Event schedulability: the attribute of performance that is the degree to which system or architectural component events and behaviors can be accurately scheduled.

[quality case] Evidence: adequate credible evidence supporting the developers' arguments.

Executable architectural representation: any architectural representation with sufficient dynamic semantics to be executable.

Executable architecture: any executable architectural representation that is a very early partial implementation of the system that (1) is used to verify aspects of the architecture and (2) may evolve into the completed system.

Extensibility: the subtype of modifiability that is the degree of ease with which a system or an architectural component can be significantly enhanced to meet specified future goals or changing requirements.

External quality characteristic: a quality characteristic that characterizes an externally visible (i.e., black box) quality of a system or architectural component when it is deployed and in service in its operational environment.

Facility: an architectural component consisting of one or more buildings and their associated equipment, instruments, and tools for facilitating the performance of an action.

Failure: an externally visible system/subsystem behavior or output that violates one or more requirements.

Failure tolerance: the subtype of tolerance that is the degree to which the system continues to provide essential mission-critical services despite the occurrence of failures, whereby a failure is the execution of a defect that causes an inconsistency between an executable work product's actual (i.e., observed) and expected (e.g., specified) behavior.

Fault: a system/subsystem state that is inconsistent with one or more architecture or design decisions. Failures often result when faults are not properly identified and handled (i.e., when the system or subsystem is not adequately fault tolerant).

Fault tolerance: the subtype of tolerance that is the degree to which essential mission-critical services continue to be provided despite the presence or execution of defects, whereby a defect (a.k.a. fault and bug) is an underlying flaw in a work product (i.e., a work product that is inconsistent with its requirements, policies, goals, or the reasonable expectations of its customers or users).

Feasibility: the internal quality characteristic that is the degree to which it is possible to develop and sustain the system or an architectural component within the practical constraints of existing budget, schedule, resources, and technology.

Focus area: the subset of one or more architectural views documenting how an architecture, architectural vision, or architectural vision component addresses a specific architectural concern.

Functional decomposition: the hierarchical decomposition of system functions into lower-level system functions; (2) the hierarchical decomposition of a system into functionally cohesive architectural components.

Functional requirement: a requirement that specifies a mandatory behavior or function that the system or subsystem must provide under specific conditions to meet a stated or implied need of a legitimate stakeholder.

Functionality: the external quality characteristic that is the degree to which the system or an architectural component implements all the functions (or functional requirements) allocated to it.

Future reusability: the subtype of reusability that is the degree of ease with which the architectural component currently being produced will be able to be incorporated with little or no modification to the architecture of other specified systems.

Government-off-the-shelf (GOTS): describing any OTS architectural element belonging to (and provided by) a governmental agency or department.

Greenfield development: the development of a system from scratch, as opposed to the development of a new version of an existing system,

Illegitimate stakeholder: synonym for system attacker or malware.

Incremental [development]: a property of a development cycle, whereby units of work are repeated to produce additional new work products or capabilities of work products. Development cycles are typically incremental because systems are too large and complex to be built all at once in a big-bang fashion.

Independence: the quality characteristic characterizing the degree to which (1) one architectural component does not depend on another, and (2) the failure of one architectural component does not cause the failure of another.

Internal quality characteristic: a quality characteristic that characterizes an internally visible (i.e., white box) quality of a system or architectural component when it is in the process of being developed, modified, or retired.

Internationalization (a.k.a. globalization or localization): the subtype of configurability that is the degree to which a system or architectural component can be (or is) configured to function appropriately in a global environment. Internationalization may include configuration in terms of:

- Native languages, language idioms, spelling, and character sets
- Formats of contact information such as name, address, and phone number
- Currencies, including real-time currency conversion
- Legal issues such as import/export laws, tariff and sales tax calculations, customs documentation, trademarks, and privacy laws culture (e.g., use of inappropriate colors, symbols, or product names)
- Country-specific security countermeasures (e.g., encryption and digital signatures)

Interoperability: the external quality characteristic that is the degree to which the system operates (i.e., interfaces *and* collaborates) effectively with specified [types of] external systems by successfully providing data, information, material, and services to those systems and using data, information, material, and services provided by those systems.

Intra-operability: the internal quality characteristic that is the degree to which an architectural component operates effectively with other architectural components within the system by successfully providing data, information, material, and services to those components and using data, information, material, and services provided by those components.

Iterative [development]: a property of a development cycle, whereby work units are repeated on existing work products to improve them (e.g., to fix defects and adapt them to changes in requirements). Development cycles are typically iterative because systems are developed by fallible humans.

Jitter: the attribute of performance that is the degree to which the variability of the time intervals between periodic actions controlled by the system or architectural component remains within its designated constraints.

Latency: the attribute of performance that is the degree to which the time that the system or subsystem takes to execute specific tasks (e.g., system operations and use case paths) from end to end is within acceptable time limits.

Learnability: the attribute of usability that is the degree to which representative users can learn to use the system or architectural component to achieve their goals (e.g., to find desired content and to perform their tasks).

Legitimate stakeholder: a stakeholder having a legitimate (i.e., valid) interest that the system meets the stakeholder's needs or expectations.

Maintainability: the subtype of modifiability that is the degree of ease with which a system or architectural component can be modified between major releases. Maintainability can be classified into the following quality characteristics: adaptive, corrective, perfective, and preventative maintainability.

Material interoperability: the subtype of interoperability that is the degree to which the system or architectural component correctly exchanges and uses materials of the appropriate types and amounts with specified [types of] external systems.

Method component selection: the task of selecting an appropriate consistent set of reusable method components from an MFESA repository.

Method component tailoring: the task of tailoring (modifying) a previously selected method component from an MFESA repository.

Method engineering: a way of creating appropriate, efficient, and effective methods for performing system or software engineering by selecting method components, tailoring them, and integrating them into a single consistent method.

MFESA: a method framework for constructing situation-specific system architecture engineering methods.

MFESA-compliant method: a method for performing system architecture engineering that has been created (instantiated) by selecting (and tailoring) an appropriate consistent set of reusable method components from an MFESA repository.

MFESA repository: any repository for storing reusable MFESA method components and MFESA methods.

Military-off-the-shelf (MOTS): describing any OTS architectural element belonging to (or provided by) a military department or organization.

Mission: a single primary usage or operation of a system to meet a cohesive set of stakeholder goals.

[system] Mode: the operational state of the system.

Model: a simplified abstraction of a part of the architecture such as a model of an architectural structure.

Modeling language: a textual or graphical language used to implement one or more related types of models.

Modifiability (a.k.a. changeability): the internal quality characteristic that is the degree of ease of changing a system, architectural component, or other work product once it has been initially developed and deployed.

Navigability: the attribute of usability that is the degree to which the system or architectural component enables users to easily move through the user interface or documentation to achieve their goals.

Observability: the attribute of testability that is the degree of ease with which the system or architectural component can be observed to:
- Be in the proper pretest state
- Provide the proper output to its clients, peers, and servers (e.g., returned values, output messages, output requests for data, and exceptions)
- Be in the proper post-test state

Off-the-shelf (OTS): describing any architectural element that is reused with little or no modification (except configuration).

Operability: the external quality characteristic that is the degree of ease of operating (as opposed to using) a system or an architectural component.

[Parallel] development: a property of a development cycle, whereby multiple teams are simultaneously (i.e., concurrently) performing work units on different work products. Development cycles are typically parallel because there is too much work to be done and too little time in which to do the work if all work is forced into a sequential waterfall development cycle.

Perfective maintainability: the subtype of maintainability that is the degree of ease with which improvements to the system or architectural component can be made (e.g., by refactoring the architecture or using a more appropriate architectural pattern).

Performance: the subtype of defendability that is the degree to which a system or architectural component operates within its designated temporal constraints; a.k.a. timeliness.

Personalization: the subtype of configurability that is the degree to which the system or an architectural component can be (or is) configured so that individual users can be presented with a unique user-specific experience.

Physical interoperability: the subtype of interoperability that is the degree to which the system or architectural component physically connects with specified [types of] interfaces with specified [types of] external systems.

Portability: the internal quality characteristic that is the degree of ease with which a software system or architectural component can be moved to specified platforms, including associated operating system, middleware, and underlying hardware.

Precision: the attribute of correctness that is the degree of dispersion of the stored and output quantitative data around their average values of a system or architectural component.

Predictability: the subtype of soundness that is the degree to which system or architectural component events and behaviors are deterministic and can therefore be predicted from a given set of inputs when in a given state and/or environment.

Preventative maintainability: the subtype of maintainability that is the degree of ease with which changes to the system or architectural component can be made to prevent future failures.

Process engineer: the role played when a person engineers (created or improves) methods to improve engineering processes (often a member of a systems engineering process group [SEPG]).

Process requirement: any requirement that constrains the development, operations, or retirement processes (e.g., as performed by the developers or operators).

Producibility: the internal quality characteristic that is the degree of ease with which a system or architectural component can be manufactured in quantity; a.k.a. manufacturability.

Product requirement: any requirement that constrains the behavior or characteristics of the system or one or more of its architectural components.

Programmatic requirement: a process requirement specifying a constraint on the process used to develop, maintain, operate, use, and retire the system.

Protocol interoperability: the subtype of interoperability that is the degree to which the system or subsystem correctly uses the interface protocols of the specified [types of] external systems.

Quality: the degree to which a system or architectural component has useful and desirable characteristics as defined by the quality characteristics and quality attributes of a quality model.

Quality attribute: a major measurable component (aggregation) of a quality characteristic or of another quality attribute.

Quality case: a cohesive collection of claims, arguments, and evidence that makes the developers' case that their work product(s) have a sufficient level of some quality characteristic.

Quality characteristic: a high-level characteristic or property of a system or architectural component that characterizes an aspect of its quality; a.k.a. quality factor and "ility".

Quality criterion: a system-specific description that provides evidence either for or against the existence of a given quality characteristic or attribute.

Quality measurement method: a method, function, or tool for making a measurement along a quality measurement scale.

Quality measurement scale: a measurement scale that defines numerical values used to measure the degree to which a system or architectural component exhibits a specific quality attribute.

Quality model: a hierarchical model for defining, specifying, and measuring the different types of quality of a system or architectural component in terms of the model's component quality characteristics, quality attributes, and associated quality measurement scales and methods.

Quality-related requirement: a requirement having ramifications regarding one or more quality characteristics.

Quality requirement: a product requirement that specifies a minimum level of quality in terms of a quality characteristic or attribute. A quality requirement can be specified in terms of one or more conditions, a quality criterion, and thresholds on one or more related quality measures.

Quality threshold: a minimum level of quality along a quality measurement scale.

Recoverability: the attribute of robustness that is the degree to which a system or subsystem automatically recovers from a failure.

[Recursive] development: a type of incremental development cycle in which a system is incrementally developed top-down in terms of one or more of its primary architectural structures. Development cycles are typically incremental because systems are too large and complex to be built all at once in a big-bang fashion.

Reference architecture: a reusable architectural vision for use on systems within a product line or application domain.

Reliability: the subtype of soundness that is the degree to which a system or architectural component *continues* to function without failure in one or more specified ways during a specified period of time under normal conditions or circumstances given its operational profile; a.k.a., continuity. Reliability is frequently measured in terms of Mean Time to First Failure (MTFF) or Mean Time Between Failures (MTBF).

Requirement: (1) an established need justifying the timely allocation of resources to develop a system capability to achieve approved goals or accomplish approved missions or tasks; (2) a *mandatory*, externally observable, unambiguous, verifiable (e.g., testable), and validatable system behavior (functional requirement), datum (data requirement), quality characteristic (quality requirement), or interface (interface requirement).

Requirements engineering: the subdiscipline of systems engineering consisting of all tasks (e.g., identification, analysis, management, and specification) directly involved with the production and maintenance of requirements and related work products.

Requirements metadata: attributes describing an individual requirement such as its priority, status, source, and whether it is architecturally significant.

Resource feasibility: the subtype of feasibility that is the degree to which the system or architectural component can be developed using existing resources (such as the availability of staffing as well development facilities, manufacturing, and maintenance facilities).

Response time: the attribute of performance that is the degree to which the time it takes for the system or subsystem to initially respond to a client request for a service is within acceptable time limits.

Retrievability: the attribute of usability that is the ease with which the system or architectural component enables users to obtain information in a form that is useful to them (e.g., print out a paper report or make a copy of a multimedia file).

Reusability: the internal quality characteristic that is the degree to which a system or architectural component supports reuse.

Robustness: the subtype of defensibility that is the degree to which the system or an architectural element tolerates potentially harm-causing events or conditions and recovers from them; a.k.a. resiliency.

Safety: the quality characteristic characterizing the degree to which the system or an architectural component prevents or reduces the probability or severity of, detects, and reacts to:
- Unauthorized, unintentional (i.e., accidental) harm to valuable assets
- Mishaps (e.g., accidents and safety incidents)
- Hazards (i.e., existence of vulnerable assets and accidental abusers)
- Safety risks (i.e., probability of accidental harm multiplied by maximum credible harm severity)

Scalability: the subtype of extensibility that is the degree of ease with which a system or an architectural component can be significantly extended to increase its existing capacities.

Schedule feasibility: the subtype of feasibility that is the degree to which the system or architectural component can be developed (initial development, initial small-scale production, or full-scale production) within schedule constraints.

Security: the quality characteristic characterizing the degree to which the system or an architectural component prevents or reduces the probability or severity of, detects, and reacts to:
- Unauthorized intentional (i.e., malicious) harm to valuable assets
- Misuses (e.g., attacks and security incidents)
- Threats (i.e., existence of vulnerable assets and civilian attackers with means, motives, and opportunities abusers)
- Security risks (i.e., probability of malicious harm multiplied by maximum credible harm severity)

Semantic interoperability: the subtype of interoperability that is the degree to which the system or subsystem communicates requests and data via specified [types of] interfaces of the specified [types of] external systems in a manner that both systems interpret the syntax in a single standard way to gain the same meanings.

Separation of concerns: the degree to which an architecture is modularized so that different architectural components do not overlap in terms of functionality and do not require the same application-domain-specific subject matter expertise to produce, use, operate, and sustain (i.e., areas of concern are separated into different architectural components).

Serviceability: the external quality characteristic that is the degree of ease of servicing a system or an architectural component (e.g., adding or replacing consumable materials such as fueling an aircraft).

Situational method engineering: a way of creating efficient and effective situation-specific methods for performing system or software engineering.

Soundness: the subtype of dependability that is the degree to which the system or an architectural component behaves properly and predictably when and for as long as needed.

Stability: the subtype of reliability that is the degree to which a system or subsystem continues to deliver *mission-critical* services during a given time period under a given operational profile regardless of any failures, whereby the:
- Failures may prevent the system from delivering less critical services
- Failures limiting the delivery of mission-critical services occur at unpredictable times
- Root causes of such failures are difficult to identify efficiently

Stakeholder: a party (i.e., person or organization) having an interest in the system meeting the party's needs or expectations.

Structure: *See* architectural structure.

Subsetability: the subtype of configurability that is the degree to which the system or an architectural component can be released in multiple variants, each of which implements either a usable partial initial subset of operational capabilities (as part of an incremental development cycle) or a different subset of a common set of requirements (i.e., each variant is a subset of a primary complete variant).

Subsystem: a system that is a component architectural element of a larger system.

Subsystem architecture: (1) the architecture of a single specific subsystem; (2) the subset of a system architecture directly related to a specific subsystem.

Subsystem requirement: a product requirement specifying a mandatory behavior or characteristic of a subsystem.

Suitability: the attribute of usability that is the degree to which users find that the system or architectural component is suitable for the performance of their tasks; a.k.a. appropriateness.

Supersetability: the subtype of configurability that is the degree to which the system or an architectural component exists in multiple variants, each of which implements a different superset of the common set of requirements (i.e., some requirements are unique to each variant).

Survivability: the subtype of defensibility that is the degree to which a system or architectural component prevents or reduces the probability or severity of, detects, and reacts to *military*:

- Unauthorized intentional (i.e., malicious) harm to valuable assets
- Misuses (e.g., attacks and survivability incidents)
- Threats (i.e., existence of vulnerable assets and military attackers with means, motives, and opportunities)
- Survivability risks (i.e., probability of malicious harm multiplied by maximum credible harm severity)

Sustainability: the quality characteristic characterizing the degree of ease with which a system or subsystem can be supported once placed into use (i.e., fielded into its operational environment).

Syntax interoperability: the subtype of interoperability that is the degree to which the system or subsystem correctly communicates data having the correct syntax (e.g., data types such as text, integer, date, and money, including associated attributes, ranges, and default values) with specified [types of] external systems.

System: a cohesive integrated set of system components (i.e., an aggregation structure) that collaborate to provide the behavior and characteristics needed to meet valid stakeholder needs and desires.

System architect: the highly specialized role played by a systems engineer when performing architecture engineering work units to produce system architectural work products.

System architecture: the set of all the most important, pervasive, higher-level, strategic decisions, inventions, engineering trade-offs, assumptions, and their associated rationales concerning how the system meets its allocated and derived product and process requirements.

System architecture engineering: the subdiscipline of systems engineering consisting of all architectural work units performed by architecture workers to develop and maintain architectural work products (including system or subsystem architectures and their representations).

System requirement: a product requirement specifying a mandatory behavior or characteristic of a system.

Technological feasibility: the subtype of feasibility that is the degree to which the system or architectural component can be developed using existing technologies.

Testability: the internal quality characteristic that is the degree of ease with which a system or subsystem facilitates the creation and execution of successful tests (i.e., tests that cause failures due to underlying defects).

Throughput: the attribute of performance that is the degree to which the system is able to complete an operation and provide a service within acceptable time limits.

Timeliness: the attribute of correctness that is the degree to which data of a system or architectural component remains current (i.e., up to date).

Tolerance: the attribute of robustness that is the degree to which a system or subsystem tolerates potentially harm-causing events or conditions.

Transportability: the attribute of usability that is the degree of ease with which a physical system or architectural component can be transported from one physical location to another.

Understandability: the attribute of usability that is the degree to which users find the human interfaces and output of a system or architectural component to be clear, legible, unambiguous, and comprehensible (especially during unusual situations).

Usability: the external quality characteristic that is the degree to which the human user interface of the system or architectural component enables a specified group of users to achieve specified goals in a specified context of use.

User satisfaction: the attribute of usability that is the degree to which users are satisfied with the system or architectural component and consider it beneficial to them.

Verifiability: the attribute of modifiability that is the degree of ease with which changes to a system or subsystem can be verified as having been correctly made and to be without unexpected and undesirable side-effects.

View: *See* architectural view.

Viewpoint: *See* architectural viewpoint.

Waterfall [development] cycle: a sequential development cycle in which phases are based on sequentially performed development activities with little or no overlap (i.e., requirements engineering precedes architecture engineering, which precedes design, etc.).

Appendix B

MFESA Method Components

As described in the previous chapters, the MFESA repository stores numerous reusable architecture engineering method components. As illustrated in Figure B.1, these method components are defined in terms of a metamodel of work products, work units that create and update the work products, and workers who perform the work units.

B.1 Architectural Work Products

As illustrated in Figure B.1, the MFESA repository contains the following types of architectural work products:

- **Architecture process work products:** architectural work products supporting the performance of architectural engineering tasks.
- **Architectures and their components:** architectural work products consisting of the system architecture and its component parts.
- **Architectural representations:** architectural work products consisting of a cohesive set of information documenting part of a system architecture.

B.1.1 Architecture Process Work Products

The MFESA repository contains the following types of miscellaneous architecture work products, listed in alphabetical order:

- **[Architectural] change control analysis report:** any report documenting the results of the analysis of the impact of a proposed change, whereby the change has significant architectural ramifications.
- **[Architectural] change request:** any change request having significant ramifications on the architecture.
- **Architectural concern:** any cohesive collection of architectural drivers.

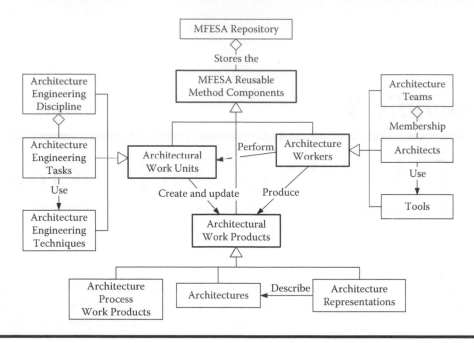

Figure B.1 Reusable method components in the MFESA repository.

- **[Architectural] discrepancy report:** any discrepancy report documenting perceived defects in any architectural work products
- **[Architectural] market survey:** any market survey of potentially reusable architectural components or technologies having significant architectural ramifications.
- **Architecture analysis report:** any report documenting the results of an analysis of all or part of the system architecture.
- **Architecture assessment report:** any report documenting the results of an architecture assessment.
- **Architecture engineering convention:** any document that constrains or guides the performance of the architecture engineering method, including checklists, examples, guidelines, procedures, standards, templates, and tool manuals.
- **Architecture engineering schedule:** any schedule documenting the actual past and planned future dates of the performance of architecture engineering tasks and the occurrence of architecture engineering milestones.
- **Architecture engineering tool evaluation report:** any report documenting the results of an architecture engineering tool evaluation.
- **Architecture engineering tool evaluation team charter:** a charter for an architecture tool evaluation team documenting team name, goals and objectives, tasks to perform, work products to produce, team profile, authority and management support, stakeholders, ground rules, success criteria, schedule and milestones, and required resources.
- **Architecture engineering training materials:** any training materials for teaching the architecture engineering method, the use of architecture engineering tools, and/or all or part of the system architecture.
- **Architecture inspection report:** any report documenting the results of an inspection of all or part of the architecture.

- **Architecture plan:** any plan for performing architectural work units to produce architectural work products.
- **Architecture prototype simulation report:** any report documenting the results of executing simulations on an architecture prototype.
- **Architecture prototype test report:** any report documenting the results of testing an architecture prototype.
- **Architecture quality assurance report:** any report documenting the results of a quality assurance evaluation of the architecture engineering process or a quality control evaluation of architectural work products.
- **Architecture-related requirements metadata:** any metadata describing requirements that are produced during the performance of architecture engineering tasks.
- **Architecture review report:** any report documenting the results of a review of all or part of the architecture.
- **Architecture team charter:** a charter for an architecture team documenting team name, goals and objectives, tasks to perform, work products to produce, team profile, authority and management support, stakeholders, ground rules, success criteria, schedule and milestones, and required resources.
- **Architecture training materials:** training materials covering the system architecture and used to train stakeholders (e.g., new members of the development team) of the architecture.

B.1.2 Architectures and Their Components

The MFESA repository contains the following types of architectures, listed in alphabetical order:

- **Architectural assumption:** any assumption that has significant ramifications on the architecture.
- **Architectural component:** any physical system element that is part of the static aggregation structure of a system.
- **Architectural decision:** any top-level, pervasive, strategic decision, the importance of which results from its significant impact on how the system and its elements [will] meet their derived and allocated requirements.
- **Architectural element:** any node in an architectural structure.
- **Architectural invention:** any top-level, pervasive, strategic invention, the importance of which results from its significant impact on how the system and its elements [will] meet their derived and allocated requirements.
- **Architectural rationale:** the rationale for making an architectural decision, invention, or trade-off.
- **Architectural structure:** any cohesive set of architectural elements connected by associated relationships that captures a set of related architectural decisions, inventions, trade-offs, assumptions, and rationales.
- **Architectural trade-off:** any top-level, pervasive, strategic engineering trade-off, the importance of which results from its significant impact on how the system and its elements [will] meet their derived and allocated requirements.
- **System architecture:** the entire set of all of the most important, pervasive, higher-level, strategic decisions, inventions, engineering trade-offs, assumptions, and their associated rationales concerning how the system meets its allocated and derived product and process requirements.

B.1.3 Architectural Representations

The MFESA repository contains the following types of architecture representations, listed in alphabetical order:

- **Architectural analysis report:** any report documenting the results of an analysis of the architecture such as cost, feasibility, performance, reliability, risk, safety, security, schedule, and trade-off.
- **Architectural concern versus vision component matrix:** a matrix documenting the degree to which each architectural vision component supports the satisfaction of the architectural concerns.
- **Architectural concern versus vision matrix:** a matrix documenting the degree to which each candidate architectural vision supports the satisfaction of the architectural concerns.
- **Architectural diagram:** Any diagram capturing a cohesive set of architectural decisions, inventions, trade-offs, and assumptions.
- **Architectural focus area:** documentation of that part of an architecture that is related to a specific architectural concern, regardless of the architectural structures in which they are found or the architectural views and models in which they are documented.
- **Architectural interfaces description document:** any document describing the interfaces between architectural components.
- **Architectural models:** any architectural representation that models a single system structure in terms of the structure's architectural elements, their relationships, and potentially their behaviors.
- **Architectural opportunity:** any recognized possibility for significantly improving the system architecture; any recognized set of architectural decisions, inventions, or trade-offs that if made will enable the system to exceed its requirements.
- **Architectural prototype:** any prototype that faithfully models one or more behaviors or characteristics of the system architecture.
- **Architectural quality case:** a document that makes the architects' case that their architecture has sufficient quality, consisting of a cohesive set of architectural *claims* (e.g., architecture sufficiently meets architectural drivers), clear and compelling *arguments* (e.g., architectural inventions and decisions, associated engineering trade-offs and assumptions, and their rationales) justifying belief in the claims, and official *evidence* supporting these arguments.
- **Architectural risk:** a measure of the expected lost (i.e., probability of loss times the severity of loss) due to defects or weaknesses within the architecture or its representations.
- **Architectural view:** any architectural representation describing a single architectural structure of a system consisting of one or more related models of that structure.
- **Architectural vision:** a conceptual overview of the architecture of the overall system or one of its architectural elements consisting of a cohesive and consistent set of architectural vision components.
- **Architectural vision component:** one or more important actual or potential architectural decisions, inventions, or trade-offs addressing one or more architectural concerns.
- **Architectural vision document:** a document describing the most suitable architectural vision.
- **Architectural vision selection report:** any report documenting the approach used to select the most suitable architectural vision and the associated evaluation results.
- **Architectural white paper:** any white paper documenting a single architectural focus area.

- **Architecture document:** any document that describes the system architecture or the architecture of one of its components, including views and models of its structures as well as other architectural decisions, inventions, trade-offs, assumptions, and their associated rationales.
- **Competing architectural visions list:** a list identifying the competing architectural visions from which a single vision is to be selected or created.

B.2 Architectural Work Units

As illustrated in Figure B.1, the MFESA repository contains the following types of architectural work units:

- **Discipline:** a work unit consisting of a cohesive collection of tasks that produce a related set of work products.
- **Task:** a work unit that models a single assigned job that may be performed by one or more workers.
- **Technique:** a work unit that models a way of performing a task.

B.2.1 Discipline

The MFESA repository contains the following reusable discipline:

- **System architecture engineering:** the subdiscipline of systems engineering consisting of all tasks performed to develop and maintain a system architecture and its associated architecture work products.

B.2.2 Tasks

The MFESA repository contains the following reusable architecture engineering tasks:

- **Task 1: Plan and resource the architecture engineering effort:** the architecture engineering task during which the overall architecture engineering effort is planned and the necessary resources to engineer the system architecture are obtained.
- **Task 2: Identify the architectural drivers:** the architecture engineering task during which the architecturally significant product and process requirements that have been derived for and/or allocated to their system or subsystem are identified and categorized into a set of architectural concerns.
- **Task 3: Create first versions of the most important architectural models:** the architecture engineering task during which the first versions of the most important architectural models are created. This typically consists of a consistent set of partial draft logical and physical, static and dynamic models of the system or subsystem based on the architectural drivers, the associated architectural concerns, and the opportunities for reuse.
- **Task 4: Identify opportunities for the reuse of architectural elements:** the architecture engineering task during which potential opportunities for the reuse of architectural decisions and inventions such as styles, patterns, and the relevant parts of existing architectural structures are identified and initially analyzed.

■ **Task 5: Create the candidate architectural visions:** the architecture engineering task during which the initial architectural models are used to create a set of competing candidate architectural visions for their system or subsystem that support meeting the derived and allocated architectural drivers and associated architectural concerns.

■ **Task 6: Analyze reusable components and their sources:** the architecture engineering task during which potentially reusable physical architectural components and their sources for reuse within candidate architectural visions are identified and evaluated.

■ **Task 7: Select or create the most suitable architectural vision:** the architecture engineering task during which either the most suitable architectural vision for the system or subsystem from the competing candidate architectural visions is selected or an even more suitable architectural vision is created by combining consistent components from multiple competing candidate architectural visions.

■ **Task 8: Complete the architecture and its representations:** the architecture engineering task during which the architecture and its representations are completed based upon the architectural vision selected or created.

■ **Task 9: Evaluate and accept the architecture:** the architecture engineering task during which the quality of the system or subsystem architecture is evaluated so that architectural risks can be managed, compliance with architecturally significant requirements can be determined, and the architecture can be accepted by its authoritative stakeholders.

■ **Task 10: Maintain the architecture and its representations:** the architecture engineering task during which the architecture and its representations are maintained and the integrity of the system architecture is ensured to not degrade over time.

B.2.3 Techniques

The MFESA repository contains the following reusable architecture engineering techniques:*

■ **Active listening:** the technique during which one or more architects or members of an architecture team carefully:
  - Listen to a stakeholder, focusing totally on what a stakeholder or subject matter expert (SME) is saying instead of what the listeners are going to say next.
  - Reiterate what the stakeholder or SME has said in order to ensure that the listeners correctly understand what the stakeholder or SME has said.

■ **Brainstorming:** the technique during which architecture stakeholders collaborate to rapidly and informally produce and then prune a list of possible architectural decisions, inventions, and trade-offs.

■ **Cost/benefit analysis:** the technique during which the relative cost and benefits of engineering an architectural work product are identified and compared.

■ **Gap analysis:** the technique during which the differences between the current architecture and the desired architecture are identified and analyzed.

■ **Incremental development:** the technique during which architectural work products are produced incrementally (i.e., during which additions are made to partial architectural work products until they are completed).

---

* Although these are general techniques having wide applicability beyond system architecture engineering, they are documented here in an architecture-specific manner to clarify how they relate to architecture engineering.

- **Iterative development:** the technique during which all or part of one or more work units are repeated in order to improve an existing architectural work product (e.g., by removing defects or improving the ability of the system or architectural component to meet its architecturally significant requirements).
- **Modeling:** the technique of producing and analyzing architectural models (i.e., simplified abstractions of a part of the architecture such as an architectural structure).
- **Parallel development:** the technique during which different workers are performing their tasks concurrently (e.g., multiple architects, multiple architecture teams, multiple teams of different types, or multiple organizations).
- **Prototyping:** the technique consisting of the production of prototypes to better understand the system or architectural component and to answer related questions.
- **Simulation:** the technique of producing an executable model of part of the system architecture in order to analyze dynamic aspects of the architecture.
- **Time-boxing:** the technique during which architectural work units are broken into small, manageable pieces that are limited in time via having set beginning and ending times (typically either large milestones or shorter "inch-pebbles").

B.3 Architecture Workers

As illustrated in Figure B.1, the MFESA repository contains the following types of architecture workers:

- **Architects:** the roles that are played by people when they perform one or more tasks of architecture engineering.
- **System architecture teams:** teams that are responsible for developing and maintaining all or part of a system's architecture.
- **System architecture tools:** tools that assist with the production, coordination, and maintenance of architectural work products.

B.3.1 Architects

The MFESA repository contains the following reusable types of architects, listed in alphabetical order:

- **Chief enterprise architect:** any system architect who leads an overall enterprise architecture team and is responsible for the architecture engineering of an entire enterprise architecture (i.e., the architecture of a system of systems, family of systems, network of systems, or a product line of systems).
- **Chief system architect:** any system architect who leads the overall system architecture team and is responsible for the architecture engineering of a single system.
- **Hardware architect:** any hardware engineer who is responsible for the hardware architecture (i.e., the hardware aspects of a system architecture).
- **Lead architect:** any system architect who leads a subordinate architecture team that is responsible for a single architectural component (e.g., subsystem, software configuration item, hardware configuration item) of the system architecture.
- **Software architect:** any software engineer who is responsible for the software architecture (i.e., the software aspects of a system architecture).

- **System architect:** the highly specialized role played by a systems engineer when performing system architecture engineering tasks to produce system architecture engineering work products.

B.3.2 System Architecture Teams

The MFESA repository contains the following reusable types of system architecture teams, listed in alphabetical order:

- **Customer architecture team:** any system architecture team that is a part of the organization that acquires the system.
- **Enterprise architecture team:** any system architecture team responsible for developing and maintaining the enterprise architecture.
- **Prime contractor/integrator architecture team:** any system architecture team that is part of the organization responsible for developing the system and integrating its major architectural components.
- **Product line architecture team:** any system architecture team responsible for developing and maintaining the common overall architecture of a product line of systems.
- **Specialty engineering architecture team:** any system architecture team responsible for developing and maintaining those aspects of the system architecture related to the area of specialty engineering (i.e., focus areas).
- **Subcontractor architecture team:** any system architecture team that is a part of the organization that develops a major architectural component for the prime contractor or integrator of the system.
- **Subsystem architecture team:** any system architecture team responsible for developing and maintaining the architecture of a single subsystem.
- **Supplier/vendor architecture team:** any system architecture team that is a part of the organization that produces off-the-shelf architectural components and that supplies them to the prime contractor/integrator for incorporation into the system.
- **System of systems architecture team:** any system architecture team responsible for developing and maintaining the overall architecture of a system of systems, network of systems, or family of systems.
- **Top-level system architecture team:** the single system architecture team responsible for developing and maintaining the entire overall top-level architecture of a single system.

B.3.3 System Architecture Tools

The MFESA repository contains the following reusable types of system architecture tools, listed in alphabetical order:

- **Business process modeling tool:** any software application that supports the modeling of the business processes that either drive the system requirements or need to be reengineered to take advantage of the new system.
- **CAD/CAM (Computer-Aided Design/Computer-Aided Manufacturing):** any combination of software applications and possibly specialized hardware platforms primarily focused on the creation and maintenance of two- and three-dimensional models of various parts of a system's physical architectural structures.

- **Configuration management tool:** any software application primarily used to establish controlling and maintaining the integrity of architectural baselines and their constituent architectural work products.
- **General-purpose drawing tool:** any software application primarily focused on the creation and maintenance of domain-independent graphical representations of ideas.
- **Graphical modeling tool (UML, SysML, and AADL):** any software application primarily focused on the creation and maintenance of constructs that model certain structural and behavioral aspects of systems architecture using a standardized representation and semantics.
- **Image capturing device:** any device used to capture and preserve an image representation of an invention, algorithm, decision, rationale, model, etc., that is generated by means other than the device itself during the creation of architectural work products.
- **Information architecting tool:** any software application used to produce logical data models of the information used by the system.
- **Mass/size/geometry modeling tool:** any software application used to model the mass, size, and geometry of the system and its subsystems.
- **Simulation tool:** any software application primarily focused on imitating specific types of architectural structures by predicting how they will behave in the context of specific scenarios.
- **Spreadsheet:** any software application primarily focused on the creation and maintenance of tabular data with associated tabular computational capabilities built in.
- **Requirements tracing tool:** any software application used to trace requirements from their sources and to the architectural components that implement them.
- **Whiteboard:** any (typically) mechanical or electronic medium for capturing graphic or text representations of architectural concepts, rationale, decisions, inventions, and/or trade-offs.
- **Word processor:** any software application primarily focused on the creation and maintenance of textual documents and typically run on general-purpose computers or workstations.

Appendix C

List of Guidelines and Pitfalls

The following provides a quick reference to the MFESA guidelines and pitfalls, listed by task:

MFESA Overview (Chapter 4)

- **Guidelines:**
  - Remember that MFESA is not a system architecture engineering method.
  - Remember that MFESA is more than a repository of reusable method components.
  - Remember that MFESA has multiple uses.
  - Remember that MFESA methods use both requirements and reusable architectural elements as inputs.
  - Use MFESA tasks to organize and understand.
  - Remember that MFESA affects more than just the architects.

MFESA Ontology of Concepts and Terminology (Chapter 5)

- **Guidelines:**
  - Remember that systems typically contain more than software and hardware.
  - Assumptions and rationales are important parts of the architecture.
  - Engineer the entire system architecture, not just the system structures.
  - Architect all important types of system structures.
  - Create other system architectural representations in addition to graphical models and views.
  - Address architectural focus areas that cut across multiple models and views.
  - Strive to keep the different types of models, views, and focus areas consistent.
  - While useful, architectural patterns and styles are insufficient by themselves.
  - Architectural concerns should center on architecturally significant requirements.
  - Architectural quality cases are critical.

- Create multiple candidate architectural visions.
- Do not forget executable architectural representations.

■ **Pitfalls:**
- Architectures are confused with designs and architecture engineering is confused with designing.
- Architectures are confused with structures or models of structures.
- Trade-offs, assumptions, and rationales are ignored.
- Models and views are engineered, but not focus areas.
- Architectural quality cases are not developed or are only developed long after the fact for assessments.

Task 1: Plan and Resource the Architecture Engineering Effort (Chapter 6)

■ **Guidelines:**
- Properly staff the top-level architecture team(s).
- Properly plan the architecture engineering effort.
- Produce and maintain a proper and sufficient schedule.
- Create or reuse appropriate MFESA method(s).
- Select appropriate architecture modeling method(s).
- Select appropriate architecture engineering tools.
- Provide appropriate training.

■ **Pitfalls:**
- Architects produce incomplete architecture plans and conventions.
- Management provides inadequate resources.
- Management provides inadequate staff and stakeholder training.
- Architects lack authority.
- Architects instantiate the entire MFESA repository without tailoring.
- Tool vendors drive architecture engineering and modeling methods.
- Planning and resourcing are unsynchronized.
- Planning and resourcing are only done once up front.

Task 2: Identify the Architectural Drivers (Chapter 7)

■ **Guidelines:**
- Collaborate closely with the requirements team.
- Notify the requirements team(s) of relevant requirements defects.
- Challenge difficult requirements.
- Consider the impact of the architecture on the requirements.
- Respect team boundaries and responsibilities.
- If necessary, clarify relevant requirements with the stakeholders.
- Concentrate on the architecturally significant requirements.
- Remember that quality attributes can be architectural concerns too.
- Formally manage architectural risks.

■ **Pitfalls:**
  - All requirements are not architecturally significant.
  - Well-engineered, architecturally significant requirements are lacking.
  - Architects rely excessively on functional requirements.
  - The architects ignore the architecturally significant functional and process requirements.
  - Specialty engineering requirements are misplaced.
  - Unnecessary constraints are imposed on the architecture.
  - Architects engineer architecturally significant requirements.
  - Requirements lack relevant metadata.
  - Architects fail to clarify architectural drivers.

Task 3: Create the First Versions of the Most Important Architectural Models (Chapter 8)

■ **Guidelines:**
  - Perform architectural trade-off analysis.
  - Reuse architectural principles, heuristics, styles, patterns, vision components, and metaphors.
  - Use a recursively incremental, iterative, parallel, and time-boxed development cycle.
  - Begin developing logical models before beginning to develop physical models.
  - Do not overemphasize the physical decomposition hierarchy.
  - Use explicitly documented system-partitioning criteria.
  - Concentrate on the interfaces.
  - Model concurrency.
  - Consider the impact of hardware decisions on usability and software.
  - Consider human limitations when allocating system functionality to manual procedures.
  - Do not start from scratch.
  - Formally manage architectural risks.

■ **Pitfalls:**
  - The architects succumb to analysis paralysis.
  - The architects engineer too few architectural models.
  - The architects engineer inappropriate models and views.
  - The architects construct views but no focus areas.
  - Some stakeholders believe that the models are the architecture.
  - Inconsistencies exist between models, views, and focus areas.
  - The architects use inappropriate architectural patterns.
  - System decomposition is performed by the acquisition organization.

Task 4: Identify Opportunities for the Reuse of Architectural Elements (Chapter 9)

■ **Guidelines:**
  - Do not start from scratch.
  - Do not be excessively constrained by the past.

- Conform to the enterprise architecture.
- Conform to the product line reference architecture.
- Consider system architecture patterns.
- Support modeling.
- Formally manage architectural risks.
- **Pitfalls:**
  - The architects start from scratch.
  - The architects ignore past lessons learned.
  - The architects overly rely on previous architectures.
  - The architects select specific OTS components too early.
  - The architects assume the reusability of immature architectural components.
  - The architects assume the reusability of immature technologies.
  - Inadequate information exists to determine reusability.

Task 5: Create the Candidate Architectural Visions (Chapter 10)

- **Guidelines:**
  - Identify an appropriate number of candidate architectural visions.
  - Complete candidate architectural visions to appropriate level of detail.
  - Prepare architectural components for OTS incorporation.
  - Formally manage architectural risks.
- **Pitfalls:**
  - The architects engineer only one architectural vision.
  - Management provides insufficient resources.
  - Management confuses the architectural vision with the completed architecture.
  - Management does not permit architects to make mistakes.
  - The architects compare the architectural visions prematurely.
  - The architects do not compare the pros and cons of the candidate visions.

Task 6: Analyze Reusable Components and Their Sources (Chapter 11)

- **Guidelines:**
  - Use appropriate decision techniques.
  - Perform task concurrently.
  - Formally manage architectural risks.
- **Pitfalls:**
  - Authoritative stakeholders assume reuse will improve cost and schedule.
  - Insufficient information exists for evaluation and reuse.
  - Stakeholders have an unrealistic expectation of "exact fit."
  - Developers have little or no control over future changes.
  - The source organization (e.g., vendor) fails to adequately maintain a reusable architectural component.

- Legal rights are unacceptable.
- Incompatibilities exist with underlying technologies.

Task 7: Select or Create the Most Suitable Architectural Vision (Chapter 12)

- ■ **Guidelines:**
  - Ensure a commensurate approach.
  - Ensure a consistent evaluation approach.
  - Ensure complete evaluation criteria.
  - Avoid unwarranted assumptions.
  - Use common sense when using decision methods to select the most suitable candidate architectural vision.
  - Take reuse into account.
  - Test reusable architectural component suitability.
  - Maintain the architectural vision.
  - Formally manage architectural risks.
- ■ **Pitfalls:**
  - Architects use an inappropriate decision method.
  - Management provides inadequate decision resources.
  - Selection is viewed as purely a technical decision.
  - Stakeholders do not understand risks.
  - The decision makers are weak.

Task 8: Complete the Architecture and Its Representations (Chapter 13)

- ■ **Guidelines:**
  - Develop quality cases as a natural part of the architecture engineering process.
  - Architect all relevant types of interfaces.
  - Work with the requirements team to provide requirements traceability.
  - Formally manage architectural risks.
- ■ **Pitfalls:**
  - Architecture engineering is finished.
  - Management provides inadequate resources.
  - The architectural representations lack configuration control.

Task 9: Evaluate and Accept the Architecture (Chapter 14)

- ■ **Guidelines:**
  - Use evaluations to support architectural milestones.
  - Evaluate continuously.
  - Internally evaluate models.
  - Perform architecture evaluation substeps.

- Collaborate with the stakeholders.
- Tailor software evaluation methods.
- Perform independent architecture assessments.
- Formally review the architecture.
- Verify architectural consistency.
- Perform cross-component consistency checking.
- Perform both static and dynamic checking.
- Set the evaluation scope based on risk and available resources.
- Formally manage architectural risks.
- **Pitfalls:**
  - Disagreement exists over the need to perform evaluations.
  - Consensus does not exist on the evaluation's scope.
  - It is difficult to schedule the evaluations.
  - Management provides insufficient evaluation resources.
  - There are too few evaluations.
  - There are too many evaluations.
  - How good is good enough?
  - Evaluations are not sufficiently independent.
  - The evaluators are inadequate.
  - Evaluations only verify the easy concerns.
  - The architectural quality cases are poor.
  - Stakeholders disagree on the evaluation results.
  - The evaluations lack proper acceptance criteria.
  - The evaluation results are ignored during acceptance.
  - The acceptance package is incomplete.

Task 10: Maintain the Architecture and Its Representations (Chapter 15)

- **Guidelines:**
  - Maintain the architectural representations to maintain architectural integrity.
  - Consider the entire scope of *Maintain the Architecture and Its Representations* task.
  - Consider the sources of architectural change.
  - Protect the architectural invariants.
  - Determine the scope of architectural integrity.
  - Train the architects and designers.
  - Formally manage architectural risks.
- **Pitfalls:**
  - The architectural representations become shelfware.
  - Architecture engineering is finished.
  - The architecture is not under configuration management.
  - The architecture is not maintained.
  - A "beautiful" architecture is frozen solid.
  - There is inadequate tool support for architecture maintenance.

Architectural Workers (Chapter 16)

- **Architect Guidelines:**
  - Provide proper authorization.
  - Empower a chief architect.
  - Develop listening and elicitation skills.
  - Encourage professional participation.
  - Wear multiple hats.
- **Architect Pitfalls:**
  - The architect knows best.
  - The architect is weak willed.
- **Architecture Team Guidelines:**
  - Develop a single shared vision.
  - Earn architecture team respect.
  - Development architects should not be replacement architects.
  - Subsystem architects should be members of the system architecture team.
  - Differentiate architecture team members from collaborators.
- **Architecture Team Pitfalls:**
  - There is only one architect.
  - The architects are isolated.
  - The architects have inadequate experience.
- **Architecture Tool Guidelines:**
  - Let the modeling method drive tool selection.
  - Use trade studies and pilot projects to evaluate new tools.
  - Provide training before tool use.
  - Obtain management commitment.
- **Architecture Tool Pitfalls:**
  - Beware of vendors selling silver bullets.
  - Architecture tools are poorly integrated and lack interoperability.
  - Architecture tools have poor support for multiple modeling languages.
  - Architecture tools have poor support for all models.
  - Low-level software design tools are used as system architecture tools.
  - Management underestimates effort and resources needed.
  - A fool with a tool is still a fool.
  - Architecture tools lack support for architectural metrics.

MFESA Metamethod (Chapter 17)

- **Guidelines:**
  - Ensure consistency.
  - Construct system architecture engineering methods early.
  - Iterate as lessons are learned.
  - Ensure the right level of formality.

- ■ **Pitfalls:**
  - There is a lack of expertise in method engineering.
  - It is difficult to keep the constructed method *internally* consistent.
  - It is difficult to keep the constructed method *externally* consistent.
  - There is a lack of adequate tool support.

Architecture and Quality (Chapter 18)

- ■ **Guidelines:**
  - Use a single common quality model.
  - Ensure proper quality requirements.
  - Develop architectural quality cases.
  - Obtain adequate resources.
- ■ **Pitfalls:**
  - The endeavor lacks a quality model.
  - The endeavor uses a poorly defined and poorly documented quality model.
  - The definitions of the quality characteristics and attributes are inadequate.
  - There is disagreement over the relative importance of the quality characteristics and attributes.
  - The quality requirements lack thresholds.
  - The quality cases contain inappropriate evidence.
  - Disagreement exists over the need to perform quality assessments.
  - Disagreement exists over the scope of the quality assessment.
  - There is no quality case index and summary.

Decision-Making Techniques (Appendix D)

- ■ **Guidelines:**
  - Independently evaluate alternatives.
  - Document the results of using the decision method.
- ■ **Pitfalls:**
  - The decision criteria are incomplete.
  - The subjective decision criteria are neglected.
  - The decision makers have tunnel vision.
  - The decision process includes systematic biases.
  - The decision makers have *a priori* preferences.

Appendix D

Decision-Making Techniques

By definition, the architecture of a system is the set of the most important, pervasive, top-level, strategic *decisions*, inventions, engineering trade-offs, assumptions, and their associated *rationales* concerning how the system and its components collaborate to meet their derived and allocated requirements. Although a few of these architectural decisions, decisions, and trade-offs can be made by an individual architect, many will require the collaboration of multiple architects within a single architecture team or the collaboration of multiple architecture teams. Finally, whereas many of these architectural decisions can be made informally as a result of brainstorming and discussion, some of these decisions will be both sufficiently complex and critical as to justify following a more formal decision-making method. While these decision-making methods can be used anywhere in the architecture engineering method, they are most commonly used for selecting among alternative:

- Architecture tools to acquire and use
- Reusable architectural components to incorporate into the architecture
- Architectural visions from the competing candidate architectural visions

As illustrated in Figure D.1, more formal decision making can be divided into six tasks. Although these tasks are documented in a strictly sequential manner for the sake of understandability, they are often performed in an incremental, iterative, parallel, and time-boxed manner.

D.1 Decision Alternatives Determination

The objective of decision-making task DM1 is to determine the set of alternatives to be evaluated so that the most appropriate can be selected. Depending on the MFESA task, this could involve identifying:

- An appropriate set of alternative architecture tools (MFESA Task 1)
- An appropriate set of alternative reusable architectural components (MFESA Task 6)
- The set of candidate architectural visions (MFESA Task 7)

449

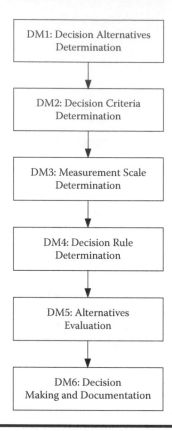

Figure D.1   A generic decision-making method.

D.2 Decision Criteria Determination

The objective of the decision-making task DM2 is to determine the criteria against which to evaluate the alternatives identified during step DM1. Examples of such decision criteria include:

- **Architecture tool decision criteria**
  - Example tool criteria include functionality, quality, interoperability, portability, ease of use, and cost.
  - Example tool vendor criteria include licensing, vendor market share, vendor longevity, and tool support.
- **Reusable architectural component decision criteria**
  - Example component criteria include functionality, quality characteristics, and cost.
  - Example source criteria include licensing, vendor market share, vendor longevity, and consulting/training support.
- **Architectural vision decision criteria**
  - Criteria include support for functional and quality requirements, simplicity and understandability, risk, and cost to implement.

The decision criteria determination task presumes that there is a means to compare the options in an objective manner. The set of architecturally significant requirements is the most direct source of evaluation criteria for reusable architectural components and architectural visions.

D.3 Measurement Scale Determination

The objective of the decision-making task DM3 is to determine the measurement scales to use in measuring how well the alternatives support the decision criteria determined during step DM2. The four main types of measurement scales include:

1. Subjective scaling
2. Quality attribute scaling
3. Requirements scaling
4. Decision-theoretic methods

If mapping exists between the architectural structures (see Figure 5.5) and architectural drivers and/or concerns, then an evaluation mechanism may be created that directly maps to the drivers and/or concerns. In the absence of such mapping, then a more explicitly subjective evaluation may be constructed. (*Note:* For the summary below, we do not treat the aggregation of the assessments of multiple advisors/evaluators to create an evaluation of an architectural vision against a particular architectural concern. This is treated in the literature, a classic reference being [Keeney and Raiffa, 1976].)

Note, however, that subjective decision criteria are sometimes appropriate, especially when a quantitative assessment of architectural complexity is infeasible — for example, because the construction of the measure itself may be as expensive or time-consuming as creating an architectural vision. In such instances, a qualitative assessment of complexity is a more appropriate approach.

D.3.1 Subjective Scaling

Subjective scaling involves the creation of a reasonable subjective nominal scale for each decision criterion. For example, when evaluating competing architectural visions against their support for the architectural concern "security," the following hypothetical nominal measurement scale could be used:

- **Excellent:** the vision sufficiently supports all security-related requirements and supports all quality attributes of security using state-of-the-art countermeasures.
- **Good:** the vision sufficiently supports all security-related requirements and supports all security attributes using better than simply adequate countermeasures.
- **Adequate:** the vision sufficiently minimally supports all security-related requirements.
- **Inadequate:** the vision currently fails to sufficiently support all security-related requirements but can easily be made to do so.
- **Unacceptable:** the vision makes it difficult or impossible for the system to sufficiently support all security-related requirements.

Note that subjective nominal scaling is highly subjective and the only formal attribute such a scale offers is that there is a total order for all values. There is no notion that "Good" is as much of an improvement over "Adequate" as "Adequate" is over "Inadequate." Any attempt to map such a scale to numerical values is fraught with potential inconsistencies and illogic.

Many people in the role of decision maker wish to have the comfort of numerical scores, whether they contribute anything to the final result. There are a few means to construct numerical scores that have some logical meaning, and all depend on additional structure (and effort). We illustrate two examples of such measures.

D.3.2 Quality-Attribute-Based Scaling

For each architectural concern, the attributes identified (see Chapter 10) may serve as a means to refine the wholly subjective model just introduced. In this case, each of the attributes of the architectural concern would be assessed for a given architectural vision, and a score might be the number of attributes assessed to be "Acceptable" or better. One should resist assuming any significance to different scores, save for "higher is better."

D.3.3 Requirements-Based Scaling

It is possible to use the architecturally significant requirements comprising each architectural concern as test cases when evaluating reusable architectural components or competing architectural visions. For each architectural concern, let us suppose that the architects and stakeholders have taken the time to construct a set of scenarios, or use cases, that address the range of potential issues relevant to the particular concern. Sharp practice would be to construct a set of vision-independent use cases that explicitly covers all the known attributes of the architectural concern. The evaluation of an architectural vision in a single dimension would be to count the number of use cases that the architectural vision successfully supports.

A numerically higher score could easily be seen as better. As an arbitrary example, an architectural vision that adequately supports 80 percent of the requirements in a given concern ought not to be thought of as twice as good as a vision that passes 40 percent of the requirements. A more appropriate approach is to establish a high threshold that serves to filter out visions with scores less than the threshold with respect to that concern. Alternatively, there may be a separation of requirements into subsets, where "passing" may require satisfying all the requirements in a subset considered *critical*, and comparison may be based on successful coverage of other, less-critical requirements.

D.3.4 Decision-Theoretic Methods

It is possible to create meaningful numerical scales where "4" is indeed twice as good as "2." The rub is that there is a rather expensive process involved that requires knowledgeable interviewers, a considerable time investment on the part of the decision makers, and nontrivial analysis. In essence, this approach constructs utility functions from a set of hypothetical wagers that compare the decision criteria as well as establish scales for individual criteria [Keeney and Raiffa, 1976]. For projects that are of considerable duration, high cost, and large impact upon stakeholders, these methods may be worthy of consideration. However, using this approach is likely to require significant expertise, experience, and training in the field of decision analysis, something that few systems engineers, architects, and managers adequately possess.

D.4 Decision Rule Determination

The construction of a decision rule is no less challenging than creating meaningful scales of measurement for each decision criterion. The following treatment of the common approaches will be in reverse order from the discussion of individual scales. In the absence of other information at hand, this order presents the types of decision rules in descending order of desirability. The filtering methods that follow the usual weighted sum are more applicable when the number of alternatives is quite large, suggesting that some kind of pre-filtering mechanism would be useful to lead to a more tractable number of alternatives subjected to detailed analysis.

D.4.1 Weighted Sum of Scores

The weighted sum of scores is an attractive and simple model:

$$WSS = \sum_{1}^{n} W_i S_i$$

where *WSS* is the weighted sum of scores, *n* is the number of architectural concerns, W_i is the weight for concern *i,* and S_i is the score for concern *i*.

A valid employment of a weighted sum of scores for individual architectural concerns is a legitimate approach under a fairly restrictive set of assumptions:

- ■ Measurements against each concern are consistent with a structure-preserving map to integers and/or real numbers. In practice, this is predicated upon the use of the decision-theoretic methods noted above.
- ■ The relative importance of each pair of architectural concerns can be agreed to, leading to a set of weights that captures this set of pair-wise comparisons.
- ■ The proportional relationships between the concerns are linear over the complete range of possible measurements.

The decision rule is to simply create a score for each of the competing architectural visions, and the maximum score would correspond to the most attractive option based on the input from decision makers.

D.4.2 Filtering Based on Relative Importance

Suppose an agreement among the decision makers can be achieved regarding the relative importance of the decision criteria. Note that the degree of difficulty in achieving this agreement should not be underestimated. With such an agreement, one can construct a total ordering of the decision criteria that represents the agreement on their relative importance: $C_1...C_n$, where C_1, is the most important decision criterion.

If we desire to avoid making inadvertent assumptions about the data at hand to support our decision, the simple weighted sum is no longer an appropriate decision rule. Let us treat the evaluation of the competing alternatives against a given decision criterion as a *filtering process*. The assessment for each alternative is compared to any thresholds or lower limits imposed for the most important decision criterion; an alternative is said to *survive the stage* if its evaluation passes the threshold on the measurement scale for the decision criterion. The set of *survivors* is then subjected to the same process for the next most important criterion, and so on. The process terminates when one alternative remains in the set or all criteria have been employed. If two or more competing alternatives satisfy all concerns, then some sort of decision process should be devised to select a single alternative; alternatively, early in a project, all surviving alternatives might be further elaborated, with the decision method repeated at a later time. If too many alternatives survive, this may be a happy occasion, or an indicator that the thresholds for individual alternatives should be made a bit less forgiving.

This filtering method is fairly robust, and mainly assumes that the assessments of each alternative are made separately, without resort to explicit pair-wise comparisons. This is not easy to accomplish in practice.

When pair-wise comparisons of the alternatives are allowed, explicitly or implicitly, the result of such a filtering method is vulnerable to changes in order. The agreement upon an ordering of importance of the decision criteria is crucial to making this method defensible.

D.4.3 Filtering When Relative Importance Is Not Available

When no agreement on the relative importance of the decision criteria is attainable, a filtering method can be usefully performed. However, we need to be careful not to allow the order in which the decision criteria are employed to contaminate our results. The order in which evidence is considered has been shown to materially affect the results of decision methods ([Madigan et al., 1996; Murphy, 2001; Schum, 2002]). Briefly, order effects are apparent when the decision method is recast as a cascaded (Bayesian) inference that generates a tree-like structure (sometimes called a belief network) where nodes indicate inferences and edges are labeled with conditional probabilities. The effect of order can be eliminated with a procedure related to the one above.

For each permutation P_k of the set of decision criteria, perform the procedure noted in the section above, resulting in a set of survivors S_k. Take the union of all such sets, S. This is the set of all alternatives that can survive consideration of the process under any ordering of the decision criteria. Again, if this set is too large, it is likely that the thresholds in individual criteria need to be adjusted. In the case of multiple survivors, there are readily available tie-breaking rules: favor the alternatives that survive under more permutations; favor the alternatives that survive the most steps over all permutations. For the context of selecting an alternative among a set of candidates, the number of alternatives and the number of decision criteria are sufficiently constrained that analyses by contemporary personal computing systems are adequate to automate the task.

D.5 Alternatives Evaluation

Evaluate each alternative against each decision criterion using the associated measurement scale. Document the results of the evaluation.

D.6 Decision Making and Documentation

Use the decision rule to select the most suitable alternative, based on the evaluation results (i.e., the associated values on the measurement scales for the different decision criteria). Document the decision and the associated results from using the decision rule.

D.7 Guidelines

Use the following guidelines (and associated rationales) regarding decision making:

- Independently evaluate alternatives.
- Document the results of using the decision method.

The following descriptions express these guidelines in more detail and provide their associated rationales:

- **Independently evaluate alternatives.** Carry out the evaluation of each alternative relative to each decision criterion independently. Unfortunately, this is more difficult to accomplish than one might otherwise assume. If there are a sufficient number of knowledgeable evaluators, there is value in having small groups dedicated to assessments of individual decision criteria.

  Rationale: This avoids contaminating evaluations with information gleaned from other evaluations. The "cleanliness" of such procedures is a prime concern when the results affect the award of contracts or have other significant effects upon people and organizations.

- **Document the results of using the decision method.** Document the results of performing each task in the decision method, including the:
  - **Alternatives.** List the alternatives evaluated and the rationale for their selection.
  - **Decision criteria and measurement scales.** Document the decision criteria, their associated measurement scales, and the rationales for their selection.
  - **Decision rule.** Document the decision rule for selecting the most suitable alternative as well as the rationale for its choice.
  - **Evaluation results.** Document the results of evaluating each alternative against its decision criteria using the associated measurement scale.
  - **Decision.** Document the final decision based on using the decision rule on the results of evaluating each alternative against its decision criteria using the associated measurement scale

  Rationale: Having a good record of the decision method used provides confidence in the decision results by enabling others to verify its adequacy and appropriateness.

D.8 Pitfalls

Avoid the following common pitfalls associated with decision making and mitigate their negative consequences when they occur:

- The decision criteria are incomplete.
- The subjective decision criteria are neglected.
- The decision makers have tunnel vision.
- The decision process includes systematic biases.
- The decision makers have *a priori* preferences.

The following express these pitfalls in more detail, describing their negative consequences and the steps that architects can use to mitigate them:

- **The decision criteria are incomplete.**
  - **Pitfall.** Incomplete decision criteria omit important architectural considerations, leading to erroneous decisions.

- **Negative consequence.** This pitfall often causes the following negative consequence:
  - Suppose a set of competing architectural visions is evaluated only relative to operational scenarios. For a long-lived system, the operational scenarios may not address maintenance issues such as the total skill set(s) required to support the to-be-developed system throughout its life cycle. This skill set may be significantly influenced by large and diverse exploitation of COTS. In this case, downstream support costs induced by architectural visions may be ignored or insufficiently recognized.
- **Mitigation.** To avoid or mitigate the negative consequences of this pitfall:
  - Selection of decision criteria should be made by a group that includes not only architects, but also subject matter experts and other engineering disciplines such as logistics and sustainment.

■ **The subjective decision criteria are neglected.**
  - **Pitfall.** Evaluation criteria are limited to those that are readily quantified.

    It is tempting to ignore any criteria that do not readily admit a quantitative measurement. However, subjective considerations can often represent important features that are not easily quantified. For example, a quantitative assessment of architectural complexity may be infeasible because the construction of the measure itself may be as expensive or time-consuming as creating an architectural vision. In such an instance, a qualitative assessment of complexity is a more appropriate approach.
  - **Negative consequences.** This pitfall often causes the following negative consequences:
    - Crucial selection criteria are ignored.
    - The wrong decision is made.
  - **Mitigations.** To avoid or mitigate the negative consequences of this pitfall:
    - Ensure that subjective evaluation criteria are considered for inclusion.
    - Avoid the temptation to ignore subjective criteria.

■ **The decision makers have tunnel vision.**
  - **Pitfall.** The decision makers are too uniform in viewpoint or experience.
  - **Negative consequences.** This pitfall often causes the following negative consequences:
    - The decision makers are in substantial agreement and are replicating one basic viewpoint.
    - The actual ratings may fail to capture the full spectrum of relevant viewpoints.
    - The wrong decision is made.
  - **Mitigation.** To avoid or mitigate the negative consequences of this pitfall:
    - Include decision makers that span the set of stakeholders for the decision in question.

■ **The decision process includes systematic biases.**
  - **Pitfall.** Vigilance is required to avoid the introduction of systematic biases into the decision procedure.

    There are a number of sources for bias in a decision procedure. Similar training/ discipline (as in the tunnel vision pitfall above) or inadvertent institutional bias can be introduced into the decision process by having all raters from the same organization.
  - **Negative consequence.** This pitfall often causes the following negative consequence:
    - The systematic biases neglect other relevant inputs, resulting in the wrong decision being made.
  - **Mitigation.** To avoid or mitigate the negative consequence of this pitfall:
    - Include evaluators that span the set of viewpoints, background, and stakeholders for the decision in question.

■ **The decision makers have *a priori* preferences.**

- **Pitfall.** The evaluation parameters are adjusted until the previously preferred decision comes out on top.

  Occasionally, one solution may be preferred over other solutions before any objective evaluation has been preformed.* However, once the evaluation method indicates a preference for another solution, many times the evaluation parameters are adjusted until the solution that was "supposed to be the best" comes out on top.

  In this case, there are multiple potential root causes, of which we discuss two here: predetermined solution and incomplete analysis factors.

  First, the predetermined solution is when management has already selected a solution and the evaluation step is *pro forma*, meaning that the evaluation should confirm what management has already decided. While there may not be a direct fix for this situation, using the evaluation to inform risk and opportunity management processes can be helpful, if the original evaluation (the objective one that had the management's solution ranked not at the top) can be used as a basis.

  Second, and less commonly occurring, is that the method may not have initially considered all the various factors needed to inform a decision. Typically, excluded factors include stakeholder acceptability, solution technical maturity, budget profile fit, and other "soft" factors that can drive a project. Eliciting these additional factors may help improve the quality and utility of the evaluations to inform the decision makers.

- **Negative consequences.** This pitfall often causes the following negative consequences:
  - The decision process may be seen as having been "rigged," placing the objectivity of the architecture process in question.
  - Prejudices of the decision makers will lead to suboptimal architectural decisions.
  - In cases where architectural choices lead to business decisions with contractual consequences, prejudices of this sort may invite litigation.

- **Mitigations.** To avoid or mitigate the negative consequences of this pitfall:
  - Establish and publish evaluation criteria, weightings of the decision criteria, and the aggregation process before initiating the actual evaluation process.
  - Include evaluators with no direct stake in the outcome of the decisions to be made.

D.9 Summary

Decision-theoretic approaches for taking decisions involving multiple criteria offer both problems and opportunities for the architect. The mechanisms offered by the decision analysis literature provide a useful framework for posing and taking decisions of architectural alternatives. There is a degree of subjectivity in the process that is unavoidable. The best we can do is to establish decision criteria, assessment scales, and decision models that make subjectivity visible to decision makers and other stakeholders. There are a number of tools that mechanize useful subsets of the approaches described here and elsewhere in the literature. We do not recommend any particular tool, yet we encourage the use of appropriate automation to facilitate the assessment process. Those who employ such tools are obligated to fully understand the assumptions and limitations

---

* This is not necessarily a problem. If a preferred solution already exists, project personnel should understand why it is preferred. It may be the case that the evaluation should not be performed because it is not necessary (or because the results will be ignored).

embodied in any such tools, taking them into account in the decision processes created for the problems at hand.

D.9.1 Steps

The generic decision-making method consists of the following steps:

1. Decision alternatives determination
2. Decision criteria determination
3. Measurement scale determination
4. Decision rule determination
5. Alternatives evaluation
6. Decision making and documentation

D.9.2 **Guidelines**

Use the following guidelines (and associated rationales) regarding decision making:

- Independently evaluate alternatives.
- Document the results of using the decision method.

D.9.3 **Pitfalls**

Avoid the following common pitfalls associated with making decisions and mitigate their negative consequences when they occur:

- The decision criteria are incomplete.
- The subjective decision criteria are neglected.
- The decision makers have tunnel vision.
- The decision process includes systematic biases.
- The decision makers have *a priori* preferences.

Annotated References/ Bibliography

[AFSTCS, 1996] *Guidelines for Successful Acquisition and Management of Software-Intensive Systems (GSAM)*, Version 2.0, Volume 1, Air Force Software Technology Support Center, June 1996, pp. 2–5. These guidelines were designed to help U.S. Air Force acquisition personnel to acquire weapon systems, command and control systems, and information systems. This document is the source of one diagram that shows the increase of software within military systems. The guidelines have been updated to version 4.0, which is available at the following URL: http://www.stsc.hill.af.mil/resources/tech_docs/gsam4.html

[Albert et al., 2002] Cecilia Albert, Lisa Brownsword, Colonel David Bentley, Thomas Bono, Edwin Morris, and Debora Pruitt, *Evolutionary Process for Integrating COTS-Based Systems (EPIC): An Overview*, CMU/SEI-2002-TR-009, Software Engineering Institute, July 2002. This technical report documents the EPIC method including its tasks and work products. EPIC is a risk-based, disciplined, spiral-engineering approach developed to better incorporate the identification, evaluation, and integration of COTS products into systems.

[Alexander, 1979] Christopher Alexander, *The Timeless Way of Building*, Oxford University Press, 1979. This is the original book about building architecture that started the patterns movement within the software community.

[ANSI/EIA, 2003] Processes for Engineering a System, ANSI/EIA-632-1999, Reaffirmed, Government Electronics and Information Technology Association, September 2003, pp. 23–26, 37, 87–90, 101. This international standard includes a requirement to produce logical solution representations and a requirement to produce physical solution representations.

[Barbacci et al., 1995] Mario Barbacci, Thomas H. Longstaff, Mark H. Klein, and Charles B. Weinstock, *Quality Attributes*, CMU/SEI-1995-TR-021, Software Engineering Institute, December 1995. An early technical report documenting a software quality model that is relatively small by today's standards.

[Beam, 1990] Walter Beam, *Systems Engineering: Architecture and Design*, McGraw-Hill, 1990. A general introduction to systems engineering concepts and approaches with emphasis on architecture and design. Architecture is understood in terms of system elements, their functions, their relationships, their performance, and other properties.

[Bengtsson and Bosch, 1998] P.O. Bengtsson and J. Bosch, "Scenario-Based Architecture Reengineering," Proceedings of the Fifth International Conference on Software Reuse (ICSR 5), 1998. This paper documents the Scenario-Based Architecture Reengineering (SBAR) software architecture evaluation method. SBAR estimates the ability of the software architecture to support the meeting of quality requirements. SBAR is based on scenarios, simulation, mathematical modeling, and experienced-based reasoning. SBAR is listed as a software architecture evaluation method that potentially may be applied as a system architecture evaluation.

[Bengtsson and Bosch, 1999] P.O. Bengtsson and J. Bosch, "Architecture Level Prediction of Software Maintenance," Proceedings of the Third European Conference on Software Maintenance and Reengineering, 1999, pp. 139–147. Architecture Level Prediction of Software Maintenance (ALPSM) is a method for predicting the average software maintenance effort based on (1) the requirements specification, (2) the architecture, (3) the expertise from software engineers, and, possibly, (4) historical data as input. ALPSM uses scenarios to both operationalize the maintainability requirements and to analyze the architecture. ALPSM is listed as a software architecture evaluation method that potentially may be applied as a system architecture evaluation.

[Bishop and Bloomfield, 1999] Peter Bishop and Robin Bloomfield, *A Methodology for Safety Case Development*, Adelard, London, U.K., 1999, p. 9, http://www.adelard.com/papers/sss98web.pdf. Describes safety cases as consisting of claims, arguments, and evidence. It primarily differs from the MFESA and QUASAR [Firesmith et al., 2006] definition of architecture quality cases in that its (1) safety cases are only for safety; (2) safety cases are not restricted to architectural claims, arguments, and evidence; and (3) it defines arguments as networks of evidence rather than architectural decisions, inventions, trade-offs, assumptions, and rationales.

[Boehm et al., 2003] Barry Boehm, Victor Basili, A. Winsor Brown, and Richard Turner, "Spiral Acquisition of Software-Intensive Systems of Systems," CrossTalk — The Journal of Defense Engineering, May 2004, pp. 4–9. This article is the source of Figure 6.2: The optimum amount of architecture engineering.

[BPMN, 2004] BPMI Notation Working Group, *Business Process Modeling Notation, Version 1.0*, Business Process Management Initiative, May 3, 2004, http://www.bpmn.org/Documents/BPMN%20 V1-0%20May%203%202004.pdf. This specification documents a graphical notation for modeling business processes in the form of business process diagrams. These diagrams can be used to document the business processes supported by or implemented by the new system. They can also be used to create logical dynamic models of logical dynamic structures within the system architectures.

[Chrissis et al., 2007] Mary Beth Chrissis, Mike Konrad, and Sandy Shrum, *CMMI Second Edition: Guidelines for Process Integration and Product Improvement*, Addison-Wesley, 2007. This book describes the Software Engineering Institute (SEI) Capability Maturity Model Integration (CMMI), which documents the need for and certain aspects of system architecture engineering as part of the engineering of the technical solution.

[CIOC, 1999] *Federal Enterprise Architecture Framework*, Version 1.1, Federal Information Officers Council, September 1999, p. 80, http://www.cio.gov/Documents/fedarch1.pdf. This document describes the United States Government Federal Enterprise Architecture Framework, which promotes shared development for common federal processes, interoperability, and sharing of information among federal agencies and other government entities.

[Clements, 2000] Paul Clements, *Active Reviews for Intermediate Designs*, CMU/SEI-TN-2000-009, Software Engineering Institute, August 2000, http://www.sei.cmu.edu/architecture/products_services/arid.html. This short technical note describes how to perform Active Reviews for Intermediate Design (ARID), a type of software design review that may be used to review software architectures. ARID is listed as a software architecture evaluation method that potentially may be applied as a system architecture evaluation.

[Clements et al., 2003] Paul Clements, Felix Bachmann, Len Bass, David Garian, James Ivers, Reed Little, Robert Nord, and Judith Stafford, *Documenting Software Architectures: Views and Beyond*, Addison-Wesley, 2003. This book documents how to use views to represent software architectures. Certain aspects of these software architecture representations may be applied to the representation of system architectures.

[DAU, 2007] Defense Acquisition University, *Naval Open Architecture (OA)*, 2007, https://acc.dau.mil/oa. The United States Navy's Naval Open Architecture (OA) is an enterprise-wide, multifaceted business and technical strategy for acquiring and maintaining national security systems that adopt and exploit open design principles and architectures. The OA core architectural principles are modular design, reusable application software, interoperable joint war fighting applications, secure information exchange, life-cycle affordability, and encouragement of competition and collaboration.

[**DODAF, 2007a**] Department of Defense, *Department of Defense Architecture Framework, Volume 1: Definitions and Guidelines*, Version 1.5, 23 April 2007, http://www.defenselink.mil/cio-nii/docs/DoDAF_Volume_I.pdf. This volume documents definitions used in the Department of Defense Architecture Framework (DODAF) and provides general guidelines for usage.

[**DODAF, 2007b**] Department of Defense, *Department of Defense Architecture Framework, Volume 2: Product Descriptions*, Version 1.5, 23 April 2007, http://www.defenselink.mil/cio-nii/docs/DoDAF_Volume_II.pdf. This volume describes the DODAF products to be produced in terms of the models comprising the all view, operational view, systems and services view, and technical standards view.

[**DODAF, 2007c**] Department of Defense, *Department of Defense Architecture Framework, Volume 3: Architecture Data Description*, Version 1.5, April 23, 2007, http://www.defenselink.mil/cio-nii/docs/DoDAF_Volume_III.pdf. This volume describes the DODAF architecture data management strategy, including data elements as well as the business rules governing the relationships between the data elements.

[**Elbert, 2006**] Christof Elbert, "Understanding the Product Life Cycle: Four Key Requirements Techniques," *IEEE Software*, Vol. 23, No. 3, May/June 2006, pp. 19–25. This article documents the results of a longitudinal study of the effects of poor requirements on project cost and schedule.

[**Elm et al., 2007**] Joe Elm, Dennis R. Goldenson, Khaled El Emam, Nicole Donatelli, and Angelica Neisa, *A Survey of Systems Engineering Effectiveness — Initial Results*, CMU/SEI-2007-SR-014, Software Engineering Institute, November 2007, p. 222. This special report documents the results of a survey that quantifies the relationship between the application of Systems Engineering (SE) best practices to projects, and the performance of those projects. This survey highlights the importance of system architecture engineering to project success.

[**Etter, 2000**] Delores Etter, "Software & Systems — Managing Risk, Complexity, Compatibility and Change," *Systems and Software Technology Conference (STC 2000)*, May 2000. This conference presentation showed the degree to which software has provided ever-increasing percentages of the functionality of military aircraft.

[**Feiler et al., 2006**] Peter H. Feiler, David P. Gluch, and John J. Hudak, *The Architecture Analysis and Design Language (AADL): An Introduction*, CMU/SEI-2006-TN-011, Software Engineering Institute, 2006, p. 145, http://www.sei.cmu.edu/publications/documents/06.reports/06tn011.html. This SEI technical note documents the Society of Automotive Engineers' standard for the Architecture Analysis and Design Language (AADL), which is a modeling language that supports the early analysis of the performance-critical properties of a system's architecture via an extensible notation, a tool framework, and precisely defined semantics.

[**FIPS, 1993**] *Integration Definition for Information Modeling (IDEF1X)*, Federal Information Processing Standards Publication Standard (FIPS PUBS) 184, December 21, 1993, p. 145, http://www.idef.com/pdf/Idef1x.pdf. This standard specifies the syntax and semantics of the IDEF1X data modeling language that can be used to produce logical data models.

[**Firesmith, 2005**] Donald Firesmith, "Quality Requirements Checklist," in *Journal of Object Technology*, Vol. 4, No. 9 November-December 2005, pp. 31–38, http://www.jot.fm/issues/issue_2005_11/column4. This paper documents the structure of quality requirements and provides a checklist for assessing their quality.

[**Firesmith, 2007**] Donald Firesmith, *Quality Assessment of System Architectures and Their Requirements*, half day tutorial at the SEPG'007 Conference, March 28, 2007, p. 158, http://www.sei.cmu.edu/programs/acquisition-support/presentations/quality-assess.pdf. This tutorial documents version 2 of the QUASAR method for assessing the quality of system architectures and their requirements.

[**Firesmith and Henderson-Sellers, 2002**] Donald Firesmith and Brian Henderson-Sellers, *The OPEN Process Framework: An Introduction*, Addison-Wesley, London, U.K., 2002, p. 330. This book documents the OPEN Process Framework, which defines a set of free, open source, reusable method components for creating system and software development methods.

[**Firesmith et al., 2006**] Donald Firesmith, Peter Capell, Joseph P. Elm, Michael Gagliardi, Tim Morrow, Linda Roush, and Lui Sha, *QUASAR: A Method for the Quality Assessment of Software-Intensive System Architectures*, Handbook CMU/SEI-2006-HB-001, Software Engineering Institute, July 2006, p. 246, http://www.sei.cmu.edu/publications/documents/06.reports/06hb001.html. This technical report documents version 1 of the QUASAR method for assessing the quality of software-intensive system architectures.

[Hatley et al., 2000] Derek J. Hatley, Peter Hruschka, and Imtiaz A. Pãrbati, *Process for System Architecture and Requirements Engineering*, Dorset House, 2000. This book documents the Hatley/Hruschka/Pirbhai method for engineering system requirements and architecture. It emphasizes the coordinated interwoven co-engineering of both requirements and architecture as part of an incremental and concurrent development cycle. It provides a good overview of the related requirements and architectural models and emphasizes the relationships between and ensuring the integrity of these models. A significant difference between the H/H/P method and this book (and other architectural engineering methods) is the restriction of architectural models to the physical models, while relegating the logical models to requirements. This book also tends to assign many requirements engineering tasks to architects rather than to requirements engineers.

[Henderson-Sellers, 2003] Brian Henderson-Sellers, "Method Engineering for OO System Development, *Communications of the ACM*, Vol. 46, No. 10, October 2003, pp. 73–78. This paper provides an accessible introduction to method engineering.

[IEEE, 1998] *IEEE Standard for a Software Quality Metrics Methodology*, IEEE Std 1061-1998, Software Engineering Standards Committee of the Institute of Electrical and Electronics Engineers (IEEE) Computer Society, December 31,1998, p. 26. This standard defines a methodology for establishing quality requirements and identifying, implementing, analyzing, and validating the process and product software quality metrics.

[IEEE, 2000] *IEEE Recommended Practice for Architectural Description of Software-Intensive Systems — Description*, IEEE Std 1471-2000, Software Engineering Standards Committee of the Institute of Electrical and Electronics Engineers (IEEE) Computer Society, September 21, 2000, p. 23. This IEEE recommended practice document addresses the creation, analysis, and maintenance of the architectures of software-intensive systems, and the recording of such architecture in terms of architecture descriptions (aka representations). It establishes a conceptual framework for and defines the contents of architectural descriptions. It includes annexes that provide the rationale for key concepts and terminology, relationships to other standards, and examples of usage.

[IFEAD, 2006] J. Schekkerman, Ed., *Enterprise Architecture Deliverables Guide*, Version 2.6, Institute for Enterprise Architecture Developments, 2006. http://www.enterprise-architecture.info/Images/Extended%20Enterprise/Extended%20Enterprise%20Architecture.htm. This guide provides an overview of the most commonly used enterprise architecture deliverables and models. It is used in combination with the *EA Implementation Guide* to implement the enterprise architectures within organizations.

[INCOSE, 2006] *Systems Engineering Handbook: A Guide for System Life Cycle Processes and Activities*, Version 3, INCOSE-TP-2003-002-03, International Council on Systems Engineering (INCOSE), June 2006. This handbook documents a standard set of system life-cycle processes and activities, including the following two architecture engineering tasks: functional analysis/allocation and system architecture synthesis.

[ISO/IEC, 2001] *Software Engineering – Product Quality — Part 1: Quality Model*, ISO/IEC 9126-1:2001, International Standards Organization (ISO)/International Electrotechnical Commission (IEC) Joint Technical Committee (JTC) 1, 15 June 2001, p. 32. This is the specification of the ISO standard quality model in terms of a set of standardized quality characteristics.

[ISO/IEC, 2002] *Systems Engineering — System Life Cycle Processes*, ISO/IEC 15288:2002(E), International Standards Organization (ISO)/International Electrotechnical Commission (IEC) Joint Technical Committee (JTC) 1, 1 November 2002, pp. 27–28. This standard establishes a common framework for describing the life cycle of artificial systems. The standard defines a standard set of processes and associated terminology that can be applied at any level in a system's hierarchical aggregation structure.

[ISO/IEC, 2005] *Unified Modeling Language Specification, Version 1.4.2*, ISO/IEC 19501:2005(E), International Standards Organization (ISO)/International Electrotechnical Commission (IEC), January 2005, p. 454. This is the ISO/IEC standard reference manual specifying version 1.4.2 of the unified modeling language (UML). Although originally developed to model object-oriented software, UML is often used to model system architectures.

[ISO/IEC, 2007] *Software Engineering: Metamodel for Development Methodologies*, ISO/IEC 24744, International Standards Organization (ISO)/International Electrotechnical Commission (IEC), 15 February 2007, p. 86. This ISO standard introduces the Software Engineering Metamodel for Development Methodologies (SEMDM), which is a comprehensive metamodel that makes use of a new approach to defining methodologies based on the concept of powertype. The intent of this standard is to ensure (via conformance to this metamodel) a consistent approach to defining each methodology with consistent concepts and terminology.

[**Johnson, 2003**] Chris W. Johnson, *Failure in Safety-Critical Systems: A Handbook of Accident and Incident Reporting*, University of Glasgow Press, October 2003, www.dcs.gla.ac.uk/~johnson/book/. This book reports the results of a large number of accident investigations that have identified the role played by requirements engineering in the failure of safety-critical software.

[**Jones, 2008**] Capers Jones, *Preventing Software Failure: Problems Noted in Breach of Contract Litigation*, Draft 5.1, Software Productivity Research LLC, March 7, 2008, p. 18, www.dcs.gla.ac.uk/~johnson/book/. This report identifies problems leading to software failures (including poor handling of requirements changes) that were noted in breach of contract litigation in which the report's author was an expert witness.

[**Kazman et al., 2000**] Rick Kazman, Mark Klein, and Paul Clements, *ATAM: A Method for Architectural Evaluation*, CMU/SEI-TR-2000-004, Software Engineering Institute, August 2000, http://www.sei.cmu.edu/publications/documents/00.reports/00tr004.html. This is the book that documents ATAM (A Method for Architecture Evaluation), which is a method for evaluating software architectures that is based on the use of scenarios to test architectural support for quality characteristics and attributes. ATAM is listed as a software architecture evaluation method that potentially may be applied as a system architecture evaluation.

[**Keeney and Raiffa, 1976**] Ralph Keeney and Howard Raiffa, *Decisions with Multiple Objectives: Preferences and Value Tradeoffs*, John Wiley, 1976. This is a classic reference on structuring and taking group decisions with multiple criteria. A highlight is a methodology for creating credible numerical scales for decision criteria.

[**Lehto and Marttiin, 2005**] Jari A. Lehto and Pentti Marttiin, "Experiences in system architecture evaluation: A communication view for architectural design," in *Proceedings of the 38th Annual Hawaii International Conference on System Sciences 2005 (HICSS '05)*, 2005. This paper describes the Architecture Evaluation Framework (AEF) for evaluating the architectures of telecommunication network elements architecture. AEF includes checking the appropriateness of architecture in relation to business drivers, suitability of external components (COTS, open source), evaluating internal coherence of a product family, and prioritizing development efforts.

[**Lions, 1996**] J.L. Lions, *Ariane V Failure Report of the Inquiry Board*, July 19, 1996, http://sunnyday.mit.edu/accidents/Ariane5accidentreport.html. The official summary report on the failure of the Ariane V rocket, which was used as an example of a system architecture failure.

[**Madigan et al., 1996**] David Madigan, Mosurski Krzysztof, and G. Almond Russel, *Graphical Explanation in Belief Networks*, 1996, http://www.stat.washington.edu/www/research/online/1994/explanation.pdf. This paper "present(s) methods for visualizing probabilistic evidence flows in belief networks, thereby enabling belief networks to explain their behavior."

[**Maier and Rechtin, 2002**] Mark W. Maier and Eberhardt Rechtin, *The Art of Systems Architecting, Second Edition*, CRC Press LLC, 2002. This book provides a good overview of architecting and architects. However, as the title implies this book treats architecting as an art rather than an engineering discipline and strongly emphasizes the distinction between architects and system engineers as well as between architecting and system engineering. Perhaps because this book describes system architecting as an art, this book strongly bases architecting on the use of a set of high-level heuristics. This is in contrast to MFESA, which considers system architects as highly specialized systems engineers and architecture engineering as a subdiscipline of systems engineering.

[**MOD, 2007**] *U.K. Ministry of Defense Architecture Framework (MODAF), Version 1.1*, U.K. Ministry of Defense, October 4, 2007, p. 222, http://www.modaf.org.uk/. MODAF is the U.K. Ministry of Defense's framework for conducting enterprise architecture and provides a means to model, understand, analyze, and specify capabilities, systems, systems of systems (SOS) and business processes. MODAF consists of the following six viewpoints: all views, acquisition views, strategic views, operational views, system views, and technical standards views.

[**Murphy, 2001**] Kevin P. Murphy, "A Brief Introduction to Graphical Models and Bayesian Networks," 2001, http://www.cs.ubc.ca/~murphyk/Bayes/bayes_tutorial.pdf. This paper introduces the reader to a graphical approach to the construction and analysis of belief networks that "marries" probability theory and graph theory.

[**NASA, 2004**] Software Engineering Requirements, NASA Procedural Requirements (NPR) 7150.2, NASA, 27 September 2004, Chapter 3, http://nodis3.gsfc.nasa.gov/displayDir.cfm?Internal_ID=N_PR_7150_0002_&page_name=Chapter3. This mandatory procedure points to poor requirements engineering as a major cause of system development project failure.

[**NATO, 2005**] ISSC NATO Open Systems Working Group, *NATO C3 Technical Architecture, Volume 2: Architecture Descriptions and Models, Version 7.0,* Allied Data Publication 34, 15 December 2005, p. 38. http://194.7.80.153/web site/home_volumes.asp?menuid=15. This volume provides the reference models making up the NATO command, control, and communications (C3) technical architecture. This volume also attempts to anticipate the potential use of new information technologies within NATO such as service-oriented architecture (SOA), component-based architecture (CBA), and net-centric technologies.

[**NAVY, 2004**] *Naval Systems Engineering Guide*, U.S. Department of the Navy, October 2004, pp. 84–94, 133–135, 189–192, 200–201, 233–234, 245. The Naval Systems Engineering Guide documents a common Naval Systems Engineering Process that has been accepted by the Naval Virtual Systems Command. This guide defines the systems engineering (SE) requirements and tasks, their implementation and products, and explains the tools and techniques used throughout a product life cycle. This guide satisfies the DOD requirement for having a documented SE process, and emphasizes the relationship between the technical management process and the SE process.

[**NCOIC, 2007**] *Netcentric Analysis Tools,* Network Centric Operations Industry Consortium (NCOIC), 11 January 2007, http://www.ncoic.org/ncat/start. The NCOIC Network Centric Analysis Tool™ (NCAT) is a Web-based tool that facilitates analysis of architectures, frameworks, and reference models against common criteria. NCAT helps network-centric operations stakeholders to measure the net-centric interoperability of systems created to meet specific operational needs.

[**Nord et al., 2003**] Robert L. Nord, Mario R. Barbacci, Paul Clements, Rick Kazman, Mark Klein, Liam O'Brien, and James E. Tomayko, *Integrating the Architecture Tradeoff Analysis Method (ATAM) with the Cost Benefit Analysis Method (CBAM),* CMU/SEI-2003-TN-038, Software Engineering Institute, December 2003. The Cost Benefit Analysis Method (CBAM) is a method for helping software architects consider the return on investment (ROI) of any software architecture decision and provides guidance on the economic trade-offs involved. CBAM is listed as a software architecture evaluation method that potentially may be applied as a system architecture evaluation.

[**Northrup et al., 2006**] Linda Northrop et al., *Ultra-Large-Scale Systems: The Software Challenge of the Future, Software Engineering Institute,* Pittsburgh, Pennsylvania, July 2006, p. 150. http://www.sei.cmu.edu/uls/the_report.html. This book is the product of a 12-month study of ultra-large-scale (ULS) systems software, which brought together experts in software and other fields to answer a question: "Given the issues with today's software engineering, how can we build the systems of the future that are likely to have billions of lines of code?" The report details a broad, multi-disciplinary research agenda for developing future ultra-large-scale systems.

[**OMG, 2007**] OMG Systems Modeling Language (OMG SysML™), Version 1.0 Specification, Object Management Group, September 2007, p. 258, http://www.omg.org/spec/SysML/1.0/PDF. The reference manual for the Object Management Group's system modeling language (SysML) that can be used to develop models (and views) of software-intensive systems.

[**OPF, 2008**] Open Process Framework Repository Organization Web site, http://www.opfro.org. This informational Web site documents the OPEN Process Framework (OPF), including its metamodel and repository of reusable method components for creating appropriate system and software engineering methods.

[**OSJTF, 2004**] Open Systems Joint Task Force, *Program Managers Guide: A Modular Open Systems Approach (MOSA) to Acquisition,* Version 2.0, The Office for the Undersecretary of Defense for Acquisition, Technology, and Logistics, United States Department of Defense, September 2004, http://www.acq.osd.mil/osjtf/pdf/PMG_04.pdf. This guide for program managers provides an introduction to modular open systems approach (MOSA), which is a both a business and technical strategy for developing new systems and modernizing existing systems. Mandated by the United States Department of Defense Directive 5000.1, MOSA is characterized by a modular architecture, the identification of key interfaces, and the use of open standards for these key interfaces. Compliance with MOSA is assessed by using the MOSA Program Assessment and Review Tool (PART) to capture the answers to a set of associated questions.

[Parnas and Weiss, 1985] David L. Parnas and David Weiss, "Active Design Reviews: Principles and Practice," presented at the *Eighth International Conference on Software Engineering,* 1985. This paper documents the Active Design Review (ADR) method for evaluating software architectures.

[Perrow, 1999] Charles Perrow, *Normal Accidents: Living with High-Risk Technologies*, Princeton University Press, 1999, p. 386. A fascinating book documenting many accident case studies and presenting the thesis that unavoidable complexity makes accidents inevitable (and therefore normal) in certain types of systems.

[Peter and Hull, 2001] Laurence J. Peter and Raymond Hull, *The Peter Principle*, Amereon Ltd., 2001. This book describes the Peter Principle, which states that "in a hierarchy, every employee tends to rise to his level of incompetence." It is relevant to system architecture engineering in that the most qualified designers and implementers tend to be promoted to the position of system architect, a position for which they may well not be qualified in terms of experience, training, and mindset. For example, architects need to be big-picture people, whereas the best designers and implementers tend to be detail people.

[Rechtin, 1991] Eberhardt Rechtin, *Systems Architecting: Creating and Building Complex Systems,* Prentice Hall, 1991. A classic book on the architecting of complex systems. A major contribution is the inclusion of a large number of heuristic rules of thumb.

[Rogers, 1962] Everett M. Rogers, *Diffusion of Innovations*, Free Press, 1962. This is a classic book that introduced the impact of organizational cultures on the introduction of innovations, including the use of new technologies. It also introduced the concept of "early adopters."

[Saint-Amand and Hodgins, 2007] David C.H. Saint-Amand and Bradley Hodgins, *Results of the Software Process Improvement Efforts of the Early Adopters of NAVAIR 4.0,* NAWCWD TP 8642, Naval Air Warfare Center Weapons Division, China Lake, California, December 2007, p. 5. This report is the source of Figure 2.3, which shows the increase in the amount of software in military aircraft over time.

[SARA WG, 2002] SARA Working Group, *Software Architecture and Review (SARA) Report, Version 1.0,* SARA Working Group, 2002, http://philippe.kruchten.com/architecture/SARAv1.pdf. Created by an international working group, this report documents industry best practices for creating and reviewing software architectures. SARA is listed as a software architecture evaluation method that potentially may be applied as a system architecture evaluation.

[Schum, 2002] David A. Schum, SAE, Alternate Views of Argument Construction from a Mass of Evidence, *Cardozo Law Review 22,* 2002, www.cardozo.yu.edu/cardlrev/pdf/225Schumalternative.pdf. This article provides applications of inference networks to complex inferences in the field of law. It also demonstrates dependence of ordering of evidence on the outcome(s).

[Standish, 2008] The Standish Group, *CHAOS Chronicles Online, Formal Methodology*, v. 10.9.6, The Standish Group, 2008. Point 3 Formal Requirements points out that all project methodologies should include a formal process of gathering and maintaining requirements.

[TOG, 2007] The Open Group, *The Open Group Architecture Framework TOGAFTM – 2007 Edition (Incorporating 8.1.1),* van Haren Publishing, 2007, http://www.togaf.org/. This book documents the Open Group Architecture Framework (TOGAF), which was developed by members of the Open Group Architecture Forum. Consisting of a detailed method and a set of supporting tools, TOGAF is a framework for developing enterprise architectures.

[van Ommering, 2004] Rob van Ommering, "Building Product Populations with Software Components," *International Conference on Software Reuse (ICSR-8),* Madrid, July 8, 2004. This conference paper documents the increasing amounts of software incorporated into high-end television sets.

[Wojcik et al., 2006] Rob Wojcik, Felix Bachmann, Len Bass, Paul Clements, Paulo Merson, Robert Nord, and Bill Wood, *Attribute-Driven Design (ADD), Version 2.0,* CMU/SEI-2006-TR-023, Software Engineering Institute, p. 43, November 2006. This technical report documents version 2.0 of the attribute-driven design method for creating software architectures.

[Womack and Jones, 1996] James P. Womack and Daniel T. Jones, *Lean Thinking — Banish Waste and Create Wealth in Your Corporation,* Simon and Schuster, pp. 162–163, 1996. This book is the source of an example of the late discovery of an architectural defect, which in this case was the selection of the Pratt and Whitney's PW2037 engine for the Boeing 757.

[WS-BPEL, 2007] WS-BPEL Technical Committee, *Web Services Business Process Execution Language, Version 2.0, OASIS Standard, Organization for the Advancement of Structured Information Standards (OASIS),* 11 April 2007, p. 264, http://docs.oasis-open.org/wsbpel/2.0/wsbpel-v2.0.pdf. The reference manual specifying version 2 of the Web Services Business Process Execution Language (WS-BPEL) from OASIS.

[Zachman, 1987] J.A. Zachman, "A Framework for Information Systems Architecture," *IBM Systems Journal,* Volume 26, Number 3, 1987. The original paper introducing the Zachman Framework for Enterprise Architecture.

[Zachman, 2008] *Zachman Framework for Enterprise Architecture,* The Zachman Institute for Framework Advancement, 2008, http://www.zifa.com/quickstart.html. The Zachman Framework for Enterprise Architecture is a model around which major organizations view and communicate their enterprise information infrastructure. Drawing upon the discipline of classical architecture, the Zachman Framework establishes a common vocabulary and set of perspectives for defining and describing the architecture for an organization's information infrastructure.

Index

A

AADL. *See* Graphical modeling tool

Acceptance of architecture, 257–277, 399
    goals, objectives, 257–258
    guidelines, 263–267, 276
    inputs, 259
    pitfalls, 267–275, 277
    postconditions, 262–263
    preconditions, 259
    steps, 259–262, 275–276
    work products, 263, 276

Acceptance package, incompleteness of, 275

Access to requirements team by architecture team, 155

Accessibility, 371

Accident prevention attributes. *See* Safety requirement

Accreditation, 108, 181, 402

Acquisition
    architecture engineering tools, 299
    cost, 361
    organization, system decomposition performed by, 187

Acronyms, 415–429

Action item list, 386–387

Activity/collaboration diagrams, 379

Actor/agent structures, 99

Adaptive maintainability, 282, 362

Address reuse incompatibilities, 195

Affordability, 361

Agenda, quality assessment initiation, 386

Aggregation structure, 90

Agile *vs.* document driven, 345

Aircraft system, 84

Airframe segment, 85

Allocation diagram, 379

Allocation structures, 97–98

Analysis
    architecture using architectural representations, 297

executable architecture prototypes, 260
    paralysis, 183
    reusable components, source, 219–231, 399, 404
        goal, objectives, 220
        guidelines, 223–224, 231
        inputs, 221
        pitfalls, 224–231
        postconditions, 222–223
        preconditions, 220–221
        steps, 221–222, 231
        work products, 223, 231

Analyzability, 363

Application domain SMEs, 311

Application layer protocols, 371

Appropriateness. *See* Suitability

AQA. *See* Architecture quality assessment

Architect systems, structures, 128

Architects, 382
    all relevant types of interfaces, 252
    architecturally significant functional, process requirements, 165
    architecturally significant requirements, 166
    authority, 148
    engineering tool, 295
    experience, 316
    focus areas, 184
    functional, process requirements, 165
    ignoring past lessons learned, 199
    inappropriate decision method, 240
    inappropriate models, views, 184
    instantiating MFESA repository without tailoring, 149
    as isolated, 316
    as knowing best, 306
    models, views, 184
    producing architecture plans, conventions, 147
    reliance on functional requirements, 164
    reliance on previous architectures, 200

reusability of immature architectural components, assumptions regarding, 201
reusability of immature technologies, assumptions regarding, 201
selecting specific OTS components early, 201
starting from scratch, 199
training of, 287
use of inappropriate traditional architectural patterns, 186
view construction, 184
Architectural analysis, reports, 123
Architectural arguments, 379
Architectural claims, 378
Architectural concerns, 66, 103, 160, 174, 193, 206, 221
determination, 174
focusing on architecturally significant requirements, 130
maturity of, 234
updated, 196
vs. candidate architectural vision matrix, 237
example, 237
vs. vision component matrix, 208, 235
Architectural constraints, 64, 105
Architectural decisions, made prematurely, 23
Architectural descriptions, 107, 123
Architectural documentation, 66
Architectural documents, 124
Architectural drivers, 102
concerns, 102–106
Architectural element, 89, 95
Architectural focus area, 121
Architectural integrity, ensuring, 298
Architectural invariants, protection of, 287
Architectural mechanisms, 102
Architectural models, 66, 109, 221, 259
current methods emphasis over other architectural representations, 36
focus areas, 109–122
treated as sole architectural representations, 26
understandability, 26–27
views, 109–122
Architectural patterns, 100
identified, analyzed, 196
Architectural prototypes, 66, 108, 124
test results, 263
Architectural quality
external quality characteristics, 363–372
internal quality characteristics, 360–363
Architectural quality cases, 124, 356, 375–380
for assessments, developed after fact, 134
components, 376–378
critical nature of, 130
development, 388
evaluation using QUASAR, 380–387
work products, 386–387
example, 378–380
importance of, 402

quality case components, 376
undeveloped, or developed long after fact for assessments, 134
Architectural relationship, 95
Architectural reliance on architectural engineering tools, 29
Architectural representations, 106–109, 202, 259
acceptance of, 262
completion of, 245–456
consistency, 282
development of, 297
maintaining, 279–292, 399
goals, objectives, 280
guidelines, 286–288, 292
inputs, 281
invariants, 283–284
pitfalls, 288–292
preconditions, 280–281
steps, 282–283, 291
work products, 284–286, 291–292
quality issues, 25–26, 247
Architectural reuse, opportunities for, 221
Architectural risks, opportunities, 156, 160, 175, 193, 206, 209, 221, 223, 235, 238, 247, 251, 259, 263, 286
management of, 177, 197, 208, 223, 237, 250, 263
Architectural simulation, 108
Architectural structures, 95–100
Architectural styles, 100–102
mechanisms, 100–102
patterns, 100–102
Architectural tools, 317–335
Architectural trade-off analysis, 177
Architectural use of multiple inconsistent architecture engineering methods, 29
Architectural views, 119–120
focus areas, 123
models, 123
Architectural visions, 125, 259
candidates, 205–217, 221
completed architecture, management confusing, 213
completion to appropriate level of detail, 211
component, 125
vs. vision matrix, 208–209
components, 125–126, 208, 235
creation of, 205–217, 246, 399
documents, 238
goal, objectives, 206
guidelines, 209–211, 216
inputs, 206
maintaining, 240
meeting requirements, 234
pitfalls, 211–217
postconditions, 208
preconditions, 206
selection report, 238

steps, 207–208, 216
work products, 208–209, 216
Architectural white papers, 124
Architectural work products, 55, 66
Architectural work unit, 55
Architectural workers, 55, 60–61
Architecturally significant constraints, 64–65
Architecturally significant requirements, 163, 285
completeness of, 155, 159
concentration on, 162
identification, 159, 174
understanding, 309
Architecture, 384–385
acceptance of, 262
architecture representations, current methods
confusing, 36
assumptions of, 127
configuration management, 289
current methods confusing, 36
design
connection between, 30
differences between, 94
maintaining, 298
maintenance, 289
never finished, 404
quality, 355–395
guidelines, 388–389
pitfalls, 389–394
quality model components, relationships,
356–360
quality requirements, 373–375
example quality requirements, 374–375
rationales of, 127
Architecture advocate, role as, 298
Architecture assessment, 402
reports, 263
Architecture completion, 250
Architecture concerns
evaluated, 159
identified, evaluated, 159
Architecture defects found during integration, testing,
24–25
Architecture development, current methods emphasis
over other tasks, 31–32
Architecture documentation, 402
Architecture engineering, 298, 301
completion, 253, 288
confusion with designing, 131
conventions, 143
conventions approved, 142
method, tailoring, 299
plans approved, 142
schedule, 144
schedules approved, 142
tool evaluation
report, 143
team charter, 143
tools, selection, acquisition, 299
training materials, 143
Architecture evaluations, 261
substeps, 264
Architecture evolution, 40, 47
Architecture maintenance, tool support, 290
Architecture patterns, 193
Architecture peer review, inspection results, 263
Architecture plan, 143
Architecture quality assessment
agenda, 387
completed, 385
follow-through task, 387
meeting task, 387
outbrief, 387
preparation task, 387
presentation materials, 387
report, 387
Architecture quality assurance, reports, 263
Architecture quality metrics, measurement of, 260
Architecture representations
architecture, current methods confusing, 36
current methods confusing, 36
Architecture risk management, 402
Architecture stakeholders
collaboration with, 299
interfacing with, 299
Architecture team
charters, 142
staffed, 142
staffing, 147
trained, 142
Architecture tools, 317, 336–337
selected, 142
support for all models, 332
support for multiple modeling languages, 332
Architecture training materials, 124
Architecture verification results:, 66
Architecture work products, 122–125
Architecture worker summary, 335–337
Architectures, as resources, 43, 48
Architectures confusion with designs, 131
Architectures confusion with structures, models of
structures, 132
Arguments, 376, 402
sufficiency of, 383
Ariane 5 flight 501, failure of, 6
Arrows, 378
Artificial system, 86
Assessor notes, 387
Assessor responsibilities, 383
Assumptions, 66–67, 80, 127
Attendee notes, 386
Attractiveness, 371
Attributes definitions, 391
Authoritative stakeholders, assumptions regarding reuse,
224

Authority, 300
Authorization, providing, 305
Automatic software generation, 108
Availability issues, 8, 105, 280, 367, 374
Available resources, setting evaluation scope based on, 266
Avionics segment, 86
Avoid unwarranted assumptions, 239

B

Baseline actual reusable components, 404
Baselined architectural representations, 251
Baselined architecture, 250
Baselined/available enterprise, reference architectures, 281
Baselined/available external system interfaces, 281
Baselines, system architectural representations as, 281
Baselining actual reusable components, 249
Blackout of August 2003, 9
Budget constraints, 344
Budgetary requirements, 65
Business process modeling tool, 324

C

CAD/CAM. *See* Computer-aided design/computer-aided manufacturing
Candidates
    architectural vision, 205–217, 221
        comparing pros/cons of, 215
        completion to appropriate level of detail, 211
        components, 208, 235
        creation of, 205–217, 399
        goal, objectives, 206
        guidelines, 209–211, 216
        inputs, 206
        meeting requirements, 234
        pitfalls, 211–217
        postconditions, 208
        preconditions, 206
        selection of, decision methods, 239
        steps, 207–208, 216
        work products, 208–209, 216
    reusable architectural elements, 197, 206, 221
Cascading network system failures, 7
Certification, 108, 181, 402
Challenges in system architecture engineering methods, 30–37
Challenges in system architecture engineering practice, 23–30
Changeability. *See* Modifiability
Changes in human-machine task division range, assessment, 281

Characteristics of quality, 391
Charter, project, 139
Chemical interoperability, 370
Chief or lead system architect, 310
Claims, 376, 402
    sufficiency of, 383
Clarification of relevant requirements with stakeholders, 162
Codification of old processes, current methods, 33
Cohesion, 180
Cohesive integrated set of system components, 90
Collaboration, 311–313
    with requirements team, 160
    with stakeholders, 264
    by system architecture teams, 23
Collaborative internal interactions, 90
Commensurate approach, ensuring, 238
Communication issues, 5, 108, 300
Compatibilities with underlying technologies, 229
Competing architectural visions list, 208, 235
Competing drivers or contradictory drivers, 158
Complete evaluation criteria, ensuring, 239
Completeness of architecturally significant requirements, 155, 159
Completeness of methods produced, 45, 48
Completion of architecture, architectural representations, 245–456
    goals, objectives, 246
    guidelines, 251–252, 255
    inputs, 247
    pitfalls, 252–255
    postconditions, 250
    preconditions, 246–247
    steps, 247–250, 255
    work products, 250–251, 255
Complex development cycles of systems, 16
Complexity, 180
Complexity of system, 406
Complexity of systems, architectures, exponential increase in, 14
Compliance, 364
Component-based development, 43, 47
Component browser, 411
Component editor, 411
Computational structures, 98
Computer-aided design/computer-aided manufacturing, 321
Computing efficiency, 369
Concurrency structure, 99
Concurrent task performance, 224
Condition/triggering event, 373
Conditional selection, potentially reusable components, 222
    identification, evaluation, 222
Conditionally passed quality assessments, 386
Configurability, 364
Configuration control, architectural representations, 254

Configuration management, 73–74
    architecture, 289
    tool, 322
Configuration management team, 312
Conforming to enterprise architecture, 198
Conforming to product line reference architecture, 198
Connection structures, 98
ConOps. *See* System concept of operations
Consensus in evaluation scope, 268
Consistency
    ensuring, 349
    system architecture, architectural representations, 282
Consistency checker, 412
Consistent evaluation approach, ensuring, 238
Constraints, 105, 159, 166
    architectural, 64, 105
    architecturally significant, 64–65
    budget, 344
    legacy architecture, 65
    legal, 65
    method, 345
    from past, avoiding, 197
    policy, 65
    regulatory, 65
Consumables efficiency, 369
Content, interpretation of external interfaces range, assessment, potential changes in, 281
Context diagram, 379
Continuity. *See* Reliability
Continuous evaluation, 264
Contract
    negotiations, 108
    project, 139
    types, 343
Contractual boundaries, 180
Contractual documentation, 108
Control flow, 17
Controllability, 363
Conventions compliance requirements, 65
Corrective maintainability, 362
Corrective maintenance, 282
Correctness, 368
Cost/benefit analysis, 236
Cost issues, 4, 8, 106
Coupling, 180
Creation of appropriate MFESA method, 145
Creation of architectural vision, 246
Creation of first versions of important architectural models, 171–189, 398
    goal, 173–174
    guidelines, 177–182, 188
    inputs, 174–175
    objectives, 173–174
    pitfalls, 183–189
    postconditions, 176–177
    preconditions, 174
    steps, 175–176, 187–188
    work products, 177, 188
Creation of multiple candidate architectural visions, 130
Credibility, 371
Criteria for acceptance, evaluations, 274
Criticality, 157
Cross-component consistency checking, 266
Cross-cutting architectural structures, 17
Current reusability, 362
Customer architecture team, 309

D

Data, 17
Data concerns, 159
Data flow, 17
Data-link layer protocols, 371
Data requirements, 64
Data structures, 98
Decision makers as weak, 242
Decision-making techniques, 449–458
Decision resources, management provision of, 241
Decision support tool, 328
Decision techniques, appropriate, 224
Decomposition performed by acquisition organization, 187
Decomposition *vs.* composition, 409
Defensibility, 365
Definitions, 296
Dependability, 365
Deployment structure, 99
Derivation, 157
Design, 71, 283
    architecture, connection between, 30
    teams, 312
Designers, training of, 287
Developers, control over future changes, 227
Development cost, 361
Development driver, 108
Development/life cycle, 345
Difficult requirements, challenge to, 161
Disposal cost. *See* Retirement cost
Disruptive technology, 18
Documentation
    incomplete, 29
    reuse, 109, 221
    sustainment, 109
Documented system-partitioning criteria, use of, 179
Domain, architectural patterns, understanding, 310
Draft architectural vision documents, 209, 235
Driver of downstream disciplines, 6
Dynamic checking, 266
Dynamic structure, 96

E

Ease of entry, 372
Ease of remembering, 372
Effort minimization, 372
Electrical connection interoperability, 370
Electronic connection interoperability, 370
Elicitation skills, development, 305
Emergent behaviors, characteristics, 91
Enabling system quality, 41, 47
Endeavor characteristics, 343
Endeavor repository, 412
Endeavor types, 343
Energy
    efficiency, 369
    interoperability, 370
Engageability. *See* Attractiveness
Engineering architectural representations, 309
Engineering method, 55
Engineering of entire system architecture, *vs.* system
        structures alone, 127
Engineering process, 54
    component, 54
    metamodel, 55
Engineering trade-off analysis, 235
Enterprise architectural representations, 285
Enterprise architectures, 65, 139, 283
    tool, 327
Environmental compatibility, 369
Environmental tolerance, 365
Error minimization, 372
Error tolerance, 365
Evaluation
    information for, 225
    need to perform, disagreement over, 268
    scheduling, 260
    of selected potentially reusable components, sources,
        222
    to support architectural milestones, 264
    too few, 270
    too many, 271
Evaluation approach, determination of, 222
Evaluation architecture
    goals, objectives, 257–258
    guidelines, 263–267, 276
    inputs, 259
    pitfalls, 267–275, 277
    postconditions, 262–263
    preconditions, 259
    steps, 259–262, 275–276
    work products, 263, 276
Evaluation needs, determination of, 259
Evaluation of architecture, 257–277, 399
    architectural representations, 298
    concerns, 159
    goals, objectives, 257–258

    guidelines, 263–267, 276
    inputs, 259
    method, process, 299
    pitfalls, 267–275, 277
    postconditions, 262–263
    preconditions, 259
    steps, 259–262, 275–276
    work products, 263, 276
Evaluation resources
    determination of, 222
    management provision of, 270
Evaluation success criteria, determination of, 260
Evaluation types, determination of, 259
Evaluation work products, development, 260
Evaluators, 272
Evidence, 376, 379, 402
Exception flow, 17
Executable architectural representation, 108, 131
    simulation results, 263
Executable architectures, 66, 108, 125
Executable models, automated analyses of, 260
Expandability. *See* Extensibility
Expectations of stakeholders, 226
Explicitly documented system-partitioning criteria, use
        of, 179
Exponential increase in, maximum size, complexity,
        architectures, systems, 14
Extensibility, 362
External consistency of constructed method, difficulty
        in, 351
External interactions with environment, 90
External quality characteristics, 358, 363–372
External reusability. *See* Future reusability
External systems, range of potential additional
        interaction assessment, 281

F

Facility requirements, 65
Failed quality assessments, 385
Failure tolerance, 366
Family of systems, 87
Fault tolerance, 17, 365
Feasibility, 109, 360
First versions of architectural models, creation of, goal,
        173–174, 221, 247
Fixability, 363
Focus area consistency, current methods' weakness,
        32–33
Focus areas, 44, 48
    architectural views, 123
    keeping various types consistent, 129
Follow-through, quality assessment initiation, 386
Formal inspections, 261

Formal review of architecture, 265
Formalism *vs.* understandability, 410
Formality level, ensuring, 350
Formally manage architectural risks, 211, 224, 240, 252, 266, 287
Fuel interoperability, 370
Functional concerns, 106, 159
Functional decomposition for logical structures, current methods overemphasis, 32
Functional decomposition structure, 97
Functional feature set concerns, 104
Functional requirements, 64
Functionality, 4, 17, 369
    to qualities, emphasis on, 408
Funding, 138, 147
Future directions
    of MFESA, 410–412
    of system architecture engineering, 405–410
Future reusability, 362

G

General-purpose drawing tool, 320
General system architecture engineering challenges, 13–23
Geographic separation, 345
Globalization. *See* Internationalization
Graphical modeling tool, 320
Graphical models
    automated analysis of, 260
    views, creation of other system architectural representations in addition to, 128
"Greenfield" development, current methods assuming, 31
Ground support system, 85

H

Hardware architects, 297, 304, 310
Hardware architecture team, 309
Hardware decisions, impact on usability, software, 182
Hardware engineering, 304
Hardware portability, 361
Hardware structure, 98
Hardware technologies, 175
Hardware to software, emphasis on, 409
Heuristics in reuse, 178
Hierarchical decomposition of systems, 19
Human factors team, 313
Human limitations, considering when allocating system functionality to manual procedures, 182
Human-machine task division range, changes in, 281
Hydraulic power interoperability, 370

I

Identification of architectural drivers, 153–169, 398
    goal, 153–154
    guidelines, 160–162, 169
    inputs, 155–156
    objectives, 153–154
    pitfalls, 162–169
    postconditions, 159
    preconditions, 154–155
    steps, 156–159, 168
    work products, 159–160, 168
Identification of architectural vision components, 206
Identification of architecture concerns, 159
Identification of potential architectural patterns, 194
Identify appropriate number of candidate architectural visions, 211
Image capturing device, 319
Impact analysis before architectural changes, 47–48
Impact of architectural drivers, determination, assessment of concerns, 297
Impact of architecture on requirements, 161
Implementation teams, 312
Importance of principles, 39
Inappropriate evidence, quality cases containing, 392
Incomplete architecture, 247
Inconsistencies between models, views, focus areas, 185
Increasing duration of system operational lifecycle period, 406
Increasing incorporation of software, 406
Incremental development, 22
Independence
    of architectural assessments, 48
    of evaluations, 272
Independent architecture assessments, 265
Independent verification, validation, 304
Individual acquired reusable components, evaluation, 404
Individual model size, complexity, 407
Individual principles, 40–47
Information architecting tool, 323
Information collector, role as, 298
Information hiding, 180
Informational web site, 410–411
Initial draft architectural models, 177, 193
Initial draft architecture, 177, 193
Initial versions of architectural models, 206
Inspection results, architecture peer review, 263
Instantiation, architecture engineering method, 299
Integration, 71, 109
    engineering, 304
    structures, 97
    teams, 312
Integrity, architectural, difficulty in maintaining, 22
Interaction analysis, 235

Interface, 283, 303–304
    concentration on, 181
    concerns, 159
    requirements, 64
    structures, 99
Interiors segment, 86
Internal consistency of constructed method, difficulty
       in, 350
Internal quality characteristics, 358, 360–363
Internal reusability. *See* Current reusability
Internally evaluate models, 264
Internationalization, 364
Interoperability, 369
    architecture tool, 331
    intraoperability, operational environments, ensuring,
       248
    technologies, 175
Interoperability white paper, 379
Interpretation of external interfaces range, assessment,
       potential changes in, 281
Intraoperability, 361
    ensuring, 248–249
    interoperability, operational environments, ensuring,
       248
    operational environments, ensuring, 248
    protocol interoperability, ensuring, 248
    semantics interoperability, ensuring, 249
    syntax interoperability, ensuring, 249
Invariants, 283–284
Iteration, 349
Iterative development, 22
    cycle, 178

J

Jitter, 367

K

Known architectural risks, opportunities, 206, 221, 235,
    247, 259, 263

L

Large decomposition structure, 90
Latency, 367
Layer diagram, 379
Layered architecture, 379
Layering structures, 97
Leading architectural activities, 298
Learnability, 372

Legacy architecture constraints, 65
Legal constraints, 65
Legal rights, 229
Legitimate stakeholders, 89
Life cycle architecture engineering, 408
Listening skills, development, 305
Local interface specifications, current methods relying
       on, 36
Localization. *See* Internationalization
Localized assembly line to globalized supply chain, 406
Logical class structure, 97
Logical data management structures, 97
Logical data structure, 97
Logical dynamic models, 112
Logical dynamic structures, 97
Logical flow structures, 97
Logical management structures, 97
Logical models before beginning to develop physical
       models, development of, 179
Logical static models, 110
Logical static structures, 97
Low-level software design tools, as system architecture
       tools, 333

M

Maintaining architecture, architectural representations,
       279–292, 399
    goals, objectives, 280
    guidelines, 286–288, 292
    inputs, 281
    invariants, 283–284
    pitfalls, 288–292
    preconditions, 280–281
    steps, 282–283, 291
    work products, 284–286, 291–292
Maintenance of architectural integrity in maintenance of
       architectural representations, 286
Maintenance of schedule, 145
Management
    availability of system architectural drivers, concerns,
       280
    baselining, 280
       availability of system architectural drivers, 280
    disciplines, 67
    provision of resources, 147
    resources, 334
Mandated compliance, 345
Manual analyses of architectural models, 260
Manufacturability. *See* Producibility
Manufacturing cost, 361
Manufacturing documentation, 109
Market surveys, 223
Mass/geometry modeling tool, 325
Material interoperability, 370

Maturity of architecturally significant requirements, 155, 159
Maximum project size, complexity, 15
Maximum size, complexity, systems, architectures, exponential increase in, 14
Maximum system size, complexity, 14
Measurements, 74–75
    metrics team, 313
Mechanical linkage energy interoperability, 370
Membership, 310–311
Metadata, 66
Metamethod, 52, 61, 339–353
    component, 55
    construction, 346–347
    determination of reuse type, 346
    documentation, 347–348
    guidelines, 348–350
    metamethod overview, 340–341
    needs assessment, 341–346
    number of methods determination, 346
    overview, 340–341
    pitfalls, 350–352
    publication, 348
    reuse, 346
    verification, 348
Metamodel, 52–56
Metaphors in reuse, 178
Method browser, 412
Method builder, 412
Method components
    architectural workers, 60–61
    construction, 353
    integration, 353
    selection, 353
    tailoring, 353
Method constraints, 345
Method construction, 346–347, 353
Method documentation, 347–348, 353
Method editor, 412
Method engineering tool support, 411
Method Framework for Engineering System Architectures, 49–80, 400–401
    architectural drivers, concerns, 102–106
    architectural models, views, focus areas, 109–122
    architectural representations, 106–109
    architectural structures, 95–100
    architectural styles, patterns, mechanisms, 100–102
    architectural workers, 60–61, 293–337
        tasks, 300–301
    architecture tools, 317–337
        example tools, 317–318
        guidelines, 328–330
        pitfalls, 331–335
        relationships, 328
        types of architecture tools, 318–328
    architecture work products, 122–125
    assumptions, 66–67, 80

authority, 300
components, 57–60, 79, 397–398, 431–439
configuration management, 73–74
defined, 51–61
design, 71
future directions, 410–412
    informational web site, 410–411
    method engineering tool support, 411
    MFESA organization, 410
goal, 51, 79
guidelines, 76–78, 80, 126–131
implementation, 71
inputs, 61–66, 79
integration, 71
mastering concepts, need for, 81
measurements, 74–75
metamethod, 52, 61, 339–353
    construction, 346–347
    determination of reuse type, 346
    documentation, 347–348
    guidelines, 348–350
    metamethod overview, 340–341
    needs assessment, 341–346
    number of methods determination, 346
    pitfalls, 350–352
    publication, 348
    reuse, 346
    verification, 348
metamodel, 52–56
methods, use of requirement, reusable architectural elements as input, 77
metrics, 74–75
more than repository of reusable method components, 77
multiple uses, 77
need for, 49–51
number, timing of system architecture engineering processes, 67
objectives, 51
ontology of concepts, 81–136
organization, 410
other disciplines, 80
    relationships with, 67–76
outputs, 66, 80
overview, 49–80, 398–399
pitfalls, 131–135, 305–307
process engineering, 72
profile, 301–304
    experience, 303
    expertise, 302–303
    guidelines, 305–307
    interfaces, 303–304
    personal characteristics, 301–302
    training, 303
project/program management, 73
quality engineering, 72
references, 140

repository, 52, 56–61, 412
requirements engineering, 68–70
responsibilities, 297–300
risk management, 74
specialty engineering disciplines, 75–76
summary, 397–399
system architects, 295–307
system architecture, 92–95
systems, 81–92
teams, systems architecture, 307–317
    collaborations, 311–313
    guidelines, 313–315
    membership, 310–311
    pitfalls, 315–317
    responsibilities, 309–310
    types of architecture teams, 307–309
terminology, 52, 81–136
testing, 71
training, 72–73
types of system architects, 296–297
vision, 125–126
    components, 125–126
Method needs assessment, 341–346, 352
Method publication, 348, 353
Method reuse, 346, 353
    type determination, 346, 352
Method selection, 353
Method tailoring, 353
Method verification, 348, 353
Metrics, 74–75
MFESA. *See* Method Framework for Engineering System
        Architectures
Middleware portability, 361
Minutes, quality assessment initiation, 387
Mode allocation structures, 98
Model concurrency, 181
Model consistency, emphasis on, 410
Model size, complexity, 407
Modeling language, 109
Models
    keeping various types consistent, 129
    views, focus areas, keeping various types consistent,
        129
Modifiability, 362
Modular architecture, 379
Most suitable architectural vision, 237
    selection, creation of, 233–244, 399
        goal, 234
        guidelines, 238–240, 244
        inputs, 234–235
        pitfalls, 240–244
        postconditions, 237
        preconditions, 234
        steps, 235–236, 243
        work products, 237–238, 243–244
Multiple architecture engineering methods, system
        architecture engineering challenges, 29

Multiple focus areas, 91
Multiple models, views
    addressing architectural focus areas cutting across,
        128
    architectural focus areas cutting across, 128
Multiple structures, 91
Multiple teams, 67
Multiple views, models, 91

N

Navigability, 372
Need for mastering concepts, ramifications, 81
Need for Method Framework for Engineering System
        Architectures, 49–51
Network layer protocols, 371
Network of systems. *See* System-of-systems systems
New architectural risks, opportunities, 177, 197, 209,
        223, 238, 251
New system architectures, 21
Notification of requirements team of requirements
        defect, 161
Number of methods determination, 346, 352
Number of system architecture engineering processes,
        67
Number of types of models, views, focus areas, 407

O

Observability, 363
Observations of architects' work, 259
Official repository, 412
Ontology, 52
    of concepts, 81–136
    current methods, 36
    underlying, current methods, 36
Operability, 371
Operating system portability, 361
Operational documentation, 109
Operational environments, intraoperability,
        interoperability, ensuring, 248
Opportunities
    architectural, 206, 221, 235, 259, 263
    goals, objectives, 192
    guidelines, 197–198, 203
    identification of, 191 204
    inputs, 193
    pitfalls, 198–204
    postconditions, 195–197
    preconditions, 192–193
    for reuse of architectural elements, 221
    risk, architectural, 206, 209, 221, 223, 235, 238, 247,
        251, 259, 263

steps, 193–195, 203
work products, 197, 203
ORD. *See* System operational requirements document
Organizational balance, 6
Organizational characteristics, 344
Organizational chief architect, 296
Organizational culture, 344
Organizational repository, 412
Organizational responsibilities, 298
Organizational separation, 344
Other system architectural representations, creation of, in addition to graphical models, views, 128
OTS components, reuse of, 180
OTS incorporation, preparation of architectural components, 211
Outbrief, quality assessment initiation, 387
Outputs, 66, 80

P

Paper or electronic documents, 348
Parallel development cycle, 178
Partial architectural concern *vs.* architectural component matrix, example, 210
Peer reviews, 260
Perfective maintainability, 363
Perfective maintenance, 282
Performance allocation structure, 99
Personal characteristics, 301–302
Personalization, 365
Pervasive design-level decisions, 94
Pervasive implementation-level decisions, 94
Physical allocation structure, logicality to, 99
Physical configuration structure, 98
Physical connections interoperability, 370
Physical decomposition for physical structures, current methods overemphasis, 32
Physical decomposition hierarchy, overemphasis on, 179
Physical dynamic
models, 119
structures, 99
Physical flow structures, 99
Physical interoperability, 369
ensuring, 248
Physical layer protocols, 370
Physical separation, 181
Physical static structures, 98, 113
Physical structure, 95
Pilot projects, to evaluate new tools, 329
Planning
architecture engineering effort, 137–152, 398
done once up front, 150
resource architecture engineering effort, objectives, 137–138

Planning and resource architecture engineering effort, 137–152, 398
goal, 137–138
guidelines, 144–146, 152
inputs, 138–140
objectives, 137–138
pitfalls, 146–152
postconditions, 141–142
preconditions, 138
steps, 140–141, 151
work products, 142–144, 151–152
Planning of architecture engineering effort, 144
Plans, 347
Points of view of different types of experts, current methods overemphasis, 37
Policy constraints, 65
Poorly defined, documented quality model, endeavor using, 390
Portability, 361
Potential changes in human-machine task division range, assessment, 281
Potentially relevant requirements, 154
Potentially reusable architectural components
descriptions, 223
list, 223
Potentially reusable components, identification, evaluation, conditional selection, 222
Power connections interoperability, 370
Power interoperability, ensuring, 248
Predictability, 369
Preexisting architectures, representations of, 193
Preference. *See* Attractiveness
Premature comparison of architectural visions, by architects, 214
Presentation layer protocols, 371
Preventative maintainability, 282, 363
Primary engineering disciplines, 67
Primary integration, subsystem intraoperability, 17
Primary responsibilities:, 297
Prime contractor or integrator architecture team, 309
Process concerns, 106, 159
Process consultant, 411
Process engineering, 72
Process engineering team, 312
Process engineering tool repository, 348
Process quality, determination of, 262
Process requirements, 65, 103
Process structures, 98–99
Processes inconsistent in practice, 25
Producibility, 362
Product line of systems. *See* Family of systems
Product quality, determination of, 262
Product requirements, 103
Production schedule, 145
Proper quality requirements, ensuring, 388
Propulsion segment, 86

Protocol interoperability, 370
   ensuring, 248
   goals, 379
   intraoperability, ensuring, 248
   requirements, 379
Prototype development cost concerns, 104
Proxies, 379

Q

Qualifications, trained, experienced architects, 28–29
Quality
   concerns, 105, 158
   criterion, 373
   engineering, 72, 304
   importance of, 401–402
Quality assessment
   initiation
      agenda, 386
      follow-through task, 386
      meeting task, 386
      minutes, 387
      outbrief, 387
      preparation task, 386
   need to perform, disagreement over, 393
   relevancy of, 385
   scope, disagreement over, 393
Quality attributes, 121, 357
   as architectural concerns, 162
   concerns regarding, 104
Quality cases, 273, 375
   appropriateness for current point in schedule, 383
   components, 376
   development as natural part of architecture engineering process, 251
   index, summary, 394
Quality characteristics, 121, 357, 391
   attribute (See Architectural claims)
   attributes, importance of, 391
   concerns regarding, 104
Quality engineering teams, 312
Quality focus area, 121
Quality issues, 27–28, 271
Quality measurement scale, 357
Quality models, 355, 357, 389
   components, relationships, 356–360
Quality-related architectural patterns, 355
Quality requirements, 64, 180, 355, 373–375
   example, 374–375
Quality threshold, 374
QUASAR
   architectural quality case evaluation using, 380–387
   assessment results matrix, 384
   materials, 386
Quebec blackout of March 1989, 9

R

Ramifications of need for mastering concepts, 81
Rationale capture, 403
Rationales of architecture, 127
Readiness. See Availability
Reasons for improved systems architecture engineering methods, 37–38
Recoverability, 366
Rectangular nodes, 378
Recursively incremental development cycle, 178
Reference architectural representations, 285
Reference architectures, 65, 139, 283
   documentation, 109
   team, 307
Regulatory constraints, 65
Relationships with other disciplines, 67–76
Relevancy of quality assessments, 385
Relevant architectural focus areas, 177, 193
Relevant architectural views, 177, 193
Relevant change control analysis reports, 285
Relevant change requests, 285
Relevant discrepancy reports, 285
Relevant process requirements, 156
Relevant product requirements, 155
Relevant requirements metadata, 174
Reliability, 8, 368
   requirement, 375
Reliability team, 313
Replacement architects, development architects, contrasted, 314
Replacement documentation, 109
Repositories, 56–61
Representations
   architectural vision components, 176
   of current architecture, 193
   of preexisting architectures, 193
   as resources, 43, 48
Request for proposal, 61, 139, 155
Requirement recommendations, 159
Requirements defect, notification of requirements team, 161
Requirements driven to evolutionary-capabilities based, 408
Requirements engineering, 68–70
   architecture engineering, current methods confusing, 33–34
   initiation of, 138
Requirements engineering team, liaison, 311
Requirements engineering tools, 323
Requirements metadata, 160, 167
   identification of architecturally significant requirements, 174
   update, 159
Requirements quality assessments completed, 385
Requirements size, 342

Requirements team, 311
    working with to provide requirements traceability,
        252
Requirements trace, 251
Requirements verification, 109
Requirements work products, 66
Resiliency. *See* Robustness
Resource architecture engineering, 137–152, 398
    goal, 137–138
    guidelines, 144–146, 152
    inputs, 138–140
    objectives, 137–138
    pitfalls, 146–152
    postconditions, 141–142
    preconditions, 138
    steps, 140–141, 151
    work products, 142–144, 151–152
Resource feasibility, 361
Resources, management provision of, 213, 253
Resourcing, done once up front, 150
Response time, 367
Responsibilities, 66, 297–300, 309–310
Responsibility for architectural work products, 299
Responsibility for safety, security, 17
Retirement cost, 361
Retirement documentation, 109
Retrievability, 372
Reusability, 362
    information to determine, 202
Reusable architectural component architectures,
        architectural representations, 285
Reusable architectural elements
    analysis of, 196
    identification, 196
        analysis, 196
Reusable architectures
    analysis, 196
    identification, 196
        analysis, 196
Reusable components
    acquisition from sources, 249, 404
    failure of source organization to adequately maintain,
        228
Reuse. *See also* Reusability; Reusable
    appropriate MFESA method, 145
    architectural elements, identification of opportunities
        for, 191–204, 398, 403
    authoritative stakeholders' assumptions regarding,
        224
    components, wrapping of, 249
    documentation, 109, 221
    effect on architecture engineering, 403–404
    heuristics, 178
    information for, 225
    metaphors, 178
    opportunities for, 221
    OTS components, 180

RFP. *See* Request for proposal
Risk management team, 313
Risks
    analysis, 236
    architectural, formal management of, 162, 182, 198
    management, 74
    opportunities, architectural, 206, 209, 221, 223, 235,
        238, 247, 251, 259, 263
    setting evaluation scope based on, 266
    understanding of, by stakeholders, 242
Robustness, 8, 365
Role structure, 99

S

Safety, 8, 181, 366
    responsibility for, 17
Safety requirement, 375
Safety team, 313
Scalability, 40, 47, 362
Scaling up current methods, 31
Schedulability, 367
Schedule, 4, 147
    concerns, 104
    constraints, 344
    feasibility, 361
    maintenance of, 145
    production, 145
    requirements, 65
Scheduling evaluations, difficulty in, 269
Scope of architectural integrity, determination of, 287
Scope of evaluation, determination of, 260
Security, 8, 181, 366
    policy, 156
    problem quality attributes, 104
    requirement, 375
    responsibility for, 17
    solution quality attributes, 104
Security team, 313
Selection as purely technical decision, 241
Selection of actual reusable components, sources, 249,
        404
Selection of architectural vision, 246
Selection of architecture engineering tools, 146
Selection of architecture modeling method, 146
Selection of candidate architectural vision, decision
        methods, 239
Selection of most suitable architectural vision, 233–244,
        399
    goal, 234
    guidelines, 238–240, 244
    inputs, 234–235
    pitfalls, 240–244
    postconditions, 237
    preconditions, 234

steps, 235–236, 243
work products, 237–238, 243–244
Selection of potential architectural mechanisms, 195
Selection of potential architectural styles, 194
Semantic interoperability, 371
Semantics interoperability
 ensuring, 249
 intraoperability, ensuring, 249
SEPG. *See* Process engineering team; Systems
  engineering process group
Service-oriented architecture, 379
Service structure, 97
Serviceability, 371
Session layer protocols, 371
Shelfware, architectural representations becoming, 288
Simulation tools, 321
Single architect, 315
Single architectural vision
 architects engineer, 212
 current methods producing, 35
Single common quality model, 388
Single shared vision, development, 314
Size efficiency, 369
SOA. *See* Service-oriented architecture
Software
 architecting, 304
 architects, 297, 310
 critical, persuasive influence on system architectures,
  16
 impact of hardware decisions on, 182
 influence on system architectures, 16
Software architecture team, 309
Software architecture tool, 326
Software engineering, 304
Software feasibility, 180
Software-intensive system, 89
Software technologies, 176
Sources of architectural change, consideration of, 287
Specialty architecting, 304
Specialty engineering
 architecture team, 308
 area subject matter experts, 311
 disciplines, 68, 75–76
 requirements, 165
Spreadsheets, 320
SRS. *See* System requirements specification
Stability requirement, 375
Staff, stakeholder training, management provision of,
  148
Staffing
 requirements, 65
 top-level architecture team, 144
Stakeholder advocate, role as, 299
Stakeholders
 disagreement on evaluation results, 274
 expectations of, 226
 need for different architectural representations, 22

systems becoming increasing critical to, 14
 understanding of risks, 242
Starting from scratch, avoiding, 182, 197
Static checking, 266
Static structure, 96
Staying at correct level, 403
Stickiness. *See* Attractiveness
Strategic design-level decisions, 94
Strategic implementation-level decisions, 94
Structure, current methods' weakness, 32–33
Styles, architectural, 100–102
Subcontractor architecture team, 309
Subordinate architects, supervision of, 299
Subsetability, 365
Subsystem architecting, 304
 software architecting, 304
Subsystem architects, as members of system architecture
  team, 314
Subsystem architecture team, 308
Subsystem lead architect, 297
Subsystem X sensors, 106
Sufficiency of evidence, 383
Sufficient architectural representations, 44, 48
Suitability, source evaluation for, 223
Summary of challenges, 38
Summary of MFESA, 397–399
Summary of principles, 47–48
Supersetability, 365
Supplier architecture team, 309
Suppliers, initiation of relationships with, 196
Support cost, 361
Support for quality characteristics, 34–35
Support for situational method engineering, 40, 47
Support modeling, 198
Supporting system architects, 310
Survivability, 367
Sustainment cost. *See* Support cost
Sustainment documentation, 109
Synchronicity, 150
Syntax interoperability, 371
 intraoperability, ensuring, 249
SysML. *See* Graphical modeling tool
SysRS. *See* System requirements specification
System architects, 2, 293, 295–307, 335
System architectural development resources, promoting
  need for, 310
System architecture, 92–95
 consistency, 282
 critical nature of, 400
 design, consistency, 284
 enterprise architecture, consistency, 284
 as essential element of success, 4–9
 external interfaces, consistency, 284
 implementation, consistency, 284
 reference architecture, consistency, 284
 requirements consistency, 284
 system architecture engineering, critical nature of, 400

System architecture engineering
  architects, trends, 407–409
    tools, 407–409
  architectural engineering tools, 29
  architectural models
    current methods emphasis over other
      architectural representations, 36
    treated as sole architectural representations, 26
  architectural reliance on architectural engineering
    tools, 29
  architectural representations, quality issues, 25–26
  architecture, architecture representations, current
    methods confusing, 36
  architecture defects found during integration,
    testing, 24–25
  architecture development
    current method emphasis over other tasks,
      31–32
    methods emphasizing over other tasks, 31–32
  architecture engineering, specialty engineering focus
    areas, 27
  challenges, 13–38
  codification of old processes, current methods, 33
  connection between architecture, design, 30
  critical nature of, 400
  current methods, scaling up, 31
  current system architecture engineering methods,
    incomplete nature of, 31
  as essential element of architecture, 9–11
  experience, 28–29
  experienced architects, 28–29
  functional decomposition for logical structures,
    current methods overemphasis, 32
  future directions, 405–410
    trends in systems, system engineering, 405–407
  general system architecture engineering challenges,
    13–23
  "Greenfield" development, current methods
    assuming, 31
  incomplete documentation, 29
  local interface specifications, current methods relying
    on, 36
  multiple inconsistent architecture engineering
    methods, 29
  "one size fits all," current methods assuming, 35
  performance of tasks for which unqualified, 28–29
  physical decomposition for physical structures,
    current methods overemphasis, 32
  points of view of different types of experts, current
    methods overemphasis, 37
  principles, 39–48
    importance of principles, 39
    individual principles, 40–47
  processes inconsistent in practice, 25
  quality issues, 27–28
  reasons for improved system architecture engineering
    methods, 37–38

  requirements engineering with architecture
    engineering, current methods confusing,
    33–34
  single architectural vision, current methods
    producing, 35
  support for quality characteristics, current methods,
    34–35
  track record of industry, architecture engineering, 24
  training, 28–29
  underlying ontology, current methods, 36
  understandability of architectural models, 26–27
  waterfall development cycle, current methods
    emphasis, 33
  weakness of current methods, 32–33
System architecture patterns, 198
System architecture team, 307, 336
System architecture worker, 293
System chief architect, 297
System complexity, 343
System concept of operations, 62, 139, 155
System criticality, 343
System decomposition performed by acquisition
  organization, 187
System design, 95
System engineering process group. *See* Process
  engineering team
System engineering trends, 405–407
System interoperability, 17
System number, 341
System-of-systems architecture team, 307
System-of-systems chief architect, 297
System-of-systems systems, 87
System operational requirements document, 139, 155
System or subsystem structures, 93
System-partitioning criteria, documented, use of, 179
System product requirements, 64
System quality, 17
System request for proposal, 61, 139, 155
System requirements evaluation results, 156
System requirements models, 156
System requirements repository, 63, 139, 156
System requirements specification, 63, 139, 156
System size, 342
  complexity, 406
System test team, liaison to, 311
System vision statement, 62, 139, 155
Systems architecture teams, 307–317
Systems engineering, 304
Systems engineering process group, 304
Systems reuse architectural components, 19

T

Tailor software evaluation methods, 265
Tailored QUASAR materials, 387

Tailoring, architecture engineering method, 299
Teams, 155, 295, 307–317
    boundaries, respect for, 162
    charter, architecture engineering tool evaluation, 143
    charters, 142
    collaboration, 160, 311–313
    configuration management, 312
    customer architecture, 309
    design, 312
    differentiation of members from collaborators, 314
    earning respect, 314
    engineering, liaison, 311
    guidelines, 313–315
    hardware architecture, 309
    human factors, 313
    implementation, 312
    integration, 312
    measurement, metrics, 313
    membership, 310–311
    multiple, 67
    notification of requirements defect, 161
    pitfalls, 315–317
    prime contractor, integrator, 309
    process engineering, 312
    project, program management, 312
    quality engineering, 312
    reference architecture, 307
    reliability, 313
    requirements traceability, 252
    responsibilities, 309–310
        respect for, 162
    risk management, 313
    safety, 313
    security, 313
    software architecture, 309
    specialty engineering, 308
    staffing, 142, 144, 147
    subcontractor, 309
    subsystem, 308, 314
    supplier architecture, 309
    system architecture, 307, 336
    system-of-systems architecture, 307
    system test, liaison to, 311
    testing, 312
    top-level architecture, 307
    training, 142, 312
    types of, 307–309
    vendor architecture, 309
Teamwork, 46, 48
Technological feasibility, 361
Technology allocation structure, 99
Temporal structures, 98
Test engineers, 304
Testability, 363
Testing teams, 312
Thresholds, quality requirements, 392
Throughput, 367

Time-boxed development cycle, 178
Time separation, 181
Timeliness, 368
Timing of system architecture engineering processes, 67
Tolerance, 365
Tools, 317–337
    acquisition, 299
    business process modeling, 324
    configuration management, 322
    decision support, 328
    enterprise architecture, 327
    evaluation, 329
        report, 143
        team charter, 143
    examples, 317–318
    general-purpose drawing, 320
    graphical modeling, 320
    guidelines, 328–330
    information architecting, 323
    integration of, 331
    interoperability, 331
    low-level software design tools, 333
    mass/geometry modeling, 325
    method engineering, support, 411
    metrics support, 333
    modeling method driving, selection, 329
    pilot projects to evaluate, 329
    pitfalls, 331–335
    process engineering, repository, 348
    relationships, 328
    reliance on, 29
    requirements engineering, 323
    selection, 142, 146, 299
    simulation, 321
    software architecture, 326
    support, 290, 351
        for all models, 332
        for architectural metrics, 333
        future directions, 411
        for multiple modeling languages, 332
    trade studies, to evaluate, 329
    training before use, providing, 330
    trends, 407–409
    types, 318–328
    vendors, as drivers of architecture engineering, modeling methods, 149
Top-level architecture team, 307
Traceability, working with requirements team to provide, 252
Track record of industry, architecture engineering, 24
Trade studies, to evaluate new tools, 329
Training, 72–73, 303
    architects, 287
    architecture stakeholders, 299
    designers, 287
    provision of, 146

for staff, stakeholder, management provision of, 148
before tool use, providing, 330
Training materials, 348
Training team, 312
Transition planning, 109
Transport layer protocols, 371
Transportability, 372
Trends in systems, system engineering, 405–407
Trustworthiness. *See* Credibility
Types of architectural components, 41, 47
Types of architecture teams, 307–309
Types of architecture tools, 318–328
Types of system architect, 296–297

U

Ultra-large systems of systems, problems with, 404–405
UML. *See* Graphical modeling tool
Understandability of architectural models, 26–27
Understanding of risks, by stakeholders, 242
Undocumented results, 237
Updated architectural concerns, 197
Updated architectural risks, opportunities, 177, 197, 209, 223, 238, 251
Updated QUASAR assessment results matrix, 387
Updated system architecture, architectural representations, 285
Updated work products, 285
User interface, 411
User satisfaction, 372

V

Various architecture analysis reports, 263
Vehicle segment, 86
Vendor-supplied technical documentation, 379
Vendors
    architecture team, 309
    initiation of relationships with, 196
    reusable architectural components, failure of, 228
    selling silver bullets, problems with, 331
Verify architectural consistency, 265
Vision. *See also* Vision of candidate
    goal, 234

guidelines, 238–240, 244
inputs, 234–235
pitfalls, 240–244
postconditions, 237
preconditions, 234
in reuse, 178
selection, creation of, 233–244, 399
steps, 235–236, 243
work products, 237–238, 243–244
Vision of candidate, 205–217, 221
    comparing pros/cons of, 215
    completion to appropriate level of detail, 211
    components, 208, 235
    creation of, 205–217, 399
    goal, objectives, 206
    guidelines, 209–211, 216
    inputs, 206
    meeting requirements, 234
    pitfalls, 211–217
    postconditions, 208
    preconditions, 206
    selection of, decision methods, 239
    steps, 207–208, 216
    work products, 208–209, 216

W

Walk-throughs, 260
Water interoperability, 370
Waterfall development cycle, current methods emphasis, 33
Web sites, 348
White box quality characteristic. *See* Internal quality characteristics
White paper
    architectural, 124
    interoperability, 379
Whiteboard, 318
Wired communication energy interoperability, 370
Wired power interoperability, 370
Wireless communication energy interoperability, 370
Word processor, 319
Work product first, 347
Wrappers, 379
Wrapping of reuse components, 249